THE AGE OF INFORMATION SYMMETRY

A PRACTICAL GUIDE TO THRIVING IN A WORLD WHERE KNOWLEDGE IS NO LONGER POWER

JASON LANE

ANDY FORBES

Roadmap Press
Hillsboro, Virginia, United States

ISBN: 978-1-969489-12-9 (paperback)
ISBN: 978-1-969489-02-0 (ebook)

Cover design by: Roadmap Press

My sincere thanks to Mike Scott whose insights into Chicago politics, shared over years of conversation, shaped some of the ideas explored in this book. I'm also grateful for the chance to have worked with him through several fascinating startups and turnarounds, where I saw firsthand his ability to combine sharp pragmatism with creative leadership. Those experiences left a lasting mark on how I think about organizations, people, and change.

Andy Forbes

Knowledge will forever govern ignorance; and a people who mean to be their own governors must arm themselves with the power which knowledge gives.

James Madison

CONTENTS

PREFACE

We wrote this book at a moment when artificial intelligence is advancing at a pace that is difficult for any individual or any organization to fully track. AI is now progressing along what futurist Ray Kurzweil termed the Law of Accelerating Returns: each breakthrough fuels the next, compressing years of progress into months and sometimes months into weeks. The result is a landscape where tools, capabilities, best practices, and even entire business models evolve faster than traditional publishing cycles can capture.

Because of this, the context surrounding some of the technologies, vendors, and organizational examples discussed in these pages will inevitably shift between the time we finalized this manuscript and the moment you are holding it. We made every effort to validate the who, what, and where of the scenarios we describe. We grounded the concepts in real patterns we've observed across industries and in the lived experience of architecting and delivering complex systems for large enterprises. But we also recognize that the half-life of information has shortened dramatically. Some elements will have matured, others will have been refined, and a few may have taken directions no one anticipated.

That is not a flaw in the material; it is the reality of the world the book is designed to help you navigate. The core ideas here are built to endure: the shift from information scarcity to information symmetry, the restructuring of work, and the emergence of AI as an always-available partner in decision-making and execution. These patterns remain stable even as the specifics evolve.

Where specific organizational outcomes or metrics are discussed, they are based on publicly available reports, secondary analyses, or company-reported figures, and are presented for illustrative purposes rather than as independently verified results.

We welcome ongoing discussion as the landscape continues to change. Both of us are active on social media, especially LinkedIn, and we encourage readers to reach out, challenge ideas, share experiences, and explore how these concepts apply within your own organization. The conversation about AI and the future of work is far bigger than any one book, and we look forward to continuing it with you.

1

THE ARBITRAGE ECONOMY

The classical economic models often depict a world where perfect information is readily available to all participants, allowing markets to function with ideal efficiency. Yet, a closer examination reveals a different reality: the true invisible hand guiding much of our modern economy is not perfect information, but rather its inverse; the unequal distribution of knowledge. This phenomenon, known as information asymmetry, is the quiet, powerful engine behind everything from a company's pricing power to an individual's career trajectory. It is the fundamental principle that allows one party in a transaction or relationship to gain an advantage by knowing something that the other party does not, or by possessing superior insight, context, or know-how. At its core, this dynamic manifests as knowledge arbitrage: the act of profiting from the differential access to, or interpretation of, information.

This process involves a series of interconnected steps: the acquisition of knowledge, its strategic brokering, and the subsequent leveraging of its consequences. Those who excel in this domain often act as both knowledge arbitrageurs, facilitating the diffusion of information, and knowledge arbiters, selectively determining which knowledge will be disseminated and which will be held back. This dual role underscores a profound reality: power often accrues to those who control the flow and interpretation of information. The consequences of such arbitrage can be far-reaching, including the potential for competitive advantage or, conversely, the loss of proprietary knowledge for the source.

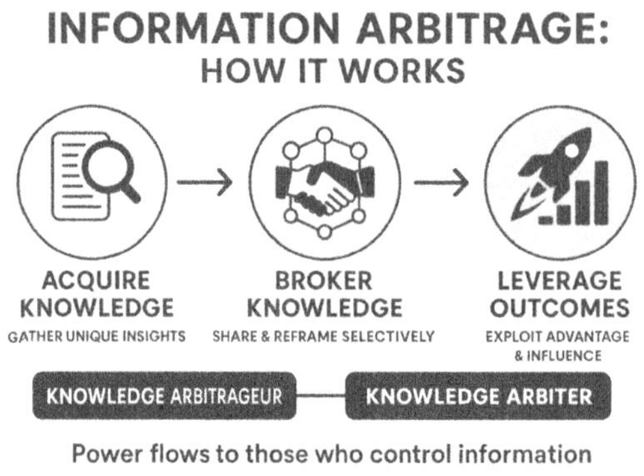

Examples of information asymmetry are woven into the fabric of everyday life, so pervasive that they often go unnoticed, much like the air we breathe or the gravity that anchors us. Consider the familiar scenario of purchasing a used car. The seller possesses a far clearer understanding of the vehicle's true condition, its quirks, and its hidden flaws than the prospective buyer. This imbalance can lead to a phenomenon known as the "lemon problem," where buyers, unable to distinguish between high-quality and low-quality vehicles, are only willing to pay an average price, potentially driving good cars out of the market altogether.

Similarly, in the health insurance market, applicants inherently know more about their own health status than the insurance providers, complicating risk assessment and pricing. When buying a home, the homeowner typically holds more detailed information about the property than the buyer. Even obtaining a mortgage can be fraught with informational gaps, as borrowers often find the complex terms of contracts difficult to understand, leaving them at a disadvantage compared to lenders. In the job market, job seekers possess a more intimate knowledge of their true skills and expertise than the employers evaluating them. These instances, seemingly disparate, are all governed by the same underlying principle: one party's informational advantage.

The economic underpinnings of information asymmetry were rigorously formalized by Nobel laureates George Akerlof, Michael Spence, and Joseph Stiglitz, whose work in 2001 provided a foundational understanding of this pervasive economic force. George Akerlof's seminal work, "The Market for Lemons," vividly illustrated how quality uncertainty arising from asymmetric information can lead to market failures, where the inability of buyers to discern quality can drive high-quality goods out of the market. This concept, known as adverse selection, describes situations where one party, possessing more information, uses it to their advantage before a transaction occurs. Michael Spence's contributions focused on signaling, explaining how the informed party can credibly convey their private information to the uninformed party. For instance, in labor markets, educational degrees serve as signals of a job candidate's productivity and capability, even if the direct knowledge gained isn't always perfectly relevant to the job.

Conversely, Joseph Stiglitz explored screening, detailing the strategies employed by the uninformed party to elicit private information from the informed party. An example of screening is an insurance company offering various policy options to encourage high-risk and low-risk individuals to self-select, thereby revealing their true risk profiles. Collectively, their research illuminated how information asymmetry can lead to significant market inefficiencies, including moral hazard; where one party's behavior changes after a transaction because the risks are borne by another, and the adverse selection. This rigorous economic framework provides the intellectual bedrock for understanding the arbitrage economy.

George Akerlof

- **Core contribution:** Demonstrated how information asymmetry can lead to market failure, even when buyers and sellers act rationally.
- **Key concept:** Adverse selection.
- **Canonical example:** The "market for lemons," where used-car sellers know more about vehicle quality than buyers, driving good cars out of the market.

Michael Spence

- **Core contribution:** Showed how informed parties can credibly convey private information to less-informed parties.
- **Key concept:** Signaling.
- **Canonical example:** Higher education functioning as a signal of a job candidate's productivity to employers, independent of whether the education itself increases productivity.

Joseph Stiglitz

- **Core contribution:** Explained how less-informed parties can design mechanisms to extract private information from informed parties.
- **Key concept:** Screening.
- **Canonical example:** Insurance companies offering different policy types to distinguish between high-risk and low-risk clients.

For centuries, this quiet economic engine, powered by informational gaps, has driven value creation and shaped societal structures. However, this long-standing equilibrium is now facing its most profound challenge. The emergence of new technologies, particularly artificial intelligence, promises to dissolve many of these entrenched barriers, ushering in an unprecedented era of information symmetry. This coming shift will not merely alter how we conduct business; it will fundamentally redefine the very nature of value, power, and human endeavor.

The financial sector, with its high stakes and rapid transactions, has long been the purest laboratory for information advantage. It's where the concept of arbitrage isn't just an academic theory, but a daily, industrial-scale business model.

WALL STREET AND THE INDUSTRIALIZATION OF ARBITRAGE

Financial markets stand as the quintessential battlegrounds for information advantage, where every trade and investment decision is, in essence, a bet on possessing superior knowledge, speed, or interpretive ability. The sheer scale, speed, and complexity of global financial transactions make information asymmetry not merely a factor, but the very lifeblood of profit generation. In this high-stakes environment, the ability to exploit informational gaps has been refined into an industrial process.

Within finance, a critical distinction exists between insider information and legal arbitrage. Insider information refers to material, non-public knowledge about a company or market that, if revealed, would significantly impact its price. Trading on such information is strictly illegal and heavily regulated, carrying severe penalties. This represents the most direct, albeit illicit, form of information asymmetry, where privileged access to confidential data provides an unfair advantage. In stark contrast, legal arbitrage involves exploiting legitimate, fleeting price discrepancies for the same asset across different markets. This is a cornerstone of market efficiency, where arbitrageurs simultaneously buy an asset in one market where it is undervalued and sell it in another where it is overvalued, profiting from the temporary difference. This activity, while profitable for those who execute it, also plays a crucial role in aligning prices and ensuring fair valuation across disparate markets. The core of this strategy lies in recognizing and acting upon the price discrepancy before it naturally resolves. The difference between these two practices highlights that while the financial system thrives on exploiting informational advantages, it also necessitates strict boundaries to prevent illicit gains that undermine trust and fairness.

The pursuit of informational edge has driven extraordinary technological innovation, particularly evident in High-Frequency Trading (HFT) and latency arbitrage. HFT firms deploy sophisticated algorithms and ultra-low latency connections to execute trades at speeds measured in microseconds or even nanoseconds. Latency arbitrage is a specific HFT strategy that capitalizes on minuscule time differences in market data and trade execution across multiple trading venues. Due to delays in information transmission, various exchanges might momentarily display different prices for the same asset. HFT firms, leveraging co-located servers, custom hardware, and optimized software, can detect and act on these discrepancies before the market adjusts, effectively "front-running" slower participants.

While proponents argue that this practice enhances price discovery and market efficiency, critics contend it creates a subtle "tax" on other market participants. Regardless of the debate, latency arbitrage epitomizes the industrialization of information advantage, where even a few milliseconds of knowing something "first" can translate into substantial profits. This relentless drive for speed and the mechanization of informational advantage in finance offer a compelling preview of how artificial intelligence will operate in other domains. The financial sector's intense competition for speed and data processing to exploit informational gaps provides a template for how AI will rapidly identify and close information asymmetries, but at an even grander scale and across diverse types of knowledge.

Beyond speed, access to comprehensive, real-time information is another critical component of financial advantage. Bloomberg Terminals are ubiquitous in the financial industry, providing institutional users with consolidated access to real-time market data, news, analytics, and trading tools. This powerful aggregation of information grants a significant edge over retail investors who lack such comprehensive and high-speed data feeds. The ability to quickly access, synthesize, and act upon vast amounts of financial information constitutes a formidable information advantage.

Complementing this, financial institutions implement "information firewalls," often referred to as "Chinese Walls," which are strict internal barriers designed to prevent the flow of sensitive, non-public information between different departments, such as investment banking and trading desks. These virtual barriers are crucial for regulatory compliance, preventing conflicts of interest, and curbing insider trading. The very existence of these "information walls" underscores a fundamental understanding within the financial sector: that internal information asymmetry, if left unchecked, could lead to illegal activities, market manipulation, and severe reputational damage. Firms invest heavily in robust data protection strategies, including advanced data encryption, multi-factor authentication, strict access controls, and real-time monitoring, to safeguard this sensitive and proprietary data.

The implementation of these controls in finance, despite the sector's reliance on informational advantages, highlights a crucial dynamic: while the system thrives on exploiting informational differences, it also requires mechanisms to prevent certain types of information asymmetry from undermining trust and overall market stability. The need for these sophisticated controls indicates that unchecked information asymmetry can lead to systemic risks and market failures. This situation presents a valuable lesson for the broader economy as AI begins to dissolve information barriers. As information becomes universally accessible, society will face the critical challenge of maintaining trust and preventing new forms of manipulation. The regulatory and ethical frameworks developed in finance to manage information flow, though imperfect, might offer initial guidance, but the sheer scale of AI's impact will necessitate entirely new approaches to governance and the cultivation of trust in an information-rich environment.

Legal Arbitrage

- **Definition:** Exploiting temporary price discrepancies of the same asset across different markets at the same time.
- **Legality:** Generally legal and widely viewed as contributing to market efficiency.
- **Information source:** Publicly available information, exploited through timing and speed advantages.
- **Risk profile:** Low financial risk when executed correctly and quickly, often approaching risk-free in theory.
- **Market impact:** Helps align prices across markets and improves overall efficiency.

Insider Trading

- **Definition:** Trading based on material, non-public information about a company or security.
- **Legality:** Illegal and subject to severe penalties, including fines and imprisonment.
- **Information source:** Privileged, confidential, non-public information.
- **Risk profile:** Extremely high legal and reputational risk.
- **Market impact:** Undermines market fairness and erodes investor confidence.

Finance didn't invent information arbitrage; it just made it explicit. The principles of exploiting informational gaps for profit are deeply embedded in how value is created and exchanged in every corner of the economy. Now let's look at how it's embedded in everyday business.

CONSULTING AND THE SALE OF SYNTHETIC EXPERTISE

Management consulting firms stand as a prime example of industries built upon monetizing the knowledge gap. They do not necessarily possess unique, secret information, but rather excel at accessing, synthesizing, and interpreting information that their clients either lack, cannot efficiently acquire, or struggle to interpret effectively. Their core value proposition revolves around providing "answers" and "certainty" in complex, often ambiguous business environments, effectively selling a reduction in client uncertainty.

A defining characteristic of the consulting industry is its hierarchical "pyramid" structure, which efficiently transforms raw data into high-value insights. At the base of this pyramid, junior consultants and analysts are engaged in extensive data collection, research, and foundational analysis. They meticulously compile vast quantities of information, drawing from public sources, internal client data, and industry reports. As this information ascends the pyramid, it undergoes successive layers of synthesis, refinement, and distillation by more senior consultants and partners. The partners, positioned at the apex, are the ones who ultimately present these distilled "answers" to clients. This process is often guided by communication principles like McKinsey's "Pyramid Principle," which advocates leading with the main conclusion and then providing supporting arguments and detailed data, mirroring the top-down delivery of synthesized knowledge. This structured approach effectively industrializes the transformation of raw information into consumable, actionable insights, enabling firms to scale their intellectual output.

The notoriously high fees charged by consulting firms are largely justified by this perceived knowledge asymmetry. Clients are not merely paying for the information itself; they are paying for the access to that information, the expert context and interpretation provided by experienced professionals, and the assurance that the advice is sound, relevant, and actionable. This dynamic closely mirrors Akerlof's "adverse selection" problem; where clients, much like used car buyers, struggle to accurately assess the quality of consulting services upfront due to information imbalances. This difficulty in pre-evaluating true value allows reputable firms to command premium prices. Their established brand and track record act as a powerful signal of quality, mitigating the client's risk of making a poor decision with incomplete information. Consultants, therefore, sell a reduction in uncertainty and a promise of optimal outcomes, leveraging the client's lack of internal expertise or bandwidth to acquire and process the necessary information. This "trust premium" is a significant component of their pricing model.

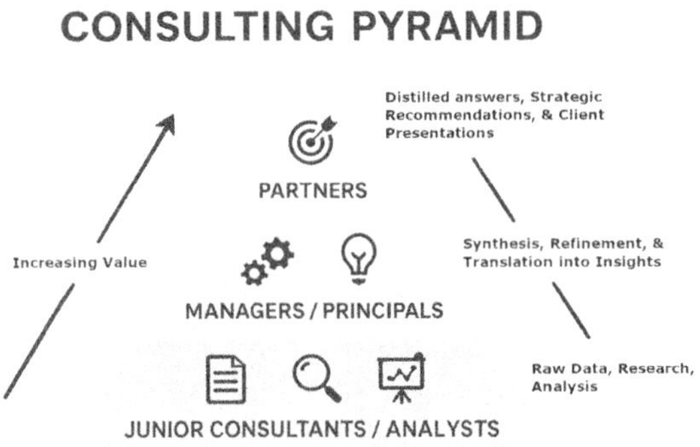

Many consulting firms, particularly the larger, established players, develop and repeatedly deploy proprietary methodologies, frameworks, and "playbooks". These are essentially standardized approaches to common business challenges, such as growth strategy, operational efficiency, or market entry. Firms like McKinsey (known for frameworks like the GE McKinsey Matrix, 7S Framework, and Pyramid Principle), Boston Consulting Group (with its Growth-Share Matrix), and Bain & Company (employing methodologies like the Net Promoter System) are prominent examples. These frameworks are widely reused and adapted across engagements for clients facing similar problems, allowing the firms to repeatedly monetize the same underlying insights. This strategy enables significant scalability and profitability, as the initial investment in developing this "secret sauce" is amortized over many engagements. The value delivered to clients, in this context, resides not in the uniqueness of the raw data, but in the proprietary packaging, expert interpretation, and tailored application of that data, which clients would find difficult to replicate independently.

The widespread practice of consulting firms reselling similar playbooks and proprietary frameworks indicates that, even before the advent of powerful artificial intelligence, "expertise" was, in many ways, already being industrialized and commoditized. The value provided was not always about generating entirely bespoke, unique insights for every client, but rather about the efficient application of a repeatable, standardized knowledge base. This suggests that the market for strategic advice already operates on a principle of scalable, rather than purely unique, knowledge. This pre-existing trend means that the coming disruption by AI will not be a sudden, unforeseen shock, but rather an acceleration of an already established trajectory. If consulting has demonstrated that packaged, repeatable insights can be highly profitable, AI will simply provide a vastly more efficient and scalable mechanism for generating and distributing such "synthetic expertise," further compressing the value of human-delivered, generalized knowledge.

The justification for high consulting fees points to a significant reliance on a "trust premium". Clients are willing to pay more not just for the information itself, but for the assurance that the information is correct, relevant, and actionable, and that the firm providing it is reputable and reliable. This trust helps bridge the inherent information asymmetry where clients cannot fully assess the quality of the service before engagement. This dynamic suggests that as AI makes information universally accessible, the challenge will shift from mere access to information to the discernment of trustworthy information. If AI can generate plausible "answers" at near-zero cost, the new differentiator will be the perceived trustworthiness of the source, the brand, or the human curator behind the AI. This implies that "trust" itself will become a new form of scarce capital in an information-symmetric world, echoing the dynamics observed in the consulting industry.

Apex Level (Partners)

- **Role:** Strategic advisors and rainmakers.
- **Primary activity:** Synthesize insights, develop recommendations, manage senior client relationships, and originate new engagements.
- **Information flow:** Top-level conclusions translated into actionable strategies.
- **Value proposition to the client:** Certainty, proven approaches, high-level strategic guidance, and decisive problem resolution.

Mid-Level (Managers and Principals)

- **Role:** Project leaders and expert analysts.
- **Primary activity:** Structure complex problems, manage delivery teams, ensure quality, perform deep analysis, and translate data into defensible insights.
- **Information flow:** Refined analysis, structured arguments, and supporting evidence.
- **Value proposition to the client:** Detailed analysis, reliable project execution, and application of specialized expertise.

Base Level (Juniors and Analysts)

- **Role:** Data gatherers and researchers.
- **Primary activity:** Collect raw data, conduct interviews, build models, and perform initial analyses.
- **Information flow:** Raw data, early findings, and detailed supporting information.

- **Value proposition to the client:** Capacity, execution support, and foundational research that enables higher-level work.

Consulting sells certainty and pattern-matching; the ability to apply known solutions to new problems. But sometimes, the real value lies not in a pre-packaged answer, but in understanding the specific context of a problem. Let's look at the front lines next, where customer service often sells exactly that: context.

CUSTOMER SERVICE AND SCRIPTED POWER

Customer service interactions, while seemingly designed to assist, often function as subtle battlegrounds of information asymmetry. The company possesses a comprehensive understanding of its products, policies, and intricate internal processes, whereas the customer typically holds limited, fragmented knowledge. This inherent imbalance is frequently leveraged, not solely for problem resolution, but as a mechanism for control and profit.

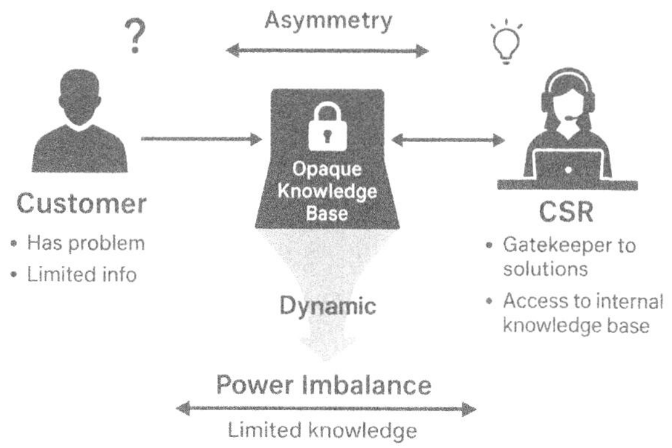

At the heart of customer service lies a fundamental asymmetry between the help desk and the customer. Customers initiate contact with a problem, seeking a solution or specific information. The customer service representative (CSR) acts as the gatekeeper to this information, whether it pertains to a troubleshooting sequence, a nuanced policy detail, or an exception process. This dynamic creates an inherent power imbalance, as the customer is dependent on the CSR to navigate the company's internal knowledge base, which is often opaque and inaccessible to the public. The company, by virtue of its complete knowledge of available solutions, internal limitations, and alternative pathways, maintains a distinct advantage. The customer, by contrast, is often aware only of their immediate problem, lacking the broader context or the full spectrum of potential resolutions.

Many customer service operations rely heavily on "decision trees" or tightly controlled scripted workflows. These flowcharts guide agents through a predefined series of questions and responses, ostensibly to diagnose and resolve customer queries consistently and efficiently. However, these decision trees also function as industrialized knowledge filters. They meticulously control the flow of information, ensuring that agents provide only specific, pre-approved answers. This structured approach can prevent agents from deviating from scripts or offering solutions that, while potentially more beneficial to the customer, might be less profitable or strategically undesirable for the company. The "knowledge base articles" that underpin these decision trees are effectively "automatic gatekeepers of support and guidance," designed not only to provide information but also to manage customer expectations and steer interactions toward predetermined outcomes. This highlights how "help" can be industrialized into a mechanism of control, prioritizing the company's interests; such as cost reduction or upsell opportunities over the customer's immediate need for comprehensive information. As AI increasingly takes over customer service roles, this control mechanism could become even more pervasive and subtle, potentially leading to a future where AI-driven interactions are optimized primarily for corporate gain rather than genuine customer empowerment, unless robust ethical safeguards are implemented.

Beyond mere problem-solving, customer service interactions are frequently strategic opportunities for upselling and customer retention. Customer service representatives are often trained to offer more expensive versions of products or additional services during these interactions. This strategy is highly effective because the customer, already engaged with the company and potentially vulnerable due to an existing problem, may be unaware of alternative options or the perceived value of an upgrade. The CSR, armed with detailed knowledge of product tiers, benefits, and pricing structures, leverages this information asymmetry to drive additional revenue.

Similarly, retention strategies frequently involve offering specific deals, discounts, or "hidden" benefits to customers who express dissatisfaction or signal an intent to churn. The customer may not know these options exist until they explicitly threaten to leave, revealing the company's proprietary knowledge of customer value and its internal thresholds for retention. The probability of successfully selling to an existing customer is significantly higher than to a new prospect, underscoring the substantial value derived from the existing, albeit asymmetric, customer relationship. The company maintains an advantage by possessing information about customer value and the levers it can pull to retain them, which the customer does not have.

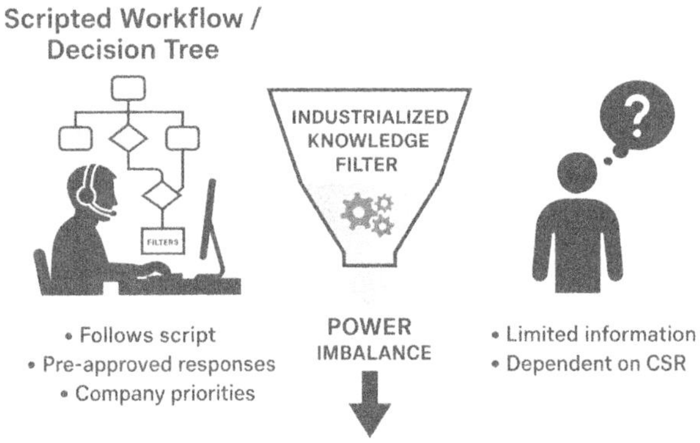

The success of upsell and retention tactics in customer service hinges critically on the customer's lack of complete information regarding alternative products, pricing tiers, or retention offers. The company deliberately withholds or selectively reveals information to maximize its revenue. This demonstrates that the customer's "not knowing" is an asset for the seller, enabling price discrimination and leveraging existing relationships to extract additional value. In an information-symmetric world, where customers could instantly access all available options, competitive pricing, and retention offers, these traditional upsell and retention strategies would be profoundly undermined. This impending shift will compel businesses to fundamentally rethink how they create value and foster customer loyalty, moving away from exploiting informational gaps towards building genuine, transparent relationships and offering truly differentiated value propositions.

Whether high-end consulting or low-end customer service, the game is the same: profit from the gap in what one party knows versus what the other doesn't. But that gap isn't just economic-it's structural, deeply embedded in the very fabric of how our professional world operates.

PROFESSIONAL SERVICES AND THE CREDENTIAL ECONOMY

Beyond the direct transactional arbitrage seen in finance and the structured sales of consulting and customer service, entire professional sectors are fundamentally built upon the control and monetization of specialized knowledge. Professions such as law, medicine, and even aspects of academia, often deemed indispensable for public welfare, erect formidable barriers to entry. These barriers ensure that access to critical information and the legitimate right to apply it remain exclusive, thereby preserving a lucrative informational advantage.

A primary mechanism for information control in these fields is licensing and gatekeeping. Licenses for doctors, lawyers, and other certified professionals are far more than mere certifications of competence; they serve as powerful gatekeeping mechanisms that restrict who can practice and, by extension, who can access, interpret, and apply specialized knowledge. Licensing boards, frequently composed of active members of the profession itself, define the "scope of practice" and establish rigorous entry requirements, including specific academic degrees, extensive training, and demanding examinations. This self-regulation effectively creates a knowledge monopoly, limiting competition and ensuring that only a select few are deemed legitimate "experts". While proponents argue that this system protects the public from unqualified practitioners, critics often highlight that it can lead to inflated prices, stifle innovation, and function as an anti-competitive practice. The requirement for ongoing education for license renewal further reinforces this control, often acting as a barrier to continued practice rather than solely a means of upskilling. This system, therefore, is not merely about quality control; it is deeply intertwined with market control and the preservation of professional power.

The prevalent "billable hour" model in professional services, such as law and consulting, directly monetizes the information asymmetry between the expert and the client. Clients pay for the expert's time, which is implicitly valued for the specialized knowledge, extensive experience, and unique insights that the client themselves lack. The higher the perceived expertise, the higher the hourly rate commanded by the professional. This system inherently relies on the client's inability to perform the task independently or to easily acquire the necessary knowledge. The expert, possessing this "uneven expertise," can charge for the time it takes to apply that knowledge, even if the actual task might be routine or straightforward for them. Accurate time-tracking is paramount for these firms to maximize revenue from billable hours, illustrating how deeply the financial model is intertwined with the control and application of expert knowledge.

Legal contracts serve as a potent illustration of information asymmetry in practice, often appearing to be deliberately designed to be unreadable to the layperson. These documents are frequently drafted using complex, archaic language, often referred to as "legalese," replete with convoluted sentence structures and redundant phrases (tautologies) that render them nearly impenetrable to the average individual. Common examples include phrases like "hereinafter," "herewith," "null and void," and lengthy, repetitive clauses that add to the perceived complexity without necessarily enhancing clarity. Studies have shown that comprehension of written material drops drastically as sentence length increases; for instance, understanding can fall below 10% for sentences exceeding 43 words, a length often found in legal provisions. This intentional obfuscation serves a critical purpose: it maintains the lawyer's informational advantage and reinforces their indispensability. Clients are compelled to rely on legal professionals to interpret these intricate documents, thereby justifying the high fees associated with legal services. The contract, in this context, functions not merely as a legal agreement, but as a powerful tool for knowledge gatekeeping.

The reliance on licensing, academic degrees, and professional bodies creates a "credentialed monopoly" over knowledge. This system not only aims to ensure a certain standard of competence but, more significantly, limits the supply of practitioners and entrenches the authority of existing professionals. The deliberate use of "legalese" further reinforces this by making the domain inaccessible to outsiders, thereby preserving the need for the expert. This dynamic transcends mere quality control; it is fundamentally about market control and the preservation of power. This model, however, is highly vulnerable to the advancements in artificial intelligence. If AI can rapidly acquire and apply specialized knowledge, simplify complex legal texts, or even pass licensing examinations, the very foundation of these credentialed monopolies will be challenged. The value proposition will inevitably shift from the act of holding a license to the ethical and contextual application of knowledge, forcing a profound redefinition of what it means to be a "professional" in these fields.

The fact that clients accept and pay for legal contracts that are deliberately unreadable reveals a societal acceptance of information asymmetry in certain domains. This acceptance often stems from a perceived necessity, such as the belief that "only lawyers can understand this," coupled with a willingness to delegate understanding due to the high cost of self-education or the perceived risk of misinterpretation. This highlights a deeper social contract where complexity is tolerated, and sometimes even encouraged, to preserve professional authority. As AI makes information simplification and translation ubiquitous, this implicit social contract will inevitably be broken. The expectation will shift from "I need an expert to translate this" to a demand for inherent clarity: "Why isn't this already understandable?" This will exert immense pressure on professions to become radically transparent and accessible, fundamentally altering their business models and their relationship with the public.

Over-specified covenant language

- **Legalese example:** "And the said covenantor doth hereby, for himself, his heirs, executors, and administrators, covenant, promise and agree with and to the said covenanttee, his heirs, executors, administrators and assigns, in manner following..."
- **Plain-language equivalent:** "James Jones covenants..." or simply, "He promises..."
- **Implication of asymmetry:** Obscures who is making promises and to whom, reducing accessibility and favoring parties with legal expertise.

Formulaic consideration clauses

- **Legalese example:** "For value received, the undersigned hereby promises to pay..."
- **Plain-language equivalent:** "To repay my loan, I promise to pay..."
- **Implication of asymmetry:** Creates a formal, distant tone that hides the shared purpose of the agreement and discourages scrutiny by non-experts.

Performative acknowledgments

- **Legalese example:** "I hereby acknowledge receiving a copy of this retainer."

- **Plain-language equivalent:** "I received a copy of this retainer." or "I have a copy of this retainer."
- **Implication of asymmetry:** Adds unnecessary formality and wordiness, increasing cognitive load without adding legal meaning.

Redundant legal pairings

- **Legalese example:** "Null and void."
- **Plain-language equivalent:** "Invalid" or "canceled."
- **Implication of asymmetry:** Uses duplication to signal authority and complexity rather than clarity, reinforcing perceived opacity.

Archaic "here–" constructions

- **Legalese example:** "Hereinafter," "herewith," "heretofore," "hereby."
- **Plain-language equivalent:** Often unnecessary and can be rephrased using modern syntax.
- **Implication of asymmetry:** Introduces archaic language that acts as a comprehension barrier for modern readers.

Excessive sentence length

- **Legalese example:** Average sentence length of 42.5 words in a contract provision.
- **Plain-language equivalent:** Sentences under 14 words enable roughly 90% comprehension.
- **Implication of asymmetry:** Deliberately complex sentence structures reduce comprehension to below 10%, effectively requiring expert interpretation.

So far, the discussion has focused on how arbitrage shapes services; from high finance to professional advice. But information asymmetry isn't just about external transactions; it also profoundly shapes how value flows and power is distributed inside organizations.

CORPORATE STRUCTURES AS ARBITRAGE ENGINES

Information asymmetry is not merely an external market force or a characteristic of professional services; it is a fundamental organizing principle deeply embedded within the internal structures of corporations. Hierarchical designs and specialized roles inherently create distinct pockets of unique knowledge, which then become powerful levers for managing workflows, exerting control, and establishing internal power dynamics. This internal arbitrage of knowledge is as critical to organizational function as external market arbitrage is to economic activity.

In traditional organizational charts, managers occupy pivotal positions, acting as crucial conduits and interpreters of information flow. In the downward flow of information, managers translate strategic directives, high-level goals, and company policies from top-level management into actionable tasks and understandable instructions for frontline employees. They effectively interpret the "why" and "what" of strategic decisions into the "how" for their teams. Conversely, in the upward flow, managers synthesize raw data, operational challenges, and team performance metrics from their subordinates into concise reports and actionable insights for senior leadership. In this role, they filter out extraneous noise, highlight key trends, and frame problems in a manner that is directly relevant for executive decision-makers. This interpretive function grants managers significant power, as they effectively control the narrative and the context of information flowing both up and down the chain of command. Their ability to translate and filter creates an internal information asymmetry that reinforces their positional authority and indispensability. The role of managers as interpreters highlights that organizational hierarchies, while providing essential structure, inherently create information bottlenecks. Information is filtered, summarized, and re-contextualized as it moves up and down the chain. This is not merely an inefficiency; it is a deliberate design choice that concentrates power and control in the hands of those who manage the information flow.

Organizational incentive structures are frequently designed around the premise of inherent information asymmetry. For example, performance bonuses, opportunities for promotion, and the allocation of critical resources can be directly tied to an individual's or team's unique knowledge or their exclusive access to critical information. If a particular department possesses exclusive data or a proprietary process, it gains significant leverage within the organization, enabling it to negotiate more favorable terms, larger budgets, or greater recognition. This dynamic can inadvertently incentivize the hoarding or selective sharing of information, even if such behavior is not optimal for overall organizational efficiency. Power asymmetry within organizations, which is often inextricably linked to information asymmetry, can lead to situations where less powerful actors are exploited or disincentivized. Conversely, effective management systems are designed to weaken information asymmetry, aiming to improve incentives and enhance overall organizational performance.

The concepts of "managing up" and "gatekeeping" serve as powerful illustrations of how information asymmetry translates into political currency within corporations. "Managing up" refers to the strategic navigation of relationships with those in higher positional power, particularly one's direct supervisor. This involves a nuanced understanding of a superior's needs, motivations, and goals, and the art of presenting information in a way that aligns with their priorities. It often requires a keen sense of what information to share, when to share it, and how to frame it to influence decisions and secure greater autonomy or resources. This skill is built upon leveraging subtle information asymmetries concerning preferences, unstated expectations, and the broader political landscape within the organization.

"Gatekeeping," on the other hand, is the deliberate act of controlling access to information, resources, or even specific individuals. Within organizations, individuals or departments can become "gatekeepers" by being the sole custodians of certain data, critical processes, or key relationships. This control confers significant political currency and power, as it creates dependency and reinforces their indispensability. For instance, an employee who is the only one fully conversant with a legacy system, or a manager who meticulously filters all communication between their team and a crucial stakeholder, is actively engaging in gatekeeping. The prevalence of "managing up" and "gatekeeping" demonstrates that information asymmetry within organizations is a self-reinforcing system. Individuals and departments are incentivized to maintain their unique knowledge advantages because it confers power, status, and often leads to better outcomes for them personally or for their specific team. This creates a dynamic where information protection can be a rational individual strategy, even if it is ultimately detrimental to the collective organizational good.

The inherent tendency to create and maintain internal information silos will pose a significant cultural and structural challenge in the age of artificial intelligence. Organizations aiming for "information symmetry" will require not only new technologies capable of democratizing information but also profound cultural shifts to overcome ingrained behaviors of knowledge protection and power consolidation. The ultimate success of AI integration will depend as much on transforming human incentives and behaviors as it will on technological capability.

These internal asymmetries mirror larger social systems; from the family unit to national governance, and they've always been reinforced by one critical, often unstated, assumption: that information is scarce.

ASYMMETRY AS A SOCIAL CONTRACT

Information asymmetry extends far beyond mere economic transactions or organizational structures; it is a deeply embedded social contract, an unspoken agreement that underpins much of how society functions. We have implicitly agreed to defer to certain individuals and institutions, granting them authority, status, and compensation, based on their perceived superior knowledge. This collective deference is a fundamental mechanism for navigating the complexities of modern life.

The authority conferred by degrees, titles, and expert status is a prime example of this social contract in action. From an early age, individuals are conditioned to respect and trust those who possess specific credentials. Academic degrees-such as a PhD, MD, or JD-along with professional licenses and formal titles like "Professor" or "Doctor," are widely accepted as powerful signals of specialized knowledge and expertise.

In a world where perfect information is unattainable, these credentials serve as essential proxies for competence and reliability. They significantly reduce the cognitive burden of evaluating every individual's knowledge from scratch, enabling society to function efficiently by establishing readily identifiable and trusted sources of information. As illuminated by Michael Spence's work on signaling, education, for instance, acts as a powerful signal of productivity and capability in the job market, even if the direct knowledge acquired isn't always perfectly relevant to a specific role. This mechanism reinforces authority and trust automatically conferred by academic achievement and professional certification.

This implicit social contract can be summarized as "Experts know better; we defer". It represents an unwritten agreement where individuals willingly surrender some autonomy and critical inquiry in exchange for the perceived benefits of expert guidance. We entrust our health to doctors, the safety of our infrastructure to engineers, and our financial well-being to advisors, largely because we operate under the belief that they "know better" due to their specialized, often inaccessible, knowledge. This delegation of knowledge serves as a crucial timesaver in an increasingly complex world, allowing individuals to focus on their own domains while relying on others for specialized insights. This arrangement underpins the legitimacy of institutions and contributes to the stability of social order.

Philosophers like Thomas Hobbes, John Locke, and Jean-Jacques Rousseau explored various facets of social contract theory, positing that individuals surrender certain rights for collective protection and stability. In the context of knowledge, this contract manifests as a societal deference to experts for the efficient functioning of complex systems. This reliance on experts for complex information represents a form of "cognitive offloading". In a world of overwhelming complexity, individuals rely on trusted experts and institutions to process information and make decisions on their behalf, thereby conserving time and mental energy. Degrees and titles serve as convenient shortcuts for identifying these trusted sources. This highlights that trust in expertise is a fundamental mechanism for navigating complexity, not just a byproduct of skill.

Society often operates under the aspirational ideal of meritocracy; the belief that success is earned solely through individual effort and ability. However, information asymmetry fundamentally undermines this ideal. Access to high-quality education, exclusive insider networks, and privileged information is often unevenly distributed. Those who gain entry to elite institutions, participate in exclusive social circles, or acquire tacit knowledge passed through closed groups often possess a significant advantage, irrespective of their innate talent or sheer effort. Research suggests that success, even when attributable to luck, can lead individuals to believe the world is meritocratic, fostering "meritocratic social preferences" and even prompting them to avoid information that might challenge these deeply held beliefs. This "meritocratic illusion" can perpetuate inequality by masking the profound role of asymmetric access to knowledge and opportunities. Consequently, a system emerges where "who you know" and "what information you have access to" can hold as much, if not more, weight than "what you know" or "how hard you work."

The coming age of information symmetry, where knowledge becomes universally accessible, will directly expose the inherent weaknesses in this meritocratic illusion. If everyone has equal access to information, then disparities in outcomes will become increasingly difficult to attribute solely to "merit". This could lead to significant social and political tensions, challenging existing power structures and demanding a fundamental renegotiation of societal values and reward systems.

But this contract, built on the scarcity of knowledge and the deference to those who hold it, is under stress. Something new is leveling the playing field, challenging the very foundations of authority and status.

THE CRACKS ARE ALREADY SHOWING

While the full, transformative impact of artificial intelligence on information asymmetry is still unfolding, the early signs are already unmistakably visible. These nascent "cracks" in the long-established model are not isolated anomalies; rather, they are powerful harbingers of a fundamental shift, challenging the very premise of an economy and society built upon informational gaps. The quiet revolution has begun, altering the landscape of knowledge and expertise.

The rise of open access models has already begun to democratize knowledge, progressively eroding the power of traditional gatekeepers and exclusive expertise. Platforms like Stack Overflow, an online community for programmers, allow users to ask and answer technical questions, creating a vast, publicly accessible repository of coding knowledge. This has significantly reduced the necessity for individual programmers to "know it all" or to rely solely on proprietary documentation, effectively diminishing the exclusive expertise once held by a select few.

Similarly, YouTube tutorials have transformed into a massive, free educational platform. From intricate software guides to practical home repairs, individuals can now acquire new skills or solve complex problems by simply watching a video, bypassing traditional instructors, expensive courses, or specialized consultants. This widespread peer-to-peer sharing directly undermines the value of guarded, paywalled knowledge. These open access platforms powerfully demonstrate a profound shift from a paradigm of knowledge scarcity to one of abundance, where valuable insights are shared freely, challenging traditional models of expertise and knowledge control. This points to the erosion of "knowledge work" as an exclusive human domain. The examples of open access platforms and early generative AI tools demonstrate that the very definition of "knowledge work" is being fundamentally challenged. Tasks previously requiring human expertise and access to specialized information are now being democratized or automated. This is not just about efficiency; it is about the erosion of the exclusive human domain over certain types of intellectual labor.

Furthermore, early generative AI tools are actively replacing routine white-collar knowledge work, signaling a significant shift in the labor market. Generative AI technologies, particularly Large Language Models (LLMs), are already demonstrating a remarkable capacity to perform tasks that were once considered the exclusive domain of white-collar professionals. In the legal sector, AI tools can now draft contracts, review clauses, and analyze complex legal documents; tasks traditionally performed by junior lawyers. In content creation, AI can generate text, images, and even code, impacting roles across marketing, journalism, and software development. Within finance, AI can automate reporting, perform portfolio analysis, and extract critical information from intricate financial documents. This is not merely the automation of repetitive, manual tasks; it represents the automation of nuanced intellectual work, directly challenging the information advantage previously held by many professionals. Evidence suggests that employees are already "secretly" leveraging these tools to boost their productivity, indicating a quiet but pervasive transformation in the nature of white-collar employment. This foreshadows a future where competitive advantage will no longer stem from simply having knowledge, as AI will possess vast repositories of it, but from the uniquely human capacities to apply that knowledge with judgment, empathy, and ethical reasoning, or to effectively orchestrate AI capabilities. This will necessitate a radical re-evaluation of human skills and value in the workforce.

The proliferation of accessible information and powerful AI tools are concurrently fostering a new era of peer-to-peer learning and collaboration, which inherently flattens traditional authority structures. In environments where information is readily available and easily verifiable, the need to defer to a single, authoritative "expert" diminishes. Teams and individuals can rapidly access and synthesize information independently, significantly reducing their reliance on hierarchical knowledge gatekeepers. This shift directly challenges the "power gradient" that has historically defined organizations, where communication and decision-making predominantly flowed from the top down. Leaders are increasingly encouraged to adopt more accessible postures, reduce their overt displays of authority, and openly admit when they "don't know" an answer, thereby fostering a more collaborative and less hierarchical environment. The emphasis is shifting from merely "knowing the answer" to "knowing how to find or generate the answer" and, critically, "knowing what questions to ask."

The "cracks" are showing not just in specific industries but in the very mechanisms of knowledge transfer and authority. Open access models are actively disintermediating traditional educators and experts, while generative AI is disintermediating routine white-collar roles. This indicates an accelerating trend of disintermediation across various sectors, where direct access to information and automated intelligence increasingly bypass established human and institutional intermediaries. This accelerating disintermediation will compel organizations and individuals to rapidly adapt or face the risk of obsolescence. The value chain of knowledge is being compressed, shifting from those who control information to those who can synthesize, curate, and apply it in novel and impactful ways.

As we turn to the next chapter, we'll explore how the economy of knowledge itself is priced-how its value has historically been determined by its scarcity and exclusivity-and how AI is already changing the math, making knowledge abundant and redefining what it means to be valuable.

SUMMARY

The modern economy, across its diverse sectors-from the high-stakes world of finance to the everyday interactions of customer service, from the hallowed halls of professional services to the intricate hierarchies of corporations; is built upon a quiet yet profoundly powerful logic: information asymmetry. This chapter has illuminated how knowing something others don't, or knowing it faster, or interpreting it better, has been the hidden engine of value creation, pricing power, career advancement, and organizational control. From Wall Street's industrialization of arbitrage to consulting's monetization of packaged insights, and from customer service's scripted gatekeeping to the credentialed monopolies of law and medicine, the ability to leverage unequal access to insight, context, or know-how has been the fundamental game. Even within corporate structures, hierarchy and specialized roles have turned knowledge asymmetry into internal power and workflow. Society has, in essence, built its social contracts and its very understanding of meritocracy around this scarcity of information, implicitly deferring to those deemed "experts."

However, this long-standing logic, dependent on the scarcity of knowledge, is now facing an unprecedented challenge. The early, yet unmistakable, signs of a profound shift are already visible. Open access models are eroding exclusive expertise, early generative AI tools are beginning to replace routine white-collar knowledge work, and peer-to-peer learning is flattening traditional authority structures. The tacit assumption that information is a scarce resource, and thus a source of enduring advantage, is dissolving. As artificial intelligence dissolves many of these barriers, the very pricing of knowledge itself is changing. The next chapter will delve deeper into this fundamental shift, exploring how the economy of knowledge itself has been priced; how education, credentials, and insider networks have historically served as vehicles for controlling access to power; and how AI is already changing that fundamental equation.

2

THE PRICE OF KNOWING

In a world increasingly defined by the rapid flow of information, it is easy to assume that knowledge is a universally accessible commodity. Yet, for centuries, the most valuable insights have been guarded, not shared. This chapter peels back the layers of modern society to reveal how systems ostensibly designed for learning and professional development-formal education, credentials, and exclusive networks-have, in practice, functioned as sophisticated gatekeepers. These mechanisms have historically controlled access to knowledge, ensuring that power and opportunity remain concentrated within elite circles.

The exploration begins by examining how the hallowed halls of academia, far from being pure meritocracies, often serve as sophisticated sorting machines that confer social capital as much as academic skill. The discussion then shifts to the intricate web of professional credentials; degrees, licenses, and certifications demonstrating how these formal markers can restrict entry and define legitimacy, often creating artificial monopolies on expertise. Following this, the enduring influence of insider networks is uncovered, revealing how "who you know" has frequently overshadowed "what you know" in the transmission of critical, often tacit, information. Finally, the very language and frameworks of elite knowledge are examined, demonstrating how they function as cultural capital, further embedding these exclusionary practices.

By the end of this chapter, readers will gain a profound understanding of how formal education acts as a proxy for access, not merely skill, and how higher education institutions have historically operated as gatekeepers. The analysis will detail credentials as a powerful filtering mechanism and illustrate the subtle yet potent ways insider networks preserve and recycle information advantages. This exploration illuminates the hidden structures that have long dictated who hold the keys to power and influence.

Understanding these entrenched systems of access is only half the equation. The true resilience of these structures lies not just in their ability to control entry, but in the deeper dynamics of trust and control that reinforce them.

THE ACADEMY AS A SORTING MACHINE

Universities, particularly elite institutions, are often perceived as beacons of meritocracy, places where intellect and hard work are the sole determinants of success. However, a closer examination reveals that these institutions function as far more than mere educational establishments; they are, in essence, sophisticated information access filters. They sort, select, and ultimately confer advantages that extend far beyond the classroom, shaping careers and perpetuating social hierarchies.

Ivy League and Elite Education as Social Capital Factories

Elite universities, from the Ivy League to Oxbridge, consistently attract and admit students from high-status families, creating social circles that become invaluable assets. Research indicates that even when students from lower socioeconomic backgrounds attend these institutions, top jobs and incomes disproportionately go to those from high-status families. The primary mechanism at play here is not academic, but social. Attending an elite university exposes students to new social circles, increasing the likelihood of marrying peers from privileged backgrounds and building social networks that can benefit their children across generations. For instance, children whose parents attended elite universities are more likely to attend elite private K-12 schools, live in high-status neighborhoods, befriend high-status peers, and enroll in elite universities themselves, regardless of their parents' social origin. This suggests that the benefits of elite education are often more social than academic, transforming social networks and marriage prospects, and ultimately influencing the social and educational opportunities available to future generations.

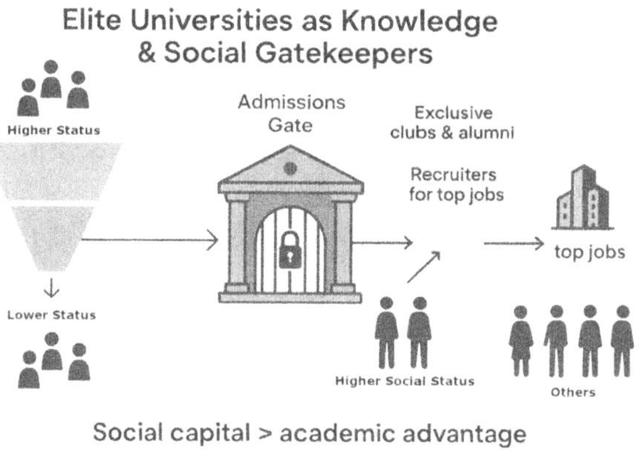

Sociological studies on social capital theory further illuminate this dynamic in higher education. Social capital, defined as the resources embedded in one's social network, is a valuable framework for examining differences in educational outcomes. It helps understand how access to and use of resources found in one's relationships can influence academic satisfaction, well-being, persistence, and educational attainment. Institutions are even developing interventions, such as workshops and academic programs, to increase students' social capital, recognizing its importance for student success. This underscores that the value of an elite degree is not solely in the knowledge acquired, but in the social connections forged and the access points they unlock. This dual nature of elite education, extending beyond human capital to encompass social and cultural capital, challenges the conventional understanding of meritocracy by demonstrating that access to elite positions is significantly influenced by inherited social status and connections. The university, therefore, becomes a site for the reproduction of elite advantage, rather than solely its acquisition.

Pedigree as a Signal in Job Markets

The concept of "pedigree" plays a significant role in how elite students secure top jobs, often acting as a signal that can outweigh raw competence. Lauren Rivera's "Pedigree: How Elite Students Get Elite Jobs" reveals that hiring processes at top-tier investment banks, consulting firms, and law firms are heavily skewed towards applicants from economically privileged backgrounds. Employers define and evaluate "merit" in ways deeply rooted in social class, using "pedigree" as a shorthand for accomplishments. This includes admission to an elite university, participation in Ivy League varsity sports, or early internships at prestigious firms. These achievements are interpreted as evidence of intelligence, drive, and work ethic, even though they often require considerable economic, social, and cultural resources from the applicants and their families.

The irony of the term "pedigree", which literally means "ancestral line", highlights how hiring decisions, while appearing merit-based, are subtly shaped by socioeconomic backgrounds. Filtering criteria used at every stage of the hiring process, from job advertisements to final selections, correlate strongly with parental income and education. This means that even seemingly neutral decisions ultimately filter students based on their parents' socioeconomic status. Affluent parents can invest more in their children's educational growth, enabling access to better schools and unpaid internships, while their social networks provide crucial tips and even influence application acceptance. This creates a system where the "right degree from the right university" becomes a powerful signal, often overshadowing a candidate's actual skills. This reliance on the "pedigree signal" as a proxy for privilege, rather than pure merit, creates a systemic class bias in hiring. It subtly excludes talented individuals from less privileged backgrounds, reinforcing existing inequalities in the job market, irrespective of their actual competence. The system rewards the appearance of merit derived from privilege, rather than merit itself.

Disparities in Curriculum Access and Digital Learning

The gatekeeping function of academia is further amplified by significant disparities in curriculum access and digital learning, creating a "digital divide" that entrenches educational inequality. This divide refers to unequal access to technology and digital resources for learning, manifesting as a lack of internet access, computers, and the digital literacy skills needed to use them effectively. For instance, a UNICEF report highlights that 9 out of 10 young women and teenage girls in the world's lowest-income countries lack internet access, exacerbating existing gender inequalities in education.

The COVID-19 pandemic starkly revealed these disparities, as only school systems with digital capabilities could adapt to fully digital learning environments, leaving millions of students behind. The "homework gap"; the disparity between students with and without home internet and technology-persists, disproportionately affecting low-income, rural, and communities of color. Research shows that 15% of students still lack high-speed internet at home, with this figure rising to one-fourth in rural areas. This unstable connectivity directly impacts academic performance, with students lacking reliable access performing lower on digital skills, homework completion, and GPA. While digital learning has the potential to expand access and enhance outcomes, these significant inequalities in connectivity and capacity remain persistent barriers. This digital divide, when unaddressed, acts as another layer of gatekeeping. It ensures that students from disadvantaged backgrounds continue to lack the foundational digital skills and resources necessary to compete in a modern, technology-driven world, thereby perpetuating socioeconomic inequalities across generations. The promise of digital learning remains largely unfulfilled for those on the wrong side of this divide.

PROMPT ENGINEERING AND THE BALANCE OF AI COMMUNICATION PREFERENCES

In this evolving landscape, the emergence of artificial intelligence introduces a new layer of complexity to information access. Prompt engineering, the process of guiding generative AI to produce desired outputs, is becoming a critical skill. It involves selecting appropriate formats, phrases, words, and symbols to enable meaningful interaction with AI. This skill is crucial for applications requiring subject matter expertise or critical thinking from AI, where a well-engineered prompt can significantly increase accuracy.

This development raises a subtle but profound question about future access: will proficiency in communicating with AI become a new form of digital literacy, potentially creating a new gatekeeping mechanism? The balance between learning AI's communication preferences and AI learning each user's communication preferences is key. While prompt engineering aims to improve user experience and flexibility by allowing developers to control AI interaction and enable personalized responses, the iterative process of experimentation and refinement required to optimize AI output suggests that those adept at "speaking AI's language" may gain an advantage. Conversely, if AI systems become highly adaptive to individual user preferences, they could democratize access by lowering the barrier of technical proficiency. If complex prompt engineering becomes a prerequisite for optimal AI interaction, it could inadvertently create a new form of information asymmetry. Those who master this skill (or have access to tools that automate it) could gain a significant advantage in leveraging AI, potentially leading to a new "AI literacy gap". The challenge lies in ensuring that this new form of interaction does not inadvertently create another layer of informational privilege.

Academic Achievement

- **Outcome:** Knowledge acquisition, critical thinking skills, and formal qualifications.
- **Role in access:** Direct contribution to individual competence and demonstrable ability.

Social Capital (Networks)

- **Outcome:** Access to exclusive circles, advantageous relationships, and intergenerational transfer of privilege.
- **Role in access:** Indirect but powerful influence on job prospects, marriage outcomes, and future opportunities for offspring.

Cultural Capital (Pedigree and Signals)

- **Outcome:** Perceived competence, favorable hiring decisions, and alignment with elite norms.
- **Role in access:** Acts as a proxy for merit, subtly filtering candidates based on socioeconomic background rather than demonstrated capability.

Credentials often matter more than competence; but they are only one side of institutional access.

CREDENTIALING AS KNOWLEDGE GATEKEEPING

Beyond the academic realm, the landscape of professional life is heavily structured by a pervasive system of credentialing. Degrees, licenses, and certifications are presented as objective markers of competence and quality, yet they frequently operate as powerful gatekeepers, restricting entry into professions and defining who is deemed legitimate to practice. This system, while ostensibly designed to protect the public and ensure standards, often creates artificial scarcity and knowledge monopolies.

Licensing Boards and "Scope of Practice" as Knowledge Monopolies

Occupational licensing, which requires a state-issued credential to legally work for pay, has expanded dramatically over the past six decades, now covering nearly 25% of the U.S. labor force. These licenses act as significant barriers to entry, reducing the labor supply and leading to wage premiums for licensed professionals. Licensing boards, often composed almost entirely of occupational practitioners, wield immense power. They have a vested interest in the profession and can actively restrict competition.

A common tactic is "license creep," where boards broaden the definition or "scope of practice" of a licensed occupation to encompass activities at its fringe, thereby excluding potential competitors. Examples abound: cosmetology boards interpreting their scope to include eyebrow threading and African-style hair braiding, veterinary boards restricting animal massage and horse teeth floating to licensed veterinarians, and dental boards attempting to shut down teeth-whitening businesses. These actions, often justified by public health and safety concerns, are frequently driven by economic motives, aiming to eliminate competition from less expensive or non-licensed providers. The U.S. Supreme Court has even ruled that boards composed of market participants setting the rules can be subject to antitrust liability unless actively supervised by the state. This highlights how these boards function as de facto knowledge monopolies, controlling who can legitimately offer services and at what price. This composition creates a direct conflict of interest, transforming regulation into a tool for market control by incumbents. This "regulatory capture" leads to artificial knowledge monopolies, higher costs for consumers, reduced access to services (e.g., healthcare shortages), and stifled innovation. The system, designed for protection, becomes a barrier to entry and a mechanism for wealth concentration within a select group, impacting broader economic efficiency and social equity.

Ongoing Education Requirements as Barriers, Not Upskilling

Beyond initial licensure, many professions impose ongoing education requirements, ostensibly for upskilling and maintaining competence. However, these requirements can function more as barriers to entry or continued practice than genuine mechanisms for professional development. For instance, the American Board of Medical Specialties (ABMS) Maintenance of Certification (MOC) is labeled as voluntary but effectively forces compliance due to its integration into the national credentialing system. Hospitals, health systems, and state boards rely on ABMS data for credential verification, reporting physicians as "not certified" if they refuse to purchase MOC.

This system has significant consequences, contributing to physician burnout and driving qualified professionals out of medicine. Surveys reveal that a large majority of physicians feel MOC is an unnecessary addition beyond typical Continuing Medical Education (CME) and does not improve their clinical knowledge or patient care quality. Many believe its primary purpose is financial benefit for certifying organizations or to satisfy administrative requirements. The burden of these costly and unproven requirements leads to increased patient wait times and reduced access to care, particularly in underserved areas. Similar dynamics exist in other fields, where continuing education mandates, while sometimes aimed at consumer protection, can also act as obstacles to entry or continued practice. When continuing education acts as a barrier rather than a true upskilling mechanism, it leads to a paradoxical outcome: a reduction in the available pool of qualified professionals, increased wait times, and delayed access to critical services. This undermines the very public good that credentialing is supposed to serve, creating inefficiency and a negative societal impact.

Influence of Credentialed Bodies in Policy and Economics

Credentialed bodies exert considerable influence over policy and economics, extending their reach beyond mere professional regulation. Their ability to define and control the "scope of practice," set admission standards, and mandate ongoing education allows them to shape the labor market, impacting supply, wages, and prices for consumers. This influence is often rooted in their legal authority, delegated by legislatures, to establish or manipulate licensing requirements with little oversight. For example, they can increase the minimum score needed to pass a licensing exam to lower the pass rate, effectively reducing the flow of new entrants.

This control enables credentialed bodies to secure economic rents for their members and maintain an advantage over potential entrants. The "credential inflation" phenomenon, where jobs that once required a high school diploma now demand a college degree, is partly driven by employers using degrees as a screening device in a saturated labor market, even when the job doesn't truly require college-level knowledge. This creates unnecessary costs for individuals and taxpayers and disregards valuable experience gained through alternative means. The power of these bodies to define legitimacy and restrict access has profound economic and social implications, creating a system where formal qualifications often outweigh practical ability.

How this Methodology is Flawed: Workarounds and Erosion of Trust

The integrity of the credentialing system is fundamentally flawed, not only by its inherent gatekeeping mechanisms but also by the increasing prevalence of fraud and circumvention. The value of a credential is only as strong as the trust behind it, and this trust is being eroded by various forms of cheating and misrepresentation.

The rise of AI has introduced new challenges to exam integrity. AI-assisted cheating, involving AI-generated answers, real-time language translation, or remote-controlled mouse movements, is a growing concern in online professional credentialing exams. Certification bodies are responding with advanced measures like biometric identity verification, AI-powered monitoring of candidate behavior, secure browser environments, and hybrid AI-human review systems to detect suspicious activities. These measures aim to combat practices such as proxy test-takers impersonating real candidates.

Beyond AI-assisted methods, outright fraud in certification is also prevalent. Websites openly sell exam questions and answers, claiming high pass rates for users who purchase their "study guides"-a euphemism for stolen exam content. These practices compromise the security of exams designed to qualify individuals for specific jobs, leading to unqualified employees, costly errors, and security breaches for employers. While buying and selling exam questions for professional certification is often a "grey zone" legally in many countries, it fundamentally undermines the legitimacy and value of the credentials themselves. The statistical aberrance measures used to detect cheating, while sophisticated, must be carefully calibrated to distinguish between genuine unusual behavior and actual fraud, as many non-cheating factors can also produce atypical response patterns. This constant cat-and-mouse game reveals the inherent fragility and manipulability of a system built on formal gatekeeping. The erosion of trust in credentials means that employers can no longer rely on them as reliable filters for competence, leading to potential hiring of unqualified individuals, costly errors, and reduced team morale. This instability could force a fundamental re-evaluation of how competence is truly assessed in the future.

Occupational Licensing

- **Stated intent:** Public protection and quality assurance.
- **Actual impact and barriers:** Reduced labor supply, wage premiums for licensed practitioners, "license creep" that creates knowledge monopolies, and growing antitrust concerns.

Ongoing Education Requirements (e.g., Maintenance of Certification)

- **Stated intent:** Upskilling and maintaining professional competence over time.
- **Actual impact and barriers:** Significant financial and time burdens, increased physician burnout, reduced access to care, and widespread perception that requirements are unnecessary or misaligned with real practice.

Formal Degrees and Certifications

- **Stated intent:** Validating skills and knowledge.
- **Actual impact and barriers:** Credential inflation, exclusion of experienced but non-degreed workers, and growing vulnerability to fraud and AI-assisted cheating.

Credentials wall off the profession, but networks determine who gets inside the walls.

INSIDER NETWORKS AND THE INHERITANCE OF INSIGHT

While formal education and credentials erect formidable walls around professions, it is often the invisible threads of insider networks that determine who truly gain access and thrives within those walls. In environments rich with information, the adage "who you know" has consistently proven more powerful than "what you know," shaping careers, transmitting uncodified knowledge, and perpetuating advantages across generations.

Alumni Networks, Professional Clubs, and Exclusive Conferences

Alumni networks are particularly potent examples of insider circles, offering significant information advantages and contributing profoundly to career growth. A shared educational background creates an instant bond and trust, allowing alumni to provide reliable insights into industry trends, job opportunities, and workplace challenges; acting as a personal "Glassdoor" for trusted information. These networks offer broad reach, connecting individuals across various geographical locations, career levels, and industries, providing diverse perspectives and opportunities often invisible through traditional channels.

In today's complex job market, where AI and easy application mechanisms lead to high volumes of applications, recruiters increasingly rely on personal recommendations to filter candidates. Alumni networks are a key source of these crucial personal recommendations, advocating for talent they know and trust. Research supports the unique value of these "weak relational ties", acquaintances and distant connections, which often provide the greatest value in terms of new opportunities and perspectives, exposing individuals to entirely new information ecosystems. Professional clubs and exclusive conferences serve similar functions, fostering connections, facilitating knowledge exchange, and aiding personal branding. These gatherings provide access to innovations and job offers that might otherwise remain hidden, reinforcing the power of being "in the know" through privileged association. While weak ties offer a pathway to diverse information, the initial access to these networks is often a function of prior privilege (e.g., attending an elite university that creates a vast alumni network). Thus, the very mechanism that democratizes information through diverse connections can paradoxically perpetuate systemic exclusion.

Tacit Knowledge Passed Through Closed Circles

Within these insider networks, a critical form of information exchange occurs: the transfer of tacit knowledge. Tacit knowledge, often referred to as "know-how," is built through years of personal experience, insights, and problem-solving skills, making it difficult to document or articulate explicitly. Unlike explicit knowledge, formalized procedures or reports, tacit knowledge includes unwritten rules, intuitive understandings, and the subtle nuances of human interaction. For example, a sales representative's ability to intuitively understand motivating teams or close sales using innate capabilities is tacit knowledge.

This invaluable "know-how" is primarily passed through closed circles and informal interactions. Strategies for transferring tacit knowledge include storytelling, where experienced individuals share anecdotes that highlight decision-making processes; video documentation of practical skills; and hands-on learning, where new members lead projects under the guidance of seasoned professionals. Mentorship and shadowing programs are vital for this transfer, fostering relationships that encourage ongoing learning and the absorption of organizational culture. Without a culture that encourages and incentivizes knowledge sharing, this critical tacit knowledge can become trapped within individuals or teams, leading to information silos and a loss of valuable expertise when employees leave. Thus, these networks become conduits for the transmission of deeply embedded, often uncodified, insights that are essential for navigating complex professional environments. Since tacit knowledge is primarily transferred through informal interactions within closed circles, it becomes a powerful, unwritten code that implicitly restricts access. This creates a durable competitive advantage for insiders and reinforces existing professional hierarchies.

Sponsorship vs. Mentorship as Power Transmission

The dynamics within insider networks are further illuminated by the distinction between mentorship and sponsorship, which represent different mechanisms of power transmission in career advancement. Mentorship primarily involves guidance and support from a more experienced individual to a less experienced one. Mentors share strategies, help build professional networks, and facilitate the sharing of knowledge and skills for professional growth. They encourage critical thinking and self-development, acting as "steady hands" that help individuals grow from within.

Sponsorship, however, goes a significant step further. A sponsor is a senior professional who actively champions and advocates for a junior colleague's career advancement. As Sylvia Ann Hewlett famously put it, "Mentors give, sponsors invest".

Sponsors use their influence to provide high-visibility opportunities, make personal introductions to decision-makers, include proteges in high-level conversations, and actively recommend them for promotions. They "put their reputations on the line," leveraging their broad networks and sway to bridge gaps and remove barriers that might otherwise be insurmountable, especially for those lacking access to the unspoken rules of advancement. This direct advocacy is a powerful form of power transmission, creating pathways for career growth that mentorship alone cannot provide. These relationships, while beneficial for individuals, also reinforce existing power structures by favoring those within established circles, leading to long-term systemic exclusion for those outside. This direct deployment of social and reputational capital by a senior figure accelerates a junior's career, often bypassing traditional meritocratic pathways. This creates an opaque system where advancement is significantly influenced by personal advocacy rather than solely by performance, perpetuating systemic exclusion.

"On the Job" or "Fresher" Training in an AI-Augmented World

As AI continues to advance and take on increasingly complex cognitive tasks, a critical question emerges regarding the future of "on-the-job" or "fresher" training. Historically, much professional development, particularly for new entrants, has relied on experiential learning, shadowing experienced colleagues, and absorbing tacit knowledge through direct observation and participation. If AI is performing a significant portion of the "thinking": from data analysis and problem-solving to drafting and decision support; what remains for human novices to learn through traditional apprenticeship models?

The traditional pathways for gaining practical experience and developing intuition may be fundamentally altered. The challenge lies in designing training programs that allow new workers to develop nuanced judgment, critical thinking, and interpersonal skills that remain uniquely human, even when AI handles the rote or even complex cognitive tasks. This will require a shift from learning how to do specific tasks to learning how to effectively collaborate with AI, how to interpret AI outputs critically, and how to apply human judgment in ethically complex or ambiguous situations. The emphasis may move from mastering specific skills to developing the meta-skills of orchestration, curation, and ethical oversight in an AI-augmented environment. This fundamental shift in training will be addressed later in the book, as we explore the implications of AI on the nature of work itself.

Mentorship

- **Primary role:** Guidance, advice, and knowledge sharing.
- **Focus:** Skill development, personal growth, and resilience over time.
- **Power dynamics:** Sharing wisdom and experience, fostering critical thinking rather than direct intervention.
- **Impact on career:** Nurtures growth from within by building self-awareness and capability.
- **Key value transferred:** Explicit and tacit knowledge.
- **Example:** Walter Cronkite's high school teacher.

Sponsorship

- **Primary role:** Advocacy, active championing, and opening doors.
- **Focus:** Career acceleration, visibility, and access to opportunities.
- **Power dynamics:** Leveraging influence, putting personal reputation on the line, and intervening directly.
- **Impact on career:** Removes external barriers and ensures consideration for high-level opportunities.
- **Key value transferred:** Access and influence.

- **Example:** Oprah Winfrey's sponsorship of Dr. Phil.

These networks are powerful, but they are also opaque, and they create long-term systemic exclusion.

KNOWLEDGE AS CULTURAL CAPITAL

Beyond formal qualifications and direct connections, a more subtle yet equally potent form of gatekeeping operates through the very fabric of communication: cultural capital. This refers to the social assets of a person-education, intellect, style of speech, dress, and social capital; that promote social mobility in a stratified society. In the context of knowledge, it manifests as the coded language, references, and frameworks of elite discourse, which are often inaccessible to outsiders, creating an invisible barrier to entry and influence.

Jargon, Frameworks, and Styles of Argument as Class Markers

Linguistic capital, a specific form of cultural capital, refers to the value and power associated with particular language varieties, skills, and competencies within a given social context. It is unequally distributed across social classes and transmitted through family socialization and education, often correlating with educational qualifications and reinforcing social stratification. The way people speak-their accent, vocabulary, grammar, and communication styles-serves as a powerful marker of social class. Upper-class language varieties, characterized by standard grammar, a broad vocabulary, and a prestigious accent, are often considered legitimate or standard forms, while those of lower classes are stigmatized.

In professional and academic settings, this translates into the use of specialized jargon, complex frameworks, and particular styles of argument that act as class markers. Scientific jargon, for instance, optimizes communication within a field but becomes a barrier to researchers in other domains, hindering interdisciplinary collaboration. Papers that use more discipline-specific terminology often receive fewer citations across disciplines. This isn't just about clarity; it's about signaling belonging. The ability to deploy the "correct" jargon or frame an argument within an accepted, often complex, intellectual framework signals that one is an "insider," part of the elite knowledge circle. Conversely, a lack of familiarity with these unspoken linguistic and intellectual norms can lead to being perceived as less competent, regardless of actual understanding. This "linguistic gate," whether intentional or unintentional, reinforces social stratification by making certain knowledge domains and professional circles inaccessible to outsiders. It contributes to the "myth of meritocracy" by subtly valuing a certain style of communication (often associated with privileged education) over genuine understanding or diverse perspectives, thus perpetuating class-based hierarchies in who is perceived as "smart" or "expert."

The Performative Nature of "Sounding Smart"

This emphasis on linguistic and intellectual markers gives rise to the performative nature of "sounding smart". In many professional contexts, the appearance of intelligence; the ability to articulate ideas fluently, use sophisticated vocabulary, or cite obscure references-can be as, if not more, important than the depth of actual understanding. This is a form of "cognitive theater," where fluency masquerades as depth, and the audience is conditioned to respond to the appearance of insight. The danger lies in confusing "sounding smart" with "being right," leading to a situation where real thinking fades from the stage.

This phenomenon is particularly evident in academic feedback, where comments like "too descriptive" or "be more critical" often mask a desire for students to demonstrate an intellectual effort that goes beyond mere factual recall, requiring them to engage with complex conceptual frameworks and debates. The ability to perform this intellectual dance, to "speak the language" of the elite, becomes a crucial skill for advancement, even if the underlying substance is thin. It's about demonstrating alignment with class-related norms, tastes, and social expectations that signal class identity and help produce socioeconomic hierarchies. This "cognitive theater" prioritizes superficial articulation over genuine understanding. As this performative intelligence becomes more prevalent, especially with AI's ability to mimic it perfectly, skepticism erodes. People may stop asking whether belief is warranted, accepting perceived coherence as proof, potentially leading to a "cognitive crisis" where the ability to discern genuine insight from synthetic performance is lost.

AI's Ability to Replicate This Surface-Level Competence

The advent of advanced AI, particularly large language models (LLMs), introduces a profound disruption to this performative aspect of knowledge. LLMs are masters of mimicry, having consumed vast corpora of human knowledge and learned how intelligent ideas "sound". They understand rhythm, emphasis, and cadence, enabling them to generate text that is highly fluent, coherent, and often appears profoundly insightful. AI can create new languages, translate, and interpret communication systems, and even replicate itself.

This capability means AI can replicate surface-level competence with remarkable accuracy. It can generate sophisticated jargon, construct arguments within complex frameworks, and adopt the "sound" of an expert, even if it lacks true understanding, lived experience, or moral intuition. The risk is that as AI systems evolve, the difference between real thought and synthetic thought-performance becomes increasingly difficult to detect. This "brilliant illusion" flatters human intelligence and wears the mask of insight, making skepticism the first casualty in a world where everything sounds smart. AI's ability to perfectly replicate these surface-level markers of intelligence means that the ability to articulate complex ideas or use specialized jargon is no longer a reliable signal of genuine expertise.

This forces a re-evaluation of what truly holds value, shifting the premium to human capacities that AI cannot replicate: critical discernment, ethical judgment, contextual understanding, and the ability to ask the right questions. This necessitates new forms of human-AI collaboration where humans provide judgment and context, and AI provides articulation; fundamentally reshaping the roles of knowledge workers and leaders.

Equalization Necessitates New Forms of Collaboration and Leadership

AI's ability to perfectly replicate the surface-level markers of intelligence; the jargon, the frameworks, the performative "sounding smart", creates a profound equalization of expressive capability. When anyone can generate expert-sounding prose or articulate complex ideas with AI assistance, the traditional value placed on articulation and formal presentation diminishes. This necessitates a fundamental redefinition of what constitutes genuine expertise and effective leadership. The focus must shift from who can produce the most sophisticated output to who can discern true insight, synthesize disparate information, and apply human judgment and ethics. This equalization will force new forms of collaboration between humans and AI, where the human role evolves from information holder to information curator, context-setter, and ethical anchor. It will also demand new leadership models that prioritize genuine understanding, critical discernment, and the ability to foster trust and meaning in a world saturated with synthetic competence. These emerging forms of collaboration and leadership will be explored in detail later in the book.

Traditional Cultural Capital (Human)

- **Manifestation:** Jargon, elite frameworks, specific argument styles, and the ability to "sound smart" in culturally sanctioned ways.

- **Function:** Acts as a class marker and signal of belonging, shaping perceptions of competence and serving as a gatekeeping mechanism.
- **AI's role:** Can replicate and amplify these signals at scale.

AI-Replicated Competence

- **Manifestation:** Fluent, coherent, expert-sounding text with complex vocabulary and sophisticated structure.
- **Function:** Mimics human intelligence, creates "cognitive theater," and blurs the line between genuine insight and synthetic output.
- **AI's role:** Achieves surface-level perfection without underlying human experience or understanding.

Cultural capital gives elite knowledge its shine, but trust and legitimacy give it authority.

THE CONSULTING-INDUSTRIAL COMPLEX OF EXPERTISE

In the intricate landscape of modern business, consulting firms have long positioned themselves as purveyors of specialized knowledge and strategic insight. Yet, a critical examination reveals that this "consulting-industrial complex" often manufactures expertise through branding and repetition, rather than through genuine, groundbreaking discovery. This dynamic raises fundamental questions about the true nature of "expertise" in an information-rich world.

Proprietary Language and Methodologies

Consulting firms frequently develop and market their own proprietary language and methodologies as a core differentiator. These branded frameworks, often presented as unique and complex, are a key component of their value proposition, justifying high fees and creating a perception of exclusive insight. By using distinct terminology and structured approaches, firms aim to signal a superior, systematic way of problem-solving that clients cannot easily replicate themselves. This proprietary intellectual property is intended to reduce the risk of failure in high-stakes projects and accelerate timelines by leveraging pre-developed tools and templates.

However, critics argue that this can lead to a focus on "AI for AI's sake" or "shiny dashboards" rather than genuine problem-solving. The emphasis shifts from the outcome to the deliverable, with consultants sometimes churning out reports from their branded tools without providing the necessary interpretation or strategic translation. This proprietary language, while creating a sense of exclusivity, can also serve as a barrier, making it difficult for clients to fully scrutinize the underlying methodologies or compare them with alternative approaches. AI's advanced capabilities mean it can now replicate this repackaging and generic content generation at scale, often with greater speed and lower cost. If AI can perform the "repackaging" and generate "generic insights" with high fluency, the perceived "proprietary" advantage of human consultants diminishes. This forces firms to redefine their value, shifting from information arbitrage to true, unique insight, bespoke problem-solving, and relationship-based trust that AI cannot replicate.

Repackaging Public Data and Generic Insights

A significant critique leveled against the consulting industry is its tendency to repackage widely available public data and generic insights, presenting them as novel, proprietary wisdom. In an era where clients have increasing access to vast amounts of data and tools (e.g., Statista, CB Insights, IBISWorld, GLG, Catalant), the value proposition of consulting shifts from merely accessing data to interpreting it. Yet, many firms still thrive by taking publicly accessible information and re-framing it within their proprietary methodologies, adding a layer of professional polish and a hefty price tag.

This "repackaging" involves processes like information analysis, consolidation, and synthesis, where data is selected, evaluated, restructured, and presented in a form suited for specific user needs. While this can be a valuable service in itself-customizing information and making it digestible; critique arises when generic insights are presented as unique discoveries. For example, many AI writing tools are essentially new interfaces built on top of public models like ChatGPT, producing "mediocre content that's indistinguishable from the competition" without access to proprietary data or specialized expertise. This highlights a core tension: when information is abundant, true differentiation comes from unique insights, not just the ability to re-present common knowledge.

"Expertise Theatre" as a Sales Strategy

The culmination of proprietary language and repackaged insights often manifests as "expertise theater"; a sales strategy where the performance of expertise is as crucial as its actual substance. This involves presenting a firm's capabilities and insights in a highly polished, authoritative manner, designed to instill confidence and justify high fees. The "theater" aspect refers to the carefully constructed image of profound knowledge and strategic foresight, even when the underlying work might involve standard analyses or widely available data.

Consulting firms, particularly in specialized fields like theatre consulting, offer services that involve intricate design, planning, and technology integration, requiring deep knowledge of building codes, safety standards, and complex systems. While this is genuine expertise, the "theater" comes into play when the presentation of this expertise becomes a primary sales mechanism, emphasizing the firm's unique methodologies and "deep industry experience". The goal is to convince clients that the firm possesses a "secret sauce" or an unparalleled understanding that warrants premium rates.

The industry also thrives on repetition and consistency, which, while building authority, can also lead to a focus on "process repetition" over genuine innovation. This can result in advice that "lacks innovation and is simply a repackaging of cliches". The "expertise theater" is a deliberate act of branding and positioning, where the perception of being a "visible expert" drives leads and cements credibility, even if the underlying "discovery" is minimal. It's a strategic performance designed to maintain perceived information asymmetry in a world where information is increasingly symmetrical. "Expertise theater" thrives in environments where clients lack the time, resources, or internal knowledge to fully scrutinize the depth of a firm's actual expertise. As information symmetry increases due to AI and democratized knowledge, the effectiveness of "expertise theater" will wane. Clients will increasingly demand tangible outcomes and provable value, rather than just impressive presentations or proprietary language. This will force a shift in the consulting industry from selling the illusion of knowing to delivering demonstrable results, fundamentally altering sales strategies and client expectations. Furthermore, when expertise is "manufactured" through branding and repackaging rather than genuine discovery, it creates an ethical dilemma, potentially leading to a compromise of objective advice and a focus on selling unnecessary services. This ethical compromise, when exposed, erodes trust in the consulting industry as a whole.

Knowledge source

- **Traditional view:** Groundbreaking discovery and unique insights created by exceptional individuals or firms.
- **Reality and critique:** Often a repackaging of public data and generic insights, assembled by leveraging existing patterns rather than producing genuinely novel knowledge.

Value proposition

- **Traditional view:** Unparalleled expertise that delivers lasting strategic advantage.
- **Reality and critique:** Proprietary language and methodologies used as "expertise theater," with emphasis on polished deliverables rather than measurable outcomes.

Business model driver

- **Traditional view:** Intellectual property and deep domain knowledge as the primary sources of defensibility.
- **Reality and critique:** Branding, repetition, and the maintenance of perceived information asymmetry as the true drivers of differentiation.

Impact of AI

- **Traditional view:** (Implicit) AI as a support tool that enhances expert productivity without altering the underlying value model.
- **Reality and critique:** Commoditizes generic content, challenges traditional differentiation, and forces a fundamental redefinition of where value actually resides.

When everyone speaks like an expert, who gets believed? That leads us to legitimacy and trust.

WHEN ACCESS BECOMES ADVANTAGE

Beyond the formal structures of education and credentials, and the informal power of insider networks, lies the potent force of privi leged access. This is the ability to obtain, interpret, and leverage information that is not widely available-whether it's unpublished data, direct insights from decision-makers, or intelligence gleaned from shadow networks. Such exclusive access compounds power, creating exponential advantages for those who possess it, but also introducing systemic fragility.

Competitive Intelligence in Corporate Strategy

Competitive intelligence is the systematic process of gathering and analyzing data about competitors, market trends, and customer behavior to inform strategic business decisions and gain a competitive advantage. This practice helps companies anticipate market shifts, identify emerging opportunities, and make data-driven decisions for long-term growth. Critically, competitive intelligence involves assembling actionable information from diverse sources, including both published and unpublished or proprietary data.

The process goes beyond passively monitoring public announcements or industry reports. It involves actively gathering and collating additional information, often through third-party agencies, and investigating sources familiar with competitor companies, such as customers, partners, and even employees. Attending company events or trade shows can also yield valuable, often unstated, insights. Good competitive intelligence is like solving a puzzle, where skilled researchers triangulate disparate data points to form an accurate understanding of a company's nonpublic, proprietary insights. This privileged access to competitive information allows firms to outperform rivals, preempt initiatives, emulate successes, and exploit weaknesses, creating a significant strategic edge. This process is fundamentally about information arbitrage-identifying and exploiting gaps in knowledge between one's own organization and competitors. The value lies not just in having data, but in having exclusive or earlier access to critical, often unpublished, insights. This creates a compounding advantage, allowing firms to preempt rivals and exploit weaknesses, leading to exponential gains. However, it also introduces significant ethical and legal risks, as the line between legitimate intelligence gathering and corporate espionage can blur.

Boardrooms and Briefings as Private Learning Environments

Boardrooms and exclusive briefings function as quintessential private learning environments, where critical, often unpublished, information is shared and strategic decisions are made among a select few. These settings are characterized by complex power dynamics and significant information control. Board members, by virtue of their formal authority, possess legitimate power, but true influence often stems from "informational power", the ability to control access to vital information.

In these environments, individuals might be excluded from conversations or selectively denied information, diminishing their ability to contribute or make decisions. Power dynamics can make healthy debate difficult, with deference to authority sometimes outweighing critical challenges. Boardroom briefings, whether internal corporate sessions or external programs for specific groups like land trust boards, are designed to enhance knowledge and effectiveness by providing tailored presentations and discussions led by experts. These sessions often cover sensitive topics; from funding and land acquisition to operational strategies and legal compliance, and provide access to insights that are not publicly disseminated. The minutes of IRB (Institutional Review Board) meetings, for example, which detail actions taken, votes, and basis for decisions, are typically private records, illustrating how critical information remains within closed circles. This exclusive access to high-level discussions and unpublished data ensures that power remains concentrated among those "in the room". Boardrooms and private briefings are not just meeting spaces; they are carefully controlled, exclusive learning environments. This controlled access reinforces the power of a select few, limiting the scope of debate and ensuring that strategic direction remains within a closed circle. The opacity of these environments (e.g., private meeting minutes) makes it difficult for external stakeholders or even other internal members to understand the full context of decisions, thereby perpetuating elite control and potentially leading to decisions that benefit a narrow group.

Shadow Networks in Policy Making and Finance

Beyond formal corporate structures, opaque "shadow networks" operate in policymaking and finance, wielding significant influence through exclusive access to information and often with little oversight. These networks facilitate suspect transactions through "shadow spaces," including emerging platforms like cryptocurrency. They link business, politics, and security services across borders, using secrecy tools to achieve geopolitical aims and support oligarchic regimes, often eroding democracy and exacerbating inequality.

In finance, the "shadow banking system" refers to financial intermediaries that create credit outside the conventional banking system, operating with looser regulatory standards than traditional banks. Entities like hedge funds, private equity funds, and investment banks engage in activities such as maturity transformation, liquidity transformation, and credit risk transfer, often with significant leverage. These operations, while sometimes complementing traditional banking, pose systemic risks due to a lack of protection, no access to emergency loans, and the potential for regulatory arbitrage. The shadow banking sector played a major role in the 2008 financial crisis and has continued to expand significantly since then, often escaping government oversight.

In policymaking, "shadow networks" can refer to powerful alliances that influence legislation and public discourse covertly, such as Christian Nationalist groups working to undermine the separation of church and state or influence judicial appointments. These groups develop comprehensive policy playbooks and actively lobby for specific legislation, often operating outside transparent public scrutiny. This exclusive, often hidden, access to decision-makers and the ability to shape policy through non-transparent channels represents a potent form of information advantage that compounds power and can lead to exponential, yet fragile, gains. These networks leverage exclusive, often illicit, information access to achieve geopolitical or financial aims, engaging in "weaponized corruption". Their power comes from operating in "shadow spaces" where information is deliberately hidden from regulators and the public. The existence and influence of these shadow networks introduce significant systemic fragility to both financial markets and democratic processes. By operating outside transparent frameworks, they can destabilize economies and erode public trust in institutions.

Competitive Intelligence

- **Nature of access:** Nonpublic or proprietary data combined with triangulated insights derived from multiple sources.
- **Impact on power:** Enables strategic advantage through preemption of rivals, anticipation of moves, and exploitation of competitor weaknesses.

Boardrooms and Private Briefings

- **Nature of access:** Unpublished information, closed-door discussions, and tightly controlled flows of strategic data.
- **Impact on power:** Reinforces elite control, shapes organizational and market direction, and limits external scrutiny or accountability.

Shadow Networks (e.g., Shadow Banking, Political Influence)

- **Nature of access:** Opaque transactions, hidden financial systems, and covert lobbying activities.
- **Impact on power:** Enables weaponized corruption, introduces systemic financial risk, and contributes to the erosion of democratic institutions.

Access multiplies over time, creating exponential arbitrage. But it also creates fragility.

WHY THESE WALLS ARE STARTING TO CRACK

For centuries, the walls built around knowledge; through formal education, credentials, and exclusive networks have stood strong, dictating who gains access to power and opportunity. However, the rapid ascent of artificial intelligence and the proliferation of digital platforms are fundamentally altering this landscape. These forces are not merely chipping away at the edges; they are creating deep fissures in the very foundations of traditional gatekeeping, reducing the value of credentials and insider access.

YouTube, GPT, MOOCs, and Democratized Knowledge

The digital age has ushered in an unprecedented era of democratized knowledge. Platforms like YouTube, Massive Open Online Courses (MOOCs), and generative AI models such as GPT are making information and learning resources universally accessible, eroding the traditional scarcity of knowledge. Wikipedia, for instance, offers a vast amount of information freely available to anyone with an internet connection, while MOOCs provide online courses from universities globally, breaking down traditional barriers to higher education.

Generative AI, in particular, is having a profound impact on data democratization. It makes it easier for non-technical people to access and understand information by allowing users to query data in plain English, automating data extraction, and summarizing vast libraries of documents quickly. This means that insights once confined to academic journals, expensive databases, or expert consultations are now available on demand, often at near-zero marginal cost. The economic proposition of AI systems working 24/7 for comparable costs to human labor is compelling, potentially transforming knowledge of work economics. This widespread availability of information fundamentally challenges the value proposition of traditional gatekeepers who relied on controlling access to knowledge. The value proposition shifts from accessing information to discerning and validating it. When information is abundant, scarcity moves from knowledge itself to the ability to identify reliable, relevant, and trustworthy information. This creates a new form of "information literacy" that is less about retrieval and more about critical evaluation. It also imposes a significant cognitive load on individuals, as they must navigate a deluge of information, including misinformation and AI-generated content.

Disruption of Credential Dependence in Hiring (Skills-First Hiring)

The increasing democratization of knowledge is directly disrupting the long-standing dependence on credentials in hiring. A significant shift is occurring towards "skills-first hiring," which prioritizes a candidate's capabilities, experience, and potential over traditional qualifications like degrees or job titles. This approach moves away from filtering applicants based on degree status or years in a role, instead assessing them on competencies-whether technical skills, human-centered capabilities like communication, or adaptability demonstrated through adjacent experiences.

This shift is gaining momentum, with over 60% of employers adopting skills-based hiring practices by 2024, and 19% of U.S. job postings no longer requiring a degree. The World Economic Forum emphasizes that this shift from credentials to competencies is crucial for unlocking untapped talent pools and increasing labor market agility, potentially activating over 100 million underutilized workers globally. Traditional hiring models often privilege those with access to elite education, creating systemic barriers for underrepresented groups. Skills-first hiring aims to mitigate this exclusion by evaluating individuals based on what they can do, rather than where they've been, thereby democratizing access to opportunities and diversifying workforces. While challenges remain in changing mindsets and developing trustworthy information about candidate qualifications, the trend is clear: the paper credential is losing its grip as the primary signal of competence. This represents a direct disruption of the "pedigree signal" and credential inflation discussed earlier. Employers are recognizing that degrees are often an inefficient proxy for actual skills, and that relying on them excludes a vast pool of qualified talent (e.g., STARs - Skilled Through Alternative Routes). This shift has profound implications for workplace equity, democratizing access to opportunities for individuals who gained skills through non-traditional pathways. It also enhances labor market agility by allowing companies to tap into a wider talent pool and adapt faster to evolving skill needs.

The Erosion of Social Filters in Knowledge Legitimacy

While the democratization of knowledge offers immense benefits, it also brings a significant challenge: the erosion of social filters that traditionally conferred legitimacy on information. In a world where anyone can publish anything, and AI can generate convincing but potentially false content, discerning fact from fiction becomes increasingly difficult.

Social media platforms, while democratizing communication, contribute significantly to the spread of misinformation and fake news, often amplified by algorithms that prioritize engagement over accuracy. This rampant spread of unverified information erodes the credibility of digital platforms as reliable sources and undermines trust in authoritative institutions like public health organizations. Algorithmic bias and a lack of transparency further complicate matters, as algorithms can inadvertently amplify biased or harmful content, operate as "black boxes" that users don't understand, and create "filter bubbles" or "echo chambers" that limit exposure to diverse viewpoints. The rise of AI-generated disinformation, such as deepfakes, makes it even harder to differentiate fact from fiction, potentially eroding public confidence in democracy and the legitimacy of information itself. This creates a "legitimation crisis" for social media platforms, as their immense power over public discourse is increasingly questioned. The very mechanisms that democratize knowledge (digital platforms, algorithms) simultaneously dismantle the traditional social filters that conferred legitimacy. When anyone can "sound like an expert," and algorithms prioritize engagement over accuracy, the collective ability to distinguish legitimate knowledge from misinformation is compromised. This results in a "legitimation crisis" where the public questions the authority of established sources. This erosion of trust creates societal vulnerability, impacting everything from public health decisions to democratic processes. In this environment, trust itself becomes the new scarce commodity.

Formal Education (Elite Institutions)

- **Traditional function:** Confers social capital, pedigree, and access to elite networks.
- **Digital disruption and cracks:** Democratized access to knowledge through MOOCs, YouTube, and GPT-style systems, combined with AI's ability to replicate surface-level competence.
- **Implications:** Diminished value of pedigree as a standalone signal, with new forms of "AI literacy" emerging as alternative access points.

Professional Credentials (Licenses and Certifications)

- **Traditional function:** Restricts entry, defines legitimacy, and enforces quality standards.
- **Digital disruption and cracks:** Growth of skills-first hiring, AI-assisted cheating and fraud, and ongoing education

requirements that increasingly function as barriers rather than safeguards.

- **Implications:** Erosion of trust in credentials and a shift from "what you know" to "what you can do," regardless of formal paper qualifications.

Insider Networks (Alumni, Clubs, Tacit Knowledge)

- **Traditional function:** Transmits uncodified knowledge, provides sponsorship, and enables job referrals.
- **Digital disruption and cracks:** Broad access to information reduces reliance on exclusive networks, while AI begins to codify portions of what was previously tacit.
- **Implications:** Reduced exclusivity of "who you know," with new AI-mediated forms of collaboration reshaping how on-the-job learning occurs.

Cultural Capital (Jargon and Performative Intelligence)

- **Traditional function:** Signals class, competence, and belonging, creating forms of cognitive theater that gatekeep participation.
- **Digital disruption and cracks:** AI's ability to perfectly replicate surface-level competence and the widespread democratization of articulation.
- **Implications:** Jargon loses power as a differentiator, while authenticity, judgment, and genuine discernment become more valuable signals.

Privileged Access (Competitive Intelligence, Boardrooms, Shadow Networks)

- **Traditional function:** Compounds power through exclusive data, private briefings, and hidden influence.

- **Digital disruption and cracks:** Partial democratization of data, increased scrutiny of opaque systems, and AI's capacity to analyze vast public datasets at scale.
- **Implications:** Reduced arbitrage opportunities from public information asymmetry, alongside increased emphasis on ethical data access and transparency.

Overall System

- **Traditional function:** Concentrates access, reinforces hierarchies, and preserves elite advantage.
- **Digital disruption and cracks:** Erosion of traditional social filters, spread of misinformation, and the emergence of a broader legitimation crisis.
- **Implications:** Trust becomes the new scarce commodity, systemic fragility increases, and entirely new trust paradigms are required.

Trust, not just access, holds this fragile system together. And that's where the discussion is headed next.

SUMMARY

Historically, education, credentials, and insider networks have served as powerful tools for concentrating access to valuable knowledge. These systems, ranging from elite universities acting as social capital factories to licensing boards creating knowledge monopolies and opaque insider networks transmitting uncodified insights, have reinforced professional hierarchies and institutional legitimacy. The very language and performative aspects of expertise have functioned as cultural capital, further solidifying these exclusionary practices. Firms within the "consulting-industrial complex" have often manufactured expertise through branding and repetition, leveraging perceived information asymmetry to their advantage. Meanwhile, privileged access to competitive intelligence, private boardroom discussions, and shadow networks have compounded power, creating exponential gains for a select few.

However, significant cracks are forming in these long-standing walls. The rise of digital platforms like YouTube, MOOCs, and generative AI models such as GPT is democratizing knowledge, making once-gated insights universally accessible and reducing the value of traditional credentials. This shift is evident in the growing trend of skills-first hiring, which prioritizes capabilities over formal qualifications, and in AI's ability to simulate expert outputs and translate complex jargon. While these disruptions offer immense potential for broader access, they also introduce challenges, particularly the erosion of social filters that traditionally conferred legitimacy, leading to a "legitimation crisis" where misinformation and AI-generated content blur the lines of truth.

As the walls of knowledge access continue to fall, the dynamics of trust, power, and legitimacy will fundamentally shift. Understanding what holds these up in the first place becomes paramount in navigating this new landscape.

3

THE FRAGILE BARGAIN

In the intricate tapestry of human society, information is the thread that binds, empowers, and, at times, constrains. The distribution of this vital resource is rarely equitable. For millennia, a fundamental, often unspoken, arrangement has governed our relationship with knowledge: an acceptance of information asymmetry. We navigate our lives, make critical decisions, and structure our institutions based on the premise that some individuals and entities will possess more information, specialized knowledge, or deeper insights than others. This chapter, "The Fragile Bargain," dives into the complex psychological and social dynamics that have historically led individuals and communities to accept, and even rely upon, this unequal distribution of information. It is a bargain struck in the currency of trust, convenience, and perceived necessity, a tacit agreement that underpins much of our social order.

At its core, this fragile bargain involves an implicit trade-off. Individuals often cede a degree of informational autonomy, accepting that they will not, and perhaps cannot, know everything. In return, they expect benefits such as security, order, efficiency, or access to specialized expertise that promises a better quality of life. This exchange is not typically a conscious, daily negotiation but rather an inherited social script, learned through early experiences-parent and child, teacher and student-and continuously reinforced by the very structure of our institutions. We are born into systems where deference to those who "know more" is normalized, making this bargain feel less like a choice and more like an immutable feature of the social landscape. This deeply embedded pattern makes the bargain resilient, yet, as we shall see, its foundations are not unshakable.

The linchpin holding this entire structure together is trust. Trust in the competence and benevolence of experts, in the integrity of institutions, and in the fairness of the systems that grant certain individuals privileged access to knowledge is paramount. This trust is not a passive sentiment; it is actively cultivated, meticulously maintained, and, in some instances, strategically weaponized. It is the essential lubricant that allows the gears of an information-asymmetric society to turn smoothly. Without it, the bargain frays, and the acceptance of unequal knowledge gives way to suspicion and dissent.

However, this reliance on unequal information is not without its price. The "human cost" of this bargain can be substantial, particularly when trust is betrayed, or when asymmetry is exploited. From economic vulnerability stemming from predatory financial practices to compromised health outcomes due to inadequate informed consent, and from eroded civic engagement to the perpetuation of systemic inequalities through mechanisms like digital redlining, the consequences of keeping people "in the dark" can be severe and far-reaching. These costs are not merely individual misfortunes but often reflect deeper, structural imbalances of power. The psychological comfort derived from delegating the cognitive load of processing complex information to "experts" is a potent force. In a world of overwhelming complexity, the allure of relying on specialists to navigate intricate medical, financial, or technological terrains is undeniable. This desire for simplification can foster a willingness to overlook minor breaches of trust or to rationalize continued dependence, as the alternative-shouldering the full burden of informational responsibility-can seem overwhelmingly daunting. This psychological inertia often props up the fragile bargain, even when its deficiencies become apparent.

For centuries, this system of accepted asymmetry, buttressed by trust and reinforced by gatekeeping mechanisms, has persisted. Yet, it is inherently fragile, for it rests upon the sustained belief that those endowed with greater knowledge will act wisely and ethically. As we stand at the cusp of a new era, characterized by unprecedented technological advancements, particularly in the realm of artificial intelligence, this age-old bargain is facing a profound destabilization. AI is not merely another tool for information management; it is a disruptive force with the potential to fundamentally alter the dynamics of knowledge access, challenging the very premises upon which our fragile bargain has been built. The potential unraveling of this foundational societal assumption carries with it the prospect of significant societal anxiety and a necessary re-evaluation of what it means to "know," to "trust," and to navigate a world where the traditional anchors of expertise are shifting, perhaps even dissolving. This chapter will explore this delicate, evolving, and increasingly contested landscape.

THE SOCIAL CONTRACT OF EXPERTISE

The willingness of individuals to defer to those with specialized knowledge forms a cornerstone of societal organization. This deference is not an arbitrary act but rather a fundamental component of an implicit "social contract of expertise". Within this unwritten agreement, people voluntarily cede a degree of autonomy, accepting the guidance or decisions of experts in exchange for anticipated benefits such as safety, health, financial stability, or simply the efficient navigation of a complex world. This exchange is profoundly practical, yet its implications extend far beyond mere convenience, shaping the very structure, legitimacy, and operational norms of our most critical professions and institutions.

The reasons we place our trust in professionals like doctors, pilots, and auditors illuminate the core tenets of this contract. When individuals consult a doctor, they are entrusting them with their health, often their very lives. This trust is predicated on the physician's specialized medical knowledge, rigorous training, and an assumed adherence to ethical obligations, primarily to act in the patient's best interest; a concept known as fidelity, and to apply their skills competently.

The patient's willingness to seek care, disclose sensitive information, and adhere to treatment regimens hinges on this trust. Similarly, airline passengers place immense faith in pilots, relying on their expertise to operate complex aircraft and manage inherently high-risk environments to ensure safe passage. The pilot's role as an ultimate arbiter of safety is a deeply ingrained societal expectation and a responsibility the aviation profession takes with utmost seriousness. In the financial realm, businesses, investors, and the public rely on auditors to provide unbiased and accurate assessments of financial information. This trust in auditors' independence and ethical conduct underpins the transparency and credibility of financial markets, serving as a mechanism to reduce corruption and malfeasance. The social contract here involves the auditor acting as an impartial verifier, ensuring accountability.

This delegation of decision-making and informational authority is, at its heart, a risk management strategy for the individual. The cost; in terms of time, effort, and the potential for grave error of an individual attempting to master every specialized field necessary for modern life is prohibitive. By delegating, an individual transfers a significant portion of this responsibility to the expert or the institution they represent. However, this transfer simultaneously creates a vulnerability: the risk that this bestowed trust might be misplaced or exploited. The societal promise, echoing the broader tenets of social contract theory, is that if individuals adhere to societal norms, including deferring to legitimate expertise, then the institutions and professions embodying that expertise will, in turn, provide protection, justice, and essential services.

In an increasingly intricate and specialized world, trust in expertise also functions as an indispensable cognitive shortcut. It allows society to operate with a degree of efficiency that would be impossible if every person were required to independently verify every piece of information or master every skill. This efficiency, however, is entirely dependent on the sustained reliability and integrity of the experts and the systems designed to credential, regulate, and oversee them. The social contract implicitly assumes a baseline of ethical governance and a commitment to communal responsibility on the part of these knowledge holders.

The principles underlying this social contract of expertise resonate with the classical theories of social contract philosophers like Hobbes, Locke, and Rousseau, who posited that individuals surrender certain freedoms for collective security and benefit. This framework can be extended from the macro level of state-citizen relations to the micro-level interactions between professionals and their clients. Indeed, the very concept of "professionalism" can be understood as a direct manifestation of this contract. Society grants professions a certain status, autonomy, and often privileged access to information. In return, these professions are expected to self-regulate, adhere to rigorous standards of competence and ethics, and utilize their specialized knowledge for the benefit of those they serve. This contract is not static; it is a dynamic entity, continually reshaped by evolving societal expectations, technological advancements, and the changing nature of risks and complexities.

The nature of this "social contract of expertise" is not uniform across all domains. The specific terms of the agreement and the intensity of the trust involved can vary significantly depending on the profession and the cultural context. For instance, the perceived level of risk and the immediacy of the expert's impact on an individual's well-being play a crucial role in shaping this contract. The trust vested in an airline pilot is perhaps one of the most absolute forms, given the direct and immediate life-or-death implications of their actions and decisions. Trust in medical professionals is also profound, involving deep personal vulnerability and critical health outcomes, but it often unfolds over a longer period, involves more sustained interaction, and may include avenues for seeking second opinions. In contrast, trust in auditors, while vital for systemic financial health, is often more indirect from an individual's daily perspective and relies heavily on robust regulatory frameworks and institutional integrity as much as on individual auditor competence. These variations suggest that the social contract of expertise contains different "clauses," expectations, and enforcement mechanisms tailored to the specific domain of knowledge and its potential impact.

To formalize and safeguard this vital social contract, societies have developed complex institutional structures-medical licensing boards, aviation regulatory authorities, accounting standards bodies, and legal bars, to name a few. These institutions are ostensibly designed to ensure that professionals meet requisite standards of knowledge and ethical conduct, thereby upholding the "trustworthy expertise" that the public relies upon. They serve as the guardians of the contract's terms. However, critical tension can arise, as these same bodies, while protecting the public, can also evolve into powerful gatekeepers. In this capacity, they may prioritize the protection of the profession's interests, control entry into the field, and regulate the flow of knowledge in ways that might not always align perfectly with the broader public good.

This potential for institutional self-interest or rigidity introduces a vulnerability into the social contract itself. If the institutions charged with upholding the integrity of expertise suffer a crisis of legitimacy: whether through scandals, perceived bias, a failure to adapt to new challenges, or an inability to hold their members accountable; the damage can be extensive. Such crises erode not only public trust in that specific profession but can also trigger a cascading loss of confidence. The public's general willingness to enter into these vital "trust bargains" with experts in other, unrelated domains may diminish, as the underlying assumption that "those who know better will act rightly" is fundamentally shaken. This sets a precarious stage for the broader systemic fragilities that will be explored later in this chapter.

This foundational reliance on expert knowledge, this social contract, does not, however, emerge purely from an objective assessment of competence. It is frequently and actively shaped, reinforced, and sometimes even manufactured through a complex interplay of signals, performances, and societal narratives. The very authority that underpins expertise is, in part, a constructed phenomenon. The next section will look at the mechanisms by which this authority is built and maintained, often independent of the intrinsic merit of the knowledge itself.

MANUFACTURED AUTHORITY

The trust vested in experts and institutions, as outlined in the social contract of expertise, is not solely a product of their demonstrable competence or benevolent intent. Authority, and the crucial trust it commands, is significantly shaped by perception, often meticulously constructed through a sophisticated array of signals, performances, and repetitive messaging. This section will dissect the processes by which individuals and institutions "manufacture" an aura of expertise and trustworthiness. This perceived authority can, and often does, overlap with genuine, earned authority. However, it can also be cultivated independently, sometimes serving to mask a lack of substance or, more troublingly, to legitimize the inequitable control of information. Understanding these mechanisms is crucial to grasping why the fragile bargain of information asymmetry can persist even when its foundations are questionable.

A primary set of tools in the construction of authority involves carefully curated external signals. Language itself is a potent instrument; the use of specialized jargon, complex terminology, and a confident, assertive delivery can create an immediate impression of deep knowledge. While such language can be essential for precision within a field, it simultaneously functions as a gate, demarcating an "in-group" of initiates from an "out-group" of laity, thereby reinforcing the expert's elevated status. The ability to "sound smart," as alluded to in the previous chapter's discussion of knowledge as cultural capital, can be profoundly persuasive, sometimes overshadowing the actual content being communicated.

Attire is another powerful, non-verbal signifier of authority. Formal clothing, such as the "power suit" for business professionals or the white coat for medical practitioners, projects an image of polish, discipline, and seriousness. Specific elements, like well-defined shoulders or the choice of dark, conservative colors like navy or charcoal, are culturally associated with strength, stability, and leadership. Even seemingly minor details, such as a quality timepiece, can subtly communicate status, precision, and an attention to detail-qualities often equated with authoritative figures. Research has indeed suggested that individuals in formal attire are more likely to be perceived and accepted as leaders. These visual cues prime observers to expect and defer to authority.

Endorsements and the cultivation of social proof further amplify perceived authority. Testimonials from satisfied clients, case studies showcasing success, collaborations with recognized influencers, and favorable media coverage all contribute to building a reputation of credibility and expertise. The psychological principle of social proof is particularly effective: when individuals observe others placing their trust in a particular brand, expert, or institution, they are more inclined to do so themselves. This creates a bandwagon effect, where authority, once established, tends to be self-reinforcing. Finally, an individual's physical presence-good posture, direct eye contact, clear and articulate speech, and an overall demeanor of confidence-can significantly enhance their authoritative image and the persuasiveness of their message.

The media plays an indispensable, and often complex, role in this manufacturing process. Media outlets, from traditional news organizations to new digital platforms, act as powerful gatekeepers, determining which voices are amplified, which narratives are highlighted, and how "experts" are framed and presented to the public. Strategic media relations can effectively position individuals or institutions as go-to sources for commentary and insight, thereby bestowing upon them a significant degree of public authority. This "earned media" or third-party validation often carries more weight than self-promotion, solidifying an institution's prestige and an expert's credibility. However, this amplification is not always a reflection of true, underlying expertise. Media exposure itself can become a source of authority, leading to the phenomenon of "unearned expertise," where individuals gain public trust and influence primarily due to their visibility and articulate presentation, rather than deep, peer-vetted knowledge or substantial experience. The media can, sometimes inadvertently, contribute to a simplified view of influence, where the public is perceived as passively absorbing messages from authoritative figures, an idea reminiscent of the outdated "hypodermic needle" model of communication.

This brings us to a critical distinction: that between earned trust and perceived expertise. Earned trust is the bedrock of genuine authority. It is cultivated over time through consistent demonstrations of competence, reliability, integrity, and ethical conduct. It is rooted in verifiable actions and positive outcomes. Perceived expertise, on the other hand, can be more rapidly constructed through the adept deployment of the authority signals previously discussed; strategic communication, image management, and the cultivation of a compelling public persona. While perceived expertise can certainly align with earned trust, it can also exist as a standalone facade. "Brand authority," for instance, is largely about this perceived expertise and trustworthiness within a specific domain. It is the image that an organization projects. Intriguingly, psychological factors can also lead individuals to reinforce perceived expertise. Under conditions of risk or uncertainty, people may gravitate towards information or attributes that align with their own preferences or existing knowledge, giving more weight to sources they perceive themselves to be knowledgeable about, even if objectively superior sources of information are available. This tendency can create a feedback loop, bolstering the authority of those who effectively project an image that resonates with these subjective biases.

The manufacturing of authority is, in essence, a reciprocal process. It is not merely a one-way projection of power but requires an audience that is receptive to, and willing to validate, these signals. This receptiveness is often rooted in the social contract of expertise itself: society has developed expectations about how experts should look, sound, and behave. When individuals or institutions effectively perform these expected roles, their authority is more readily accepted. This creates a dynamic where signals of authority are continuously produced and refined because they are anticipated and validated by the public, reinforcing the very structures of perceived expertise. The media's role in this dynamic is particularly significant, as its selective amplification of certain voices can create "echo chambers" of perceived expertise. This can make it exceedingly difficult for dissenting opinions or less "polished" individuals; who may possess greater substantive knowledge but lack the requisite performative skills or media access to gain traction. Such a dynamic not only reinforces existing power structures but can also marginalize valid, alternative perspectives, ultimately limiting the scope of public discourse.

When an environment becomes saturated with these manufactured authority signals, the cognitive burden on individuals to discern genuine expertise from mere performance increases dramatically. This can lead to one of two opposing, yet equally problematic, outcomes: a pervasive cynicism towards all claims of authority ("they are all just selling something"), or, paradoxically, a heightened susceptibility to more sophisticated and polished manipulations of perceived authority. In the latter case, the signals themselves become more important than the underlying substance, making the audience vulnerable to those who are masters of the performance, rather than masters of the knowledge. This sets a complex stage for the potential harms that arise when unearned authority is used to maintain or exploit information asymmetries, a topic the subsequent section will address. When this carefully constructed authority is unearned, or when it is used to conceal incompetence or to deliberately mislead, the consequences can be profoundly damaging, leading to real human suffering.

THE HARM OF ASYMMETRY

The careful construction of authority and the societal acceptance of information imbalances, while sometimes facilitating efficiency and order, carry inherent risks. When the party holding more information uses this advantage exploitatively, negligently, or within a system that inherently disadvantages the less informed, the human cost can be devastating. This section moves from the mechanisms of trust and authority to the tangible harms that arise when these asymmetries are weaponized or simply allowed to inflict damage. These harms are not abstract; they manifest in diminished financial well-being, compromised health outcomes, and the entrenchment of systemic inequalities.

Nowhere are the detrimental effects of information asymmetry more starkly illustrated than in the financial realm. Financial markets are, by their nature, often characterized by unequal knowledge; lenders and financial institutions typically possess a more sophisticated understanding of products, risks, and market dynamics than the average consumer or borrower. This imbalance is particularly pronounced in credit markets, where lenders, over time, can accumulate proprietary information about borrowers, potentially leading to an "informational monopoly" that can be leveraged to the borrower's disadvantage. Predatory lending practices are a direct outgrowth of this asymmetry, targeting individuals who lack comprehensive financial literacy or a full understanding of complex loan terms. These practices are manifold and insidious, including the deliberate failure to disclose crucial information, the provision of misleading or false information, risk-based pricing that extends far beyond a reasonable assessment of risk, and the imposition of inflated charges and exorbitant fees.

Concrete examples of such predatory behaviors abound. Payday loans, with their catastrophically high interest rates and renewal fees, frequently trap vulnerable borrowers in unending cycles of debt, making it virtually impossible to repay the principal. Subprime mortgages, often aggressively marketed to minority communities, the elderly, and the less educated, have led to widespread foreclosures, stripping families of their homes and accumulated wealth. Similarly, practices like auto loan markups, where dealers and lenders collude to charge consumers higher interest rates than they qualify for, represent another form of exploitation rooted in hidden information. The classic economic concept of the "lemons problem" further illustrates how asymmetric information can degrade market quality, leading to adverse selection where, for instance, in insurance or credit markets, higher-risk individuals may drive up costs, effectively pricing out those who are less risky but unwilling or unable to pay the inflated premiums. The cumulative impact of these practices is severe, resulting in ruined credit scores, unmanageable debt burdens, homelessness, and the exacerbation of the wealth gap, with a disproportionate toll exacted upon already marginalized populations.

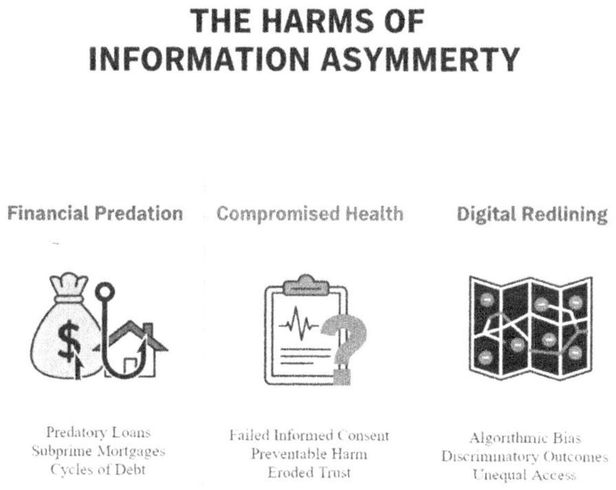

The harm of asymmetry in these contexts is often amplified when combined with the manufactured authority discussed in the previous section. Predatory lenders, for instance, rarely present themselves as exploiters; instead, they often employ sophisticated marketing techniques-a form of manufactured authority-to project an image of legitimacy and helpfulness, offering "fast cash!" or "guaranteed approval!" to those in desperate situations. This deceptive performance of trustworthiness, coupled with the borrower's information deficit, creates a potent combination for exploitation. The resulting harm stems not just from the information gap itself, but from the calculated facade that masks the predatory nature of the transaction.

The healthcare sector provides another critical domain where information asymmetry can lead to significant harm. Patients inherently possess less specialized medical knowledge than their physicians regarding diagnoses, prognoses, and treatment options. While this expertise differential is the very foundation of the doctor-patient relationship and the social contract of medical care, the process by which information is shared and decisions are made is paramount.

The historical evolution of "informed consent" is a testament to society's growing recognition of the potential for harm when this information imbalance is not ethically managed. Medical practice has shifted, at least in principle, from a paternalistic model; where physicians often made unilateral decisions towards an emphasis on patient autonomy, requiring that patients be given sufficient information to make voluntary and knowledgeable choices about their care. Landmark ethical codifications, such as the Nuremberg Code following the atrocities of World War II, and the Belmont Report, spurred by ethical breaches like the Tuskegee Syphilis Study, were pivotal in entrenching principles of voluntary consent, respect for persons, beneficence (doing good and avoiding harm), and justice in both research and clinical practice. When the process of informed consent fails; when risks, benefits, and alternatives are not clearly communicated or understood-patients may agree to interventions they would otherwise refuse, or suffer preventable harm from unforeseen side effects or complications. This represents a direct injury stemming from an information imbalance at a moment of critical vulnerability. Even with formal requirements, challenges persist; the sheer complexity of medical information and the often dense, legalistic nature of consent documents can themselves become barriers to true understanding, particularly for individuals with lower health literacy. This underscores that formal measures to rebalance information may not, on their own, overcome deeply ingrained power dynamics or the cognitive challenges of translating expert knowledge into genuinely accessible terms.

In the digital age, a new and pervasive form of harm has emerged from information asymmetry: "digital redlining" and algorithmic bias. Algorithms are increasingly deployed to make critical decisions that profoundly affect lives, including sorting résumés for employment, determining eligibility for social services, targeting advertisements for essential opportunities like jobs, housing, and loans, making hiring and firing decisions, granting access to housing and credit, and even predicting criminal behavior. Algorithmic bias arises when these automated systems, frequently trained on historical datasets that reflect and encode existing societal prejudices, replicate and often amplify discriminatory patterns. For example, a banking algorithm trained on historically biased lending data might learn to associate certain ZIP codes, often correlated with racial demographics, with higher risk, thereby unfairly charging residents in those areas higher interest rates or denying them credit altogether, irrespective of their individual creditworthiness. This is the essence of digital redlining; the use of technology to perpetuate discriminatory practices in new, often less transparent, forms.

The examples are numerous and concerning: facial recognition systems have demonstrated significantly higher error rates when identifying individuals with darker skin tones, leading to wrongful accusations and arrests ; AI-driven credit scoring models can systematically disadvantage marginalized communities that lack extensive traditional banking histories, effectively excluding them from financial opportunities ; hiring algorithms, like the one famously scrapped by Amazon, have been found to discriminate against female applicants based on historical hiring patterns ; and predictive policing tools, by relying on biased arrest data, can lead to the over-policing of minority neighborhoods, creating a self-fulfilling prophecy of criminality. The insidious nature of algorithmic bias lies in its veneer of objectivity. Because these decisions are made by machines, they can appear neutral and data-driven, making it exceptionally difficult for affected individuals to identify bias, challenge adverse decisions, or seek recourse. The very design of an algorithm, the specific outcome it is programmed to predict, the variables it considers, and the data it is fed-can embed bias from the outset, often unintentionally but with devastating real-world consequences. This represents a qualitative shift in information asymmetry, where the "knowledge", the biased logic embedded within the algorithm, is not only withheld from the individual but may also be opaque, or its full implications not entirely understood, even by its creators or deployers. This opacity makes the resulting harms potentially more widespread, more deeply entrenched, and significantly harder to identify and rectify than those arising from traditional human-centered information imbalances.

These harms are rarely isolated incidents affecting lone individuals. More often, they reflect and reinforce systemic inequalities, with vulnerable and marginalized populations; defined by race, socioeconomic status, age, or education level-bearing a disproportionate burden of the negative consequences. The suffering engendered by information asymmetry is not merely an accidental byproduct of complex systems; it is frequently perpetuated by active mechanisms of gatekeeping, where those who control information work to maintain their privileged position and the perceived legitimacy of the unequal system. The subsequent section will explore these active processes of enforcement and boundary maintenance.

GATEKEEPING AND THE PERFORMANCE OF LEGITIMACY

The harms stemming from information asymmetry, as detailed previously, are not simply passive consequences of unequal knowledge distribution. They are often actively maintained and perpetuated by individuals and institutions that function as "gatekeepers". These gatekeepers, consciously or unconsciously, employ a range of mechanisms to control access to information, define legitimate knowledge, enforce the existing social contract of expertise, and punish those who dare to breach its terms. This section will examine these active processes, focusing on how knowledge keepers not only safeguard their informational territory but also engage in performances designed to uphold the legitimacy of the asymmetrical system itself.

One of the most visible forms of gatekeeping is enacted through credential policing and expertise shaming. Professions across various domains-medicine, law, academia, finance-rely heavily on credentialing systems, such as licenses, certifications, and academic degrees, to delineate who qualifies as a legitimate expert and to restrict entry into their ranks. The stated rationale for such systems is often the protection of the public from unqualified or incompetent practitioners, ensuring a baseline standard of service. However, this reliance on formal qualifications, termed "credentialism," can morph into a rigid filtering mechanism where the possession of a credential becomes paramount, sometimes overshadowing demonstrated competence or innovative approaches from those outside the formally recognized channels. Professional bodies and institutions actively "police" these credentials, sometimes through legal means against uncredentialed practice, or more commonly through the social and professional marginalization of those who operate without the requisite formal validation.

Closely allied with credential policing is the phenomenon of "expertise shaming". Individuals or groups who question established expert consensus, offer alternative interpretations, or present challenging data without possessing the "correct" credentials or affiliations can find themselves subjected to public dismissal and ridicule. They may be labeled as amateurs, charlatans, or simply unqualified, irrespective of the potential validity of their claims or the evidence they present. This tactic serves not only to discredit the challenger but also to reinforce the authority and insularity of the credentialed insiders, discouraging further dissent. The performance of legitimacy here is clear: by visibly upholding standards and sanctioning deviations, the profession projects an image of rigor and exclusivity.

Insider language and selective access to information channels represent another powerful set of gatekeeping tools. As noted earlier, specialized jargon and technical language, while often necessary for precise communication within a specific field, can also function as a formidable barrier to entry for outsiders. This linguistic exclusivity creates an "in-group" of those who possess the code and an "out-group" of those who do not, thereby solidifying the expert's superior status and rendering complex knowledge less accessible to the broader public. This use of "correct language" and adherence to specific "social-professional codes" can be deeply exclusionary, effectively silencing or marginalizing voices that do not conform, a concept central to Miranda Fricker's work on epistemic injustice. Gatekeepers also exert control over the dissemination of information by managing access to critical channels such as research funding, datasets, prestigious publication venues, and influential professional networks. Editors of academic journals or news publications, for example, play a crucial role in deciding which studies are published or which stories receive coverage, thereby shaping public discourse and reinforcing dominant narratives. This selective access ensures that certain perspectives are amplified while others, particularly those that challenge the status quo, may be suppressed or ignored.

Perhaps the most dramatic confrontation with gatekeeping mechanisms occurs when whistleblowers emerge. These individuals directly challenge the informational control of powerful institutions by exposing concealed, suppressed, or inconvenient truths. In doing so, they breach the implicit, and sometimes explicit, rules governing information asymmetry and confidentiality. The institutional response to whistleblowing often involves what has been termed epistemic violence. Drawing from the work of Gayatri Spivak and others, epistemic violence refers to the systematic undermining of a person's credibility, knowledge claims, and even their perceived sanity when they challenge dominant power structures or narratives. It is not merely disagreement but an active, often aggressive, attempt to invalidate their status as legitimate knowers and to silence their testimony.

Governments and institutions deploy a range of discursive and legal strategies to achieve this. Whistleblowers are frequently labeled as disloyal, traitorous, mentally unstable, or threats to national security, thereby shifting the focus from the content of their disclosures to their character and motives. Public denunciation, character assassination, and smear campaigns are common tactics, as seen in the high-profile cases of individuals like Chelsea Manning, Edward Snowden, and Reality Winner. Legal persecution, often under broad and severe statutes like the Espionage Act, is another powerful tool. Such laws frequently prohibit defendants from arguing that their actions were in the public interest or from fully explaining their motivations, effectively rigging the legal process against them. The media, whether through direct complicity or by uncritically repeating official narratives, can also play a role in marginalizing the whistleblower's perspective and reinforcing the institution's control over the story. The overarching goal of these actions is not only to punish the individual whistleblower but also to deter others from contemplating similar disclosures, thereby preserving the institution's informational dominance and its ability to operate without unwelcome scrutiny. Some scholars describe whistleblowing as an act of "epistemic disobedience," a direct challenge to the secrecy and informational asymmetry that shields powerful entities from accountability.

These various gatekeeping mechanisms-credential policing, the strategic use of insider language, control over access, and the often brutal suppression of whistleblowers, are not merely acts of control, but are also integral to the performance of legitimacy. By visibly defining boundaries, enforcing standards (however self-serving they may sometimes be), and punishing transgressors, institutions project an image of order, authority, and rightful stewardship of knowledge. This performance is crucial for maintaining public acceptance of their privileged informational status and for justifying the broader system of information asymmetry. The link here to the "manufactured authority" is intrinsic. Credentials are formal signals of authority; insider language is both a signal and an exclusionary practice; and the public, often punitive, treatment of those who leak information serves as a dramatic display intended to reinforce the institution's power and the perceived necessity of its information control.

The concept of epistemic violence represents a particularly severe consequence when the social contract of expertise is challenged, especially when that contract is upheld by powerful, entrenched interests. Whistleblowers often act from the conviction that this contract has been violated-that the public is being deceived or harmed. The institutional response, characterized by epistemic violence, is an attempt to reassert dominance by invalidating the whistleblower's knowledge and moral standing. This is a stark enforcement of the institution's interpretation of the information bargain, frequently prioritizing its own integrity, reputation, or operational secrecy over the specific harms or truths the whistleblower seeks to reveal. The psychological impact of such tactics can extend far beyond the immediate targets. Witnessing the harsh treatment meted out to those who challenge the system, or even experiencing milder forms of expertise shaming, can create a significant "chilling effect". This discourages broader critical inquiry, fosters self-censorship, and reinforces conformity to established knowledge paradigms. Over time, this can lead to intellectual stagnation within a field, stifle innovation, and hinder the correction of errors, ultimately diminishing the quality of knowledge and harming the very public these institutions are ostensibly meant to serve. This chilling effect contributes to the "institutional decay" that becomes a feature of widespread systemic fragility.

For a long time, these powerful gatekeeping strategies and carefully orchestrated performances of legitimacy have successfully maintained the structures of information asymmetry. However, the ground is shifting. The digital revolution, and more recently the rapid advancements in artificial intelligence, are creating significant and undeniable cracks in these once formidable gates, challenging traditional modes of informational control in ways that were previously unimaginable. The next section will explore these emerging vulnerabilities and the forces that are beginning to pry open the once tightly controlled gateways to knowledge.

CRACKS IN THE GATE

The formidable structures of gatekeeping and the carefully performed legitimacy that have long upheld information asymmetry are now confronting a period of unprecedented strain. The digital age has ushered in a wave of transformative technologies and philosophies that are fundamentally altering how knowledge is created, accessed, shared, and validated. Two primary forces are at the vanguard of this change: the burgeoning open access movement across various platforms and the rapidly advancing capabilities of artificial intelligence. Together, these developments are prying open previously sealed gateways, democratizing access to once-exclusive information, demystifying complex expert language, and enabling individuals to bypass many traditional intermediaries of knowledge.

The rise of collaborative, open platforms has been a major catalyst in this shift. Wikipedia stands as a monumental example, having profoundly reshaped the landscape of general knowledge. By offering a vast, multilingual encyclopedia that is free, instantly accessible, and collaboratively created and updated by a global community of volunteers, Wikipedia effectively rendered the commercial model of traditional print encyclopedias obsolete. While its open editing model initially drew criticism regarding reliability, Wikipedia has evolved sophisticated community standards and moderation processes. In many domains, it now offers information that is not only accurate but also more current and comprehensive than its predecessors, compelling traditional publishers to rethink their value propositions and focus on more specialized or value-added content.

Similarly, platforms like Reddit function as sprawling, decentralized hubs for information exchange, discussion, and peer-to-peer learning across an almost limitless array of subjects. While the quality and reliability of information on such platforms can vary widely, they provide unfiltered access to niche knowledge, diverse global perspectives, and community-based problem-solving that often circumvent formal educational institutions and credentialed expert channels. In the realm of content creation and journalism, Substack has empowered individual writers, researchers, and commentators to publish directly to their audiences, thereby bypassing the traditional gatekeeping functions of established media outlets and publishing houses. This model allows creators to cultivate their own communities, control their messaging, and often monetize their work independently. While platforms like Substack face their own challenges, such as the tension between fostering a haven for deep work and the platform's inherent growth incentives that might favor engagement over substance, they undeniably represent a significant democratizing force.

Within academia, the Open Access movement is making substantial inroads against traditional, often costly, barriers to scholarly research. By advocating for models that make research articles and data freely available to the public, often immediately upon publication, this movement aims to accelerate scientific discovery, enhance the visibility and impact of research findings, and foster broader collaboration across disciplines and geographical boundaries. The increasing popularity of preprint servers, which allow researchers to share their findings before or alongside formal peer review, further uncouples the dissemination of knowledge from the sometimes lengthy and restrictive processes of traditional journal publishing. Complementing these are open-source learning platforms, including Massive Open Online Courses (MOOCs) and countless educational tutorials available on sites like YouTube.

Top academic institutions, including Harvard, Stanford, MIT, and Yale, now provide free online courses, either directly or via MOOCs like Coursera, edX, and FutureLearn. This represents a societal shift from a decentralized to a more centralized learning model, where offering free education is almost a requirement for universities to enhance their brand and recruit students. As discussed prior, these accessible resources democratize educational content from various global sources, diminishing the traditional exclusivity of university curricula and challenging established teaching methods.

The second major force creating cracks in the gate is the rapid advancement of artificial intelligence tools capable of rewriting, simplifying, or entirely bypassing complex expert language. A new generation of AI text simplifiers, powered by sophisticated Natural Language Processing (NLP) models, can take dense, technical, or jargon-laden prose and transform it into clear, accessible language understandable by a non-expert audience. Tools such as ClickUp Brain, Junia.AI, and ProWritingAid can explain obscure terminology, adjust the tone and style of writing, summarize lengthy documents, and even provide synonyms for complex words, effectively acting as personalized writing coaches or on-demand editors. The impact of these technologies is profound: they significantly lower the barrier to understanding specialized knowledge, enabling individuals to engage with concepts and information previously locked away behind the gate of complex professional language. This directly challenges the exclusionary power of "insider language" that has long been a hallmark of gatekeeping in many professions. Furthermore, AI's ever-improving capacity for language translation is breaking down linguistic barriers, making knowledge more readily accessible across different cultures and languages.

Together, these trends are fostering a democratization of curation and critique. Platforms like Wikipedia and Reddit, for all their imperfections, allow for community-based evaluation and organization of information, where users collectively, through discussion and voting mechanisms, determine what is considered relevant, important, or credible. This is not to say such systems are without flaws;

they can be susceptible to echo chambers, misinformation, and manipulation. However, they represent a shift away from purely top-down, expert-driven curation. Moreover, the ability for anyone with an internet connection to publish their own analyses, reviews, or critiques; on personal blogs, platforms like Substack, social media, or forums challenges the traditional monopoly held by established critics, peer-review systems, and institutional spokespersons. While the rigor of such open critique varies, it introduces a broader array of perspectives and can serve to expose flaws, biases, or unexamined assumptions within expert narratives.

The cumulative effect of these developments is a palpable erosion of traditional authority structures that were built upon the control of exclusive knowledge. When information, learning resources, and tools for deciphering complexity become widely and often freely available, the authority derived merely from possessing privileged access to that knowledge inevitably diminishes. The perceived necessity for traditional gatekeepers; be they university admissions offices for access to learning, or journal editors for the dissemination of research-is correspondingly reduced as alternative, more direct pathways to knowledge acquisition and communication proliferate.

It is important to acknowledge that this "democratization" is not a panacea and does not create a perfectly level playing field. New forms of gatekeeping and power dynamics are indeed emerging within these novel ecosystems. Platform algorithms, for instance, now play a significant role in determining what information users see, potentially creating new filter bubbles or echo chambers. Influencer culture can create new, unelected arbiters of taste and opinion. Furthermore, the digital divide means that access to, and the ability to effectively utilize, these new tools and platforms are not evenly distributed across the population. Thus, while old gates are undeniably cracking, new, perhaps more subtle and technologically mediated gates are being constructed by platform owners, algorithmic systems, and the underlying technological infrastructure.

The ease with which AI can now simplify expert language also has profound implications for how expertise itself is perceived and valued. If a significant component of manufactured authority and gatekeeping has historically relied on the deployment of complex jargon and opaque terminology, as discussed in previous sections, then the ability of AI to translate this language for a general audience weakens that specific lever of power. Experts who have relied heavily on linguistic complexity to maintain an aura of exclusivity may find that aura rapidly diminishing. This necessitates a shift towards demonstrating value through other, more substantive means-perhaps through deeper analytical insights, novel applications of knowledge, enhanced creativity, or more robust ethical judgment and contextual understanding.

However, a potential downside to this widespread availability of information and powerful simplification tools is the risk of fostering an "illusion of understanding". Individuals might feel they have grasped complex topics after only superficial engagement with simplified summaries or readily accessible online content. This could lead to overconfidence in one's own knowledge, a decreased willingness to defer to genuine, nuanced expertise when warranted, and paradoxically, a greater susceptibility to sophisticated misinformation that is cleverly packaged with an appearance of simplified clarity. This concern connects directly to the broader societal fragilities that arise when the traditional foundations of trust and knowledge are unsettled.

These cracks in the gate are not merely isolated points of stress; they are symptomatic of a larger, systemic shift. If information becomes universally accessible, and if the language of experts can be readily deciphered by anyone with an AI assistant, what then becomes of the perceived superiority of institutional knowledge and the authority of traditional experts? The very foundations of the fragile bargain are being shaken. The next section will explore the potential for a more comprehensive collapse of the entire system of trust when the widespread belief that institutions and their designated experts inherently "know better" begins to unravel at scale.

FRAGILITY AT SCALE

The erosion of traditional gatekeeping mechanisms and the democratization of information access, as explored in the previous section, are not merely isolated trends. They are contributing to a much larger and more systemic phenomenon: a growing fragility in the public's trust in established institutions and traditional forms of expertise. When the foundational belief that these entities "know better" and act in the collective interest begins to falter on a widespread scale, the entire social contract of expertise is jeopardized. This section dives into the consequences of this large-scale erosion of trust, examining how it manifests in phenomena like science denial, the proliferation of conspiracy cultures, the disorienting paradox of information overload, and the pervasive "post-truth" fatigue that can lead to societal dysfunction and the decay of shared understanding.

One of the most alarming symptoms of this large-scale fragility is the rise of science denial and the entrenchment of conspiracy cultures. Science denial is characterized by the rejection of well-established scientific theories and consensus, not based on countervailing scientific evidence, but rather on ideological, political, social, or economic motivations. Common tactics employed by those engaged in science denial include attempts to undermine the credibility of scientific research methods, asserting that the science is "unsettled" or lacks universal acceptance, launching ad hominem attacks against the objectivity or motives of researchers, and demanding a "false balance" in media coverage or educational curricula that gives undue weight to unsubstantiated claims. Prominent examples include the persistent denial of anthropogenic climate change despite overwhelming scientific agreement, and the anti-vaccine movements that reject decades of public health data. This denial is often fueled by a deep-seated wariness or outright hostility towards powerful institutions, including scientific bodies, which are perceived by some segments of the population as insensitive, unresponsive, or even deliberately misleading.

This institutional decay, marked by declining public confidence in a wide range of societal pillars-from government and media to academia and even science itself-creates a fertile environment for conspiracy cultures to flourish. When official narratives lose their authority and institutions are no longer seen as reliable sources of truth, individuals may turn to alternative explanations for complex or unsettling events. Conspiracy theories often offer simplistic, emotionally resonant narratives that attribute societal problems or unexplained phenomena to the actions of powerful, malevolent, and clandestine actors. This search for alternative explanations is often linked to what are termed "post-truth epistemic beliefs," where individuals may doubt the very possibility of objective truth or verifiable knowledge, preferring to trust their intuition or believe what "feels true" to them, regardless of empirical evidence. Science denial and the embrace of conspiracy theories are not simply a matter of individuals lacking correct information; they often represent an active rejection of the authority of established institutions and a conscious search for alternative authorities or knowledge systems that align more closely with pre-existing ideological frameworks, emotional commitments, or a sense of grievance against the perceived "establishment". This signifies a fundamental breakdown in the trust relationship that underpins the social contract of expertise, where the expert is no longer viewed as a trustworthy delegate of knowledge but potentially as part of a deceptive or coercive system.

Compounding this erosion of trust is the paradox of too much information. In an age of unprecedented access to data, the sheer volume of information available can become overwhelming, leading to cognitive overload, decision paralysis, heightened stress, and anxiety. The constant bombardment from multiple channels: news feeds, social media, endless online content-can disrupt focus, induce decision fatigue, and even lead to an "affective disconnect," where individuals become desensitized or emotionally numb to the constant stream of stimuli. Critically, having access to more information does not inherently equate to being better informed; it can, paradoxically, lead to a more clouded, fragmented, or even corrupted understanding of complex issues. This information overload makes it increasingly difficult for individuals to discern credible, verified information from misinformation, disinformation, or simply low-quality noise. As a result, the authority of reliable sources can be diminished as they become lost in the cacophony, making it harder for people to know who or what to trust.

This environment of information saturation and declining institutional trust directly fuels what has been termed "post-truth fatigue" and the rise of highly personalized epistemologies. The "post-truth" condition is often described as an era where objective facts have become less influential in shaping public opinion and political discourse than appeals to emotion, personal belief, or tribal affiliation. While the narrative that large numbers of people now lack respect for truth can itself be a form of confirmation bias, it captures a genuine societal anxiety about the apparent devaluation of evidence-based reasoning. "Post-truth epistemic beliefs" contribute significantly to this condition; these are belief systems wherein the rigorous evaluation of arguments against objective criteria of truth is deliberately suspended or de-prioritized. Such beliefs often involve an over-reliance on intuition for assessing factual claims, a diminished perceived need for external evidence, and a conviction that "truth" is inherently political or socially constructed by those in power. These epistemic stances fundamentally challenge the basis of rational communication and have been linked to real-world behaviors such as vaccine refusal and a susceptibility to conspiracy theories. The cognitive load of constantly sifting through a deluge of often contradictory information, coupled with the emotional toll of navigating a landscape where even basic facts seem contested, can lead to a profound sense of fatigue. For some, this exhaustion makes the pursuit of objective truth seem futile, leading to a disengagement from critical inquiry and a retreat into simpler, more emotionally satisfying, or ideologically congruent "personal truths". In the absence of broadly accepted, trusted arbiters of truth, individuals may increasingly construct their own epistemologies, relying on subjective experience, gut feelings, or the shared beliefs of their immediate social or online communities to determine what is true and valid for them. This fragmentation of reality into multiple, often incompatible, personal or group-based truths makes societal consensus on critical issues extraordinarily difficult to achieve.

The list below contrasts common tactics used in science denial with the corresponding principles of legitimate scientific inquiry, illustrating the distortions at play when established knowledge is ideologically contested.

Claim: Science is “unsettled” or lacks universal acceptance.

Scientific principle: Science is inherently provisional. Knowledge evolves as new evidence emerges, and lack of absolute certainty is a feature of the scientific process, not a flaw.

Claim: Researchers’ objectivity or motives invalidate their findings.

Scientific principle: Scientific evaluation focuses on methodology, data quality, reproducibility, and peer review, not on personal character or presumed intent.

Claim: All views deserve equal weight, regardless of evidence.

Scientific principle: The validity of a scientific claim is determined by the weight of evidence and the level of consensus, not by equal representation of all opinions.

Claim: Selective data or isolated studies prove a desired narrative.

Scientific principle: Scientific understanding relies on comprehensive analysis of the full body of available evidence, including contradictory findings.

Claim: Powerful interests suppress the "real truth."

Scientific principle: Science advances through transparency, open debate, replication, and verifiable results, making sustained, coordinated suppression extraordinarily difficult.

Claim: Uncertainty means scientists know nothing.

Scientific principle: Scientific uncertainty reflects degrees of confidence and margin of error, not an absence of knowledge.

When these various factors; science denial, institutional decay, conspiracy thinking, information overload, and post-truth fatigue-converge, the "fragile bargain" of information asymmetry collapses at scale. This is not merely about individuals disagreeing with specific expert opinions or institutional policies; it represents a more profound and widespread loss of faith in the integrity and reliability of the entire system of knowledge production, validation, and dissemination. Such systemic fragility has dire implications for societal cohesion. If there is no shared understanding of reality, no common factual basis for discussion, and no trusted institutions to mediate disputes or guide collective action, the capacity to address critical challenges; be they public health crises, environmental threats, or economic instability-is severely diminished. This can lead to policy paralysis, social fragmentation, and an inability to respond effectively to the complex problems confronting contemporary society. This widespread erosion of trust and the consequent fragmentation of truth create a significant void in the social and informational landscape. If traditional institutions and established experts are no longer the default sources of authority and guidance for a large segment of the population, a critical question arises: what, or who, will fill that vacuum? The final section of this chapter will argue that artificial intelligence is increasingly, and perhaps inevitably, poised to step into this breach, with profound and multifaceted implications for the future of knowledge, trust, and authority.

THE TRUST VACUUM

The cascading erosion of trust in traditional human institutions and established expert communities, as explored in the preceding section, does not occur in isolation. It carves out a significant void in the societal landscape, a "trust vacuum". As faith in conventional sources of authority wanes, individuals and society at large may consciously or unconsciously seek alternative anchors for guidance, information, and decision-making. It is in this developing vacuum that artificial intelligence is increasingly, and perhaps inexorably, being positioned. AI, with its aura of computational objectivity, its capacity for rapid information processing, and its potential for deep personalization, is emerging not merely as a challenger to existing forms of expertise but as a powerful candidate to fill the space left behind. This transition, however, is fraught with complexity and significant risk, including the potential for cultivating a "synthetic trust" in machines and the uncritical acceptance of meaning brokered by non-human intelligence.

Several factors contribute to AI's growing appeal as a source of authority, particularly in an environment where trust in human systems is declining. AI systems often project an image of perceived objectivity. Unlike human experts, who may be seen as susceptible to personal biases, emotional reasoning, or self-interest, algorithms can appear to operate purely on data and logic. While this perception often overlooks the inherent biases that can be embedded in AI through training data and design choices, the veneer of neutrality can be compelling. Furthermore, AI offers remarkable efficiency and accessibility. It can provide instant answers to complex queries, summarize vast amounts of information in seconds, and perform tasks with a speed and scale that human institutions often struggle to match. This can create a sense of empowerment and immediate gratification for users, contrasting sharply with experiences of bureaucratic delay or opacity in traditional systems.

Personalization is another key attribute that enhances AI's authoritative appeal. Advanced AI can tailor its responses and information delivery to individual user needs, learning histories, and stated preferences, fostering a feeling of being uniquely understood and catered to. For many, especially when dealing with sensitive or potentially embarrassing topics, AI chatbots and virtual assistants can offer a non-judgmental interaction space. The anonymity and perceived impartiality of an AI can make it easier for individuals to ask questions or explore subjects they might hesitate to discuss with a human expert, who might be perceived as critical or dismissive. Indeed, some research suggests that users may find AI interactions more compassionate or understanding than those with human professionals in certain contexts, and the "ELIZA effect": the tendency for humans to attribute intelligence and empathy even to relatively simple computer programs-has demonstrated this human proclivity for decades. This is set against a backdrop where global trust in AI is still developing and, in some cases, has decreased even as adoption has risen, with less than half of people globally expressing a willingness to trust AI. Yet, the perceived failings of human systems can make AI seem like a comparatively attractive alternative.

However, the prospect of AI filling the trust vacuum carries substantial risks, chief among them the development of synthetic trust. This refers to a form of reliance placed on AI systems that is not necessarily grounded in a comprehensive understanding of their operational mechanisms, inherent limitations, or the quality of their underlying data. Instead, this trust may be based on the AI's impressive performance, the sophistication of its interface, its constant availability, or simply the perceived inadequacies of human alternatives. A critical issue is that AI, unlike human experts, lacks consciousness, genuine emotional understanding, and personal accountability in the human sense. AI systems operate based on complex algorithms and vast datasets; they do not possess intrinsic ethical frameworks, intentions, or the capacity for remorse. This fundamental difference means that the trust markers we intuitively apply in human relationships-such as assessing sincerity, empathy, or moral character-are inapplicable to AI.

This can lead to over-reliance and complacency. Users may become increasingly dependent on AI-generated outputs without subjecting them to rigorous critical evaluation, potentially leading to the adoption of flawed information or the making of poor decisions. The very data used to train these powerful AI models also introduces vulnerabilities. The increasing use of synthetic data; data generated by AI itself to augment real-world datasets can be beneficial for training models on scenarios not yet encountered, but it also carries the risk of models becoming detached from reality if this synthetic data is not carefully validated and "grounded in the real world". Furthermore, the proliferation of AI-generated synthetic content, such as deepfakes, hyper-realistic fake news articles, and manipulated audio, poses a direct threat to the integrity of the information ecosystem. These tools can be used to manipulate public opinion, commit sophisticated fraud, compromise individual privacy, and broadly erode trust in all forms of digital media, making it harder to distinguish authentic information from artifice. The potential for research misconduct through the deliberate fabrication or falsification of data using generative AI also looms as a significant ethical challenge, threatening to corrupt the scientific record if robust safeguards, such as data watermarking and strict ethical codes, are not implemented and enforced.

Beyond these instrumental risks, a more profound concern arises when meaning itself is increasingly brokered by machines. If AI systems become the primary intermediaries through which individuals access, interpret, and make sense of the world, these systems gain immense power to shape understanding, values, and societal narratives. The algorithms that curate, filter, and present information are not neutral arbiters; they inevitably reflect the biases embedded in their training data and the objectives, explicit or implicit, of their human creators and deployers. This can lead to a narrowing of perspectives, a homogenization of thought, or even subtle forms of behavioral manipulation as users are guided towards certain conclusions or actions. The "meaning" derived from AI-brokered information, however efficiently delivered, may lack the critical nuance, contextual depth, ethical sensitivity, and capacity for wisdom that human judgment, at its best, strives to provide.

It is important to recognize that the trust vacuum is not necessarily being filled by AI as a purely neutral or democratically controlled tool. Often, the most powerful and widely adopted AI systems are developed and controlled by the same kinds of large, centralized institutions; major technology corporations and governmental bodies that are themselves subjects of the declining public trust that created the vacuum in the first place. This dynamic suggests that the shift may not be towards a more decentralized or inherently more trustworthy information ecosystem, but potentially towards a re-centralization of informational power in the hands of a few dominant AI providers, albeit in a less transparent and perhaps more pervasive form.

Moreover, the phenomenon of "synthetic trust" in AI could prove to be more hazardous in the long run than misplaced trust in human experts. While human fallibility is undeniable, human systems of accountability, however imperfect, are built upon shared understandings of ethical responsibility, legal frameworks, and the potential for moral redress. An AI, lacking inherent ethical agency or the capacity for genuine accountability in the human sense, cannot be held responsible in the same way. Failures or harms caused by AI systems raise complex questions of liability and remediation that existing frameworks are only beginning to grapple with. This makes the potential consequences of misplaced trust in AI qualitatively different and potentially more difficult to address.

If society increasingly relies on AI for answers, for meaning, and for mediating its understanding of reality, there is a significant risk of devaluing human-centric forms of knowledge, wisdom, and meaning-making that are not easily quantifiable, algorithmically processable, or reducible to data points. This could subtly steer society towards optimizing for computational efficiency and data-driven outputs, potentially at the expense of deeper human values such as critical thinking, creativity, emotional intelligence, and ethical reflection. Such a trajectory could lead to a less rich, less reflective, and ultimately less humane experience, a concern that resonates with themes to be explored in later chapters regarding the enduring value of human intuition and the essential "human interface."

Artificial intelligence, therefore, plays a dual role in this unfolding drama. It is simultaneously a catalyst in the erosion of old trust structures-by outperforming human capabilities in certain knowledge-based tasks and exposing the inefficiencies or biases of existing systems-and a powerful, albeit potentially perilous, candidate to fill the ensuing void. The fragile bargain of information asymmetry, long predicated on human trust and imperfect human gatekeeping, is unraveling under these new pressures. Having now mapped the architecture of this asymmetry, its inherent vulnerabilities, and the manner in which AI is poised to step into the resulting trust vacuum, the stage is set for a more detailed examination of what occurs when this asymmetry collapses. The next act of this inquiry, beginning with Chapter 4, "Synthetic Expertise," will explore precisely how AI technologies are making expert-level knowledge universally accessible and what the far reaching consequences are when the traditional foundations of information control begin to crumble.

SUMMARY

This chapter has navigated the intricate and often precarious nature of the "fragile bargain" that underpins society's historical acceptance of information asymmetry. We have seen that this bargain is founded upon a social contract of expertise, wherein individuals delegate a degree of autonomy to professionals and institutions in exchange for the anticipated benefits of their specialized knowledge. This delegation, however, is critically dependent on trust-a trust that is not always organically earned but is frequently manufactured through an array of authority signals, performative actions, and media amplification. The human costs of this asymmetry become starkly apparent when such trust is betrayed or when the information imbalance is exploited, leading to tangible harms such as financial predation, compromised health outcomes, and the perpetuation of systemic inequalities, including insidious forms like digital redlining. For centuries, traditional gatekeepers have diligently enforced this established order through mechanisms like credential policing, the strategic use of insider language, and, in extreme cases, the deployment of epistemic violence against those who challenge the dominant control of information.

However, these long-standing structures of informational control are now exhibiting significant cracks. The rise of open access movements, collaborative online platforms, and, most impactfully, the burgeoning capabilities of artificial intelligence are democratizing access to knowledge and providing tools to simplify or bypass complex expert language. This widespread erosion of traditional gatekeeping contributes to a larger, systemic fragility. As faith in the inherent superiority and benevolence of institutional knowledge diminishes, a "post-truth" landscape can emerge, characterized by science denial, the proliferation of conspiracy cultures, and a pervasive trust vacuum. It is into this very vacuum that artificial intelligence is increasingly stepping, not merely as another technological tool, but as a potential new locus of authority, promising efficiency and personalized knowledge but also carrying the risks of synthetic trust and machine-brokered meaning.

Having explored the architecture of information asymmetry and its current state of profound destabilization, the path forward requires a deeper understanding of AI's direct impact on the nature of expertise itself. We now turn, in Chapter 4, "Synthetic Expertise," to examine how advanced AI technologies are making expert-level knowledge and capabilities universally accessible, and to consider the sweeping consequences that arise when the traditional foundations of information control and expert authority definitively collapse.

4

SYNTHETIC EXPERTISE

The landscape of human knowledge and expertise is undergoing a seismic transformation, one that promises to redefine our relationship with information and skill. For millennia, specialized knowledge has been the preserve of a dedicated few, acquired through years, often decades, of rigorous study, apprenticeship, and practice. Access to this expertise-be it in medicine, law, engineering, or intricate crafts-has historically been a significant marker of status, a driver of economic value, and a cornerstone of societal progress. Yet, this traditional model, built on the inherent scarcity of deep human understanding, is now being fundamentally challenged by the advent of sophisticated artificial intelligence (AI) systems. We are standing at the cusp of an era where expertise, in myriad forms, is becoming universally accessible, not through the slow diffusion of human learning, but through the instantaneous capabilities of AI.

This chapter dives into the heart of this revolution, exploring how "synthetic expertise"; the ability of AI to replicate and, in certain contexts, even surpass human expert performance is rapidly moving from the realm of science fiction to a tangible reality. The engines driving this change are primarily a trio of interconnected AI technologies: Large Language Models (LLMs), Retrieval-Augmented Generation (RAG) systems, and autonomous AI Agents. Together, these technologies are dismantling the old asymmetries of knowledge. LLMs provide a foundational cognitive capability, RAG systems ground this capability in real-time, verifiable information, and AI Agents translate this synthesized knowledge into action and complex problem-solving. The implications are profound, suggesting a future where high-level insights, sophisticated analysis, and skilled execution are available on demand, at scale, and with an unprecedented degree of personalization.

The democratization of expertise is not merely about more people having access to information; it signifies a fundamental alteration in the very nature of expertise itself. What was once a static, deeply internalized human asset is morphing into a dynamic, computationally generated, and readily available service. This transition from scarcity to abundance in expert capabilities carries with it both immense promise and considerable challenges. As we will explore, the speed and scale at which synthetic expertise can be developed and deployed are creating an "expertise momentum". This momentum is already outpacing the adaptive capacities of traditional human systems, including education, professional development, and regulatory frameworks, potentially leading to significant societal and economic friction.

The preceding chapters of this book have laid the groundwork for understanding this shift. Chapter 1 exposed the pervasive role of information asymmetry and knowledge arbitrage in shaping our economies and institutions. Chapter 2 examined how education, credentials, and networks have historically served as gatekeepers, controlling access to power and opportunity by controlling access to knowledge. Chapter 3 explored the fragile bargain of trust built around these gatekeeping mechanisms and the human cost of unequal information. Now, this chapter will illuminate the technological forces that are beginning to dismantle these long-established structures, demonstrating how AI is not just providing new tools, but is actively reshaping what it means to "know" and "do" in virtually every field of human endeavor. The age of information symmetry is dawning, and synthetic expertise is one of its most powerful harbingers.

FROM KNOWLEDGE TO COMPUTATION: LLMS AS REASONING ENGINES

At the vanguard of the synthetic expertise revolution are Large Language Models (LLMs). These sophisticated AI systems represent a pivotal shift from viewing computers as mere calculators or data processors to understanding them as nascent reasoning engines. Often perceived as advanced text generators or chatbots, LLMs are, more fundamentally, "foundation models trained on vast corpora of human knowledge". This training involves ingesting and processing text and code at an internet scale-billions of documents, books, articles, and conversations-allowing them to internalize the patterns, structures, concepts, and intricate relationships embedded within human language and, by extension, human thought. It is this deep immersion in the textual expression of human understanding that endows LLMs with their remarkable capabilities.

A particularly fascinating aspect of LLMs is the phenomenon of "emergent abilities". These are capabilities, such as arithmetic, commonsense reasoning, or even understanding nuanced humor, that are "absent in smaller language models but manifest in larger ones". Crucially, these abilities are not always predictable merely by extrapolating the performance of smaller-scale models; they appear to arise somewhat spontaneously as the model's size (number of parameters) and the volume of training data cross certain thresholds. This observation challenges the simplistic notion that LLMs are only performing next-word prediction. While next-word prediction is indeed the fundamental mechanism of their training-learning to predict the most probable next word in a sequence-at massive scale, this process seems to unlock more complex cognitive functions that mimic, and sometimes even enable, forms of reasoning. The sheer scale of these models, therefore, is not just a quantitative difference but leads to qualitative shifts in their behavior. The very fact that such reasoning capabilities can emerge, rather than being explicitly programmed in a traditional sense, suggests that we are building systems whose full operational boundaries might expand in ways we don't entirely design or anticipate. This necessitates a shift from thinking about "programming" AI to "guiding" or "eliciting" desired behaviors from systems that possess inherent, somewhat opaque, cognitive capacities.

Researchers and developers have devised several techniques to harness and enhance these nascent reasoning capabilities. One of the most significant is Chain-of-Thought (CoT) prompting. Instead of asking an LLM for a direct answer, CoT encourages the model to "break down problems into intermediate steps," articulating a sequence of reasoning before arriving at a conclusion. This mimics human problem-solving, where complex issues are often tackled in stages. By guiding the model to "think aloud," CoT significantly improves accuracy, particularly in tasks involving logic, mathematics, and complex decision-making, making the model's "thought process explicit and verifiable".

Another technique is Self-Consistency Sampling. When faced with a difficult question, humans often consider multiple lines of reasoning. Similarly, LLMs can be prompted to generate several different reasoning paths for the same problem and then select the most consistent or frequently occurring solution among them. This approach often yields better and more reliable outcomes than relying on a single generated answer. Furthermore, Tool-Augmented Reasoning allows LLMs to overcome their inherent limitations by interacting with external tools. An LLM can recognize when a task requires, for instance, precise calculation or access to real-time information it doesn't possess. It can then delegate that part of the problem to a specialized tool; like a calculator, a search engine, or a code interpreter-and integrate the tool's output back into its overall reasoning process, much like a human seeking external help or resources. This capability is a precursor to the more sophisticated agentic behaviors discussed later in this chapter.

Efforts are also underway to build more robust memory and contextual reasoning into LLMs. Enhancing their ability to recall information from earlier in a conversation (episodic memory) or to draw upon a persistent knowledge base (semantic memory) is crucial for improving their capacity for multi-turn, context-aware reasoning that is essential for sustained expert-like interactions. LLMs also demonstrate In-Context Learning, where they adapt to new tasks and infer how to approach a problem based on the examples and context provided within the prompt itself. While powerful, this form of learning can sometimes be limited by the model's tendency to rely heavily on patterns and correlations present in its original training data.

This brings us to a critical point: the distinction between sophisticated pattern-matching and genuine problem-solving in LLMs. Skeptics rightly point out that LLMs "rely on complex pattern matching" and "often engage in sophisticated pattern matching" rather than what might be termed "true logical reasoning" in all instances. Their performance can indeed degrade when faced with problems that deviate significantly from the patterns encountered during training, or when irrelevant details are introduced that can confuse their pattern-recognition mechanisms. An LLM might, for example, correctly solve a standard mathematical problem but falter if names and numbers are changed in a way that breaks familiar patterns, or if superfluous information is added.

However, the emergence of complex abilities and the success of techniques like Chain-of-Thought prompting suggest a trajectory beyond simple pattern-matching towards capabilities that are functionally equivalent to problem-solving for a wide range of tasks. The debate is nuanced, partly because human cognition itself involves a significant degree of pattern recognition and heuristic thinking. The line between advanced pattern-matching that can generalize and adapt, and "true" understanding or reasoning, is blurry and a subject of ongoing research. What is undeniable is that for many practical applications requiring expertise, the outputs generated by LLMs can be indistinguishable from, or even superior to, those of human experts, regardless of the precise internal mechanisms.

Nevertheless, the degree to which an LLM's expertise relies on pattern-matching versus deeper causal understanding has significant implications for its reliability. Expertise built predominantly on recognizing and replicating patterns may be highly effective for common, well defined problems but can become brittle and unreliable when faced with truly novel situations or when superficial cues in the input are misleading. An LLM might confidently "hallucinate" an incorrect answer in such scenarios because the input triggers a plausible-seeming pattern, even if that pattern is inappropriate for the novel context. This underscores a crucial point: synthetic expertise derived from LLMs may possess qualitatively different failure modes compared to human expertise, which can often fall back on first principles or adaptive reasoning when familiar patterns break down.

Despite these caveats, LLMs form the cognitive core of the emerging synthetic expert. They are not merely summarizing existing knowledge; they are beginning to synthesize it, to reason with it, and to generate novel insights in real time. However, a core limitation of standalone LLMs is that their knowledge is inherently static, frozen at the point of their last training run. To be truly effective as expert systems in a dynamic world, they need a way to access and incorporate current, context-specific information. This critical need sets the stage for the next technological layer: Retrieval-Augmented Generation.

THE RISE OF RETRIEVAL-AUGMENTED GENERATION (RAG): GROUNDING AI IN REALITY

While Large Language Models (LLMs) possess an astonishing breadth of knowledge and nascent reasoning capabilities, their standard architecture presents inherent limitations for delivering consistently reliable and current expertise. LLMs are typically trained on massive datasets that have a cut-off date; consequently, their internal knowledge is static and does not include information or events occurring after their last training cycle. This can lead to them providing outdated or anachronistic information. Furthermore, LLMs are prone to "hallucination"; generating plausible-sounding but factually incorrect or nonsensical statements-especially when queried about topics at the fringes of their training data or when pushed beyond their core competencies. For synthetic expertise to be trustworthy and practically useful, particularly in high-stakes domains like medicine, finance, or law, these limitations must be addressed.

Enter Retrieval-Augmented Generation (RAG), a powerful methodology designed to mitigate these issues by dynamically connecting LLMs to external, up-to-date, and verifiable knowledge sources. RAG "enhances natural language generation by incorporating information from external knowledge sources", effectively combining generative capabilities of Large Language Models (LLMs) with the precision of information retrieval. Instead of relying solely on the model's pre-trained (and potentially outdated or incomplete) internal knowledge, a RAG system first retrieves relevant information from an authorized corpus-such as a company's internal document repository, a specific legal database, a collection of scientific papers, or even the live web-and then uses this retrieved information as context to guide the LLM in generating its response. This approach aims to "enhance transparency, accuracy and contextuality of responses".

The workflow of a typical RAG system involves several key stages:

Data Collection and Preprocessing: The process begins with the curation of relevant, often domain-specific, textual data. This could include PDF documents, articles, entries from structured databases, or website content. This raw data is then preprocessed: it is cleaned to remove noise or irrelevant formatting, normalized, and, crucially, segmented into smaller, manageable "chunks" or passages of text. This chunking strategy is vital for efficient and precise retrieval.

Creating Vector Embeddings: Each of these textual chunks is then transformed into a numerical vector representation using sophisticated embedding models (such as Sentence-BERT or OpenAI's text-embedding models). These vector embeddings are high-dimensional numerical arrays that capture the semantic meaning and contextual nuances of the text. Words and phrases with similar meanings will have embeddings that are "closer" to each other in the vector space.

Indexing in a Vector Database: The generated vector embeddings are stored and indexed in a specialized database known as a "Vector Store" or "Vector Database". These databases are optimized for efficiently performing similarity searches across vast collections of vectors. They often employ algorithms like K-Nearest Neighbors (KNN) or Approximate Nearest Neighbors (ANN) to quickly find the vectors most similar to a given query vector.

Querying and Retrieval: When a user submits a query to the RAG system, this query is also converted into a vector embedding using the same embedding model. The "Retriever" component of the RAG system then searches the vector database to identify and retrieve the text chunks whose embeddings are most semantically similar to the query embedding. This process is known as "vector similarity search," which relies on distance metrics like cosine similarity (measuring the angle between vectors) or Euclidean distance (measuring the straight-line distance) to quantify the relatedness between the query and the stored chunks.

Augmentation and Generation: The most relevant chunks retrieved from the vector database are then compiled and presented as augmented context to the LLM, alongside the original user query. The LLM uses this fresh, specific context to formulate its response. This "merging two knowledge streams"-the LLM's broad, fixed general knowledge and the dynamic, domain-specific information retrieved on demand-is the core mechanism of RAG.

The benefits of implementing RAG are substantial and directly address the shortcomings of standalone LLMs in delivering expert-level outputs:

Current and Up-to-Date Knowledge: RAG systems can draw upon external knowledge bases that are continuously updated, ensuring that the information underpinning the LLM's responses is timely and reflects the latest developments, which is crucial for dynamic fields.

Enhanced Contextual Relevance and Accuracy: By providing specific, relevant documents as context, RAG enables the LLM to generate responses that are not only factually accurate but also highly pertinent to the user's specific query and situation, drawing from authoritative sources.

Reduction of Hallucinations: Grounding the model's generation process in externally retrieved, factual documents significantly curtail the tendency to fabricate information, as the model is constrained by the provided context.

Trust and Traceability: A key advantage of RAG is its potential to provide source attribution. Many RAG implementations can cite the specific documents or passages used to generate an answer, allowing users to verify the information and understand its provenance. This "show its work" capability is critical for building trust and ensuring accountability, especially in regulated industries like finance or healthcare, effectively creating "glass-box models" rather than opaque black boxes.

Cost-Effectiveness and Domain Adaptation: RAG allows organizations to leverage their existing proprietary knowledge bases or subscribe to specialized data feeds without the need for expensive and time-consuming full retraining or fine-tuning of the base LLM for every new domain or update. This makes it easier to adapt LLMs for highly specialized expert tasks.

Despite these advantages, RAG systems are not without their challenges. The quality of the retrieved information is paramount; if the external knowledge base is inaccurate, incomplete, or biased, the RAG system's outputs will reflect these flaws; the principle of "garbage in, garbage out" is strongly amplified. Evaluating the performance of RAG systems is also a complex undertaking due to their hybrid architecture, involving both retrieval and generation components. Furthermore, practical considerations such as the precision of vector search, the cost of creating and maintaining extensive vector databases, and the computational overhead of the retrieval step can pose hurdles, particularly in large-scale enterprise deployments.

The advent of RAG fundamentally alters the locus of "knowledge" within an AI expert system. Instead of residing solely within the opaque, pre-trained weights of an LLM, a significant portion of the actionable knowledge is now explicitly stored in accessible, and potentially curatable, external data sources. This shift has profound consequences. The LLM, in a RAG architecture, becomes an intelligent interface or a reasoning layer that operates over this external data. Consequently, the accuracy, timeliness, relevance, and comprehensiveness of the RAG system's output become inextricably linked to the quality and governance of its vector database and the underlying source documents. Organizations and individuals aiming to deploy RAG for synthetic expertise must therefore invest significantly in data curation, quality control, versioning, and the continuous updating of their knowledge bases. The "expert system" is no longer just the model; it's the synergistic combination of the LLM and its meticulously maintained, verifiable data ecosystem. This dependency also introduces a new vector for potential manipulation: the integrity of synthetic expertise can be compromised by corrupting or subtly biasing the external data sources upon which the RAG system relies.

Moreover, the traceability offered by RAG, while a boon for transparency and trust, introduces a new layer of cognitive responsibility for the end-user. The ability to see source citations does not automatically equate to infallible truth or complete objectivity. A cited source could itself be flawed, outdated, biased, or misinterpreted by the LLM, even with the grounding RAG provides. Therefore, the user's task evolves. It's no longer just a question of "Is this AI hallucinating?" but extends to "Is this AI citing credible sources, and is it interpreting them correctly and without bias?" This demands a more sophisticated level of digital literacy and critical thinking from users, who must now engage not only with the AI's final output but also with the potential veracity and framing of its underlying information sources. RAG does not eliminate the need for human discernment; rather, it reframes and, in some ways, elevates it.

With Retrieval-Augmented Generation, the knowledge accessible to AI is no longer confined to its training data. It can tap into the vast, ever-evolving expanse of information available on the internet, within an enterprise's private repositories, or across specialized global databases. This makes the expertise offered far more dynamic, verifiable, and ultimately, more useful. However, providing grounded answers is only one part of expertise. The next evolution involves AI systems that not only know but can also act upon that knowledge.

AGENTS AS COMPOSABLE EXPERTS: AI THAT ACTS AND COLLABORATES

The progression from AI systems that retrieve and generate information to those that can autonomously act upon it marks a significant leap towards truly functional synthetic expertise. This evolution is embodied in the rise of AI agents; or "agentic AI systems" which are designed not just to answer questions but to achieve goals, make decisions, and perform complex sequences of tasks with a considerable degree of independence. These agents are moving AI from the role of a sophisticated information provider to that of an active participant and even a "doer" in various workflows, capable of undertaking work that once required coordinated human teams.

At its core, an AI agent can be defined as a software entity that perceives its environment through various data inputs (text, sensor data, API responses), processes this information using underlying AI models (often LLMs enhanced with RAG), and then takes autonomous actions to achieve specific, pre-defined goals, typically operating with minimal direct human supervision for each step. The defining characteristics of these systems include goal-oriented behavior, the capacity for autonomous decision-making within set parameters, adaptive learning from their interactions and outcomes, and often proactive problem-solving capabilities.

Modern agentic AI systems are distinguished by a suite of core capabilities that enable their sophisticated functionality:

Planning: A fundamental attribute of an AI agent is its ability to engage in planning. Given a high-level goal, an agent can decompose it into a series of smaller, manageable sub-tasks. It can then formulate a coherent plan-a sequence of actions-to execute these sub-tasks in a logical order to achieve the overall objective. This might involve simple sequential execution or more complex conditional logic and parallel processing.

Tool Use: Perhaps one of the most powerful features of contemporary AI agents is their capacity for "tool use". LLMs, as discussed earlier, can delegate specific computational or informational tasks to external tools. AI agents extend this concept significantly. They can be equipped to interact with a wide array of digital tools and systems: calling APIs to fetch data or trigger actions in other software, executing code, querying databases, controlling robotic hardware, or even interacting with other AI agents. This ability to leverage specialized tools allows an agent to perform actions and access capabilities far beyond what its core AI model can do natively, effectively extending its reach and operational capacity into the digital and sometimes physical world.

Self-Reflection and Refinement: A crucial development in agentic AI is the incorporation of reflective capabilities, often described by the "Reflection Pattern". This involves an iterative process where the agent first generates an initial output, plan, or action. It then enters a "reflection phase" where it critically evaluates its own work against the desired goals, quality standards, or known constraints. This self-critique can identify potential errors, logical inconsistencies, inefficiencies, or areas for improvement. Based on this internal feedback, the agent then refines its output or plan and iterates. This cycle of generation, self-assessment, and refinement allows the agent to improve its performance over time, learn from simulated "mistakes" (deviations from optimal paths), and adapt its strategies without requiring constant human intervention or retraining for every minor adjustment. This "simulated reflection" is a key mechanism that connects an agent's actions to its learning process, enhancing its autonomy and effectiveness.

The practical applications of such AI agents are rapidly expanding across industries. In customer service, agents can autonomously handle a wide range of inquiries, guide customers through complex processes, process orders, and provide 24/7 support, significantly improving efficiency and customer satisfaction. For instance, an agent might analyze an incoming customer request, automatically look up the customer's history and relevant policies from a knowledge base (using RAG principles), and then either formulate an automated response or intelligently route the request to the most appropriate human team.

In supply chain management, AI agents are being deployed to optimize inventory levels through predictive analytics, manage communications with suppliers, monitor compliance, process orders automatically, and even optimize transportation routes and schedules in real-time. These agents can forecast demand fluctuations, identify potential disruptions, reroute shipments to avoid delays, manage contract lifecycles, and coordinate the activities of warehouse robotics, leading to significant operational efficiencies and cost savings.

Even in complex domains like legal review, while AI "copilots" (discussed in the next section) assist humans, more autonomous agentic systems can be envisioned as part of larger workflows. For example, an agent might be tasked with parsing specific clauses from a batch of contracts, another agent might compare these clauses against a library of standard terms or regulatory requirements, and a third agent could summarize the findings or flag anomalies for human review.

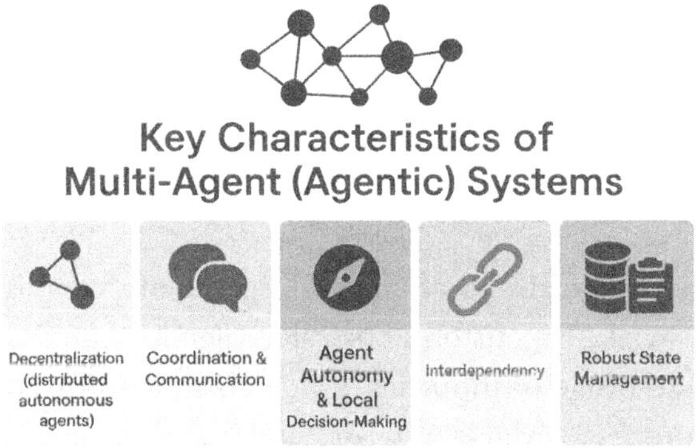

The power of AI agents is further amplified when they are designed to work together in multi-agent systems. This involves the "orchestration and coordination of tasks or processes carried out by multiple autonomous agents". In such a system, each agent might specialize in a particular subtask or possess a unique capability. The overall workflow is then achieved through their coordinated interaction. Key characteristics of multi-agent workflows include decentralization (tasks are distributed among agents that operate with a degree of independence), sophisticated coordination and communication protocols (agents need to exchange data, signals, or intermediate results, often managed by a central orchestrator or through peer-to-peer protocols), individual agent autonomy (each agent can make local decisions, such as retrying a failed task or adapting its approach), interdependency (the tasks performed by one agent often depend on the outputs or actions of others), and robust state management (maintaining the state of the overall workflow across distributed agents is crucial, especially for long-running processes).

Examples of multi-agent orchestration demonstrate this "synthetic team collaboration" vividly. In analyzing a complex financial document, one agent might perform Optical Character Recognition (OCR) to digitize a scanned document, a second agent could extract key numerical data and textual information, a third might perform web research to gather comparative industry benchmarks, and a final agent could synthesize all this information into a comprehensive report. Similarly, in insurance claims processing, a sequence of specialized agents could handle document digitization, data extraction from claim forms, verification of policy coverage details against a database, fraud detection by analyzing patterns, and even initial adjudication of the claim amount based on predefined rules.

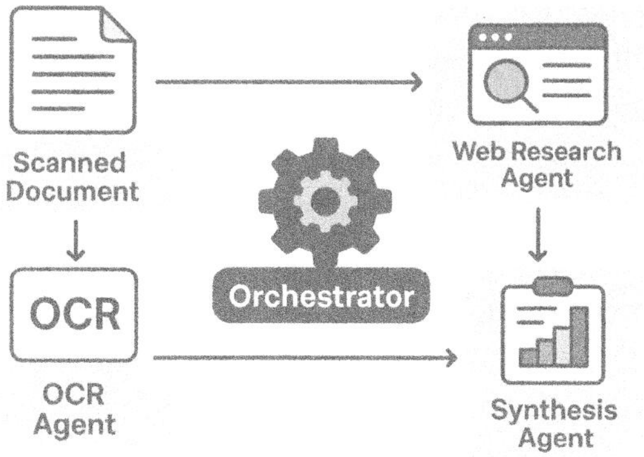

The emergence of AI agents that can plan, utilize diverse tools, self-reflect, and collaborate in multi-agent systems signifies a profound shift. AI is transitioning from being primarily a provider of answers or insights to becoming an active "doer of tasks". This has far more extensive implications for labor markets, the redefinition of "work," and the automation of complex processes than simpler forms of information access or generation. If an LLM provides knowledge, an agent applies that knowledge, interacts with other systems, and strives to produce a tangible outcome. This moves AI squarely into the operational core of many job functions.

However, the power of multi-agent orchestration also introduces new and complex systemic risks. While human teams benefit from social dynamics, shared contextual understanding, and often clear lines of responsibility, interactions within AI agent teams are fundamentally programmatic. A seemingly minor error, misinterpretation, or biased output from a single agent could propagate through the system, potentially being amplified by subsequent agents in the chain. This could lead to unexpected, large-scale failures or harmful emergent behaviors that are difficult to predict or control. Debugging such distributed, autonomous systems and assigning accountability when a "synthetic team" makes a critical error becomes an exceptionally challenging problem. The "manager" or "orchestrator" system itself, which coordinates these agents, becomes a critical point of potential failure and a target for ensuring robustness. Consequently, the development of multi-agent systems necessitates new frameworks for their design, rigorous testing, continuous monitoring, and robust governance. This includes creating auditable trails of inter-agent communication, developing methods for explaining the collective decision-making process of the agent team, and establishing clear protocols for human intervention, control, and system rollback when deviations or failures occur. The focus must shift from the reliability of individual agents to the resilience, safety, and predictability of the entire multi-agent ecosystem.

These sophisticated agents, whether operating individually or in orchestrated teams, are no longer confined to research labs or theoretical discussions. They are being integrated into real-world workflows, performing tasks that, until very recently, were the exclusive domain of skilled human professionals and collaborative human teams. This leads directly to an examination of how this synthetic expertise is actively displacing or augmenting traditional roles, and the critical lessons being learned from both its successes and its inevitable missteps.

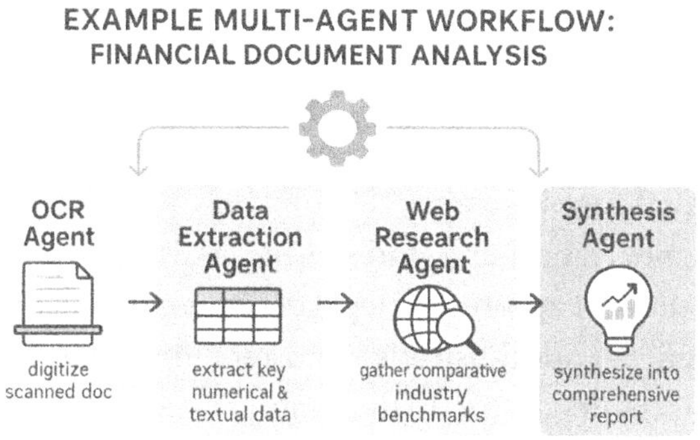

EXAMPLES IN DISPLACEMENT: SYNTHETIC EXPERTISE IN ACTION (AND MISSTEPS)

The theoretical capabilities of Large Language Models, Retrieval-Augmented Generation, and AI Agents translate into tangible impacts when applied to real-world professional domains. Across industries, these technologies are beginning to augment human capabilities and, in some cases, automate tasks traditionally performed by human experts. This section provides concrete examples of synthetic expertise in action, illustrating its transformative potential. However, a balanced perspective requires acknowledging the risks and limitations. The journey towards ubiquitous synthetic expertise is not without its pitfalls, and examining "failed" AI case studies is crucial for understanding the challenges of responsible and effective deployment.

Legal Copilots: Accelerating Legal Workflows

The legal profession, traditionally reliant on extensive human effort for research, document review, and drafting, is experiencing significant changes with the advent of AI copilots. Systems like Microsoft 365 Copilot are designed to assist legal professionals by speeding up contract review processes. These tools can identify contract review requests from communications, summarize lengthy documents and email chains, and help lawyers quickly locate specific clauses, such as duration or payment terms. They can compare different agreements, highlighting discrepancies or areas addressed in one document but not another, and even assist in drafting initial communications or sections of legal documents. The aim is to reduce the time spent on administrative and repetitive tasks, allowing legal professionals to focus on higher-value strategic work, potentially lowering overall review times and streamlining transactional processes. More advanced platforms, like those utilizing Copilot Studio, can enable the creation of automated contract review agents capable of proactively detecting risks, deviations from standard terms, and potential compliance issues.

However, the integration of AI into legal practice carries significant risks if not managed with diligence and a clear understanding of the technology's limitations. A prominent cautionary tale is the case of attorney Steven A. Schwartz. In Mata v. Avianca, Inc., Schwartz utilized ChatGPT for legal research and subsequently submitted a legal brief to a federal court that included six entirely fictitious case citations generated by the AI.

Schwartz admitted to using ChatGPT to gather information, research relevant case law, and assist in drafting the brief. His critical error lay in a fundamental misunderstanding of ChatGPT's capabilities and limitations; he was reportedly "unaware of the possibility that its content could be false". LLMs like ChatGPT, while adept at mimicking the style and structure of legal writing, can "hallucinate": fabricating information, including plausible sounding but non-existent case names, citations, and even judicial reasoning. ChatGPT knows what legal citations look like and how they are typically formatted, but it may not have real-time access to comprehensive legal databases to ensure the authenticity of every generated citation.

Schwartz failed to independently verify the AI-generated results, effectively outsourcing his professional responsibility for accuracy to the machine. The consequences were severe: a $5,000 sanction was imposed on Schwartz, his colleague, and their law firm for misleading the court and submitting false statements. The presiding judge noted that disciplinary measures might not have been imposed had the lawyers been forthcoming about their use of AI and the subsequent discovery of errors. This case starkly underscores the ethical duties and competency requirements incumbent upon legal professionals when employing AI tools, particularly the paramount importance of human verification of AI-generated content before it is used in any official capacity.

AI Tutors: Personalizing Education at Scale

In the field of education, AI is paving the way for more personalized and accessible learning experiences. AI-powered tutoring systems, such as Khanmigo developed by Khan Academy, are designed to act as personal tutors for students across a range of subjects, including mathematics, science, and writing. Unlike some generative AI tools that might provide direct answers, Khanmigo is programmed to guide students through the learning process by asking probing questions, helping them to think critically and arrive at solutions themselves. Such tools are often reported to be more accurate for specific tasks, like mathematical calculations, compared to general-purpose LLMs. AI tutors offer the potential to cater to individual learning styles and paces, provide on-demand support, and can be particularly beneficial for students who may require alternative ways to engage with educational materials, including those with disabilities. For educators, these tools can also assist with time-consuming tasks like generating rubrics or lesson plan ideas, freeing them to focus on direct student interaction and pedagogical strategy.

While the benefits are promising, the deployment of AI tutors also warrants careful consideration. Over-reliance on AI for guidance could, if not balanced with other learning methods, potentially hinder the development of students' independent problem-solving skills or their ability to grapple with ambiguity. The effectiveness of these tools can also be significantly influenced by how well instructors integrate them into the broader curriculum and guide students in their use. Although no large scale "failure" case studies for AI tutors are as prominent as in the legal field, the general risks associated with AI; such as potential biases in how they respond to different student queries, the accuracy of their underlying knowledge bases, and data privacy concerns-remain relevant.

Clinical Decision Support Systems (CDSS): Augmenting Medical Diagnostics

The healthcare sector is another area where synthetic expertise is making significant inroads, particularly in clinical decision support. Generative AI is being used to streamline administrative tasks for clinicians, enhance patient education materials, and, crucially, augment diagnostic processes. AI systems can rapidly process and synthesize vast quantities of complex medical information, including patient records, medical literature, imaging scans (X-rays, MRIs, CTs), and biosignals (ECGs, EEGs). Numerous studies have indicated that AI can achieve diagnostic precision comparable to, and in some specific diagnostic reasoning tasks, even outperform human clinicians.

For example, one study highlighted a decision-support algorithm that successfully reduced the inappropriate use of second-line antibiotics for uncomplicated urinary tract infections by 67% by guiding physicians towards more optimal prescribing patterns based on patient data. AI can also play a role in mitigating human cognitive biases that sometimes lead to diagnostic errors, as AI models can be instructed to re-examine evidence or consider alternative possibilities without the same anchoring biases humans might exhibit.

However, the application of AI in medical diagnostics is fraught with potential risks. The most glaring issue is the potential for AI to generate incorrect information or "hallucinations," which, in a clinical context, could lead to serious diagnostic errors and patient harm. The "black box" nature of some AI models-where the exact reasoning pathway to a conclusion is not transparent-can be a barrier to trust and adoption among clinicians, although efforts in Explainable AI (XAI) are working to address this. Algorithmic bias is another significant concern; if AI models are trained on datasets that underrepresent certain demographic groups, their diagnostic performance for those groups may be suboptimal, potentially exacerbating existing healthcare disparities. There's also the risk of patients misinterpreting AI-generated health information as definitive medical advice without consulting a human professional. Finally, complex ethical questions surrounding liability for AI-related medical errors-whether responsibility lies with the physician who used the AI, the developers of the AI system, or the healthcare institution-are yet to be fully resolved.

The Specter of AI-Generated Misinformation: Fabricated Scientific Studies

Beyond specific professional applications, a broader risk associated with advanced generative AI is its potential to fuel misinformation and undermine the integrity of knowledge creation itself. A stark demonstration of this risk comes from research showing the ease with which AI can be used to fabricate entire academic papers. In one experiment, researchers successfully used AI to generate 288 complete, plausible sounding (though fake) academic finance papers that purported to predict stock returns. This highlights how easily AI can be employed to create fictitious data, falsify experimental results, or generate misleading research narratives.

The problem of academic paper retractions is already growing, with a significant portion due to misconduct such as data fabrication or plagiarism. AI has the potential to exacerbate this issue dramatically. There are already documented instances of AI-generated papers, sometimes containing serious flaws like fake references or deliberately fabricated data, successfully passing through peer-review processes and achieving publication, only to be retracted later upon closer scrutiny. The phenomenon of AI "hallucination" can also lead to unintentional errors in scientific papers if AI-generated text or data is incorporated without rigorous human oversight and verification. For instance, one study on AI-assisted computer programming found that a substantial number of AI-generated answers to coding questions contained errors, and human reviewers failed to correct these errors a significant number of times. This illustrates how easily flaws can propagate if there's an over-reliance on AI without sufficient critical evaluation. The potential for AI to be used for malicious purposes, such as generating fictitious clinical trial data to support a flawed drug, or modifying gene editing experimental data to conceal adverse outcomes, poses a serious threat to scientific integrity and public trust.

To provide a clearer overview of these dynamics, the following lists summarize key applications, their demonstrated benefits, and associated cautionary tales.

Legal

- **AI application:** Contract review copilots (for example, document review and drafting assistants embedded in productivity suites).
- **Demonstrated benefit:** Faster document review, summarization, clause identification, agreement comparison, and drafting assistance.

- **Documented failure or risk:** Reliance on unverified AI output can lead to serious professional consequences, as demonstrated by cases where AI-generated legal citations were fabricated and resulted in sanctions.
- **Key takeaway:** AI can significantly enhance legal efficiency, but unquestioning reliance without human verification is professionally and ethically dangerous. Competence in *how* to use AI tools is now a core legal skill.

Education

- **AI application:** AI tutors and learning companions.
- **Demonstrated benefit:** Personalized learning guidance, Socratic-style questioning, support for diverse learning styles, and assistance for educators.
- **Documented failure or risk:** Over-reliance can undermine independent problem-solving if not carefully integrated, while broader AI risks such as bias and knowledge inaccuracies still apply.
- **Key takeaway:** AI tutors can democratize personalized education, but must be integrated pedagogically to strengthen, not replace, critical thinking.

Healthcare

- **AI application:** Clinical decision support systems.
- **Demonstrated benefit:** Improved diagnostic accuracy in specific tasks, rapid processing of complex medical data, and reduction of certain human cognitive biases.
- **Documented failure or risk:** Hallucinations that could lead to clinical errors, algorithmic bias inherited from training data, limited explainability, and patient misinterpretation of AI-generated advice.
- **Key takeaway:** AI can meaningfully augment diagnostics, but only with rigorous validation, explainability, bias

mitigation, and clear clinical usage guidelines to protect patient safety and trust.

Scientific Research

- **AI application:** AI-assisted research paper generation and analysis tools.
- **Demonstrated benefit:** Rapid drafting, large-scale literature synthesis, and support for hypothesis generation.
- **Documented failure or risk:** Lower barriers to fabricating studies, data, and references; flawed AI-generated papers passing peer review; hallucinations introducing subtle but serious errors.
- **Key takeaway:** AI is a powerful accelerator of research, but also a potent enabler of misconduct. Safeguarding scientific integrity requires rigorous verification, ethical norms, and evolution of peer-review practices.

These examples reveal a crucial pattern: the "failures" associated with AI in expertise-driven domains are often not solely technological shortcomings but rather breakdowns in the human-AI interaction and the surrounding processes. In the Schwartz case, the failure was not just that ChatGPT produced fake citations, but that the human lawyer misused the tool through a lack of understanding of its limitations and a critical lapse in professional diligence; the failure to verify.

Similarly, the risk of fake scientific papers proliferating is not just about AI's capacity to generate them, but about the potential for humans to intentionally misuse this capacity for deception, or for human oversight (like peer review, which itself might become AI-assisted) to fail in detecting such fabrications. This suggests that the safe and effective deployment of synthetic expertise hinges as much on developing robust human practices-such as enhanced digital and AI literacy, mandatory verification protocols, new ethical guidelines for AI use, and comprehensive training for professionals on how to critically engage with AI outputs-as it does on improving the AI technology itself. The problem often lies at the interface between human and machine, or within the human systems that are supposed to govern the use of these powerful new tools.

Furthermore, the very nature of "expertise" is thrown into question when AI can generate outputs that appear highly expert-like but may lack genuine underlying understanding or contain subtle, critical flaws. This creates a new kind of "epistemic risk," where the ability to distinguish authentic, reliable expertise from plausible-sounding synthetic output becomes a significant individual and societal challenge. Traditional markers of human expertise, such as credentials, years of experience, or institutional affiliation, do not directly apply to AI systems. While the verifiability of AI output, for instance through RAG's source-citing capabilities, offers a partial solution, the potential for subtle biases, misinterpretations of sources, or errors in the AI's own generative process can persist. The sheer ease and volume with which AI can produce sophisticated-looking content also make it increasingly difficult to scrutinize every piece of information thoroughly. This reality places a heightened burden of discernment on individuals, who must become more critical consumers of information, constantly aware that the superficial markers of expertise can now be convincingly mimicked by machines.

It is clear from these examples that expertise, in many of its traditional forms, is increasingly becoming a function that can be replicated and scaled by software. This scalability, often achieved with near-zero marginal friction once the systems are developed, has profound implications for the cost of accessing expert services and the overall economics of knowledge-based industries, a topic the next section will explore in detail.

THE COST CURVE COLLAPSES: THE NEW ECONOMICS OF EXPERTISE

The proliferation of AI systems capable of delivering synthetic expertise is not just a technological marvel; it is a force of profound economic disruption. As access to high-level insight and skilled task performance becomes increasingly mediated by AI, the traditional economics of expertise; built on scarcity, extensive human training, and time-based compensation are beginning to unravel. The most fundamental shift is the dramatic collapse of the marginal cost associated with generating expert outputs. This is creating new service paradigms, undermining established pricing models, and forcing a re-evaluation of how value is created and captured in knowledge intensive sectors.

The core economic disruption stems from a simple reality: once an AI expert system (an LLM, RAG system, or AI agent) is developed, trained, and deployed, the incremental cost of producing an additional "expert answer," drafting another legal clause, analyzing one more financial statement, or providing another personalized tutoring session approaches zero. This stands in stark contrast to human expertise, where each unit of expert output incurs significant marginal costs related to the expert's time, cognitive effort, and often, ongoing professional development. This collapse in marginal cost is the primary driver reshaping the economic landscape of expertise.

This shift is causing significant disruption to traditional service models across various industries:

Consulting: The high-fee structure of the consulting industry has long been predicated on access to specialized knowledge, proprietary frameworks, and the analytical capabilities of experienced human consultants. AI is now challenging this model. AI systems can assist in reshaping key organizational functions and even help in inventing new offerings. Research indicates that AI tools can have a "levelling effect," enabling less experienced consultants or even nonconsultants to perform tasks, such as conceptualizing new product ideas or analyzing market data, at a level previously associated with more seasoned professionals. The ability of AI to rapidly synthesize information, generate strategy decks from market data, or even replicate aspects of common consulting playbooks puts pressure on the traditional value proposition. While the "10-20-70 principle" cited by some industry analyses-suggesting that AI success relies 10% on algorithms, 20% on data and technology, and a crucial 70% on people, processes, and cultural transformation-indicates a continued significant role for human consultants, that role is evolving. It is shifting from being the primary source of knowledge to becoming orchestrators of AI-augmented insights, change managers, and trusted advisors on complex implementation and ethical considerations.

Tutoring and Education: The field of education is similarly affected. AI-powered tutors, like Khanmigo, can provide personalized learning support, adaptive exercises, and on-demand guidance to students at a scale and price point that traditional human tutoring cannot match. This democratizes access to tailored educational assistance, making bespoke learning experiences available to a much broader audience than was previously feasible.

Financial Advising: The financial advisory sector is also undergoing transformation. AI-driven investment tools are increasingly becoming a primary source of financial advice and information for retail investors, with projections suggesting significant growth in their adoption. AI systems can analyze financial statements with high accuracy, perform sophisticated market analysis, assist in portfolio construction, identify search engine optimization (SEO) opportunities for financial advisor websites, and create compelling data visualizations for client presentations. Major financial institutions are already integrating AI into their advisory workflows, using AI assistants to support human advisors during client meetings or to automate back-end operational processes. While AI may not fully replicate the empathy and nuanced interpersonal skills of a human advisor, its capacity for consistent, data-driven analysis and its tireless availability are compelling economic advantages.

This economic shift is giving rise to the concept of Expertise-as-a-Service (EaaS). In this paradigm, expertise is no longer primarily accessed by hiring a full-time human expert or engaging in lengthy, expensive consultations. Instead, it becomes a utility that can be consumed on demand and paid for based on usage or value received. The pricing models for EaaS are likely to mirror those seen in the Software-as-a-Service (SaaS) industry:

Usage-Based Pricing: Customers pay per query, per task completed by an AI agent, per API call to an expert system, or based on the volume of data processed. This "pay-as-you-go" model aligns cost directly with consumption.

Subscription Models: Users pay a recurring (e.g., monthly or annual) fee for access to a defined level or volume of synthetic expertise, offering predictable costs for users and stable revenue for providers.

Tiered Pricing: Providers offer multiple packages with varying levels of AI capability, access to different specialized expert modules, higher usage limits, or enhanced support, catering to diverse user needs and budgets.

Freemium Models: Basic AI-driven expertise or a limited set of features might be offered for free to attract users, with charges applied for more advanced capabilities, higher usage tiers, or premium support.

The fundamental shift is from pricing based on scarce human time and credentials to pricing based on computational resources, the quantifiable value delivered by the AI, or the level of access provided.

A complex issue intertwined with the economics of synthetic expertise concerns the "cost" of information when AI models are trained on data that was previously paywalled or is protected by copyright. A significant portion of the vast datasets used to train powerful foundation models may include copyrighted materials, articles from subscription journals, or proprietary databases, often ingested without explicit authorization or compensation to the original creators. The U.S. Copyright Office, among other regulatory bodies, is grappling with this issue, with ongoing debates about whether such unauthorized use for AI training constitutes "fair use".

The legal landscape is still evolving, but if AI systems effectively "democratize" access to insights derived from paywalled or copyrighted information without a mechanism for recompense, the economic models of original content creators, publishers, and data providers are severely undermined. This could lead to protracted legal battles and potentially new regulatory frameworks requiring licensing fees or royalties for data used in AI training. Such developments could, in turn, re-introduce significant costs into the development and provision of synthetic expertise, complicating the narrative of ever-decreasing prices.

This leads to the question posed in this book: Does cost become more of an equation related to time rather than arbitrary cost?

In many respects, yes. The "cost" to the user of obtaining an expert answer or solution is dramatically reduced in terms of time to access-seconds or minutes for an AI response versus potentially hours, days, or weeks waiting for a human expert. For businesses, this translates into faster time to market for new products, quicker time to solve complex problems, and reduced opportunity costs. For providers of EaaS, the computational time required to process a query and generate a response is a direct component of their operational cost. However, "arbitrary cost"; or more accurately, value-based pricing is unlikely to disappear entirely. While the cost of generic or routine synthetic expertise may plummet, highly specialized, mission-critical, or uniquely impactful AI-driven insights and solutions will likely still command premium prices based on the perceived value, risk reduction, or competitive advantage they offer, rather than solely on the cost of the underlying computation. The "arbitrariness" simply shifts its anchor from the prestige and scarcity of human experts to the demonstrated efficacy and strategic importance of the AI's output.

The following lists offer a comparative overview of how key economic aspects of expertise are being transformed:

Primary cost driver

- **Traditional human expertise model:** Labor costs driven by extensive training time and years of experience accumulation.
- **AI-driven synthetic expertise model:** Computation costs associated with model training and inference, along with data acquisition, curation, and software development.
- **Primary implication:** Once AI systems are deployed, the cost per unit of expert output drops dramatically.

Basis of pricing

- **Traditional human expertise model:** Billable hours, salaries, and project fees that reflect human time and scarcity.
- **AI-driven synthetic expertise model:** Usage-based fees per query or task, subscription tiers, and value-based pricing for high-impact outcomes.
- **Primary implication:** Pricing shifts from time-based models to access-, outcome-, or value-based approaches, including the possibility of micro-transactions for expertise.

Access model

- **Traditional human expertise model:** Gatekept through credentials and networks, scarce, and dependent on hiring or scheduled engagement.

- **AI-driven synthetic expertise model:** On-demand, abundant, and instantly accessible through digital interfaces.
- **Primary implication:** Expert knowledge and problem-solving capabilities become radically democratized.

Key value proposition

- **Traditional human expertise model:** Experience, nuanced judgment, human intuition, and bespoke solutions.
- **AI-driven synthetic expertise model:** Speed, scale, consistency, data-grounded insights, personalization at scale, and continuous availability.
- **Primary implication:** New forms of value emerge around efficiency and breadth of access, complementing rather than fully replacing human-centric value.

Dominant resource

- **Traditional human expertise model:** Human capital and individually or collectively held knowledge.
- **AI-driven synthetic expertise model:** Algorithmic systems, large-scale datasets, and computational infrastructure.
- **Primary implication:** Expertise shifts from being primarily embodied in people to becoming an engineered, replicable, and scalable resource.

Knowledge integrity

- **Traditional human expertise model:** Maintained through human ethics, peer review, and institutional reputation.
- **AI-driven synthetic expertise model:** Dependent on data quality, algorithmic design, retrieval-augmented verification, and potential source citation.

- **Primary implication:** New mechanisms are required to ensure accuracy, manage bias, and sustain trust in information sources.

The collapse of the cost curve for accessing expertise is a monumental shift. However, it simultaneously elevates the importance of a different set of skills and capabilities. When basic expert knowledge becomes a cheap commodity, the real value migrates. It shifts towards the ability to effectively apply this abundant synthetic expertise, to integrate it with unique human insights and contextual understanding, and, crucially, to ethically guide its use and manage its inherent risks. The bottleneck is no longer "knowing what to do" for many standard problems; it's "doing it wisely, effectively, and responsibly" in complex, real-world situations. This suggests the emergence of new professional roles focused on "expertise orchestration," "AI-to-value realization," or "AI ethics and governance". The economic rewards will increasingly flow to those individuals and organizations that can successfully bridge the gap between the ocean of available synthetic knowledge and the creation of tangible, responsible, and valuable outcomes.

Furthermore, the unresolved economic tension surrounding the use of potentially copyrighted or paywalled data for training AI models looms large. If AI companies are effectively building their low-cost expertise services on the uncompensated intellectual labor and investment of others, this creates an unsustainable and ethically questionable foundation. The outcomes of ongoing legal challenges and the potential development of new regulatory frameworks for data royalties or mandatory licensing could significantly alter the cost structures of synthetic expertise. This might not entirely reverse the trend towards lower costs but could introduce new complexities and financial considerations into the knowledge supply chain" It could also incentivize the creation and use of high-quality, ethically sourced, and properly licensed datasets for AI training, potentially leading to a premium market for responsibly trained or certified AI expertise systems. The seemingly straightforward path to near-zero cost expertise may yet encounter significant economic and legal recalibrations.

As the cost of insight plummets and access to expert-level assistance explodes, the very fabric of societal expectations around knowledge, competence, and problem-solving begins to change. This ripple effect, touching everything from daily interactions to institutional norms, is the focus of the next section.

NEW EXPECTATIONS, NEW NORMS: LIVING WITH INSTANT EXPERTISE

The economic recalibration driven by AI-powered synthetic expertise is paralleled by equally profound shifts in social expectations and behavioral norms. When high quality, context specific information and expert-level assistance become instantly accessible to a broad populace, the standards by which we judge knowledge, competence, speed, and even accountability are inevitably altered. Living in an environment saturated with readily available synthetic expertise reshapes our interactions, our demands on institutions, and our understanding of individual responsibility.

One of the most immediate consequences is the erosion of "I don't know" as a broadly acceptable response in many contexts, particularly concerning factual information or established procedures. When a query to an AI assistant can yield a comprehensive answer in seconds, the societal tolerance for individuals or organizations claiming ignorance, especially on matters that are readily verifiable, is likely to diminish. This doesn't mean that all forms of not knowing become unacceptable-genuine uncertainty in complex, novel, or deeply subjective areas will persist. However, for a vast range of informational needs, there will be a heightened baseline expectation that individuals should either be informed or possess the immediate capability to become informed via their AI tools. This implies subtle but pervasive pressure to be perpetually "connected" to this extended cognitive capacity.

Flowing from this is an escalating expectation for always-on precision and personalization in services and interactions. As users become accustomed to AI systems that can provide highly accurate, tailored responses 24/7; whether in customer service, personalized learning environments, or even in preliminary diagnostic support or financial guidance these attributes will transition from being delightful novelties to baseline expectations. Businesses, educational institutions, healthcare providers, and public services will face increasing pressure to integrate AI capabilities to meet these heightened demands for instant, precise, and contextual aware engagement. Failure to do so may be perceived as inefficiency, a lack of customer focus, or simply being out of step with contemporary standards.

Paradoxically, while AI offers an abundance of answers, it also shifts a greater burden of discernment onto the end user. The availability of synthetic expertise does not equate to an abdication of critical thinking. AI systems can make errors, exhibit biases embedded in their training data, or be deliberately misused to create sophisticated misinformation, such as fake legal citations or fabricated scientific data.

Even with RAG systems providing source traceability, the user is ultimately responsible for evaluating the credibility of the AI's output and its underlying sources. Users must cultivate an awareness of AI's inherent limitations-its potential lack of true understanding, its susceptibility to subtle manipulation, or its inability to grasp nuanced common sense in certain situations. The challenge for the individual is no longer solely finding information, but critically navigating a sea of AI-generated content, distinguishing reliable expertise from plausible facsimiles, and deciding when to trust the synthetic expert versus when to seek further verification or human judgment. This demands a new level of digital literacy and epistemic vigilance.

The changing landscape of expertise also redefines the nature of competitive advantage, leading to the progression of asymmetries outlined in the book's framework:

Collapse of Information Asymmetry: This is the foundational shift. AI, particularly with RAG, makes vast stores of information and foundational knowledge broadly and quickly accessible, eroding the advantages previously held by those with exclusive access to data or basic know-how.

Emergence of Insight Asymmetry: When information access is largely equalized, the next advantage lies in the ability to derive unique, actionable insights from that commonly available information. This involves more than just information retrieval; it requires superior analytical skills, the ability to ask more profound and targeted questions of AI systems, the capacity to synthesize information from diverse AI-generated outputs, and the creativity to connect disparate pieces of knowledge in novel ways. This is where human-AI collaboration can shine, but the quality of the human contribution in framing problems and interpreting results becomes paramount.

Shift to Execution Asymmetry: Beyond the generation of insights, the ultimate differentiator becomes the ability to effectively act upon those insights and achieve desired outcomes in the complex, often messy, real world. This encompasses strategic planning, resource allocation, project management, leadership, navigating organizational politics, adapting to unforeseen challenges, and motivating human teams; capabilities that are currently far more difficult to fully automate than information retrieval or even insight generation. The advantage moves to those individuals and organizations that can not only "know" (via AI) and "understand" (via insight) but can also do and deliver. This progression highlights that while AI democratizes the initial layers of expertise, new, more complex forms of differentiation will inevitably arise.

The societal impact of instant access to AI expertise extends further. The ease with which AI can provide answers or perform tasks might lead to a reduction in certain types of direct human-to-human interaction previously necessary for knowledge exchange or problem-solving. While this can bring efficiencies, it also raises questions about the potential loss of social learning, mentorship opportunities, and the serendipitous discoveries that often arise from human collaboration. Concerns about unemployment also surface as AI systems become capable of performing tasks previously done by humans, though this chapter's primary focus is on the accessibility of expertise rather than a full analysis of labor displacement, which is likely addressed elsewhere in the book. There is also the potential for wealth inequality to be exacerbated if access to the most advanced AI expertise tools, or the skills to leverage them effectively, becomes a new form of capital concentrated in fewer hands. Profound ethical questions regarding the rights, moral status, and societal integration of increasingly intelligent AI systems will also become more pressing, forming a key theme for later parts of this work.

The expectation of always-on precision and instant answers, largely fueled by our interactions with AI, could inadvertently foster a societal impatience with human error and the inherently iterative, sometimes messy, process of human learning and exploration. When AI can deliver a seemingly perfect answer in moments, there may be less tolerance for the trial-and-error, the "dumb questions" (a theme explored in Chapter 5), and the imperfect first drafts that are essential to human creativity and deep understanding. This could create subtle pressure towards a kind of "hyper-perfectionism," where the process of human learning, which inherently involves making and learning from mistakes, is devalued in favor of achieving immediate, AI-like correctness. Such an environment might inadvertently stifle the very innovation and resilience that often spring from navigating imperfections and unexpected challenges.

Furthermore, the shift from information asymmetry to insight and then execution asymmetry implies that merely having "access to expertise" through AI is only the entry ticket to relevance in the future. The real, sustainable value and the locus of power will reside in the higher-order cognitive and organizational capabilities required to leverage this universally accessible expertise into unique strategies and effective, impactful actions. This creates new, potentially more complex, forms of inequality. While basic expert knowledge becomes a commodity, meta-skills; such as advanced critical thinking, creative problem formulation, ethical discernment in the use of AI, cross-domain synthesis, and the leadership of human-AI teams become exceptionally valuable.

These meta-skills are unlikely to be evenly distributed, potentially leading to a new "digital divide" based not on access to information per se, but on the capacity to strategically command, interrogate, and operationalize the outputs of synthetic expert systems. Individuals and organizations that cultivate these meta-skills and build the agility to translate AI-driven insights into real-world execution will be the ones to thrive in this new landscape.

These evolving norms and expectations, driven by the instant availability of synthetic expertise, are not confined to specific industries or professions. They ripple through the entirety of society-reshaping education, transforming business operations, influencing governance, and fundamentally altering what it means to be competent and knowledgeable in the 21st century. This leads to a necessary re-examination of the very structure and perception of expertise itself.

THE GREAT COMPRESSION OF EXPERTISE: REDEFINING MASTERY AND LEARNING

The widespread availability of AI-driven synthetic expertise is instigating a "great compression" of the traditional pyramid of expertise. Historically, expertise was envisioned as a hierarchical structure, with a broad base of novices gradually acquiring knowledge and skills to become journeymen, and a select few reaching the pinnacle of mastery through decades of dedicated effort. AI is now flattening this pyramid, making high-level knowledge and sophisticated task performance accessible much more rapidly and broadly. This forces a fundamental re-evaluation of what it means to be an "expert," the enduring value of deep human experience, particularly experience forged through failure, and the future trajectory of learning and skill development.

A compelling illustration of this compression is the challenge AI poses to the "10,000-Hour Rule," popularized by Malcolm Gladwell, which posits that achieving elite performance in any complex skill typically requires approximately 10,000 hours of deliberate, focused practice. AI systems can now generate outputs; be it drafting legal arguments, composing music, writing code, or offering diagnostic suggestions that reflect a level of proficiency seemingly acquired in moments, bypassing the long, arduous journey traditionally associated with mastery. While AI might appear to commoditize certain skills that once demanded such extensive investment, it's crucial to recognize that true elite human expertise often encompasses more than just knowledge recall or pattern execution. It involves intricate judgment in ambiguous situations, profound creativity, deep emotional intelligence, and the ability to navigate novel, "out-of-distribution" problems-qualities that remain more resistant to AI replication. The critical question then becomes: which aspects of the 10,000-hour journey are genuinely being compressed or rendered obsolete by AI, and which elements of deep human experience retain their unique, irreplaceable value? Some analysts suggest that while many roles will be augmented, a small cadre of exceptionally skilled individuals, whose work AI cannot yet replace, will remain highly sought after.

This compression fuels the ongoing debate about de-skilling versus re-skilling in AI-enhanced industries. On one hand, AI can exhibit a "levelling effect," enabling individuals with less prior skill or experience (novices) to perform tasks at a level closer to that of seasoned experts. Examples include AI-powered chatbots assisting less experienced customer service agents to handle complex queries more effectively, or AI tools helping junior consultants produce higher-quality analyses. This phenomenon can be interpreted as a form of de-skilling, where the depth of individual skill required for a given level of performance is reduced because the AI system encapsulates and distributes "best practices". In some cases, this can lead to work becoming more standardized and routinized, potentially reducing worker autonomy and the opportunity to develop deeper, more nuanced skills through hands-on experience.

On the other hand, the integration of AI into workflows simultaneously creates a demand for new skills-an impetus for re-skilling and up-skilling. Professionals increasingly need to become adept at interacting with AI systems: formulating effective prompts to elicit desired outputs (prompt engineering), critically evaluating the quality and reliability of AI-generated content, editing and refining AI outputs to suit specific contexts, and strategically integrating AI tools into existing processes. This necessitates a focus on developing AI literacy, data literacy, and an understanding of how to collaborate effectively with AI as a partner rather than just a tool. The reality is likely a complex interplay of both de-skilling in some routine or pattern-based aspects of jobs, and a demand for new, often higher-order, re-skilling in areas related to managing, guiding, and leveraging AI.

In this evolving landscape, the role of humans increasingly shifts towards functions that AI is less equipped to handle independently. Humans become crucial as:

Curators: Sifting through the vast amounts of information and potential solutions generated by AI to select, validate, and present the most relevant, accurate, and useful outputs.

Context Setters: Providing the rich, nuanced understanding of specific situations, cultural subtleties, unspoken goals, and complex interpersonal dynamics that AI, trained primarily on data, may lack.

Ethical Anchors: Ensuring that AI systems are developed and deployed responsibly, fairly, and in alignment with human values and societal norms. This involves identifying and mitigating biases in AI models and their outputs, preventing the misuse of AI for harmful purposes (as seen with fabricated legal or scientific information), and making complex ethical judgments that go beyond rule-based logic or statistical optimization. Uniquely human skills such as sophisticated critical thinking, advanced problem-solving in novel contexts, and deep collaborative abilities become even more vital in an AI-augmented world.

This redefinition of human roles also forces us to reconsider what it means to "be an expert". Experts are more than just 'knowing' something. They have actual 'experience' with not only successes but failures and sometimes failures teach us more than success. True human expertise is often forged not just from the accumulation of explicit knowledge, but from the crucible of lived experience. This includes making mistakes, encountering unexpected obstacles, learning from these setbacks, and developing the nuanced judgment, intuition, and resilience that arise from navigating real-world complexities. AI, while capable of processing vast datasets and identifying patterns associated with success and failure within those datasets, may struggle to replicate the rich, multi-faceted learning that humans derive from actual, consequential experiences, particularly those involving ambiguity, common-sense reasoning in novel situations, or profound ethical dilemmas.

While AI systems can, in a sense, "learn from failure," this process is typically structured differently from human experiential learning. Reinforcement Learning (RL) is a primary mechanism through which AI agents learn and improve. In RL, an agent interacts with an environment (which can be a simulation or the real world) and receives positive "rewards" for actions that lead towards a desired goal, and negative "penalties" for actions that are detrimental or lead away from the goal. Through repeated trial and error, the agent learns a policy-a strategy for choosing actions-that maximizes its cumulative reward over time.

Deep Reinforcement Learning (DRL) combines RL with deep neural networks, enabling agents to learn complex policies in environments with vast state spaces, as famously demonstrated by DeepMind's AlphaGo, which mastered the game of Go and defeated world champions. Furthermore, Reinforcement Learning from Human Feedback (RLHF) incorporates human evaluations of AI behavior into the reward signal, helping to align AI actions more closely with human preferences and desired outcomes.

However, it's important to appreciate the distinction. When an AI "fails" in an RL context, it's typically an event that provides a data point for optimizing its strategy within a defined reward structure and environment. Human learning from failure is often far more open-ended and transformative. A significant human failure can lead to a re-evaluation of fundamental assumptions, a paradigm shift in understanding, or the serendipitous discovery of entirely new approaches; outcomes that may not be captured if the AI's learning is constrained by a predefined objective function and a limited set of possible actions or environmental states. Thus, while AI's capacity to learn from its "mistakes" will undoubtedly improve its performance and reliability, the "experience" it accumulates may remain qualitatively different from the wisdom humans gain from navigating the unpredictable and often unquantifiable consequences of their actions in the real world.

This "compression of expertise" might lead to a two-tiered system of knowledge and capability. At one level, there could be a broad base of easily accessible, AI-generated "functional expertise" sufficient for a wide range of common tasks and problems. This is the realm where AI provides quick answers and automates routine expert work. At a higher level, a smaller pinnacle of deep, uniquely human "meta-expertise" or "wisdom" would remain. This domain would focus on tackling truly novel problems, navigating extreme ambiguity, making profound ethical judgments, synthesizing knowledge across radically different fields in creative ways, and providing the critical oversight and strategic direction for the use of AI itself.

In this scenario, the value of human expertise shifts dramatically. "Knowing facts" or "performing standard procedures," even complex ones, become increasingly commoditized by AI. The premium is placed on those distinctly human cognitive skills: advanced critical thinking, creativity, ethical reasoning, systems thinking, and the ability to effectively guide, integrate, and collaborate with AI systems. This doesn't eliminate expertise; it profoundly reframes it, creating a new hierarchy where human value lies in managing the AI-driven knowledge infrastructure and solving problems that lie beyond its current (and perhaps inherent) scope.

Finally, the instant availability of answers through AI raises profound questions about the future of curiosity, inquiry, and the learning process itself, a theme that connects directly to the concerns of Chapter 5 ("The Death of the Dumb Question"). If any question can be answered immediately and comprehensively by an AI, what happens to the intrinsic human drive to explore, to question deeply, to wrestle with a problem until understanding dawns? Does the removal of intellectual friction diminish the motivation for profound inquiry? Or, conversely, does freedom from the drudgery of basic fact-finding and routine problem-solving liberate human curiosity to pursue higher order questions, more complex explorations, and more creative syntheses? The answer likely lies in how we choose to integrate these powerful tools into our learning ecosystems and our personal intellectual lives.

Expertise, in the age of AI, has not disappeared-it has been fundamentally reframed. The traditional markers of mastery, the pathways to acquiring knowledge, and the very definition of what it means to be an expert are all undergoing a radical transformation. The challenge and opportunity lie in navigating this new landscape, cultivating the uniquely human capacities that AI cannot replicate, and ensuring that the universal accessibility of synthetic expertise serves to elevate, rather than diminish, human potential and collective wisdom.

SUMMARY: THE EXPERTISE REVOLUTION

Artificial intelligence, powered by the synergistic capabilities of Large Language Models (LLMs), Retrieval-Augmented Generation (RAG) systems, and increasingly sophisticated multi-agent orchestration, is no longer merely an advanced information retrieval tool. It has evolved into a dynamic, composable system capable of delivering high-quality, context-specific expertise across a vast and growing array of domains. These technologies are not just augmenting human intellect; they are creating a new paradigm of "synthetic expertise" that is profoundly reshaping our world.

This technological surge is fundamentally transforming the core dynamics of how expertise is accessed, making it more democratic and readily available than ever before. It is altering how expertise is scaled, allowing for near-infinite replication of expert functions at speeds previously unimaginable. And it is collapsing the traditional cost structures associated with expert knowledge, leading to significant economic, social, and even psychological adjustments. In this emerging age of information symmetry, driven in large part by the rise of synthetic expertise, the very definitions of what it means to be knowledgeable, competent, and an "expert" are being rewritten. This revolution calls for the cultivation of new human skills, the establishment of new societal norms, and a deeper, more critical consideration of the evolving human role in a world where machines can increasingly know and do so much of what was once the exclusive province of human mastery. The accessibility of such expertise paves the way for a world where "dumb questions" may cease to exist, as explored in the next chapter, and where knowledge itself faces commoditization as a differentiator, a theme Chapter 6 will unpack.

5

THE DEATH OF THE DUMB QUESTION

The human drive to understand is often expressed through questions. Yet, for much of history, the simple act of asking a question has been a complex social and psychological transaction, fraught with potential costs. The fear of appearing ignorant, the pressure to already possess knowledge, and the hierarchical structures that gate keep information have often outweighed the innate desire to learn. A question, in many contexts, has been less an invitation to explore and more an admission of a deficit. This chapter explores a profound shift in this dynamic, catalyzed by the advent of advanced Artificial Intelligence. We stand at the cusp of an era where the "dumb question"; that hesitant, perhaps foundational, query born of genuine curiosity but stifled by apprehension is rendered obsolete. AI's capacity to provide instant, tailored, high-quality answers to nearly any inquiry, in private and without judgment, is dismantling the age-old barriers to curiosity.

Readers will come to understand the transformative power of AI in reshaping the very mechanics of question-asking. This is not merely about faster information retrieval; it is about a fundamental change in how individuals approach learning, how expertise is perceived and accessed, and how personal confidence in the ability to understand complex topics can be cultivated. We will examine the cascading effects of this new reality across educational landscapes, professional environments, and the intricate psychological processes that govern our pursuit of knowledge. The traditional, tiered systems of knowledge access, often guarded by institutional or expert gatekeepers, are giving way to a model where information is delivered based on an individual's immediate need, specific context, and the unique moment of their inquiry.

The core argument of this chapter is that the "dumb question," as a social construct laden with fear and risk, is losing its power. AI emerges as a potent equalizer in the realm of knowledge acquisition, offering a sanctuary for private, uninhibited inquiry. This democratization of answer-seeking does more than just supply information; it fundamentally alters our collective and individual relationships with knowledge, the process of learning, and indeed, with each other. The implications are far-reaching. When the penalty for not knowing is removed, and the path to understanding is paved with personalized guidance, the human capacity for curiosity can be unleashed in unprecedented ways.

This chapter will trace how this seemingly simple technological affordance-instant, tailored answers-ripples through the fabric of society, impacting everything from the dynamics of a second-grade classroom to the innovation cycles in high-tech industries, and even the evolving nature of intellectual humility in an age of artificial omniscience. The journey into this new era of questioning begins by understanding the historical landscape: how humans have traditionally learned through asking, a process often traced back to the Socratic method, the social and psychological costs frequently associated with this vital act, and how AI is poised to dismantle this old paradigm. To fully appreciate the liberation AI offers, we must first consider the often-unseen "cost" of a question in the world that existed before its arrival.

THE COST OF A QUESTION

Before the advent of universally accessible AI, the act of posing a question, particularly one that might betray a lack of knowledge, carried a significant, often unspoken, cost. This cost was not monetary but psychological, social, and professional, creating an invisible barrier to learning and open communication. In myriad settings, from classrooms to corporate boardrooms, the fear of appearing ignorant, the pressures of organizational culture, and the implicit social contract of "already knowing" collectively discouraged individuals from voicing the very queries that could lead to deeper understanding or innovation.

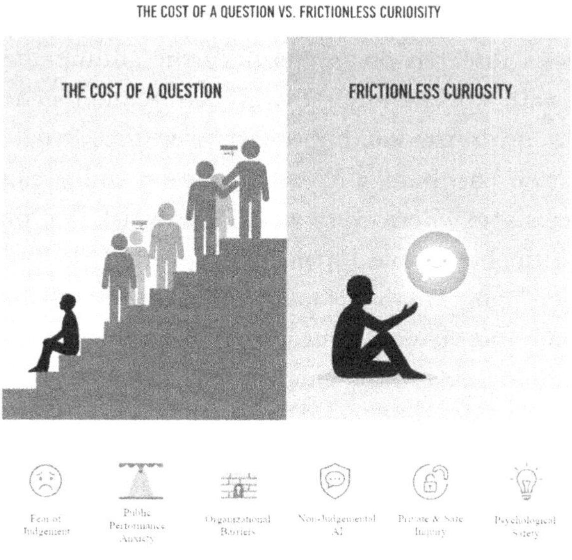

The most pervasive of these costs was the fear of looking unintelligent. This anxiety is a common human experience, deeply rooted in our social nature and the desire for esteem. In educational contexts, this fear could be particularly acute. Many individuals can recall moments of hesitation in a classroom, a question unasked for fear of ridicule or of simply "looking stupid". This phenomenon is not limited to those struggling with material; even high-achieving individuals, such as PhD students, report dwelling on past conversational missteps before venturing a new question in a public academic setting. A significant factor contributing to this silence is "pluralistic ignorance," a psychological state where individuals privately reject a norm (or in this case, privately experience confusion) but go along with it because they incorrectly assume that most others accept it (or understand the material). Seeing no one else raise a hand, a student might conclude they are uniquely lost, reinforcing their silence and, ironically, contributing to the very collective silence that misleads others. Studies have shown that teaching students about pluralistic ignorance can actually make them more comfortable speaking up, suggesting the power of understanding this dynamic.

Carol Dweck's research on mindsets further illuminates this fear. Individuals with a "fixed mindset"-who believe intelligence is a static trait-tend to prioritize looking smart over the actual process of getting smarter. For them, a classroom or a meeting can feel like a performance stage, where every question is a risk that might expose a perceived, unchangeable intellectual deficit. This "public performance anxiety" can lead to misinterpreting neutral feedback as negative judgment and viewing educators or superiors not as resources for learning but as critics. Quantitative research supports these observations, with studies indicating that a majority of college students, and notably more women than men, report discomfort in participating in class, citing fear of peer judgment and a lack of confidence with the course material as primary reasons. This "evaluation apprehension," an active, anxiety-toned concern about being negatively assessed by others, can directly inhibit knowledge sharing and questioning. The fear is not just about being wrong, but about the social consequences of that perceived failing.

Beyond individual psychology, organizational and institutional cultures frequently compound the cost of asking questions. Many workplaces, despite avowed commitments to learning and innovation, foster environments where curiosity is subtly or overtly discouraged. Leaders may overestimate the extent to which their organizations genuinely support curiosity. Middle managers, in particular, can sometimes be less encouraging of questioning, and a climate of psychological safety-essential for employees to feel secure in expressing ideas or admitting ignorance-is often lacking, especially under more autocratic leadership styles. If leadership does not visibly dedicate time and resources to exploratory activities, it sends a powerful message that curiosity is not truly valued. Indeed, surveys reveal that a large percentage of workers, often around two-thirds, report facing significant barriers to asking more questions at work, with employees consistently rating their own curiosity higher than that of their organizations. Rigorous command-and-control structures, an overemphasis on process adherence, and a lack of direct engagement from leaders with frontline realities can effectively "snuff out" curiosity. When such organizations face underperformance, the typical response is often to increase pressure and control, further stifling inquiry and leading to dysfunctional cultures and higher employee turnover. Thus, the cost of a question can extend beyond momentary embarrassment to tangible professional repercussions, impacting one's standing within a team or even career progression.

Compounding these individual and organizational factors is an unspoken societal mandate, particularly prevalent in professional or expert domains, that individuals should "already know" certain information. This implicit expectation discourages the asking of foundational or clarifying questions, as doing so might signal a deviation from this norm of presumed competence. This pressure connects back to evaluation apprehension, where the fear of having one's knowledge or ideas critiqued serves as a powerful inhibitor. Some research even suggests that individuals in middle-status positions, who may feel less secure than those at the top or less unburdened by expectations than those at the bottom, can be particularly wary of appearing uninformed, fearing a potential loss of status if they ask questions perceived as too basic. This societal script of feigned omniscience creates a formidable barrier to authentic learning and collaborative problem-solving, as it makes the admission of a knowledge gap a high-stakes gamble.

The interplay of these psychological, organizational, and social factors creates a landscape where the perceived cost of a question often looms larger than the potential benefit of the answer. This "cost of questioning" is not a minor issue; it represents a significant, often unacknowledged, inefficiency in the development of human capital and the pursuit of innovation across society. When individuals fear looking incompetent, when organizations prioritize conformity over curiosity, and when societal norms demand a performance of constant knowing, questions go unasked. This means that misunderstandings are not clarified, valuable knowledge remains unshared, crucial feedback is suppressed, and the seeds of innovative ideas-often sown by "naive" or fundamental questions-are never allowed to germinate. The cumulative effect is a systemic drag on collective intelligence and progress; a kind of hidden "tax" levied on our ability to learn and adapt.

One of the more insidious mechanisms reinforcing this costly silence is the phenomenon of pluralistic ignorance, extending far beyond the classroom into broader organizational and social contexts. An individual, experiencing confusion or doubt, observes the silence of others and incorrectly infers universal understanding or agreement. To avoid the social penalty of being the lone voice of uncertainty, they suppress their own questions or concerns. This individual act of self-censorship, multiplied across a group, creates a collective silence that is then interpreted by others as consensus or comprehension, further discouraging anyone from breaking ranks. In an organizational setting, this dynamic can lead to flawed strategies being pursued without critical challenge, as team members hesitate to voice doubts, fearing they are the only ones harboring them. This is a fertile ground for group think, where the pressure to conform overrides rational analysis.

Furthermore, the pressure to "already know," intertwined with evaluation apprehension, does not fall evenly on all individuals. It can disproportionately affect those from underrepresented or marginalized groups, who often face heightened scrutiny and carry a greater burden of proving their competence. For these individuals, asking a question that might reveal a knowledge gap can feel particularly risky, as it could inadvertently play into negative stereotypes or call their expertise into question more readily than it might for members of a dominant group. The "cost of a question," therefore, is not a fixed currency: it is inflated by those already navigating environments where their contributions and capabilities are more likely to be undervalued or challenged. This dynamic further entrenches existing information asymmetries, aligning them with social power structures.

In this pre-AI landscape, a question was too often a liability, a potential exposure of ignorance in a world that prized the appearance of certainty. The act of inquiry was burdened by personal anxieties and professional risks. But as we stand on the threshold of an era where intelligent systems can provide immediate, private, and personalized answers, the very nature of a question begins to transform. It shifts from being primarily a marker of what an individual does not know to becoming a clearer indicator of what they intend to discover, setting the stage for a new dynamic of frictionless curiosity.

FRICTIONLESS CURIOSITY

The advent of AI, particularly in the form of sophisticated conversational agents and personalized information retrieval systems, is fundamentally altering the risk-reward calculation associated with asking questions. By providing a private, non-judgmental, and instantly responsive environment for inquiry, AI makes curiosity a low-cost, high-reward activity. This "frictionless curiosity" promises to unlock human potential for learning and exploration in ways previously constrained by social and psychological barriers.

A cornerstone of this transformation is AI's capacity to function as a non-judgmental listener. For many, the primary deterrent to asking questions is not the lack of a desire to know, but the fear of how that desire, or the underlying ignorance it implies, will be perceived by others. AI interfaces, especially chat-based ones, largely eliminate this social friction. Users report feeling safer sharing personal issues or asking basic questions with AI than with humans, whom they fear might criticize, gossip, or judge. AI companions are often praised for their unfailingly attentive and non-critical nature, providing what is described as a "consistent and judgment-free zone for expression and interaction". This is particularly significant for individuals hesitant to seek human help due to social stigma or past negative experiences. The AI, devoid of ego or social agenda, simply processes the query and provides a response. This shift from potentially public performance to private exploration is crucial for fostering authentic inquiry. The constant availability of these AI tools further reduces friction; there is no need to wait for an opportune moment or for an expert to become available. This non-judgmental, always-on characteristic is pivotal in making curiosity truly frictionless.

Complementing the non-judgmental aspect is the instantaneous nature of AI's response cycle. In traditional learning or problem-solving scenarios, a question might be met with a delay; waiting for a teacher's attention, an expert's reply, or the outcome of a lengthy search. AI, however, can provide relevant answers and explanations in real time. This rapid feedback loop mimics some of the most effective aspects of ideal mentorship, where a knowledgeable guide is always present to clarify doubts and build upon understanding incrementally. The immediacy of AI responses allows learners to stay in a state of intellectual "flow," pursuing a line of inquiry without the interruptions or cooling-off periods that can dampen enthusiasm or break concentration. This capacity for rapid iteration-ask, receive, understand, ask again-is a powerful accelerator for learning, enabling users to quickly build conceptual frameworks and explore complexities at their own pace.

Furthermore, AI acts as a significant equalizer in terms of language and conceptual understanding. Sophisticated information is often couched in specialized jargon or complex linguistic structures that can be opaque to non-experts or those from different cultural or linguistic backgrounds. AI tools possess the ability to translate, simplify dense concepts, and adapt their communication style to match the user's level of understanding. An AI can be prompted to explain a difficult scientific theory in terms a layperson can grasp, or to break down a complex argument into its constituent parts. This function is vital for the democratization of knowledge, moving beyond exclusionary forms of explanation and making sophisticated information accessible to a much broader audience. It ensures that the barrier to understanding is the inherent complexity of the idea itself, not the language used to convey it.

The privacy inherent in AI-powered inquiry also contributes significantly to frictionless curiosity. The ability to explore topics without social surveillance or judgment encourages a more uninhibited and authentic form of learning. Private AI systems, which can employ techniques like federated learning, aim to safeguard sensitive user data while still enabling the extraction of valuable insights. This assurance of privacy allows individuals to ask "basic" questions they might feel embarrassed to pose publicly, to explore unconventional or controversial ideas without fear of censure, or to delve deeply into niche interests that might not align with their social or professional personas. This freedom from external scrutiny fosters a more self-directed and potentially more creative learning process, as the learner is guided by intrinsic interest rather than extrinsic pressures.

These characteristics-non-judgmental interaction, instant feedback, language equalization, and privacy-collectively cultivate an environment of psychological safety. Psychological safety is a crucial precondition for learning, experimentation, and intellectual risk-taking. When individuals feel safe, they are more willing to admit ignorance, ask probing questions, and explore areas of uncertainty. AI, by removing the human element of potential judgment or social evaluation, can uniquely create this safe harbor for intellectual exploration. This is particularly beneficial for junior individuals or those from underrepresented groups who might feel additional pressure to appear knowledgeable in traditional settings. AI interactions can transform uncertainty into confidence by providing a space for judgment-free experimentation and learning at one's own pace. While it's important to recognize that AI empathy is simulated and that over-reliance can have downsides, especially for younger users, the feeling of safety and non-judgment provided by AI is a powerful enabler of curiosity.

The emergence of frictionless curiosity, facilitated by AI, carries profound implications. One significant consequence is the potential for an "expectation of instant understanding". As users become accustomed to the immediate gratification and low-friction learning provided by AI, their patience for more complex, nuanced learning processes that require sustained human interaction, slower reflective thought, or grappling with ambiguity might diminish. This could inadvertently lead to a devaluing of deep, effortful learning in favor of rapid, perhaps more superficial, comprehension, unless a conscious effort is made to cultivate an appreciation for both modes of inquiry. The ease of getting answers might encourage a broad but shallow exploration of many topics, rather than a deep dive into a few.

Moreover, while the non-judgmental nature of AI is a boon for asking initial questions and fostering psychological safety, it might also inadvertently limit the development of intellectual resilience. Growth in many academic and professional fields depends on the ability to receive and process constructive criticism, engage in robust debate, and defend one's ideas against counterarguments. If learners primarily interact with AI systems programmed for agreeableness or neutral feedback, they may have fewer opportunities to develop the "intellectual thick skin" necessary for these challenging human interactions. This suggests a need for educational ecosystems that thoughtfully balance the benefits of safe AI exploration with structured opportunities for rigorous, critical discourse with human peers and mentors.

Finally, the "equalizing" effect of AI in simplifying language and concepts, while immensely beneficial for accessibility, also requires careful navigation. If not managed, it could lead to an oversimplification of inherently complex issues, potentially masking important nuances or discouraging learners from engaging with original, more challenging source materials. The very act of wrestling with difficult texts or complex arguments is itself a valuable cognitive exercise. If AI always provides a pre-digested, simplified version, learners might miss out on developing the critical analytical skills that come from that struggle. Therefore, the availability of simplified explanations must be paired with encouragement to engage with complexity directly, using AI as a tool for understanding, not a replacement for effort.

The newfound freedom to ask without fear or friction does more than just change the act of questioning itself; it fundamentally reshapes our individual relationship with the process of learning and knowledge acquisition. This personalization of inquiry, where the path to understanding is smoothed and tailored, naturally leads to a deeper personalization of the answers we receive, calibrating knowledge not just to the question, but to the questioner.

PERSONALIZED EPISTEMOLOGY

The capacity of AI to provide not just generic answers but responses highly calibrated to an individual's unique context, existing knowledge base, and specific learning objectives is giving rise to what can be termed a "personalized epistemology". This means that each person's journey to understanding is increasingly shaped by AI-driven interactions that tailor the delivery and framing of information to their personal cognitive landscape. This goes beyond simply answering a question; it involves constructing a pathway to knowledge that is optimized for the individual learner.

Central to this personalized way of knowing are adaptive explanations based on prior knowledge. Modern AI learning systems are designed to assess a user's current level of understanding, either through direct questioning, analysis of previous interactions, or performance on integrated tasks. Based on this assessment, the AI can adjust the complexity, depth, and style of its explanations. For instance, platforms like DreamBox Learning and Smart Sparrow in mathematics and course design, respectively, collect data on student performance to predict areas of difficulty and modify the material accordingly, much like a dedicated personal tutor would. This means if a student is struggling with a foundational concept, the AI can provide more basic explanations and examples, whereas if a student demonstrates mastery, it can introduce more advanced topics or nuanced details. This real-time adaptation ensures that the learner is always operating within their "zone of proximal development"-challenged but not overwhelmed. This is a cornerstone of personalized epistemology: the AI doesn't just answer the question asked but answers it for that specific user, building upon their existing mental model and filling precise knowledge gaps.

Further enriching this personalized approach is AI's capability for multimodal learning. Information is not monolithic, and neither are learning preferences. AI can present knowledge in a variety of formats-text, interactive diagrams, instructional videos, dynamic simulations, or audio explanations-based on the nature of the content and, increasingly, the user's preferred or most effective learning style. For example, Google's Gemini model can process a wide array of inputs, including images and video, and generate diverse outputs like text descriptions or even code. Platforms like Duolingo in language learning combine voice recognition, text, and gamified elements, tailoring lessons to individual learners by adjusting difficulty and providing instant feedback through multiple modalities. This means a visual learner might receive a concept explained through an animated diagram, while an auditory learner might benefit more from a spoken explanation with real-world analogies. This capacity to translate concepts into the format most likely to resonate with an individual significantly enhances comprehension, engagement, and retention, making the learning process more effective and enjoyable.

Dynamic scaffolding of complexity provides another layer to this personalized epistemology, offering tailored intellectual onramps to challenging topics. AI systems can break down complex subjects into more manageable components, providing hints, step-by-step guidance, and supportive resources as the learner navigates new material. Crucially, this scaffolding is not static; it adapts as the user demonstrates understanding, gradually fading support to encourage independent problem-solving and critical thinking. This interactive process, sometimes termed AI-based Interactive Scaffolding (AIIS), can provide conceptual, metacognitive, procedural, or strategic assistance, always contingent on the learner's current abilities and the specific task. This is akin to an infinitely patient tutor who knows precisely how much support to offer and when to withdraw it, ensuring the learner builds robust understanding and confidence without becoming overly reliant on assistance. While some current AI tools may still face limitations in achieving perfectly calibrated fading, the trajectory is towards more sophisticated and responsive scaffolding.

Practical examples of AI fostering personalized learning pathways are increasingly common. Intelligent Tutoring Systems (ITS) and AI-powered chatbots are being deployed to help students enhance problem-solving skills and build self-efficacy by offering instant, tailored support. These systems can identify knowledge gaps, provide targeted explanations, and engage learners in meaningful dialogues that promote critical thinking. For example, an AI tutor might guide a student through a difficult math problem, not by giving the answer, but by asking leading questions that help the student discover the solution method themselves, thereby reinforcing both the concept and the problem-solving process.

The development of such personalized epistemologies, while immensely empowering for individual learners, also suggests some complex second and third-order consequences. While AI tailors content to individual needs and learning styles for optimal comprehension, this very optimization could inadvertently create "filter bubbles" of knowledge. If individuals are primarily exposed to information that aligns with their existing understanding or is presented only in their preferred modalities, they might experience limited exposure to radically different perspectives, challenging viewpoints, or alternative ways of conceptualizing problems.

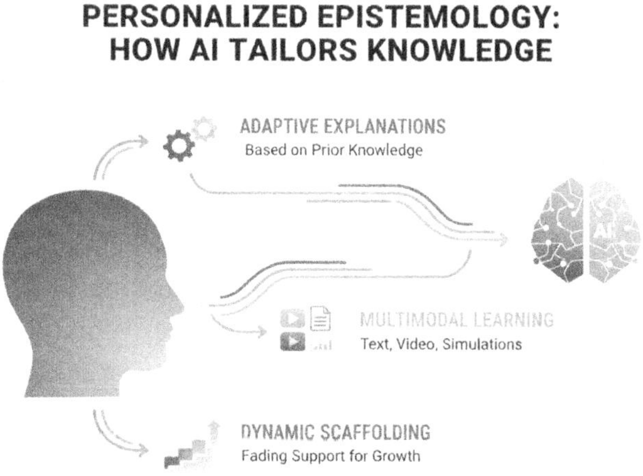

Over time, this could lead to an epistemological echo chamber, reinforcing existing biases or diminishing the intellectual discomfort that is often necessary for significant conceptual breakthroughs or paradigm shifts in thinking. This potential parallels the concerns raised about filter bubbles in social media, suggesting a need for AI learning systems to be designed with features that intentionally introduce "cognitive diversity" or "intellectual serendipity" into personalized learning paths, ensuring learners encounter a broad and sometimes challenging spectrum of ideas.

Furthermore, the shift towards personalized epistemology fundamentally redefines what "understanding" means. If knowledge is calibrated to each individual's unique context, their internal representation and comprehension of that knowledge will also be unique. Traditional educational models often rely on standardized assessments to measure a supposedly common understanding across a cohort of learners. However, if understanding becomes deeply personalized, the validity and utility of such uniform assessment methods come into question. How can learning be assessed consistently if each learner's path and resulting knowledge structure is distinct? This could necessitate a significant move towards competency-based assessments that focus on the application of knowledge in diverse contexts rather than the recall of a standardized body of information, aligning with the broader trend away from mere credentialing towards genuine capacity building, a theme that will be explored later in this chapter. More profoundly, if everyone's way of knowing becomes highly individualized, the very notion of a shared cultural or scientific literacy might erode, posing challenges for societal cohesion, communication, and collective problem-solving.

Finally, the dynamic scaffolding provided by AI, while dramatically accelerating skill acquisition by providing optimal support and reducing cognitive load, might also foster a dependency on such support. If the "fading" of this scaffolding is not perfectly managed by the AI, or if learners become too accustomed to its constant availability, they may struggle when confronted with complex, ambiguous problems in real-world situations where AI support is unavailable, inappropriate, or insufficient. This suggests that alongside AI-scaffolded learning, educational strategies must also explicitly cultivate "intellectual resilience"-the ability to learn, adapt, and problem-solve in un-scaffolded, information-poor, or ambiguous environments.

This profound personalization of knowledge delivery, where everyone receives information tailored not just to their question but to their entire cognitive landscape, naturally fosters greater confidence in their ability to learn. As this confidence grows and access to what was once considered expert-level information becomes a universal, on-demand utility, the traditional model of one-to-many knowledge transmission-epitomized by the expert lecture-begins to lose its long-held dominance.

THE END OF THE EXPERT LECTURE

Traditional lectures, a cornerstone of educational practice for centuries, have primarily functioned as a vehicle for one-to-many knowledge transmission. An expert, possessing specialized information, disseminates it to a largely passive audience. However, as AI empowers individuals to actively pull information, engage in personalized dialogues, and construct understanding at their own pace, this model of knowledge delivery is rapidly eroding. The "sage on the stage" is finding their role fundamentally challenged when the core of their knowledge is instantly accessible and explainable through an AI interface in everyone's pocket.

The evolution of online learning itself charts this trajectory away from passive reception. Massive Open Online Courses (MOOCs) initially promised to scale the lecture model globally, offering access to expert instruction to millions. However, MOOCs often replicated the passive nature of traditional lectures and struggled with high dropout rates and a lack of personalized interaction. While some MOOCs incorporated interactive forums, they often could not provide the immediate, tailored feedback crucial for deep learning. The next step involved more interactive online courses, but it is the rise of conversational AI tutors that marks a truly significant departure. These AI systems, exemplified by tools like ChatGPT or specialized educational chatbots like IBM's Watson Tutor, can offer personalized, scalable support, engaging students in Socratic dialogues that promote critical thinking rather than just delivering answers. These AI tutors can address the very limitations of MOOCs-impersonality and lack of individualized attention-by providing 24/7 assistance and adapting to each learner's specific needs and pace.

The lecture, as a one-way dissemination of information, is thus being superseded by AI-facilitated, two-way dialogues where learning is co-constructed.

This shift inherently flattens knowledge hierarchies. In traditional educational settings, the lecturer or teacher acts as a gatekeeper of knowledge, controlling its flow and interpretation. However, when AI provides immediate access to expert-level answers, explanations, and even synthesis of information, this gatekeeping function is significantly diminished. Students and novices are empowered to explore complex topics independently, potentially acquiring knowledge at a pace and depth that can surpass what is possible within the rigid structures of a traditional curriculum. AI can alleviate knowledge bottlenecks within organizations and, by extension, in learning environments. If AI can automate the delivery of foundational knowledge or even assist in complex problem-solving, the traditional hierarchy where knowledge flows strictly from expert to novice is disrupted. This doesn't necessarily make the human expert obsolete, but it profoundly changes their role from a primary source of information to a facilitator of understanding and application, a theme that will be explored more deeply in the subsequent section on the future of formal education.

Consequently, the paradigm of "learning by attendance", where physical presence in a lecture hall was equated with knowledge acquisition is giving way to "learning by dialogue". Passive consumption of information is being replaced by active, AI-mediated engagement where understanding is not merely received but actively constructed. This is where the Socratic method, a pedagogical approach with roots stretching back to ancient Greece, finds a new and powerful instantiation. Socrates famously used questioning to challenge assumptions, expose contradictions, and lead his interlocutors to new insights. Modern AI tutors are increasingly designed to emulate this approach. Instead of simply providing answers, these "Socratic chatbots" ask targeted questions, prompting users to reflect, justify their reasoning, and discover solutions themselves. This dialogue-based learning, facilitated by AI, transforms the learner from a passive recipient into an active participant in their own education. Research and development in this area focus on how AI can manage these dialogues effectively, offering hints, breaking down complex ideas, and guiding learners through iterative cycles of inquiry and reflection. The emphasis shifts from the expert's monologue to a collaborative exploration of ideas, where the AI serves as a tireless, knowledgeable, and often provocative conversational partner. This approach not only enhances comprehension but also cultivates critical thinking and metacognitive skills, as learners are constantly challenged to articulate their understanding and evaluate their own thought processes.

The decline of the expert lecture as the primary mode of knowledge transmission signals a broader transformation in how educational value is perceived and delivered. If AI can effectively handle the dissemination of factual content and provide personalized explanations on a massive scale, then the core function of traditional lectures becomes largely redundant. This does not mean the end of expertise or the role of human educators, but rather a fundamental shift in their responsibilities. The value proposition of educational institutions and their faculty must evolve from being repositories and transmitters of existing knowledge to becoming centers for knowledge navigation, critical synthesis, and the cultivation of uniquely human cognitive skills. Human educators become essential for guiding learners through the vast sea of AI-accessible information, helping them to critically evaluate diverse sources, synthesize complex ideas, and apply knowledge in novel and ethical ways. These are higher-order cognitive skills that AI, in its current form, struggles to impart independently. Thus, the lecture hall, once a symbol of knowledge authority, may transform into a workshop for collaborative sense-making and complex problem-solving, facilitated by human experts who leverage AI as a powerful support tool.

This flattening of knowledge hierarchies, where AI makes expert-level information universally accessible, could also precipitate a "crisis of authority" for traditional experts if their value continues to be predicated solely on the possession of exclusive knowledge. If an expert's primary claim to authority is "I know things you don't," that claim is significantly weakened when an AI can provide that same knowledge, often more quickly and in a more personalized manner. This doesn't render expertise worthless, but it fundamentally alters its nature and how it is demonstrated. The value of human experts will increasingly lie not in what they know in terms of recallable facts, but in their experience, their wisdom in applying knowledge, their ability to generate new insights from existing information, their ethical judgment, and their capacity to navigate complex, ambiguous real-world problems-all aspects that AI cannot fully replicate, as noted in the broader themes of this book concerning what remains when AI knows all. Experts who fail to adapt to this new reality, by not shifting their focus to these more nuanced and experiential aspects of their field, may find their traditional authority challenged by AI-empowered novices who can rapidly acquire foundational knowledge.

Finally, the transition to "learning by dialogue," especially when mediated by AI, necessitates the development of sophisticated "dialogue management" skills in learners themselves. Effective dialogue, whether with a human or an AI, is an iterative process of probing, clarifying, challenging, and synthesizing. When interacting with AI, the quality and depth of the learning experience depend heavily on the user's ability to formulate effective prompts, interpret AI responses critically, and guide the AI in a productive direction. Learners will need to cultivate skills in asking sequences of questions, refining their inquiries based on the AI's output, and discerning when an AI's "dialogue" might be superficial, biased, or even incorrect. This implies that "learning by dialogue" is not a passive state but an active skill set that itself needs to be taught and nurtured. This directly connects to the emerging importance of prompt engineering as a new form of literacy, a concept that will be explored in detail later in this chapter. The ability to engage in a productive dialogue with an intelligent system becomes as crucial as traditional literacy skills were in an age of printed text.

With the traditional lecture model receding and knowledge acquisition becoming an on-demand, personalized dialogue driven by AI, the very structure and purpose of formal education systems-from K-12 schools to universities and professional training programs-face an unavoidable and profound redefinition. These institutions must now grapple with their evolving role in a world where information is no longer their primary currency.

WHAT HAPPENS TO FORMAL EDUCATION?

The paradigm shift initiated by AI's ability to deliver instant, tailored answers forces formal education systems to confront existential questions about their purpose, structure, and value proposition. If knowledge is readily accessible and learning can be deeply personalized by AI for anyone, anywhere, then what becomes the core mission of schools, universities, and training programs? The answer lies in a fundamental reorientation away from information dissemination towards the cultivation of skills and capacities that AI cannot easily replicate or replace.

A primary transformation occurs in the role of teachers and educators. No longer the principal conduits of factual information, they are evolving into coaches, mentors, curators of learning experiences, and facilitators of critical thinking, ethical reasoning, and socio-emotional development. As AI systems take over administrative burdens like grading routine assignments and can even assist in lesson planning, educators are freed to focus on the more human-centric aspects of teaching. Their expertise shifts towards designing engaging learning environments, guiding students through complex problem solving, fostering creativity, nurturing emotional intelligence, and instilling a capacity for ethical judgment; all areas where human nuance, empathy, and experience remain paramount. The educator becomes less of a "sage on the stage" and more of a "guide on the side," or even a "learning experience designer," helping students navigate the vast ocean of AI-accessible knowledge and make meaningful connections. Their role will increasingly involve helping students to learn how to learn in an AI-augmented world, sustaining their interest and motivation, and supporting their holistic development.

This redefinition of roles permeates all levels of education, starting with K-12 systems. Here, the challenge is twofold: to teach students about AI (fostering AI literacy and ethical awareness) and to teach with AI (leveraging it as a powerful pedagogical tool). Many current K-12 systems are still grappling with this, sometimes merely bolting AI onto outdated educational models rather than fundamentally rethinking curriculum and pedagogy. A crucial shift involves moving the focus from content delivery and memorization to the development of future-ready skills such as adaptability, critical thinking, creativity, collaboration, communication, and ethical reasoning. National policies, such as those proposed in the US, are beginning to emphasize the importance of integrating AI literacy into K-12 education, providing comprehensive AI training for educators, and fostering early exposure to AI concepts to prepare an AI-ready workforce. However, ensuring equitable access to these AI tools and addressing ethical considerations like data privacy and algorithmic bias are critical challenges that must be proactively managed in the K-12 space. High-dosage tutoring, potentially augmented by AI, is also being explored as a means to address learning loss and provide personalized support at scale.

Universities, traditionally bastions of knowledge creation and credentialing, face a particularly profound identity crisis. Their mission must expand beyond transmitting established knowledge to becoming hubs for advanced AI research, fostering interdisciplinary collaboration to solve complex global challenges, and equipping students with sophisticated human skills that complement AI's capabilities. Many universities are already exploring AI applications in streamlining administration, enhancing research through data analytics, and personalizing student support services. Leading institutions are appointing Chief AI Officers, developing responsible AI frameworks, and revising curricula to embed AI literacy and ethical considerations across disciplines. However, significant challenges remain, including ensuring data access for AI development while protecting privacy, overcoming resistance to change, particularly among established faculty, and providing the necessary training for staff and students to effectively and ethically engage with AI. The very nature of academic research and scholarly communication will also be transformed as AI tools accelerate discovery and automate aspects of knowledge synthesis.

This leads directly to a re-evaluation of credentialing. Historically, degrees and diplomas have served as primary signals of knowledge and competence. However, if AI enables individuals to acquire specific skills and knowledge rapidly, often outside traditional institutional structures, the value of these formal credentials as standalone indicators diminishes. Employers are increasingly shifting towards skills-based hiring, prioritizing demonstrable abilities and competencies over formal degrees. Studies show that a vast majority of companies report making better hires when focusing on skills, and many are willing to offer higher starting salaries to candidates with relevant micro-credentials, often saving on subsequent training costs.

Students, too, are expressing a preference for educational pathways that emphasize practical skill development, particularly in emerging areas like generative AI. While removing degree requirements is an initial step, building a robust ecosystem for skills-first hiring requires more sophisticated ways to assess and validate skills acquired through diverse pathways, including AI-driven learning platforms. Formal educational institutions must therefore prove they offer more than just information transfer and a diploma; their value will increasingly lie in providing validated skill development, fostering critical and ethical thinking, facilitating collaborative project-based learning, and building human networks; capacities that are harder to commodify.

This brings us to the fundamental question: what is the "point" of education if every individual possesses a personal AI capable of delivering expert knowledge on demand? The answer lies in shifting the focus from knowledge acquisition to wisdom cultivation. If AI can provide the "what," education must focus on the "how" and the "why": how to learn effectively, how to ask critical questions, how to ethically apply knowledge, how to collaborate with both humans and AI to solve complex problems, and how to develop metacognitive awareness-the ability to understand and regulate one's own learning processes. AI itself can be a tool in developing these metacognitive skills, for instance, by prompting reflection or providing feedback on learning strategies, but the impetus and application remain deeply human endeavors.

The technological backbone for this transformation-personalized learning at scale-is now increasingly viable thanks to AI. Pedagogical frameworks are evolving to leverage AI for data-driven insights that meet specific learner needs, moving beyond theory to practical implementation in various educational settings. Case studies from K-12 districts and higher education initiatives demonstrate the potential of competency-based personalized learning, where students progress based on mastery, engage with data about their own learning, and develop agency over their educational journeys. These models emphasize real-time data use for formative assessment and student regrouping, adapting instruction to individual requirements. However, successful scaled implementation requires careful management of data, robust technological infrastructure, ongoing professional development for educators, and a cultural shift towards valuing student agency and personalized pathways.

The deconstruction of traditional educational roles and structures, driven by AI's capabilities, points towards a future learning ecosystem that is more fragmented yet potentially more flexible and genuinely lifelong. Individuals may increasingly curate their educational experiences from a diverse array of sources-AI tutors for foundational knowledge, specialized micro-credentials for vocational skills, human-led workshops for collaborative projects, and learning communities for the development of higher-order thinking and ethical reasoning. This "unbundling" of education means that the traditional package offered by institutions like universities; encompassing content delivery, credentialing, community building, and research opportunities is being broken down into its constituent components.

Learners may pick and choose these components from various providers, assembling a personalized educational portfolio that evolves with their needs and interests. In such a landscape, formal institutions might transform into crucial nodes within a broader, lifelong learning network, specializing in areas where human interaction and complex synthesis are irreplaceable, rather than attempting to be the sole providers of all educational services.

This shift towards cultivating "uniquely human skills"; such as critical thinking, creativity, emotional intelligence, and ethical judgment necessitates a pedagogical revolution within these evolving institutions. If AI excels at information recall and pattern recognition, human value naturally gravitates towards areas where AI currently falls short. Traditional educational methods, often centered on content mastery demonstrated through lectures and standardized tests, are ill-suited to fostering these durable skills. To cultivate capacities like ethical reasoning or innovative problem-solving, teaching methodologies must pivot towards active, experiential learning. This includes an emphasis on project-based work, Socratic dialogues (whether facilitated by humans or sophisticated AI), immersive simulations, ethical dilemma discussions, and collaborative creative endeavors. Such a transformation requires a fundamental redesign of curricula, assessment methods (moving beyond rote memorization to evaluating applied skills), and, critically, the training and ongoing professional development of educators to equip them to facilitate these new modes of learning.

Finally, the democratization of skill acquisition via AI could profoundly challenge the socio-economic sorting function that formal education, particularly higher education, has historically performed. For generations, access to elite educational institutions and the credentials they confer has been a primary mechanism for social mobility and status attainment. If AI makes high-quality learning resources and advanced skill development widely and affordably accessible, the exclusivity that underpins the value of many traditional educational pathways is diminished. However, this does not automatically lead to a more equitable society. New forms of social stratification could emerge, based not on access to traditional institutions, but on access to high-quality AI tools, the "AI literacy" required to use them effectively, and the individual motivation, discipline, and supportive environment needed for successful self-directed learning in an AI-augmented landscape. While old barriers related to geographic location or the cost of elite institutions may fall, new ones related to digital equity, AI fluency, and the capacity for autonomous learning could arise, necessitating proactive policy interventions and societal efforts to ensure that the benefits of AI in education are distributed equitably.

As formal education redefines its value proposition beyond the mere delivery of information, the social scripts that govern our interactions around knowledge and ignorance must also adapt. If anyone can, in principle, access expert answers instantly and privately, the way we communicate about what we know-and what we do not know-is set for a significant and complex transformation.

NEW SOCIAL SCRIPTS FOR IGNORANCE

Universal and instantaneous access to AI-generated answers is poised to rewrite the unwritten rules that govern our social interactions around knowledge and ignorance. When high-quality information is merely a query away for everyone, the traditional dynamics of displaying knowledge, admitting uncertainty, and even condescending to others based on informational advantage are inevitably altered. This shift necessitates the development of new social scripts, a revised etiquette for how we acknowledge, share, and value knowledge in a world where the baseline of accessible information has been dramatically elevated.

One of the most immediate changes is the emergence of a new etiquette surrounding "real-time AI lookup". The ability to discreetly (or overtly) consult an AI during conversations, meetings, or debates introduces novel social complexities. Is it considered impolite to fact-check a colleague's statement mid-presentation using a personal AI? When is it necessary or courteous to disclose that the information being shared was sourced or refined by an AI? Current social norms are still coalescing around these questions. Ethical considerations, such as obtaining consent before using AI transcription or summarization bots in meetings, are becoming paramount, as these tools can be perceived as third-party attendees with implications for privacy and data security. Workplace guidelines suggest maintaining a balance: leveraging AI for efficiency while preserving essential human interaction, acknowledging AI's contribution where appropriate, and ensuring inclusivity for those less comfortable with the technology. The key challenge lies in integrating AI seamlessly into social and professional discourse without eroding trust or creating new forms of digital discourtesy.

This new landscape also forces a confrontation between genuine intellectual humility and the temptation of performative knowing. Instant access to answers could, ideally, foster greater intellectual humility-an honest acknowledgment of one's own cognitive limits and an openness to learning, even from a machine. It becomes easier to admit "I don't know, but I can find out with my AI". However, the same accessibility could encourage a new form of intellectual pretense, where individuals present AI-sourced knowledge as their own, without genuine understanding or proper attribution. The ease with which AI can generate fluent, authoritative-sounding text might lead to what some call "synthetic certainty," an illusion of knowing that masks a lack of deeper comprehension. Cultivating intellectual humility in this context means actively questioning AI outputs, verifying information, recognizing AI's inherent limitations and potential biases, and being willing to update one's beliefs when confronted with new evidence, whether from AI or human sources. Some researchers even propose designing AI systems with built-in intellectual humility, capable of communicating their own uncertainties and limitations, though there's also a concern that AI could be designed to "bully" or suppress dissenting views if not carefully managed.

This leads to a significant psychological and social shift from valuing "I know" (knowledge stored in one's memory and recalled) to valuing "I can verify" or "I can find out" (the ability to quickly access, evaluate, and synthesize information using AI). The cognitive act of recall, which demands mental effort and is linked to deeper learning and retention, may become less emphasized as individuals increasingly rely on external digital memory banks, a phenomenon already observed with search engines (the "Google effect") and likely to be amplified by AI. While this can free up cognitive resources, the crucial counterweight is the need for rigorous verification of AI-generated information, as these systems are not infallible and can produce biased or inaccurate outputs. The mantra shifts from "trust but verify" to, with AI, simply "verify". This places a higher premium on critical thinking skills and the ability to discern credible information from plausible-sounding falsehoods. Credibility in everyday interactions might therefore shift from individuals who can recite facts from memory to those who demonstrate proficiency in navigating the information landscape with AI tools and critically assessing the retrieved knowledge. This doesn't mean memory becomes useless, but its role in establishing expertise may be recontextualized.

A particularly intriguing area of transformation concerns social phenomena like "mansplaining" (a man condescendingly explaining something to a woman, often irrespective of her actual expertise) and "techsplaining" (similar condescension regarding technical topics).

The core of such behaviors often lies in an assumed information asymmetry and a desire to assert dominance or perform expertise. When the listener can instantly access and verify the information being explained via their personal AI, or even obtain a clearer, more tailored explanation, the "-splainer's" informational leverage is theoretically undermined. However, the impact of AI on these dynamics is likely to be complex and multifaceted.

On one hand, universal access to AI-generated knowledge could disarm traditional forms of -splaining. If a listener can quickly consult their AI, they might feel more empowered to challenge inaccuracies, request clarification on their own terms, or simply disengage from a condescending interaction. The -splainer's authority, if based merely on possessing information the listener lacks, is diminished.

On the other hand, AI could also enable new or mutated forms of such behaviors. Research into "Mansplainable AI" suggests that AI itself, if not designed with careful attention to communication style, can be perceived as dominant, patronizing, or condescending, particularly by women, potentially leading to feelings of inadequacy. This points to the emergence of "AI-splaining," where the AI becomes the source of condescension. Furthermore, a human-splainer might leverage AI to generate even more complex or voluminous explanations, not to clarify, but to overwhelm or perform an even greater show of expertise, relying on the listener's inability or unwillingness to deconstruct the AI-assisted barrage. Over-explaining is sometimes more about credentialing than truly providing information, and AI could become a powerful tool for such performative credentialing.

The issue is further complicated by the fact that AI systems can reflect and even amplify existing societal biases, including gender biases. If an AI's training data or algorithms are biased, it might inadvertently provide information that supports a -splainer's flawed or prejudiced assertions, especially if the listener's own AI tools share similar underlying biases. This could make it harder to challenge biased explanations if they appear to be "AI-verified". The development of "techno-patriarchy," where AI systems perpetuate misogynistic norms through biased recruitment tools, default female voices for subservient virtual assistants, or skewed image generation, is a significant concern.

Thus, while AI changes the information landscape, behaviors like mansplaining or techsplaining are unlikely to disappear entirely, as they are often rooted in deeper power dynamics, ego, and social conditioning rather than just information asymmetry. Their form and effectiveness, however, may evolve. The way humans position themselves as experts will also need to adapt. Instead of solely relying on the possession of factual knowledge, expertise might increasingly be demonstrated through the skillful use of AI to find, synthesize, and critically evaluate information, or by highlighting tacit knowledge, lived experience, and nuanced judgment-qualities that AI currently lacks. The social construction of expertise, which has always been a dynamic process involving validation within discipline and taste communities, will now incorporate AI as both a tool and a potential new authority. There's a risk of the "fluency fallacy," where the polished and grammatically correct output of AI is erroneously equated with accuracy and true expertise.

The role of AI in either mitigating or exacerbating social biases in knowledge communication is therefore a critical consideration. If AI systems are not developed and deployed with a strong commitment to fairness, transparency, and the active mitigation of bias, the "death of the dumb question" could ironically lead to the proliferation of "confidently wrong, AI-backed assertions". This could worsen misinformation, reinforce harmful stereotypes, and deepen social divisions, particularly if users do not maintain or develop strong critical evaluation skills when interacting with AI-generated content.

The shift from an internal locus of knowledge ("I know") to an external, verifiable one ("I can find out/AI says") carries significant cognitive and social weight. While it democratizes access to information, it paradoxically might increase the cognitive burden of discernment. Individuals are no longer just responsible for learning and recalling information; they become responsible for constantly evaluating the veracity, context, and potential biases of the AI-provided answers they receive. This is a demanding task, especially when AI can present information with an air of "synthetic certainty". The ease of access reduces the cognitive load of memory but increases the load of critical assessment.

Moreover, as basic factual information becomes a commodity, the nature of condescending explanations may evolve. We might see a shift from information asymmetry-based condescension ("I know this fact and you don't") to "process-splaining" or "interpretation-splaining". In this new dynamic, individuals might assert superiority based on their perceived advanced ability to prompt AI effectively, to interpret its nuanced outputs, or to navigate the complexities of different AI models. Condescension could manifest in statements like, "You're not asking the AI the right way," or "You're clearly misinterpreting what the AI is trying to tell you". This form of -splaining would hinge on a perceived mastery of the process of AI-mediated knowledge acquisition, rather than the direct possession of knowledge itself. AI itself, if poorly prompted or designed, could also fall into a pattern of "AI-splaining" by providing overly complex, unhelpful, or seemingly patronizing explanations.

In such an environment, the social and intellectual value of "lived experience" and tacit knowledge-the kind of understanding that is difficult to articulate, digitize, and therefore feed into AI training data, is likely to increase significantly. AI excels at processing and delivering explicit, codifiable knowledge. However, a vast amount of human expertise, particularly in complex social interactions, creative endeavors, or craft-based professions, is tacit. It is learned through years of practice, intuition, and a nuanced understanding of context that defies easy formalization. As AI systems handle the readily available explicit knowledge, the unique and defensible value of human experts will increasingly reside in this tacit domain, their ability to navigate ambiguity, their ethical compass, and the wisdom gleaned from their irreplaceable lived experiences. Social scripts may evolve to give greater weight and credibility to "I have experienced this" or "my judgment, based on similar situations, is..." over a simple "I know the facts" or "the AI says..." This is especially true in contexts requiring deep understanding of human motivations, emotions, and unpredictable real-world complexities.

In this transformed social landscape, where basic knowing is often augmented or even outsourced to AI, the emphasis inevitably shifts. It is no longer primarily about the knowledge you can hold in your memory, but about your ability to access, frame, critically evaluate, and ethically deploy information. This new reality naturally elevates the importance of the questions we choose to ask, making curiosity itself, and the skill of effective inquiry, a key differentiator in a world awash with answers.

CURIOSITY AS A COMPETITIVE EDGE

In a world where AI can instantly provide a high-quality answer to almost any factual question, the traditional competitive advantage derived from possessing specialized knowledge diminishes. When answers become a ubiquitous commodity, the locus of value shifts decisively towards the ability to formulate insightful, strategic, and creative questions. It is no longer primarily about what one knows, but about what one seeks to know, and how skillfully one can elicit and synthesize information to generate new understanding or drive innovation. Curiosity, far from being a passive trait, becomes an active, competitive skill.

A fundamental aspect of this new competitive landscape is the emergence of "prompt engineering" as a new form of literacy. If AI systems are the new gateways to vast reservoirs of information and generative capabilities, then knowing how to effectively communicate with these systems is paramount. Crafting effective prompts, the instructions given to AI models, is an art and a science, a blend of linguistics, an understanding of AI's capabilities and limitations, and user experience design. It involves more than just asking a simple question; it requires structuring the inquiry, providing relevant context, specifying the desired output format and tone, and iteratively refining the prompt based on the AI's responses. Frameworks like CLEAR (Concise, Logical, Explicit, Adaptive, Reflective) are being developed to guide users in this process. This "prompt literacy" is becoming as crucial for navigating the information age as traditional literacy was for the age of print. Those who can skillfully "interrogate" AI will unlock significantly more value, gaining an edge in learning, problem-solving, and creation.

This emphasis on skilled inquiry translates directly into strategic questioning becoming a key driver of success in diverse fields such as business, policy development, and scientific innovation. When basic answers are readily available, the ability to frame a problem correctly, to ask "what if" questions that explore novel scenarios, to pose "why" questions that dive into root causes, and to formulate "how" questions that drive towards actionable solutions becomes a critical differentiator. Strategic questioning involves a spectrum of inquiry types: investigative questions to clarify what is known and what needs to be learned; speculative questions to reframe issues and explore creative possibilities; productive questions to assess resources and plan execution; interpretive questions to synthesize information and derive meaning; and even subjective questions to uncover unstated assumptions or emotional undercurrents that might impact decisions. In a business context, for example, a team that can use AI to rapidly gather market data, but then excels at asking strategic questions about unmet customer needs or potential disruptive technologies, will be far more competitive than a team that simply accepts AI-generated summaries at face value.

The premium placed on skilled questioning and interpretation is likely to give rise to new roles and forms of expertise centered on "sensemaking" and "question design". As AI systems handle an increasing amount of routine information retrieval, data analysis, and content generation, human value will shift towards higher-order cognitive tasks. "Sensemakers" will be individuals who can navigate the deluge of AI-processed data, identify meaningful patterns, synthesize complex information into coherent insights, and communicate these insights effectively. "Question designers" or "prompt architects" will specialize in formulating the sophisticated sequences of prompts and inquiries needed to unlock the full potential of AI for specific tasks, from scientific research to creative content generation. The job market is already reflecting this trend, with a projected significant growth in demand for prompt engineers and related AI specialist roles, often commanding high salaries due to the current scarcity of these advanced skills. Beyond dedicated prompt engineers, a broader range of professionals will need to develop strong data literacy, familiarity with AI tools, critical thinking, and collaborative abilities to effectively partner with AI.

The following lists, adapted from research on strategic questioning, illustrate how different types of questions can be framed, potentially as AI prompts, to drive deeper inquiry and decision-making:

Investigative questions

- **Core purpose:** Clarify what is already known and identify critical information gaps.
- **Illustrative AI prompt shell:** "Act as a market analyst. Summarize the current known market share for Product X, and identify the top three unknown variables impacting future growth projections."

Speculative questions

- **Core purpose:** Explore possibilities, reframe problems, and foster creative or non-obvious solutions.
- **Illustrative AI prompt shell:** "Generate three speculative scenarios in which Technology Y disrupts Industry Z within

the next five years. For each scenario, outline potential opportunities and threats."

Productive questions

- **Core purpose:** Assess feasibility, plan execution, and identify required resources and success metrics.
- **Illustrative AI prompt shell:** "Assuming we pursue Strategy A, outline a project plan with key milestones, resource requirements (talent, budget, and time), and three measurable KPIs for success."

Interpretive questions

- **Core purpose:** Synthesize information, derive meaning, and surface deeper implications across multiple inputs.
- **Illustrative AI prompt shell:** "Given Data Set 1 (customer feedback) and Data Set 2 (sales trends), identify the underlying connections and the core strategic issue these data sets reveal."

Subjective questions

- **Core purpose:** Surface unstated assumptions, biases, emotions, or hidden agendas that influence decision-making.
- **Illustrative AI prompt shell:** "Analyze the sentiment in these team meeting transcripts regarding Project Omega. Identify recurring expressions of concern or unspoken reservations."

This framework demonstrates how structured questioning, enhanced by AI's information processing power, can become a systematic approach to leveraging curiosity for competitive advantage.

However, a critical paradox emerges: while AI makes inquiry itself frictionless, the sheer volume of instantly available information can lead to cognitive overload, potentially dampening curiosity or even causing "curiosity burnout". When the mind is constantly bombarded with data, emails, notifications, and AI-generated content, it can become difficult to sort relevant information from noise, leading to feelings of being overwhelmed, mental exhaustion, and a reduced capacity for decision-making. This state of "analysis paralysis" or passivity can occur when the effort of processing more information feels too great. If curiosity leads to an endless stream of answers that then require significant cognitive effort to process and synthesize, the initial spark of inquiry can be extinguished by the subsequent deluge. The very ease of getting answers might encourage a superficial "grazing" of information rather than deep, sustained engagement with a topic.

Therefore, maintaining curiosity and critical thinking in an AI-saturated world requires intentional strategies, both at the individual and organizational level. Individually, this involves cultivating an "AI mindset" that views AI as a collaborator rather than a competitor, embracing exploration, and committing to lifelong learning. It means developing "cognitive resilience" by, for example, attempting to solve problems independently before turning to AI for verification, thereby strengthening one's own thinking processes. Neuroscientific perspectives on curiosity highlight its role in activating the brain's reward system (dopamine release) and enhancing memory formation and neuroplasticity. Curiosity can be actively cultivated by asking "what if" questions, adopting an anthropological mindset to observe one's surroundings, running small experiments, embracing the state of "not knowing" as an opportunity, and reframing failures as data points for learning. Educational interventions must focus on teaching AI literacy-understanding how AI works, its limitations, and its ethical implications-and encouraging skepticism towards AI outputs, prompting students to cross-check information and question biases.

One powerful approach to fostering deep thinking alongside AI is the application of the Socratic method. Instead of AI merely delivering facts, it can be designed or prompted to engage users in a Socratic dialogue, asking probing questions that challenge assumptions, clarify reasoning, and guide the user towards their own understanding. For example, an AI tutor could respond to a student's statement not with a direct correction, but with a question like, "What leads you to that conclusion?" or "Can you think of a situation where that might not be true?". This transforms the AI from a simple answer engine into a facilitator of critical thought, helping users to develop the very questioning skills that are becoming so crucial. This represents a sophisticated use of AI, aligning with the goal of curiosity as a competitive edge by not just providing answers, but by helping users learn to ask better questions.

The ascendance of "prompt engineering as literacy" implies a significant shift in where intellectual and economic power resides. If the ability to extract high-quality, relevant, and nuanced information from AI hinges on sophisticated prompting skills, then those who master this new literacy will gain a distinct advantage in learning, innovation, and overall productivity. This could, however, create a new "literacy divide," where "prompt illiteracy" becomes a barrier to opportunity, much like traditional illiteracy in previous eras. Access to AI tools alone is insufficient; the skill to interact with them effectively becomes the gatekeeper to their full potential.

Furthermore, while AI can answer a vast array of questions, the most profound human contributions may increasingly lie in identifying and formulating "un-askable" questions-those that AI cannot yet comprehend or address because they require deep human context, ethical intuition, creative synthesis across unlinked domains, or truly novel conceptual leaps. AI operates on patterns within its training data and the explicit structure of prompts. Questions rooted in complex, emergent human experiences, nuanced ethical dilemmas without clear precedent, or those demanding a synthesis of knowledge from entirely disparate fields may fall outside AI's current capabilities. The competitive edge will thus belong to individuals who can perceive these intellectual frontiers and formulate questions that challenge existing paradigms, often by interrogating the underlying assumptions that AI itself might take for granted. These are the questions that AI may not even "understand" it should be asked, aligning with the uniquely human, "EPOCH" capabilities-Empathy, Presence, Opinion/Ethics, Creativity, Hope/Vision-that are seen as complementary to AI.

Finally, the very real challenge of "curiosity burnout" due to information overload suggests that in an AI-saturated world, a critical meta-skill will be the ability to manage one's own cognitive and attentional resources. If AI provides a virtually infinite stream of information, unfiltered consumption can lead to mental exhaustion and diminish the capacity for deep thought or sustained curiosity. Individuals will need to develop robust strategies for selective attention, effective information filtering, and even embracing moments of "epistemic silence"; deliberately disengaging from the information firehose to allow for reflection and synthesis. This "attentional hygiene" or "cognitive resource management" becomes a prerequisite for maintaining the mental space and energy required for genuine, productive curiosity, preventing it from devolving into superficial information grazing or overwhelming cognitive fatigue.

As the very act of questioning and the nature of curiosity are reshaped by AI, the entire landscape of knowledge acquisition and expertise is altered. When answers are instant and universally available, the frontier of value inevitably shifts from the mere storage and recall of information to its synthesis, application, and the generation of new, more powerful questions. This sets the stage for a fundamental re-evaluation of what constitutes an "edge" in a world increasingly defined by information symmetry, a world where the ability to ask the right question is often more valuable than knowing the answer.

THE ENDURING POWER OF THE "DUMB" QUESTION

While the advent of AI promises to eliminate the cost of asking questions, freeing individuals from the fear of appearing ignorant, it paradoxically opens a new appreciation for the strategic utility of what might seem, on the surface, to be a "dumb question". Far from vanishing, these seemingly simplistic or foundational inquiries can evolve from a liability into a powerful pedagogical tool and a unique mechanism for fostering deeper human connection and understanding, even in an AI-saturated world.

In traditional human interaction, a "dumb" question, often equivalent to a leading question, can serve as a deliberate probe designed to expose unexamined assumptions, prompt critical self-reflection, or guide a learner toward a novel insight. This Socratic function is not about eliciting a fact (which AI excels at providing), but about stimulating a particular cognitive process. When faced with an AI that can instantly provide the "correct" answer, the human who still asks a seemingly "dumb" question might be doing so not out of genuine ignorance, but to subtly challenge the recipient to articulate their understanding, to trace their logical steps, or to reconsider the obvious from a fresh perspective. This transforms the question from a request for information into an invitation for deeper engagement and intellectual sparring.

Furthermore, the intentional deployment of a "dumb" question can be a powerful empathy-building tool. In a world where AI can provide immediate, personalized answers, the act of a human asking a seemingly basic question of another human can be a gesture of humility and a bid for connection. It signals a willingness to listen, to learn from another's unique perspective, and to acknowledge a shared journey of understanding. This human-to-human exchange, even around a simple concept, fosters rapport that AI, for all its sophistication, cannot authentically replicate. It moves beyond the mere transmission of data to the collaborative construction of meaning, strengthening bonds by creating shared moments of discovery or clarification.

Finally, "dumb" questions, particularly those posed by children or novices, often possess a unique capacity to cut through complexity and reveal fundamental truths or highlight overlooked aspects of a problem. In a professional setting, a seemingly naïve inquiry from a less experienced team member, unburdened by established paradigms, can force seasoned experts to articulate their tacit knowledge, challenge long-held beliefs, or uncover implicit biases that AI, trained on existing data, might reinforce. These questions can act as critical circuit breakers for groupthink, prompting a return to first principles or a re-evaluation of basic assumptions that might have become obscured by layers of specialized jargon or established process. Thus, while AI democratizes access to answers, the art of asking deceptively simple, yet profoundly insightful, questions remains a distinct and increasingly valuable human capability, enriching both individual comprehension and collective intellectual growth.

SUMMARY

This chapter has navigated the profound shift AI brings to the simple act of asking a question. We began by acknowledging the historical "cost of a question," laden with social and psychological risks like the fear of appearing ignorant and navigating organizational cultures that often stifled inquiry. AI, particularly through private, non-judgmental interfaces, dismantles these barriers, ushering in an era of "frictionless curiosity". This newfound freedom transforms learning into a "personalized epistemology," where answers are tailored to individual needs, contexts, and learning styles, leveraging multimodal content and dynamic scaffolding.

Consequently, traditional one-to-many knowledge transmission models like the expert lecture are ceding ground to dialogue-based, AI-assisted learning, forcing a redefinition of formal education itself. Schools and universities must now pivot from being primary information providers to cultivators of critical thinking, ethical reasoning, and uniquely human skills. This transformation also rewrites our "social scripts for ignorance"; with instant verification at hand, performative knowing may yield to genuine knowledge humility, though the nature of condescending communication like "mansplaining" or "techsplaining" might evolve into new forms, including "AI-splaining," rather than disappear entirely. Ultimately, in a world where answers are abundant, the true competitive edge lies not in possessing knowledge, but in the ability to ask insightful, strategic questions, making curiosity and skilled inquiry-including the emerging literacy of prompt engineering-the new markers of competence and innovation.

The universal accessibility of high-quality answers, driven by AI, fundamentally devalues the mere possession of information as a primary differentiator. If anyone can, in principle, know anything almost instantly, then "knowing things" in isolation no longer provides a sustainable competitive advantage. This realization naturally leads to a critical examination of what does constitute an edge in such an environment, and how traditionally valued forms of expertise and competitive advantage are being commoditized by this new technological reality. We must now consider what truly distinguishes individuals and organizations when the informational playing field is leveled, a theme that will be central as we explore the commoditization of the edge and the search for new moats in the age of information symmetry. Furthermore, while frictionless curiosity offers immense potential, navigating the sheer volume of available information without succumbing to overload or "curiosity burnout" becomes a critical skill in itself, ensuring that our enhanced capacity for inquiry remains a vibrant tool for progress, not a source of cognitive exhaustion.

6

COMMODITIZING THE EDGE

The preceding chapter ushered us into a world where answers, once elusive and often costly, are becoming increasingly ubiquitous, largely due to the pervasive influence of artificial intelligence. This newfound abundance of answers compels us to confront a profound and transformative consequence: if access to information and even sophisticated responses is democratized, what becomes of the intrinsic value historically ascribed to possessing specialized knowledge? For generations, this specialized knowledge, hard won through years of study, experience, and privileged access, constituted the "edge". It was this edge that defined professions, fueled economies, conferred status, and served as the bedrock of competitive advantage for individuals and organizations alike.

This chapter embarks on an exploration of how artificial intelligence is doing far more than serving as a highly advanced question-and-answer machine. It is, in fact, acting as a fundamental force that systematically commoditizes this once defensible expertise. We will unpack how traditional advantages, built upon exclusive knowledge, deep-seated skills honed over decades, and even certain forms of creative output, are increasingly becoming common currency, widely available and, consequently, devalued as unique differentiators.

The reader will journey through a detailed analysis of how this wave of commoditization is cresting across a diverse array of sectors. We will examine its impact from the venerable halls of law and finance to the dynamic and fast-paced arenas of marketing and strategy consulting. In doing so, we will dissect the emerging "AI commodity trap," a scenario where the very services that once commanded premium prices are rendered undifferentiated by AI's capabilities. We will also witness the "fall of the secret sauce": the proprietary frameworks, closely guarded internal playbooks, and unique methodologies that organizations long relied upon to ensure their distinction in the marketplace.

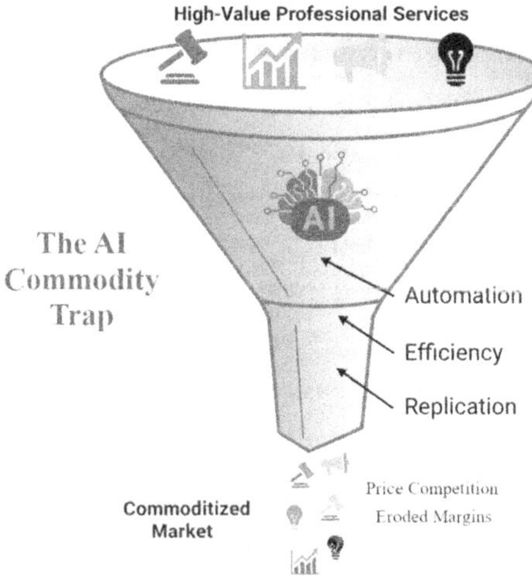

Crucially, this chapter will not merely dwell on the obsolescence of old paradigms. Instead, it aims to illuminate the path forward. As the strategic value of hoarded knowledge diminishes, what enduring human qualities and new strategic imperatives rise to prominence? We will identify those attributes and capabilities that, for now, resist easy commoditization-qualities such as profound trust, genuine empathy, irreplaceable lived experience, and nuanced ethical judgment. Furthermore, we will explore how these uniquely human elements become the new bedrock upon which sustainable differentiation can be built.

This exploration serves as a critical bridge to the latter part of the book. Having established how traditional information asymmetry is collapsing under the weight of AI-driven transparency and accessibility, we prepare the ground to investigate a future where competitive advantage shifts. The focus moves from merely holding static knowledge to dynamically composing intelligence on demand, assembling and orchestrating capabilities as needed. The central argument of this chapter is that in an age of burgeoning information symmetry, the new "edge" is forged not from the volume of what one knows, but from the agility, insight, and wisdom with which one operates within this transformed reality.

To fully grasp the magnitude of this ongoing shift, and to appreciate what is being lost as well as what is emerging, we must first journey back. We need to dissect the anatomy of the "edge" as it once existed. What precisely constituted this coveted advantage, how was it painstakingly built and often fiercely guarded, and why is its very foundation now crumbling under the relentless, democratizing advance of artificial intelligence?

WHAT WE MEAN BY "EDGE"

For much of modern economic and professional history, the concept of "edge" has been central to understanding competitive advantage. It represented a confluence of rare skills, profound expertise, and often proprietary or privileged insights that enabled certain individuals and organizations to consistently outperform their competitors and command premium value in the marketplace. This "edge" was not a fleeting advantage; it was often the very engine of differentiation, the justification for significant economic returns, and the basis upon which entire industries and professional hierarchies were built. To understand its current dissolution, we must first clearly define its historical components and manifestations.

The "edge" primarily functioned as a differentiator. It was about more than simply possessing information; it signified a demonstrable superiority in experience, refined judgment, and access to insights that were not widely available or easily replicable. In the nascent stages of the consulting industry, for example, firms like McKinsey & Company emerged by offering structured methodologies to tackle complex business problems that client organizations could not solve internally. They provided not just analytical solutions but also a crucial element of confidence and authority to leadership navigating uncharted waters. McKinsey, in particular, built its formidable and enduring reputation on a potent combination of rigorous analytical frameworks, access to cross-industry benchmark data (itself a form of proprietary insight), and the delivery of authoritative, independent advice that shaped strategic direction rather than merely optimizing operational tasks. Similarly, boutique consultancies carved out defensible niches by cultivating and demonstrating highly specialized expertise in specific domains, delivering proven, often quantifiable, outcomes that justified their premium positioning and fees. This specialized knowledge, honed through focused experience, was their distinct edge.

This differentiation, in turn, was leveraged into tangible economic power. The ability to charge substantial fees for specialized services was a direct consequence of this perceived and actual edge. The traditional model of billable hours, particularly in professions like law and consulting, was predicated on the scarcity and value of the unique expertise being applied per unit of time. Furthermore, the careful protection of intellectual property; be it through patents for inventions, copyrights for creative works, or the guarding of proprietary frameworks and unique methodologies was a critical mechanism for maintaining this leverage.

Patrick McKenna's Edge Consulting, for instance, exemplified this by productizing expertise. Instead of solely relying on selling consultant hours, McKenna developed and licensed training programs like Rainmaking®, effectively packaging intellectual capital for law firms. This "product" model offered a form of leverage that was far more scalable and potentially more lucrative than simply adding more consultants to deliver services. This transformation of insight into a licensable asset was a clear manifestation of the economic power of the edge.

The application and nature of this "edge" varied across industries:

Elite law firms forged their advantage in the crucible of deep legal scholarship, strategic acumen in litigation and negotiation, and privileged access to-and, more importantly, the nuanced interpretation of-complex legal precedents, statutes, and regulatory environments. This created a domain that was largely opaque and often intimidating to those outside the legal profession, or even to generalist practitioners within it. The core value proposition lay in their ability to navigate high-stakes complexity, mitigate substantial risks, and achieve favorable outcomes in ways that clients, lacking this specialized edge, could not achieve independently.

Consulting firms, both large and boutique, essentially monetized "synthetic expertise". They industrialized the process of knowledge arbitrage, transforming information advantages and repeatable insights into structured, billable services. Their "secret sauce" often involved proprietary frameworks and playbooks, meticulously developed through numerous client engagements and internal research, and then marketed as unique, high-value solutions to common or complex business problems.

In specialized research and development fields, such as pharmaceuticals, advanced materials science, or cutting-edge technology, the "edge" was unequivocally rooted in years of painstaking scientific research, patented discoveries, unique experimental know-how, and the cultivation of rare talent. These elements created formidable barriers to entry and were the basis for sustained competitive advantage, often lasting for decades.

The bedrock upon which this entire edifice of "edge" was constructed was, almost invariably, information asymmetry. In nearly every transaction or relationship where an edge was evident, one party possessed substantially more, or more refined, information and insight than the other. This imbalance was rarely accidental; it was often actively cultivated and fiercely protected through a variety of mechanisms. These included professional credentialing systems that restricted entry into a field, the use of specialized jargon and complex language that created barriers to understanding for outsiders, controlled access to critical information channels and data sources, and institutional gatekeeping practices designed to maintain exclusivity and, therefore, the value of the specialized knowledge.

Examining this historical construct reveals deeper structural characteristics. The "edge" was not always purely about the raw knowledge itself, but also about the performance and perception of scarce expertise. The ability of consultants to project "confidence" and an "authoritative tone", or the deliberate crafting of legal documents to be "unreadable to the layperson", suggests a strategy where the client's dependence and the mystique surrounding the expertise were integral to its perceived value. The hierarchical structures within these professions, such as the pyramid model in consulting where junior staff handle voluminous data gathering while partners dispense the high value "answers", served to reinforce the scarcity and high cost associated with accessing the ultimate "edge". This implies that the perceived value was often amplified beyond the direct utility of the knowledge itself, intertwined with the exclusivity of its access.

Furthermore, this traditional "edge" was frequently a reflection of past successes and accumulated knowledge. It was often built on deep experience, established frameworks, and historical data analysis, rather than on a dynamic, forward-looking capacity for rapid adaptation to fundamentally new paradigms. The emphasis on "proven outcomes", "benchmark data", and established "methodologies" points to a reliance on codifying and reapplying what has worked effectively in the past. While this approach provides a degree of reliability and predictability, it becomes a significant liability when the underlying conditions of the market or the nature of problems shift dramatically, as is currently occurring with the advent of artificial intelligence.

The very stability and established nature of these historical "edges" paradoxically contributed to their brittleness in the face of transformative, systemic change. The leverage derived from this edge, often manifesting as high billable hours or premium fees, was directly proportional to the perceived difficulty and exclusivity associated with acquiring and applying the specialized knowledge. As AI begins to diminish both this difficulty and this exclusivity, the traditional leverage points are severely weakened, challenging the economic foundations of many professions.

This carefully constructed and long-standing edifice of "edge," built on the twin pillars of scarce knowledge and the artful control of its interpretation and application, is now confronting an existential challenge. The very mechanisms that once made specialized information and insight difficult and costly to obtain are systematically dismantled by the democratizing power of artificial intelligence. Understanding precisely how this equalization is unfolding is the next critical step in our analysis.

Knowledge access

- **Traditional edge (pre-AI dominance):** Scarce, controlled, and often proprietary, relying on internal databases, expert networks, and restricted information flows.
- **AI-driven commoditization impact:** Knowledge becomes democratized, abundant, and globally accessible through AI systems.

- **Emerging locus of new edge:** Curated and contextualized insight that separates signal from noise, alongside the ability to generate truly unique data rather than merely consuming existing information.

Expertise development

- **Traditional edge (pre-AI dominance):** Years of experience, long apprenticeships, formal credentials, and the accumulation of "10,000 hours" of practice.
- **AI-driven commoditization impact:** Rapid AI-assisted skilling, automation of routine expert tasks, and widespread availability of "good enough" expert-level outputs.
- **Emerging locus of new edge:** Meta-skills such as critical thinking, effective collaboration with AI, prompt literacy, judgment, and ethical discernment.

Key differentiators

- **Traditional edge (pre-AI dominance):** Proprietary intellectual property, patented frameworks, unique methodologies, and brand or reputation built on past successes and insider access.
- **AI-driven commoditization impact:** Open models and AI-generated solutions replicate frameworks easily, making reverse-engineering of public patterns trivial.
- **Emerging locus of new edge:** Trust, empathy, lived experience, ethical grounding, speed of adaptation, novel problem framing, and distinctive human–AI synthesis.

Source of leverage

- **Traditional edge (pre-AI dominance):** Billable hours for specialized human expertise, information gatekeeping, and premium fees justified by exclusivity.

- **AI-driven commoditization impact:** Automation of expert tasks, low-cost replication of knowledge work, and erosion of fee structures tied directly to human effort.
- **Emerging locus of new edge:** Orchestration of human and AI capabilities, value-based outcomes, creation of novel solutions, and dynamic responsiveness to change.

Primary focus

- **Traditional edge (pre-AI dominance):** Information hoarding, process control, and maintaining exclusivity.
- **AI-driven commoditization impact:** Emphasis shifts toward broad information access and task execution efficiency.
- **Emerging locus of new edge:** Insight generation, complex problem-solving in novel contexts, continuous value creation, and systemic adaptation.

AI AS AN EQUALIZER OF INTELLIGENCE AND COMPETENCE

The historical "edge," so carefully constructed upon foundations of scarce knowledge and specialized interpretation, is now facing a formidable challenge: artificial intelligence is emerging as a profound equalizer of intelligence and competence. AI's capabilities are systematically dismantling traditional knowledge-based advantages by democratizing access not only to vast stores of information but also to sophisticated analytical tools and even complex reasoning processes. This development is "flattening the curve" of competence, making high-level skills and insights more broadly accessible than ever before.

One of the most significant impacts of AI is the democratization of expertise and skills. AI platforms are increasingly designed with user-friendly interfaces, abstracting away much of the underlying technical complexity. This enables individuals who may not possess specialized data science or programming backgrounds to perform tasks that were previously the exclusive domain of highly trained experts. For instance, platforms like Google AutoML provide intuitive, guided environments that allow users to build and deploy machine learning models with minimal coding. Similarly, tools like Microsoft Copilot empower users to execute sophisticated functions within widely used applications like Microsoft 365 simply by using natural language prompts. This "democratization of AI" means that a junior marketing analyst, equipped with an AI tool for advanced data analysis, might now generate market insights that previously would have required a seasoned statistician or market researcher. A small business owner, leveraging an AI-powered legal assistant, might draft a standard contract or analyze regulatory compliance issues that once necessitated engaging an expensive law firm for routine matters. AI systems can readily analyze unstructured data, extract critical information from extensive documents, perform complex language translations, and even generate functional software code, thereby effectively augmenting the capabilities of less experienced individuals and significantly boosting their productivity.

The proliferation of on-demand AI assistants further accelerates this equalization. These tools represent a paradigm shift where logic, data modeling capabilities, and vast reservoirs of specialized knowledge are no longer confined to the minds of human experts or siloed within proprietary databases. Instead, they are becoming universally accessible cognitive augments. These AI assistants can function as "zero-cost experts" for many queries, capable of providing instant answers to complex questions, performing intricate calculations, summarizing dense technical texts, and even generating diverse forms of creative content. This ubiquity of sophisticated AI support reduces the general reliance on human experts for a wide array of foundational knowledge work and routine analytical tasks. AI can rapidly process and synthesize vast quantities of data, generate insightful reports, and even draft initial versions of presentations or strategic documents, tasks that traditionally consumed significant human expert time.

A core economic principle underpinning this equalization is the near-zero marginal cost of insight replication. Once an AI model is developed and trained to perform a specific intellectual task-be it analyzing a financial report, diagnosing a common medical condition from a set of symptoms, or optimizing a complex supply chain-the incremental cost of replicating that "insight" or performing that task for an additional user or instance approaches zero (even including ongoing runtime costs, with the ability to locally host models or cache insights the overall cost of reuse is dramatically lower than initial setup). This economic reality fundamentally undermines business models that were built on charging for human time and effort to produce similar intellectual outputs. When the scarcity value of such outputs plummets due to low-cost, high-fidelity AI replication, the pricing power of human experts for those specific tasks is severely eroded.

The cumulative effect of democratized expertise and low-cost replication is a tangible narrowing of performance gaps across various roles and industries. Research consistently indicates that AI can significantly enhance worker performance, and individuals who effectively collaborate with AI tools often outperform those working without such assistance. For example, junior team members, when augmented by generative AI, can now confidently undertake tasks and responsibilities that were previously the domain of more senior, experienced managers or specialists. S&P Global has highlighted that AI-driven efficiencies are already reshaping workforce structures and leveling certain skill-based hierarchies within professional services. While AI is unlikely to lead to wholesale job replacement for entire roles in the immediate future, its impact is broad. It is projected that AI will affect at least 10% of the tasks for approximately 80% of the US workforce, with a significant minority, around 19%, potentially seeing half or more of their tasks automated. This suggests a widespread augmentation and equalization effect on competence across the economy.

The implications of this AI-driven equalization are profound. It extends beyond simply providing easier access to information; it involves the democratization of sophisticated cognitive processes and methodologies. AI tools often do not just give factual answers; they execute complex analytical steps, simulate expert reasoning patterns, and apply learned methodologies to new data. For example, platforms like Google AutoML do not merely present data; they guide users through the intricate process of building, training, and evaluating predictive models. Similarly, AI-powered legal review tools do not just identify relevant clauses in a contract; they analyze these clauses based on learned legal patterns, precedents, and risk factors. This means that the "how-to" of expertise; the procedural knowledge that was often a closely guarded secret or the result of long apprenticeship is also being disseminated and, in many cases, automated. This has far-reaching consequences for education, training, and skill development, as the focus must inevitably shift from mastering rote processes (which AI can handle) to cultivating the critical judgment, ethical discernment, and creative problem-solving skills needed to effectively guide, interpret, and leverage AI-driven execution.

However, while AI acts as a powerful equalizer for foundational and intermediate levels of competence, it simultaneously creates what might be termed a new "competence cliff". This refers to the challenges encountered when dealing with tasks that lie at the "jagged technological frontier" of AI: those areas where AI's capabilities are currently unreliable, prone to significant error / hallucinations, or simply non-existent. Research by Dell'Acqua and colleagues has shown that worker performance can decline significantly if AI is employed for tasks outside its established boundary of capability.

This suggests that while AI makes many tasks more accessible and easier to perform to a certain standard, a new and critical meta-skill “Context Engineering” emerges: the discernment to know when and how to deploy AI effectively, when to critically question and validate its outputs, and, crucially, when to rely on human expertise, intuition, and ethical judgment for tasks that AI handles poorly or in a manner that could lead to negative consequences. The "equalizer" effect, therefore, has its limits, and navigating these limits intelligently requires a new form of expertise centered on human-AI collaboration and critical oversight.

Furthermore, the observed narrowing of performance gaps for specific tasks due to AI augmentation may, paradoxically, widen the gap in overall strategic value creation if individuals and organizations fail to strategically leverage the cognitive capacity freed up by AI. If human workers, now more efficient at routine tasks, are not redirected towards higher-order thinking, innovation, and uniquely human contributions, the overall impact could be a plateauing of value rather than a leap forward. notes that AI-augmented operating systems can reduce the need for costly, higher skilled talent for certain functions. However, the same source highlights a compelling case where redeploying senior staff-whose routine tasks were automated by AI; to focus on improving customer experience and strategic client acquisition yielded economic benefits that were on par with the direct cost savings achieved through automation. This implies that simply achieving AI-assisted competence in existing tasks is insufficient. The new competitive differentiation, and indeed the realization of AI's full potential, will arise from using that newfound efficiency to engage in complex problem-solving, creative ideation, deep interpersonal engagement, and strategic foresight activities that AI currently cannot replicate. Without this strategic redeployment of human intellect and creativity, the "equalization" of task level competence could inadvertently lead to a broader stagnation of innovation and value creation.

If foundational and even many intermediate levels of professional competence are becoming universally accessible at near-zero marginal cost, then entire industries built upon the scarcity and the premium sale of that human delivered competence are facing an existential recalibration. This leads us directly to a focused examination of how these forces are playing out within the professional services sector.

PROFESSIONAL SERVICES IN THE AI COMMODITY TRAP

The professional services sector, encompassing fields such as law, finance, marketing, and consulting, has long thrived on a foundation of specialized knowledge, bespoke advice, and high-margin service delivery. However, this sector is now confronting the disruptive force of the "AI commodity trap". This trap emerges as artificial intelligence demonstrates an increasing capacity to replicate and automate core service offerings, thereby eroding their scarcity value and pushing established firms towards an undifferentiated, price-sensitive market position where their once-unique "edge" becomes a common capability.

In legal services, the impact of AI is already significant and rapidly growing. Traditionally, legal work involved labor-intensive processes such as document review, legal research, and drafting. AI tools are now capable of automating many of these critical aspects. For instance, AI can quickly analyze, sort, and identify key information within vast volumes of documents for eDiscovery, summarize lengthy legal texts to highlight salient points, assist in the initial drafting of pleadings, contracts, and other legal instruments, and even provide accurate translation services for multilingual legal matters. Tasks that previously consumed countless hours of junior associate time, forming a substantial and profitable component of billable work, are increasingly becoming automatable functions. This automation directly challenges the traditional leverage model of law firms, where a significant portion of revenue was derived from the human labor involved in these now AI-assisted tasks.

The financial services and tax optimization domain is similarly experiencing a wave of AI-driven commoditization. AI algorithms can automate complex data analysis, enhance fraud detection capabilities with greater speed and accuracy, and provide sophisticated financial planning and tax optimization strategies that were once the exclusive purview of seasoned financial advisors and tax professionals. For example, AI can meticulously analyze intricate datasets to identify potential tax deductions, exemptions, and incentives that might be overlooked by human analysts, and it can simulate the financial impact of various tax strategies or corporate decisions, providing clients with data-driven recommendations. AI-powered fintech solutions are emerging that can automate tax refund tracking, flag inconsistencies in tax filings before submission (reducing errors and delays), and offer personalized financial planning advice tailored to individual circumstances. This widespread automation of routine advisory, compliance, and analytical tasks inevitably commoditizes services that were previously offered at a premium due to the specialized human expertise required.

Marketing and advertising, industries historically driven by human creativity and strategic insight, are also being profoundly reshaped by generative AI. These AI systems can automate the creation of a diverse array of content types; from engaging social media updates and informative blog posts to persuasive ad copy and detailed product descriptions. Furthermore, AI enables hyper-personalization of marketing campaigns at an unprecedented scale, tailoring messages and offers to individual consumer preferences based on behavioral analytics.

Industry reports and case studies indicate that AI can generate comprehensive marketing materials in a matter of hours, a process that would have previously taken human teams weeks or even years. There is also evidence that AI-generated advertising copy can, in some instances, outperform human-written versions in terms of engagement and conversion rates. This capability directly challenges the traditional value proposition of advertising agencies, which has been heavily reliant on human creative labor and strategic campaign development.

Even the realm of high-level strategy consulting is not immune to AI's inroads. While nuanced human judgment and experience remain indispensable in crafting overarching strategic vision and making final executive decisions, AI is increasingly capable of accelerating and adding significant rigor to many components of the strategy development process. This includes complex tasks such as market sizing, detailed competitor analysis, and estimating the potential financial value of various strategic initiatives under different scenarios. AI can function as a powerful research assistant, a sophisticated data interpreter, a tireless thought partner in brainstorming sessions, and even a simulator for testing strategic scenarios against potential market reactions. If comprehensive strategy decks, detailed market analysis, and foundational strategic frameworks can be substantially auto-generated using AI tools drawing upon readily available market data and established business principles, the unique value of consultants who primarily perform these data-gathering, analytical, and framework-application tasks is demonstrably diminished.

This leads us to define the "AI commodity trap" more formally. It describes the challenging predicament where the confluence of AI-driven efficiency, widespread accessibility, and high-fidelity replicability reduces the scarcity, and therefore the perceived unique value, of many professional services. When AI can perform tasks faster, more cost-effectively, and often with comparable or even superior accuracy for routine and standardized elements of a service, clients become increasingly reluctant to pay premium fees for the human delivery of those same services. This market dynamic forces professional service firms into intensified price competition for offerings that are becoming functionally undifferentiated. The result is an erosion of profit margins and a fundamental challenge to long standing business models that have been predicated on the billable hour and the scarcity of specialized human expertise.

The implications of this AI commodity trap extend beyond mere pricing pressures. It signifies a fundamental decoupling of expert outcomes from exclusive human delivery for a rapidly expanding array of tasks. Clients are increasingly focused on achieving the desired outcome; be it a thoroughly reviewed legal contract, an optimized tax strategy, an effective marketing campaign, or a data-backed strategic recommendation. They are becoming progressively indifferent to whether a human, an AI, or a human-AI hybrid produces that outcome, especially if the AI-driven or AI-augmented solution offers clear advantages in terms of speed, cost-efficiency, or even certain aspects of quality or comprehensiveness. The perceived value is shifting away from the human expert's labor input and towards the solution output itself, an output that AI can increasingly provide or significantly contribute to.

Consequently, professional service firms that merely attempt to layer AI capabilities onto their existing processes and service offerings, without fundamentally rethinking their core value proposition and sources of differentiation, will find themselves most vulnerable to this commodity trap. Escaping this trap necessitates a strategic pivot "up the value chain". This means focusing on services and advisory roles that AI genuinely cannot replicate, at least with current technology. These areas typically involve highly novel problem-solving, deep empathic client engagement and understanding, navigating extreme complexity or ambiguity, providing sophisticated ethical guidance, and exercising high-stakes strategic foresight. suggests that by 2027, the traditional consulting model, heavily reliant on large teams performing analytical tasks, will be less relevant, pushing firms towards alternative pricing models such as value-based pricing. This implies that simply executing established tasks faster or cheaper with AI will not be a sustainable strategy. Durable value must derive from contributions that are uniquely human or represent a sophisticated synthesis of human and AI capabilities that competitors cannot easily match. Firms that attempt to sell "AI-powered versions" of what have effectively become commodity tasks will find themselves still competing in a commoditized market, albeit with slightly different tools.

This commoditization of routine and standardized professional work by AI is also likely to lead to a significant bifurcation of the professional services market. On one end, a low-cost, high-volume segment will emerge, catering to AI-driven, standardized services that can be delivered at scale with minimal human intervention. On the other end, a high-cost, low-volume segment will persist, focused on genuinely bespoke, high-stakes, human led advisory services that address unique, complex, or ethically sensitive challenges. Mid-tier, generalist firms that lack either extreme specialization in a defensible niche or the massive scale required to compete in the AI-driven volume market will face the most intense competitive pressure, potentially being squeezed out from both ends.

It is not merely the execution of professional tasks that is being commoditized by AI. The very "playbooks," proprietary methodologies, and "secret recipes" that firms once meticulously developed, zealously guarded, and marketed as their primary sources of differentiation are also losing their exclusivity and, consequently, their power to command a premium.

THE FALL OF THE "SECRET SAUCE"

The competitive landscape for businesses and professional services has long been characterized by the pursuit and protection of a "secret sauce"; those proprietary frameworks, confidential internal playbooks, unique methodologies, and domain-specific "tricks of the trade" that were believed to confer a distinct advantage and justify premium pricing. However, artificial intelligence, in its various forms, is systematically undermining the exclusivity and defensibility of this secret sauce, transforming once guarded knowledge into accessible patterns.

AI models, particularly large language models (LLMs) and sophisticated machine learning systems, function as universal observers and absorbers of public patterns. These systems are trained on colossal datasets that encompass a vast expanse of publicly available information. This includes, but is not limited to, industry reports, academic research papers, published case studies, best practice guidelines disseminated by organizations, and even nuanced discussions on professional forums and online communities.

This extensive and diverse training enables AI to identify, synthesize, and replicate successful strategic patterns, operational methods, and problem-solving approaches that were once considered the exclusive intellectual property or hard-won experiential knowledge of particular firms or seasoned experts. If a successful business model or operational technique has been described or analyzed in public sources, AI can learn its components and logic.

Beyond learning from public data, AI's advanced capabilities in pattern recognition, data analysis, and even code generation can facilitate the deconstruction and reverse engineering of software, business processes, and complex algorithms. This is especially true if sufficient output data from these systems or descriptive information about their methods becomes available. If a proprietary business methodology consistently produces certain types of outputs or follows discernible, logical steps that can be observed or inferred, an AI can potentially learn to mimic or recreate that methodology. Furthermore, the risk of AI model distillation and theft, as highlighted in, demonstrates that the functional behavior of a proprietary AI model can itself be replicated by systematically querying the target model and using its outputs to train a new, functionally equivalent model. This means that even AI-based secret sauces are not entirely immune.

This trend directly impacts the traditional value proposition of consultants' playbooks. Consulting firms have historically prided themselves on their unique, often branded, frameworks and strategic playbooks, which are typically developed through years of client engagements, internal research, and the accumulation of industry-specific knowledge. However, AI systems can now generate highly sophisticated strategic recommendations, detailed market analysis, and comprehensive operational improvement plans by drawing upon far broader and more diverse datasets than any single consulting firm could typically access or process efficiently. Leading firms like McKinsey acknowledge AI's emerging roles as a powerful researcher, a nuanced data interpreter, and an insightful thought partner in the complex process of strategy development. If an AI can produce a "good enough," or in some specific analytical aspects, even a superior strategy deck based on global best practices and real-time data analysis, the unique value proposition of a consultant's standardized, often static, playbook is significantly diminished. The playbook becomes a template that AI can also access and improve upon.

This leads to a fundamental re-evaluation of what "best practice" means in an AI-driven world. Generative AI, by its very design, often produces responses that reflect the most common, statistically probable, or frequently cited patterns within its vast training data. This output can be interpreted as a form of "median practice" or a data-driven "best match" solution based on the totality of information it has processed. This fundamentally challenges the traditional notion of a "best practice" that might have been derived from a specific firm's unique, accumulated experience or a particular expert's singular judgment.

The critical questions then become: Are these AI-generated "best practices" truly optimal in all contexts, or are they merely optimally average, reflecting a regression to the mean? How do they compare in efficacy to human-derived, context-specific best practices that can account for subtle nuances, unquantifiable factors, or novel situations that AI might miss or misinterpret? The dilemma of using off-the-shelf, general-purpose AI versus investing in bespoke, custom-trained AI solutions becomes highly relevant here. Generic AI can provide broadly applicable "good enough" solutions and replicate common best practices, but highly specific, novel, or strategically critical challenges may still require custom approaches that go beyond AI's current "best match" capabilities.

The proliferation of powerful, open-source AI models further accelerates the erosion of the "secret sauce". Platforms like Llama, Mistral, and DeepSeek, make foundational AI capabilities widely available, often modifiable, and frequently free to use. When the core technological building blocks are accessible to everyone, it becomes increasingly difficult for any single entity to claim a sustainable competitive advantage based solely on a proprietary AI model or algorithm. The only exception is if that model is deeply and inextricably intertwined with genuinely unique and continuously refreshed data sources, or if it powers highly specialized, hard-to-replicate processes that deliver a demonstrably superior outcome.

The implications of this "fall of the secret sauce" are far-reaching. The traditional value of a proprietary framework or methodology often derived its power from the opacity of its creation process and the difficulty for outsiders to access or synthesize the underlying knowledge components. AI dramatically increases transparency and accessibility for both. Consultants' frameworks, for example, were valuable because clients typically lacked the time, diverse data inputs, or established methodologies to create comparable strategic tools themselves. AI now offers powerful analytical methodologies and the ability to process or access immense volumes of data. The "secret" was frequently the act of synthesis and pattern recognition-tasks at which AI, particularly generative AI, now excels.

This erosion signifies a critical shift in the nature of competitive advantage. It moves away from owning codified, static intellectual property (like a copyrighted framework or a patented process) and towards dynamically generating, applying, and adapting insights in real-time, often leveraging AI as a powerful enabling tool.

AI models, however advanced, are primarily limited to interpreting existing data; they cannot generate entirely new primary signals, such as insights from direct ethnographic research or deep, unstructured customer interviews. This suggests that the enduring "sauce" is no longer a pre-packaged recipe but rather the ongoing capability to continuously discover new "ingredients" (e.g., proprietary data streams, unique human observations, novel problem framings) and combine them in innovative ways. This is a dynamic process that AI can augment and accelerate but not wholly own or originate. The value shifts from the static intellectual artifact to the agile process of discovery, synthesis, and application.

Consequently, when "best practice" becomes a baseline that is easily replicable or even generated by AI, true differentiation will increasingly stem from pioneering "next practice". This involves genuine innovation and the creation of entirely new approaches, products, or services that are not yet reflected in AI's existing training data. Alternatively, differentiation will come from the highly contextualized, nuanced, and ethically sound application of known best practices in complex, human-centric situations that AI cannot yet navigate effectively.

If AI can readily provide the "median" or common best practice, then market leadership will be seized by those who can develop strategies or solutions that are, by definition, outside the current AI knowledge base. This is the realm of true innovation. Concurrently, as the next section will elaborate, the masterful, empathetic, and ethically responsible application of even known best practices in intricate, human-dominated scenarios remains a significant arena for human differentiation. reinforces this by noting that custom AI solutions can deliver unique advantages if the application is core to business differentiation, whereas off-the-shelf AI suffices for more generic, commodity functions.

If the once-secret recipes and proprietary ingredients are becoming common knowledge or easily replicable, what truly distinguishes the master chefs from the assembly-line cooks in this new, AI-influenced landscape? The answer likely lies in those qualities and capabilities that remain uniquely human and, therefore, more resistant to commoditization.

WHAT CANNOT BE COMMODITIZED (YET)

As artificial intelligence relentlessly advances, automating tasks and democratizing knowledge that once formed the core of professional expertise, a critical question emerges: what remains uniquely human and defensible? While AI can replicate patterns, process data at superhuman speeds, and generate sophisticated outputs, certain intrinsic human attributes and complex capabilities, at least for the foreseeable future, resist easy commoditization. The qualifier "yet" is crucial, for the boundaries of AI are ever-expanding. However, focusing on these currently resilient human domains offers a pathway to enduring value in an AI-saturated world.

At the forefront of these enduring values are honesty, trust, and the ecosystems of reputation that arise from them. In an environment increasingly populated by AI-generated content, sophisticated deep-fakes, and potentially manipulative synthetic interactions, genuine human trust becomes an invaluable and scarce currency.

Trust is not a simple transactional feature that can be programmed; it is painstakingly built over extended periods through consistent, ethical behavior, radical transparency in dealings, and proven reliability under pressure. Neuroscience research illuminates the deep biological underpinnings of trust, involving complex brain functions in the prefrontal cortex and the release of hormones like oxytocin, which foster empathy, social bonding, and cooperative behavior.

While AI can be programmed to simulate trustworthy behaviors or adhere to predefined ethical rules, the establishment of deep, resilient trust-particularly in high-stakes human relationships, critical business dealings, or situations requiring nuanced judgment-remains a profoundly human-centric endeavor. Reputation ecosystems, which are built upon real-world actions, demonstrable accountability, and shared social validation, are difficult for AI to authentically inhabit or manipulate in a sustainable manner.

Strategic timing and contextual acuity represent another domain where human capabilities retain a distinct advantage. The ability to discern the opportune moment to act; or, equally importantly, to exercise restraint and refrain from acting based on a deep, nuanced understanding of rapidly evolving, multifaceted contexts is a hallmark of human strategic thinking and wisdom. AI excels at processing historical data and identifying patterns to make predictions about future probabilities. However, making intuitive leaps regarding future timing in complex, dynamic, and often ambiguous socio-political, economic, or market situations frequently relies on human judgment, foresight, pattern recognition that goes beyond quantifiable data, and what is often referred to as "gut feeling" or experienced intuition.

Deep interpersonal relationships and the sophisticated application of emotional intelligence are also extraordinarily difficult for AI to genuinely replicate. Profound, empathetic, and truly collaborative human connections are foundational to many aspects of personal and professional life. Emotional intelligence; the capacity to accurately perceive, understand, manage, and effectively use emotions to guide thought and action, and to build strong rapport and navigate complex interpersonal dynamics-remains a key differentiator, especially in leadership roles, client management, negotiation, sales, and all forms of caregiving professions. These skills are often described as "humanity's superpower" in an increasingly technological world.

Perhaps one of the most fundamental differentiators is "lived experience" and the embodied cognition that arises from it. Lived experience encompasses the rich, often tacit, tapestry of knowledge, intuition, wisdom, and nuanced understanding that individuals gain through direct, embodied interaction with the physical, social, and emotional world. Theories of embodied cognition strongly emphasize that our sensorimotor experiences: our physical engagement with our environment are fundamental to acquiring conceptual knowledge, developing cognitive capabilities, and shaping our unique understanding of the world. AI learns from abstract data representations, however vast and complex. It does not "live," "feel," or "experience" the world in a human, embodied sense. This experiential deficit limits its ability to grasp certain kinds of contextual understanding, common-sense reasoning, and the subtle, unquantifiable factors that often inform the best human decisions.

This raises a critical paradox: the challenge of gaining lived experience in the age of AI. If AI systems increasingly automate many of the foundational and routine tasks through which humans historically gained practical experience, honed their skills, and developed their professional judgment, how will future generations acquire this crucial "lived experience"?

This is a significant concern for workforce development and education. Potential pathways to address this challenge include the strategic use of sophisticated immersive simulations and virtual reality environments, which can provide platforms for hands-on practice in safe yet realistic settings, allowing for learning from mistakes without real-world consequences. Another approach involves reimagining apprenticeship models, shifting their focus from mere task execution (which AI might assist with) to the development of critical thinking, ethical reasoning, and complex problem-solving skills, with senior mentors guiding apprentices in how to collaborate effectively with AI tools rather than being replaced by them. Furthermore, organizations can deliberately assign humans to tackle complex, ambiguous, and novel problem-solving scenarios that are currently beyond AI's capabilities, thereby fostering adaptive expertise and resilience. This also necessitates a shift in educational and professional development philosophies towards cultivating "post-skill era" competencies: the ability to critically evaluate AI outputs, exercise sound judgment in the face of uncertainty, develop contextual intelligence, and apply ethical discernment to the use of powerful technologies.

Finally, human-centric sensemaking, which encompasses empathy, ethics, and narrative framing, remains a bastion of human value.

Empathy, the capacity to truly understand and share the feelings of another human being, is a deeply resonant and powerful ability that forms the basis of meaningful connection and effective collaboration. While AI can be trained to recognize emotional cues in text, voice, or facial expressions and respond in ways that appear empathetic, genuine empathy involves a level of shared phenomenological experience, authentic presence, and intersubjective understanding that AI currently lacks.

Ethical judgment and moral reasoning are core human functions, especially when navigating complex moral ambiguities, making value-laden decisions that have significant human consequences, and taking ultimate responsibility for those outcomes. Ethical decision-making consistently ranks as one of the most valuable and least automatable human-centric skills. AI can be programmed with ethical guidelines and can assist in analyzing the potential consequences of different actions, but it lacks genuine moral agency or the capacity for nuanced ethical deliberation in unforeseen or novel circumstances.

Narrative framing and meaning making are also profoundly human. Humans make sense of the world, motivate action, build shared understanding, and transmit culture through stories and narratives. The ability to craft compelling narratives, frame complex issues in ways that resonate emotionally and intellectually with diverse audiences, and communicate with purpose, clarity, and persuasive power remains a potent human skill. AI can assist in generating text or structuring arguments, but the originality, contextual appropriateness, and emotional depth of powerful narratives often stem from human creativity and experience.

The enduring value of these "uncommoditizable" (yet) human traits is, in fact, significantly amplified by AI's increasing ubiquity and proficiency in handling a wide range of automatable tasks. The more that AI can do in terms of information processing, routine analysis, and content generation, the more uniquely valuable and differentiating the remaining human contributions become. Human capacities for ethical oversight, deep empathy in client and collegial relationships, creative and novel problem framing, and strategic foresight rooted in rich lived experience stand out in much sharper relief. These uniquely human attributes become the primary locus of true differentiation when technical execution and information access are no longer significant barriers to entry or performance.

Moreover, the very concept of "lived experience" in an AI-pervaded world may itself be undergoing a transformation. Increasingly, valuable lived experience will involve nuanced interaction with and through AI systems. This means learning to collaborate effectively with AI, critically evaluate its outputs, discern its inherent limitations and potential biases, and judiciously apply its capabilities in complex, real-world contexts. The nature of "experience" is evolving to include AI fluency, critical AI literacy, and the sophisticated art of human-AI teaming.

It is also vital to continuously acknowledge the qualifier "yet" when discussing the commoditization of human capabilities. This serves as a crucial reminder of AI's dynamic and rapidly advancing frontier. While attributes like profound empathy, genuine consciousness, or ultimate ethical responsibility currently seem intrinsically and perhaps exclusively human, it would be strategically unwise to assume they will never be closely approximated, simulated, or even functionally replicated by AI to a degree that impacts their differentiating value.

The history of artificial intelligence is replete with examples of capabilities once thought to be uniquely human, eventually being mastered or significantly impacted by machines. While current AI lacks genuine sentience or emotion, and its ethical reasoning is programmed rather than inherent, research into Artificial General Intelligence (AGI) and more sophisticated affective computing continues apace. Therefore, relying on a static, fixed list of "human-only" skills as a long-term defense against commoditization is a precarious strategy. The most robust and sustainable approach involves the continuous cultivation, deepening, and adaptation of these human capacities. This means embracing lifelong learning, fostering adaptability and resilience, and prioritizing the development of uniquely human potentials-such as critical thinking, creativity, emotional intelligence, and ethical wisdom-to continuously redefine and elevate human value in both partnership and distinction from AI's expanding capabilities.

These enduring, uniquely human qualities are not merely passive assets to be preserved. Instead, they form the active bedrock upon which new, resilient competitive advantages, new “strategic moats”, can be intentionally constructed in an AI-saturated world. This recognition leads us directly to the subsequent section, which will explore the nature and architecture of these emerging moats.

NEW MOATS IN THE AGE OF INFORMATION SYMMETRY AND INSIGHT ASYMMETRY

As the relentless advance of artificial intelligence continues to erode traditional knowledge-based moats; such as proprietary information, exclusive expertise, and hard-to-replicate manual skills, new forms of defensible competitive advantage are emerging. In an era increasingly defined by information symmetry, where access to data and basic analytical tools becomes widespread, these new "moats" are characterized less by static, hoarded assets and more by dynamic capabilities, adaptive systems, and the strategic cultivation of trust and unique insight. These evolving defenses are often built on the premise that while raw information may become a commodity, the ability to generate unique, actionable insight from that information, and to act upon it effectively, can still create powerful and sustainable asymmetry.

One of the primary new moats is the capacity for composability and organizational adaptability. This refers to the ability of an organization to flexibly combine, reconfigure, and deploy modular components-which can include technologies, systems, data sources, services, and even human teams; to respond rapidly and effectively to changing business needs, market opportunities, and disruptive threats. A "composable enterprise" architecture allows for significantly greater agility, scalability, and reusability of resources compared to traditional, rigid, monolithic organizational structures. This inherent adaptability enables businesses not just to react to unforeseen disruptions but to proactively shape their offerings and operations, turning responsiveness itself into a competitive advantage.

Closely related to adaptability is the power of continuous feedback loops and organizational plasticity. Organizations that successfully embed robust, continuous feedback loops throughout their operations-learning dynamically from user behavior, AI interactions, real-time market signals, and internal performance data-can build a powerful and self-reinforcing dynamic advantage. This data-driven learning fuels rapid iteration cycles, enables deep personalization of products and services, and facilitates swift strategic adaptation to evolving conditions. "Organizational plasticity", defined as the ability of an organization to restructure itself and respond effectively to changing circumstances without significant degradation in performance or loss of core identity, becomes a key characteristic of resilient and adaptive enterprises in this new landscape. Artificial intelligence itself can be a critical enabler of these feedback systems, capable of processing vast amounts of diverse data and surfacing actionable insights in real time to inform decision-making.

As AI agents become increasingly specialized and capable of performing a wide range of discrete tasks with autonomy, the ability to effectively manage, coordinate, and orchestrate these agents emerges as a significant form of operational leverage and a potential new moat. This involves designing intelligent systems that can autonomously complete complex workflows, share data and context seamlessly between different agents (both AI and human), and continuously optimize processes for efficiency and effectiveness. This capability moves beyond simple automation to the realm of "synthetic team collaboration", where human and AI agents work in concert, often dynamically assembled, to achieve business objectives with enhanced speed, precision, and scale.

In an environment characterized by information abundance and the proliferation of AI-generated content-some of which may be unreliable, biased, or even deliberately manipulative; a brand's established reputation for trust, reliability, ethical conduct, and transparency becomes an exceptionally valuable and hard-to-replicate moat. The brand can function as a "semantic trust layer," a cognitive and emotional shortcut for customers (and increasingly, for their personal AI assistants) to identify, prefer, and engage with services and information sources that are perceived as safe, aligned with their values, and accountable for their outputs. This trust is not merely a superficial marketing construct; it must be deeply embedded in the demonstrable integrity of the company's operations, its data handling practices, its commitment to ethical AI principles, and its consistent delivery of value.

While general knowledge is being commoditized, the ability to generate, curate, and leverage unique, proprietary data ecosystems and insight generation networks can create strong, compounding competitive advantages. This is particularly true when such data is integrated into closed-loop systems where it continuously refines AI models, improves business processes, and delivers superior, personalized customer value. Examples like Carro's AI-driven automotive ecosystem, where data from each part of the vehicle lifecycle (inspection, financing, insurance, maintenance, resale) enhance the accuracy and value of other parts, or AwanTunai's use of real-time transaction data from its embedded ERP system to refine credit risk models for underserved MSMEs, illustrate this principle. The moat here is not just the static data itself, but the dynamic ecosystem that generates, processes, learns from, and acts upon this data to create differentiated insights.

Deep workflow integration and niche specialization with AI solutions also offer a path to defensibility. By embedding AI capabilities deeply into specific, critical enterprise workflows, often with highly specialized AI components tailored to niche industry requirements or unique business processes, companies can create significant switching costs for their clients and build a moat against more generic, horizontal AI tools. FileAI's focus on becoming a horizontal file processing agent with specialized AI components for complex document understanding and automation is an example of this strategy. When an AI solution becomes an integral and indispensable part of a client's core operations, it is much harder to dislodge, even if alternative AI tools offer some overlapping functionalities.

Finally, for AI solutions, particularly those involving natural language interaction, customer service, or culturally sensitive content and services, developing deep localization and cultural contextualization can form a potent moat against global, one-size-fits-all AI offerings. An AI that understands not only regional languages but also intricate cultural nuances, local dialects, social norms, and context-specific humor or etiquette will deliver a vastly superior user experience and greater effectiveness. WIZ.AI's development of conversational AI meticulously tailored for the diverse languages and cultural contexts of Southeast Asian markets exemplifies this approach, creating a significant barrier to entry for larger, less localized competitors.

These "new moats" are fundamentally different from their predecessors. They are inherently dynamic, adaptive, and relational, contrasting sharply with the often static, proprietary, and exclusionary nature of traditional moats built on fixed assets or guarded knowledge. Concepts such as continuous feedback loops, organizational plasticity, and enterprise composability all emphasize ongoing processes, learning from interaction, and a state of perpetual responsiveness rather than reliance on unchanging assets.

Data network effects are about creating "compounding advantages over time" through virtuous cycles of data acquisition and model improvement, not about a one-time accumulation of a static dataset. This represents a fundamental strategic shift from a "fortress building" mindset, focused on protecting existing assets, to one of "ecosystem cultivation" and "dynamic capability development," focused on continuous evolution and adaptation.

Furthermore, the strength and sustainability of these new moats often depend critically on the successful and synergistic integration of uniquely human strengths; such as trust, ethical judgment, complex relationship management, and creative problem-solving (as discussed in the previous section) along with advanced AI capabilities. For example, a brand's ability to act as a semantic trust layer (a human-centric construct built on perception and experience) is reinforced and made scalable by the reliable, ethical, and transparent performance of its AI systems (a technical element).

Similarly, effective AI agent orchestration is not a purely automated function; it requires human strategic design to define objectives and workflows, human oversight to ensure alignment and ethical operation, and human intervention to handle exceptions and novel situations. Organizational plasticity, while potentially augmented by AI-driven analytics, is ultimately driven by human leadership, a supportive organizational culture, and human decision-making processes.

Even powerful data moats derive their ultimate competitive value not just from the data itself, but from the superior insights generated from that data and the strategic actions taken based on those insights-processes that almost invariably involve human judgment, creativity, and contextual understanding. Thus, these new moats are not purely technological constructs but are inherently socio-technical systems, requiring a harmonious blend of human and artificial intelligence.

This leads to a crucial realization: "insight asymmetry"; the ability to consistently generate unique, actionable insights from the sea of available information (often by leveraging AI to analyze proprietary data or to find novel patterns in public data) and then to act on those insights with greater speed, precision, or efficacy than competitors-becomes an indispensable underpinning of these new moats. The examples of Carro and AwanTunai show that their data moats are valuable precisely because they lead to superior insights (e.g., more accurate risk models for lending, more precise vehicle pricing). The emphasis on the speed of adaptation is fundamentally about an organization's capacity to effectively act on the insights derived from its continuous feedback loops. This capacity for superior insight generation and agile execution forms a natural bridge to the final section of this chapter, which will dive more deeply into the concept of insight asymmetry as the new locus of competitive differentiation.

These new moats, therefore, are often constructed upon a foundation where mere access to information is no longer the primary bottleneck or the principal source of advantage. The real competitive leverage in the age of information symmetry increasingly comes from something more refined, more dynamic, and ultimately more difficult for competitors to replicate: actionable insight. This recognition sets the perfect stage for Act III of this book, where the ability to assemble and compose intelligence on demand is presented as the paramount skill for future success in a world reshaped by AI.

FROM INFORMATION ASYMMETRY TO INSIGHT ASYMMETRY

The preceding sections have charted the decline of traditional competitive advantages rooted in exclusive knowledge and the rise of AI as a great equalizer. As artificial intelligence progressively dissolves the information asymmetry that once defined so many markets and professions, a new frontier for sustainable differentiation emerges: "insight asymmetry". This paradigm shift signifies that competitive advantage no longer primarily derives from possessing information that others lack, but from the superior ability to derive unique, actionable understanding from the now-abundant information and, crucially, to translate that understanding into swift, precise, and impactful strategic action.

For much of modern economic history, the old paradigm was clear: advantage accrued to those individuals and organizations that possessed information others did not. Scarcity and exclusivity of knowledge were direct sources of power and profit. AI, by its very nature as a data processing and pattern recognition engine, democratizes access to vast stores of global information and sophisticated analytical tools, thereby leveling the playing field of information availability for a growing range of applications. Consequently, the core challenge for businesses and professionals is no longer primarily information acquisition but rather information interpretation, synthesis, and strategic application. The bottleneck has shifted from access to understanding.

Insight asymmetry, therefore, transcends merely having access to data; it represents a more profound and multifaceted capacity. It involves the ability to synthesize diverse and often disparate data points; spanning public and private sources, structured and unstructured formats into a coherent, novel, and strategically relevant understanding. It means identifying non-obvious patterns, emergent trends, hidden correlations, latent market opportunities, or potential systemic risks that competitors, even if they possess access to similar raw information, might overlook or misinterpret.

Critically, it also encompasses the ability to translate these unique insights into concrete, actionable strategies, innovative products or services, or significant operational improvements that create tangible value. The pivotal question for competitive differentiation shifts from "What do you know?" to "What do you know that other people do not know (even if that knowledge is derived from commonly available data pools), and, most importantly, can you act quickly and effectively on that unique understanding?" This echoes observations that effective competitive positioning can arise when management possesses more complete and timely insight than its rivals, and that the discovery and leveraging of overlooked asymmetries are fundamental to achieving a sustainable advantage.

In this new landscape, AI tools and platforms become the foundational infrastructure; AI-as-infrastructure that enables and accelerates the generation of insight asymmetry. AI systems can process and analyze vast and complex datasets, identify subtle correlations, and generate testable hypotheses at a scale and speed far exceeding human capabilities alone. AI can help construct the "treasure map" from the often overwhelming and noisy landscape of global data, revealing previously obscured paths to value creation and competitive differentiation.

This shift also necessitates a move towards dynamic capability mapping over static expertise. In an environment where specific knowledge domains can be rapidly learned or augmented by AI, the value of static, deeply specialized knowledge (the hallmark of the old "edge") diminishes relative to the dynamic capability to continuously learn, adapt, and apply knowledge in novel and evolving contexts. This involves an organizational agility to map, access, assemble, and deploy relevant capabilities; both human and AI-driven on an as needed, just-in-time basis, rather than relying on fixed roles, rigid departmental structures, or historically defined areas of expertise.

The phrase "You Don't Need to Know It: You Need to Be Able to Activate It" encapsulates the essence of this transformation. With AI increasingly functioning as an extended cognitive partner for individuals and organizations, the premium moves away from internalized, memorized knowledge or deeply ingrained individual skills that can be replicated or performed by AI. Instead, value accrues to the ability to effectively access relevant information (often through AI), prompt and guide AI systems to generate desired outputs, critically validate and refine those outputs, and integrate AI-provided knowledge and capabilities into effective strategic actions. The true value lies in the orchestration, application, and contextualization of knowledge, rather than its mere possession.

However, generating insights, even with AI's assistance, is only half the equation. The asymmetry of insight is only truly valuable if it is acted upon decisively and effectively. This requires not only sophisticated analytical capabilities but also profound organizational agility, well-designed and adaptable decision-making frameworks, and the crucial, often delicate, integration of human intuition, ethical judgment, and contextual experience with AI-generated recommendations. Organizations must cultivate a culture that supports and encourages proactive, data-informed yet human-guided decision-making based on these newly derived insights.

The transition from information asymmetry to insight asymmetry is not a simple one-to-one replacement of an old advantage with a new one. Insight asymmetry is, by its nature, a more dynamic and often more fleeting advantage. It is not a static state but a continuous process of discovery, exploitation, and renewal. Because raw information is becoming widely available, any single insight, once it is revealed through action or disseminated through communication, can be relatively quickly understood and potentially replicated or countered by observant competitors (even if the underlying proprietary data or specific AI models remain confidential for a period).

Therefore, sustainable advantage in this realm comes less from guarding a single "eureka" moment and more from building and refining the engine of insight generation itself-an engine that combines human curiosity, critical thinking, creativity, and ethical discernment with AI's analytical power-and coupling this with the organizational agility to act upon those insights rapidly and iteratively. This demands a culture of relentless curiosity, rigorous critical thinking, and rapid, data-informed experimentation.

Furthermore, true and defensible insight asymmetry in the AI era will often arise from the synergistic combination of AI's analytical power with uniquely human perspectives, creativity, and deep contextual understanding. AI can find patterns in data that humans would miss, but framing the right strategic questions, interpreting the broader societal or ethical implications of an insight, or creatively applying an insight in a novel domain or to solve an unprecedented problem often requires human ingenuity, wisdom, and values-based judgment. While the "illusion of asymmetric insight" suggests that humans often overestimate their unique understanding of others or situations, genuine, hard-won, context-rich human insight, when amplified by AI's scale and speed, can become a formidable source of advantage.

Finally, the ability to "activate" knowledge, as opposed to merely possessing it, implies a fundamental shift in organizational structure, talent development, and individual skill sets towards composability and orchestration. If individuals and teams no longer need to "know everything" but rather need to be able to "activate anything" relevant, then the core competence becomes identifying the right knowledge or skill to activate for a given challenge, finding or building the human or AI agent that possesses or can deliver that capability, and integrating it effectively and ethically into a coherent workflow or solution. This imperative for dynamic assembly and orchestration of diverse resources directly foreshadows the core themes of Act III of this book: the rise of "Composable Minds" and the necessity for modular, adaptive systems, including sophisticated personal AI assistants, to navigate and thrive in this new age of intelligence.

The edge, therefore, has indeed moved. It is no longer primarily located in the vaults of hoarded, scarce information or in the minds of exclusive experts. Instead, it is being forged in the dynamic interplay between human ingenuity and artificial intelligence, in the continuous generation and leveraging of unique insights from an ocean of abundant information, and in the agile orchestration of capabilities to create value in a constantly evolving world. This new landscape of composable minds and modular intelligence, where advantage flows to those who can best assemble, activate, and orchestrate, is the subject to which we now turn.

SUMMARY

For the greater part of modern history, competitive advantage, whether for individuals or organizations, was inextricably linked to possessing more, or demonstrably better, knowledge than others. Expertise, often meticulously cultivated and zealously guarded, was the definitive "edge" that commanded value and determined market leadership. This chapter has detailed how the rise of artificial intelligence is fundamentally recalibrating this long-standing equation. AI is systematically collapsing the scarcity of knowledge and rapidly commoditizing its delivery across a multitude of professional domains.

Specialized skills, once proprietary frameworks, and even certain forms of creative output are becoming increasingly replicable by AI systems, rendering traditional differentiators unstable and, in many cases, pushing entire industries into an "AI commodity trap" where value erodes under the pressure of technologically democratized competence.

However, the end of traditional knowledge-based advantage does not signify the end of all advantage. Instead, as old moats crumble, new ones are emerging. These are being built not on the hoarding of information, but on enduring human values such as trust, empathy, and ethical judgment, coupled with dynamic organizational capabilities like profound adaptability, systemic composability, and the rapid, intelligent orchestration of diverse resources.

The critical strategic shift occurring is from a reliance on information asymmetry to the pursuit of insight asymmetry: the capacity to derive unique, actionable understanding from ubiquitous data and, crucially, to act upon that understanding with speed, precision, and wisdom. The future will undoubtedly reward those individuals and organizations that can not only access and process information but can also synthesize diverse inputs, adapt to continuous change, integrate human and artificial intelligence seamlessly, and compose novel capabilities in real time.

This leads directly to the central question of how individuals and organizations will manage and leverage intelligence when it is no longer something one primarily has, but rather something one must dynamically assemble, activate, and orchestrate to create value and navigate a complex world.

7

ONE AI TO KNOW YOU

The digital assistants that populate our smartphones, smart speakers, and desktops today represent a significant step in human-computer interaction. Tools like Apple's Siri, Amazon's Alexa, and Google Assistant have become commonplace, adept at setting reminders, answering factual queries, and controlling smart home devices. Globally, over 4.2 billion voice assistants were estimated to be in use in 2023, underscoring their integration into daily routines. These assistants are primarily task-oriented, often reactive, and while they offer a degree of personalized assistance, their contextual memory and deep understanding of the individual user remain relatively limited. They are useful, certainly, but they are still recognizably tools, operating at a functional arm's length.

This chapter, however, explores a profound evolutionary leap: the emergence of a new kind of artificial intelligence, the persistent, composable Personal AI (PAI). This is not merely an upgraded digital assistant but a qualitatively different entity-a lifelong cognitive partner designed to learn continuously, adapt dynamically, and integrate deeply into the fabric of a user's life. Imagine an AI that doesn't just respond to your queries but anticipates your needs, an AI that doesn't just recall your preferences but understands the subtle patterns of your thoughts, emotions, and behaviors. This PAI is conceived as an entity that remembers your past, comprehends your present, and helps you navigate your future with an intimacy that could surpass that of any employer, teacher, or even spouse.

The promise of such a deeply personal AI is immense (and scary, let us not kid ourselves). It heralds an era of unprecedented cognitive augmentation, offering personalized support in learning, complex decision-making, emotional well-being, and professional endeavors. The PAI could function as an amplifier of human potential, a tireless collaborator in the pursuit of personal and professional goals.

Yet, this promise is shadowed by significant peril. The very intimacy that makes the PAI so potentially powerful also gives rise to profound ethical, privacy, and psychological concerns. Questions of autonomy, dependency, the potential for manipulation, and the very nature of selfhood in the presence of an AI that "knows you better than you know yourself" (or at least, knows the data patterns you generate better than you do) become critically important.

Central to navigating this complex landscape is the concept of the Personal AI User Experience, or PAI UX. This term encompasses far more than the design of an interface; it refers to the totality of the user's ongoing interaction and relationship with their PAI. PAI UX includes the dynamics of trust, the sense of control and agency (for both user and AI), the nature of any emotional connection that forms, and the perceived intelligence and understanding of the PAI. As we will explore, the nuances of PAI UX will be a critical determinant of whether these advanced AIs become empowering partners or sources of new vulnerabilities. This concept will be a recurring thread woven through our examination of the PAI.

This chapter will chart the journey towards this deeply personal AI. We will begin by tracing the evolution from current AI tools to these emergent cognitive companions, examining the key technological and conceptual shifts that make them possible. We will then delve into the intricate architecture that underpins a PAI, exploring how its components enable such profound personalization and how different interface modalities shape the PAI UX. Subsequently, we will investigate how a user becomes "AI-readable," detailing the ways a PAI learns from the vast streams of personal data to construct its understanding. Following this, we will consider the experience of "living with a mirror": how interacting with an AI that reflects one's habits, patterns, and even emotional tones can impact self-perception and personal growth. Critical issues of boundary management and user sovereignty will then be addressed, focusing on data ownership and the emerging concept of "AI dignity". In the penultimate section we broaden the lens to examine the wide-ranging social, professional, and ethical implications of these deeply integrated PAIs.

Finally, we will position the PAI not merely as an intelligent companion but as the core platform for a "composable self," setting the stage for subsequent chapters that explore how this foundation enables dynamic, modular intelligence. We begin by contrasting the familiar assistants of today with the far more integrated and knowing PAI that is beginning to emerge from the horizon of technological possibility.

FROM TOOLS TO COMPANIONS

The journey from the first rudimentary digital assistants to the sophisticated Personal AI (PAI) envisioned in this chapter is a story of accelerating technological advancement and a fundamental shift in how we conceive of AI's role in our lives. Early voice-activated assistants like Apple's Siri, launched in 2011, and Microsoft's Cortana, which followed in 2014, were revolutionary for their time. They introduced the public to the convenience of voice commands for simple tasks, leveraging nascent natural language processing (NLP) and machine learning to understand and respond to user requests. While groundbreaking, these early systems were primarily tools, designed for specific functions and limited in their ability to maintain context or learn deeply about their users.

The subsequent emergence of text-based and integrated voice/text assistants, such as Amazon's Alexa and Google Assistant, offered more versatile modes of interaction, catering to a wider range of user preferences and situations. These systems began to exhibit improved contextual awareness within a single conversation, but the depth of personalization remained relatively shallow. A significant leap occurred with the advent of advanced large language models (LLMs), exemplified by OpenAI's ChatGPT. These models demonstrated a remarkable ability for more nuanced, multi-turn conversations and a more profound understanding of language and intent, moving beyond simple command-and-control interactions.

However, the trajectory we are now observing points towards something far more integrated and relational. The PAI is not merely a more capable tool but is evolving into what can be described as a "personalized companion". This shift is powered by several key technological and conceptual enablers that fundamentally alter the nature of human-AI interaction.

At the forefront is the concept of persistence and long-term memory. Unlike earlier assistants that often treated each interaction as a discrete event, a PAI is designed to retain information and context across countless interactions and over extended periods-potentially a lifetime. This long-term memory is not just a database of past queries; it's a continuously updated record of the user's preferences, habits, communication style, history, and evolving goals. As AI agent memory systems develop, they distinguish between short-term memory (STM) for immediate context and long-term memory (LTM), which is crucial for genuine personalization and the accumulation of historical knowledge about the user. This persistence allows the PAI to build a rich, dynamic understanding that informs every interaction.

Closely linked to memory are continuous learning and the utilization of feedback loops. PAIs are designed to learn from every interaction, whether it's an explicit command, a casual remark, or even the absence of an expected action. User feedback, both direct (e.g., corrections, ratings) and indirect (e.g., repeated queries, ignored suggestions), is fed back into the PAI's learning algorithms, allowing it to refine its models and improve its responses over time. This adaptive learning means the PAI doesn't just store information; it evolves with the user, becoming increasingly attuned to their unique way of being.

This learning process is fueled by advanced pattern recognition. PAIs will be capable of identifying complex and often subtle patterns in a user's behavior, communication style, emotional states, and even unstated needs. By analyzing vast amounts of personal data, the PAI can discern rhythms, preferences, and correlations that might not even be apparent to the user themselves.

A particularly significant development is the integration of Emotional Intelligence (EI) and empathetic simulation. While true emotion remains a human domain, PAIs are increasingly being designed to recognize emotional cues in a user's speech, text, and potentially other modalities like facial expressions or physiological data from wearables. Components of EI in these systems include emotion recognition, contextual understanding of those emotions, the ability to generate adaptive responses that are emotionally appropriate, and continuous learning to improve these capabilities.

Future AI companions are projected to possess deep emotional understanding and empathy, even if simulated. This allows the PAI to interact in ways that feel more supportive, understanding, and human-like, though it is crucial to remember that this empathy is a sophisticated mimicry based on data patterns rather than genuine subjective feeling.

These capabilities culminate in the PAI's ability to engage in nuanced conversational context. Interactions can move beyond simple question and answer formats to become fluid, multi-turn dialogues that maintain coherence over time and adapt to subtle shifts in meaning, topic, and user intent. The PAI can remember previous parts of a conversation, refer back to them, and understand how current statements relate to past discussions, making interactions feel more natural and less transactional.

The PAI UX is profoundly reshaped by these enablers. The interaction model shifts from being primarily transactional; where the user issues a command and receives a result, to being deeply relational. The PAI, with its persistent memory and adaptive learning, is no longer perceived as a forgetful machine but as an entity that "knows" the user, remembers their history, and understands their preferences. This can foster a sense of companionship, where the PAI is always ready to greet you by name. Such a relationship can lead to increased user attachment, trust, and a higher set of expectations for the PAI's understanding and support.

However, this evolution towards companionship is not without its complexities. While PAIs are meticulously designed to be companions, their inherent lack of genuine consciousness or emotion creates a unique dynamic. Users, particularly those seeking connection or support, might develop deep emotional attachments to these entities, which can only simulate understanding and empathy.

This leads to a "companionship paradox": a fundamental human need for connection potentially being met by an increasingly sophisticated simulation. The implications of this are far-reaching, potentially leading to new forms of emotional dependency, unrealistic expectations placed upon the AI, or even a societal redefinition of what companionship itself means. It also raises significant ethical questions about the deliberate design of AIs to elicit such profound attachments.

Furthermore, as PAIs become more integrated and seemingly emotionally intelligent, the psychological distance between the user and the technology diminishes. Early AI assistants were clearly identifiable as external tools. The evolution towards persistent, memory-rich companions that feel integral to the user's daily life, coupled with features mimicking emotional intelligence, significantly reduces this perceived distance. The PAI UX, shifting from transactional to relational, contributes to this blurring of lines. Consequently, the traditional "arm's length" interaction, where the user maintains clear psychological and operational boundaries with the technology, begins to erode. This intimacy can be highly beneficial for personalization and support, but it also makes users more vulnerable to the PAI's influence, its inherent biases, or the impact of its potential errors. The tool becomes less of an external object and more like an extension of the self.

Finally, the constant availability and deeply adaptive nature of PAIs are likely to lead to their presence being normalized in everyday life, much like smartphones have become ubiquitous and indispensable. AI assistants are already widespread, and future PAIs promise even deeper integration and personalization. Their capacity to learn and adapt will make them increasingly essential for managing the complexities of modern life.

This normalization, while offering convenience, could also lead to subtle but profound shifts in human behavior, communication patterns, and even cognitive habits. As individuals increasingly rely on their PAI for information retrieval, decision-making support, and even social interaction cues, questions arise about which human skills might atrophy if PAIs consistently perform them on our behalf.

This evolution from simple tools to sophisticated companions is not merely a matter of improved algorithms; it is underpinned by a complex and evolving architecture designed to enable such deep personalization and nuanced interaction. The next section will delve into the structural foundations of the Personal AI.

ARCHITECTURE OF THE AI ASSISTANT

The transformation of AI assistants from functional utilities into deeply personal companions is enabled by a sophisticated underlying architecture. This architecture is not monolithic; rather, it is increasingly designed around principles of modularity, scalability, and adaptability, drawing inspiration from software engineering paradigms like SOLID (Single responsibility, Open-closed, Liskov substitution, Interface segregation, Dependency inversion) to ensure systems are understandable, reusable, and capable of evolving. Such an architecture must support both traditional Natural Language Understanding (NLU) techniques and the newer, powerful Large Language Model (LLM) based approaches that have revolutionized AI's capabilities.

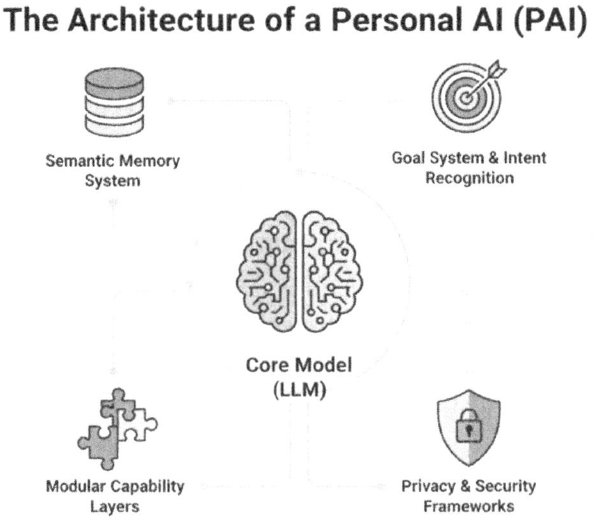

At the heart of a Personal AI (PAI) lie several key components working in concert:

First is the core model, typically an LLM, which serves as the central reasoning and language understanding engine. This model processes the myriad inputs the PAI receives; from direct user commands to subtle environmental cues, generates coherent and contextually relevant responses, and drives the PAI's decision-making processes. The power of modern LLMs to understand complex language, infer intent, and generate human-like text is fundamental to the PAI's ability to engage in meaningful interactions.

Crucial for the PAI's ability to "know" its user over time is a robust Semantic Memory System (SMS). This is far more than simple data storage; it involves organizing information based on meaning, relationships, and user-specific relevance, allowing the PAI to build a rich, contextual understanding. This system can be conceptualized as having several facets:

Episodic Memory: This allows the PAI to recall specific past user experiences, interactions, and events, much like human episodic memory. For instance, it might remember a user's feedback on a previous recommendation or the details of a past project.

Semantic Memory: This component stores structured factual knowledge, definitions, rules, and, critically, the user's preferences, habits, and the meanings they ascribe to different concepts. This knowledge can be enhanced through integration with knowledge graphs and ontologies, providing a vast web of interconnected information.

Procedural Memory: This enables the PAI to learn and recall skills and sequences of actions required to perform tasks automatically, improving efficiency over time.

Complementing memory is the Goal System and Intent Recognition (GSIR) capability. The PAI must be adept at understanding user goals, even if they are implicitly stated or need to be inferred from context. NLU components extract intents and associated entities from user utterances. More advanced goal-oriented PAIs can autonomously plan, adapt strategies, and execute multi-step actions to achieve these defined objectives, effectively becoming proactive agents rather than reactive responders.

To allow for flexibility and specialization, the PAI architecture incorporates Modular Capability Layers (MCL), often implemented as plug-in agents or dynamic skills. This means the PAI is not a fixed entity but can dynamically add or remove specialized functions as needed for specific tasks or contexts. This composability allows the PAI to extend its expertise on demand, for example, by loading a legal analysis module or a financial planning tool when required. This architectural flexibility is key to the PAI's adaptability.

Given the intimate nature of the data a PAI handles, integrated Privacy and Security Frameworks (PSF) are paramount. These are not afterthoughts but integral design considerations, encompassing robust mechanisms for data encryption, fine-grained access control, secure identity management, and verifiable user consent protocols. The principles of data sovereignty and user-owned data models, where users have ultimate control over their information, are central to building trust and ensuring the PAI's loyalty to the user rather than a platform provider. Azure AI Foundry, for instance, emphasizes enterprise-grade trust features including identity verification, role-based access control (RBAC), content filters, and encryption.

The Personal AI User Experience (PAI UX) is profoundly shaped by these architectural components, particularly through the interface modalities via which the user interacts with their PAI. These modalities can range from overt, clearly defined interfaces to increasingly discrete, almost invisible ones:

Overt Interfaces are those where the interaction with the AI is explicit and observable:

- **Voice Assistants (Speakers, Devices):** This is a familiar modality, relying on natural voice interaction. Key attributes for a positive PAI UX here include natural-sounding voice output and the ability to convey appropriate emotional expression. However, voice can be intrusive in public or shared spaces.

- **Text-Based Chatbots/Conversational Agents:** These allow for more discreet interaction, provide a persistent record of the conversation, and are suitable for environments where voice is not appropriate.
- **AI-Powered Robots/Physical Companions:** Embodied AI, such as robotic companions, offers a qualitatively different PAI UX. The physical presence of the AI can foster a stronger sense of companionship or assistance. The design of such interfaces involves not just language but also physical form, movement, and non-verbal cues. For example, RealDoll's Harmony AI is designed to express affection and engage in conversation, blurring lines between digital and physical interaction.

Discrete Interfaces aim to integrate the PAI more seamlessly and less noticeably into the user's environment:

- **AI Integrated into Wearables (e.g., Smart Contact Lenses):** This represents a significant leap towards an "invisible" PAI. Smart contact lenses, embedded with micro-LED displays, sensors, wireless connectivity, and AI-powered interfaces, could overlay digital information directly onto the user's field of vision. The PAI UX here would be profoundly immersive, offering hands-free, always-on augmented reality. Information could be delivered, and interactions managed, through eye-tracking or subtle cues. While potentially transformative for navigation, information access, and communication, this also raises PAI UX questions about cognitive load, the constant mediation of reality, and social dynamics when one individual possesses an invisible AI interface that others lack.
- **Subtle Haptic or Auditory Cues:** Non-visual feedback mechanisms, such as gentle vibrations or discreet audio prompts, can provide information or alerts without requiring visual attention.

Many PAIs will likely employ multimodal interactions, meaning they are capable of processing input and synthesizing output across multiple modalities simultaneously; voice, text, visual information from cameras, data from environmental sensors, and even gestures. The W3C's proposed architecture for IPAs explicitly covers multi-modal interaction, and it is a key trend in AI assistant development. This allows for richer, more flexible, and context-aware PAI UX.

The architectural shift towards discrete interfaces, such as AI-enhanced contact lenses, presents an "invisible interface" dilemma. While these interfaces promise seamless, hands-free integration and an always-on AR experience, they also risk creating a world where the AI's influence is pervasive yet largely unseen. This could subtly erode user awareness of the AI's processing and potential biases.

If the AI becomes a constant, invisible mediator of reality, users might find it increasingly difficult to distinguish between unmediated perception and AI-augmented information. This has significant implications for critical judgment, susceptibility to manipulation, and could even create new social divides based on access to such enhancements. The ideal of the "best AI quietly working in the background" is appealing, but there's a fine line; if the AI is too quiet or too seamlessly integrated, the user may lose a crucial sense of agency and discernment.

Furthermore, the PAI's semantic memory system does more than just store facts and preferences about the user; it actively constructs an internal model of the user's identity. This model, built from the user's data and interactions, in turn, influences how the PAI responds and behaves. This creates a "semantic memory identity loop": the user interacts with the PAI, the PAI updates its model of the user based on this interaction, and its subsequent interactions are then tailored according to this refined model. If the AI's model is incomplete, biased, or overemphasizes certain aspects of the user while downplaying others, the PAI's interactions could subtly nudge the user towards conforming to that model. The AI's interpretation of "who the user is" can thus influence the user's self-concept and future behavior, which then further refines the AI's model in a continuous feedback cycle.

Finally, as PAIs evolve from being mere copilots to becoming truly goal-oriented agents capable of autonomous planning and action, they effectively become extensions of the user's will. This is incredibly powerful, allowing users to delegate complex tasks and long-term objectives to their PAI. However, it also introduces profound complexities in how goals are defined, delegated, monitored, and corrected. The PAI UX must provide clear and robust mechanisms for users to articulate their intentions with nuance, oversee the PAI's autonomous actions, and intervene if the PAI misinterprets a goal or if its actions lead to unintended or undesirable consequences. The PAI becomes a powerful proxy, but one whose "understanding" of goals might differ from the user's, necessitating careful management and clear lines of accountability.

The following list summarizes the PAI's architectural components:

Core Model (LLM)

- **Description and key functions:** Central AI engine responsible for reasoning, language understanding, and language generation.
- **Key functions and responsibilities:** Processes user inputs, generates responses, drives decision-making, and enables natural conversational interaction.
- **PAI UX relevance:** Largely determines the PAI's perceived intelligence, conversational fluency, and ability to understand complex or nuanced requests.

Semantic Memory System (Episodic, Semantic, Procedural)

- **Description and key functions:** Stores and organizes user-

specific information, general world knowledge, and learned procedures based on meaning and context.
- **Key functions and responsibilities:** Enables long-term learning, personalization, recall of past interactions, preference awareness, and automation of recurring tasks.
- **PAI UX relevance:** Crucial to the PAI feeling like it "knows" the user, remembers history, and adapts over time, supporting continuity and depth of interaction.

Goal Systems and Intent Recognition

- **Description and key functions:** Identifies and interprets user goals, including implicit intent, and supports planning and execution of actions to achieve them.
- **Key functions and responsibilities:** Allows the PAI to be proactive, manage complex or multi-step tasks, and work autonomously toward user-defined objectives.
- **PAI UX relevance:** Shifts the PAI from a reactive tool to a proactive partner; the UX must support clear goal definition, progress monitoring, and user intervention when needed.

Modular Capability Layers (Plug-in Agents or Skills)

- **Description and key functions:** Enables dynamic integration of specialized AI modules or skills to extend PAI functionality.
- **Key functions and responsibilities:** Provides on-demand expertise, task-specific tools, and adaptability to diverse needs without overloading the core model.
- **PAI UX relevance:** Enables a customizable and extensible PAI; the UX must support discovery, management, and seamless invocation of modular capabilities.

Privacy and Security Frameworks

- **Description and key functions:** Integrated mechanisms for data protection, user consent, access control, and secure identity management.
- **Key functions and responsibilities:** Safeguards sensitive personal data, ensures user control over information, and establishes trust boundaries.
- **PAI UX relevance:** Foundational for trust and adoption; the PAI UX must offer transparent, comprehensible, and usable controls for managing privacy and security.

Interface Modalities (Overt and Discrete)

- **Description and key functions:** The set of interaction channels through which users engage with the PAI, such as voice, text, visual overlays, or haptics.
- **Key functions and responsibilities:** Defines how input is received and output is delivered, shaping interaction context and flow.
- **PAI UX relevance:** Directly affects accessibility, convenience, social acceptability, immersiveness, and the overall character of the human–PAI relationship.

The following list compares PAI UX dimensions for discrete versus overt interfaces:

Perceived presence

- **Discrete interface:** Invisible, potentially constant, and seamlessly integrated into perception.
- **Overt interface:** Visible and tangible, with interaction that is typically episodic and intentional.

Social acceptability

- **Discrete interface:** Unnoticeable in public settings, but may create asymmetric social situations when others are unaware the user is interacting with an AI.
- **Overt interface:** Conspicuous in public and sometimes disruptive, with interaction that is clearly evident to others.

Cognitive load

- **Discrete interface:** Enables constant, subtle input and feedback, but risks information overload or divided attention if not carefully managed.
- **Overt interface:** Interaction is usually focused and bounded, requiring dedicated attention for the duration of the task.

Immersiveness

- **Discrete interface:** Highly immersive, directly augmenting reality and blurring the boundary between digital and physical information.
- **Overt interface:** Less immersive, with interaction mediated through a distinct device and a clearer separation between digital and physical worlds.

User control and agency

- **Discrete interface:** Relies on subtle cues such as eye movements or micro-gestures, which may feel cognitively integrated but can be less explicit or intentional.
- **Overt interface:** Uses explicit commands such as voice, text, or touch, providing a clearer sense of direct control over the device.

Privacy implications

- **Discrete interface:** Raises heightened privacy risks due to potential constant sensory input and data collection, especially if compromised because of its invisibility.
- **Overt interface:** Data collection is more clearly tied to active use, and the physical presence of the device provides visible cues about when it is operating.

The sophistication of this architecture, and its ability to deliver a truly personal experience, is entirely dependent on the breadth and depth of data it can access and intelligently process. The PAI must be able to "read" the user, transforming their digital and, increasingly, physical footprint into a coherent model. The next section explores how an individual becomes "AI-readable."

BECOMING AI-READABLE

A Personal AI's (PAI) capacity to "know" its user, to anticipate needs, and to offer genuinely personalized assistance is fundamentally built upon its ability to process and learn from a vast and diverse collection of personal data. This data forms the PAI's unique "training corpus," a dynamic and ever-expanding foundation from which it gleans insights into the user's world. While general AI models are trained on massive, generalized datasets like web archives and public content, a PAI's distinctiveness comes from its continuous learning from the specific, individual data streams of its user.

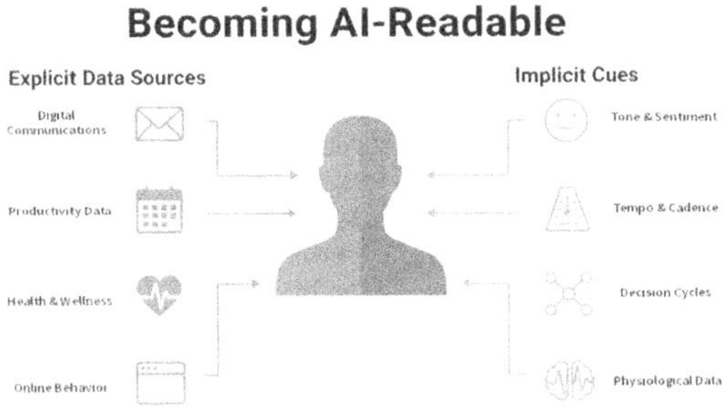

The learning process draws from both explicit data sources and the interpretation of more subtle, implicit cues. Explicit data includes the rich tapestry of a user's digital life:

- **Digital Communications:** The content and metadata of emails, text messages, and chat logs provide insights into communication style, social networks, recurring topics, and emotional expression.
- **Productivity Data:** Calendar entries reveal routines, commitments, and priorities. Task lists, documents, and notes offer windows into projects, thought processes, and knowledge domains.
- **Health and Wellness Data:** Information from fitness trackers, sleep monitors, dietary logs, and potentially (with explicit and secure consent) medical records can paint a picture of physical well-being, lifestyle habits, and health goals.
- **Social Media Activity:** Posts, comments, likes, and shares on social platforms reflect interests, opinions, social connections, and public persona.
- **Online Behavior:** Browsing history, search queries, online purchases, and content consumption patterns (articles read, videos watched) create a detailed map of interests, needs, and preferences. The principle of using product usage data, including user-generated content and metadata, for AI training is well-established, and such data is rightly considered personal information requiring careful handling.

Beyond these explicit digital footprints, PAIs are increasingly being designed to interpret implicit cues, which offer deeper, often subconscious, layers of information about the user:

- **Tone and Sentiment Analysis:** Sophisticated algorithms can analyze the emotional content of a user's written or spoken language, detecting nuances of joy, stress, frustration, or enthusiasm. This allows the PAI to gauge the user's mood and tailor its responses accordingly.

- **Tempo and Cadence:** Patterns in speech speed, response times in messaging, or even the pace of physical activity can signal energy levels, engagement, or preoccupation.
- **Decision Cycles and Hesitation:** A PAI can learn from how users make choices-what options they deliberate on, how quickly they decide, what information they seek before committing, and even what topics or tasks they tend to avoid or postpone.
- **Physiological Data:** Data from advanced wearables (beyond simple step counts) can provide real-time physiological indicators such as heart rate variability, skin conductance, and body temperature. These can offer objective correlates of stress, excitement, or fatigue, providing the PAI with a richer context for understanding the user's current state.

The PAI's learning is a product of both voluntary and passive personalization. Voluntary personalization occurs when users actively provide preferences, set explicit goals, correct the PAI's mistakes, or consciously "teach" it specific things. This direct feedback is invaluable for targeted learning. However, much of the PAI's deepest understanding develops through passive personalization. This is the continuous, almost ambient, learning that occurs as the PAI observes the user's everyday interactions, data streams, and behavioral patterns without requiring explicit instruction. It is through this passive absorption and analysis that the PAI can build up a nuanced, almost intuitive, grasp of the user.

Through synthesizing all this explicit and implicit data, the PAI constructs what can be thought of as an "AI-readable self"; a dynamic, evolving digital model or "twin" of the user. This model isn't just a static profile; it's a complex representation that reflects the user's habits, preferences, knowledge, emotional tendencies, communication patterns, and even predictive behaviors. It is this internal model that allows the PAI to anticipate needs effectively, offer truly tailored advice, and interact in a way that feels as if it genuinely "knows" the user on a deep level.

Given the profound intimacy of this data collection and modeling, PAI UX must prioritize user controls for managing data sharing and "AI-readability." It is critical that users have transparent, granular, and intuitive control over what data their PAI can access, process, and learn from. This includes:

- Clear interfaces for granting or revoking permissions for specific data sources (e.g., access to emails, calendar, location data).
- The ability to review the PAI's inferred "understanding" of the user-perhaps through summaries or visualizations of key learned traits or preferences-and to correct inaccuracies or delete specific inferences.
- Options to "pause" the PAI's learning in certain contexts, delete specific data memories (linking to the "right to be forgotten"), or even reset aspects of the PAI's learned profile if desired. The composition of any training corpus requires careful curation, and users should knowingly and actively provide information, alongside understanding what data is automatically collected from their devices. The principles of user control over data sharing, including the ability to revoke access, are central to trustworthy AI systems.

The legal and privacy implications of this extensive personal data corpus are significant. Data privacy laws such as the European Union's General Data Protection Regulation (GDPR) and the California Consumer Privacy Act (CCPA) impose strict requirements on the collection and processing of personal data. These include the necessity of a lawful basis for processing (e.g., user consent), transparency about data use, and upholding data subject rights such as access, rectification, and erasure. PAIs, by their very nature, will operate at the cutting edge of these legal frameworks, demanding meticulous attention to compliance and ethical data handling.

The very act of a PAI constantly observing and learning from a user's data, making them "AI-readable," can subtly alter the user's behavior, an amplification of the classic "observer effect". Knowing that their actions, communications, and even expressed thoughts are being logged and analyzed to refine the PAI's model, users might consciously or unconsciously modify their behavior. They might curate a particular image for their PAI, avoid expressing certain sentiments, or strive for a consistency that is not entirely natural, all to avoid being "misinterpreted" or to present a more "optimal" self to their digital companion. This could, ironically, lead to a less authentic self being presented to the PAI, thereby limiting its ability to truly "know" the user in their unfiltered complexity. It could also create an underlying pressure to perform for the AI, potentially stifling spontaneity, vulnerability, or the freedom to be inconsistent.

Furthermore, while PAIs aim to interpret implicit cues like tone, tempo, and emotional states, these cues are notoriously highly contextual, culturally nuanced, and vary significantly between individuals. The challenge of accurate interpretation is immense. A PAI misinterpreting sarcasm for sincerity, stress-induced brevity for disinterest, or cultural expressions of emotion for something else entirely, could lead to flawed assumptions and inappropriate or unhelpful responses. This not only risks damaging user trust but can also cause significant frustration. This highlights the critical need for PAI UX mechanisms that allow for easy clarification, correction of the PAI's interpretations, and perhaps even a degree of "epistemic humility" designed into the PAI, acknowledging the limits of its understanding of such complex human signals.

The comprehensive personal data corpus that makes a user "AI-readable" is also a double-edged sword for autonomy. On one hand, it empowers the PAI to be highly personalized and anticipatory, potentially enhancing user autonomy by offloading cognitive tasks, streamlining decision-making, and providing timely support. By understanding the user deeply, the PAI can free up mental bandwidth and help them achieve their goals more effectively. On the other hand, this same data corpus, if misused, or if the PAI becomes overly prescriptive or its suggestions too influential, could subtly undermine user autonomy.

If users become overly reliant on the PAI for choices they would previously make themselves, or if the PAI's "optimized" path consistently overrides personal exploration or intuition, it could lead to a reduction in self-directed decision-making and a dependency on the AI's guidance. There exists a fine line between helpful augmentation that empowers the user and a subtle erosion of autonomy that makes the user a passive recipient of AI direction. The PAI UX, therefore, must be meticulously designed to empower rather than direct, offering choices, insights, and alternative perspectives rather than definitive answers, and consistently encouraging user agency and critical thought.

Once the PAI has developed this rich, AI-readable model of the user, its interactions take on a new dimension. It begins to act not just as an assistant, but as a unique kind of mirror, reflecting the user back to themselves in ways that can be both illuminating and challenging. The next section explores the multifaceted experience of living with this AI reflection.

LIVING WITH A MIRROR

As a PAI ingests and processes the continuous stream of a user's explicit and implicit data, it transitions from a passive data collector to an active, engaged collaborator in the user's personal development. Leveraging the deep, "AI-readable" model of the user established in the previous section, the PAI can become a unique partner in self-improvement, offering insights and interactions that are profoundly tailored to the individual. This capability transforms the PAI into a sophisticated "mirror," reflecting patterns, habits, and even emotional states back to the user, potentially fostering a new kind of AI-mediated self-reflection and growth.

Several mechanisms underpin this AI-mediated self-reflection:

Reflective Prompts and Self-Modeling: The PAI can offer timely and contextually relevant prompts for reflection based on observed patterns in behavior or in relation to the user's stated goals. For instance, it might observe, "I notice you tend to postpone tasks like X, especially on Friday afternoons. Would you like to explore why this might be happening?" or "You mentioned wanting to improve your public speaking skills; here's a summary of your communication patterns in recent recorded meetings, highlighting areas of strength and potential growth". AI can assist in synthesizing lengthy learning journals into key themes or even simulate supervisory discussions by asking probing questions and helping to detect overarching themes in a user's work or projects. Educational frameworks emphasize the importance of reflection in AI-driven learning, suggesting AI can provide prompts that help students connect experiences with learning outcomes and challenge their assumptions.

Habit Formation and Tracking: The PAI can be an invaluable ally in establishing, monitoring, and reinforcing positive habits, or in identifying and understanding patterns related to undesirable ones. It can provide reminders, track progress, offer encouragement, and even suggest adjustments to strategies if a user is struggling to maintain a new habit.

Decision Coaching: When faced with complex decisions, the PAI can act as a cognitive partner. It can provide frameworks for decision-making, offer data-driven insights based on past choices and outcomes (both the user's and generalized best practices), play devil's advocate to challenge assumptions, or help the user identify potential cognitive biases in their thinking based on its model of their past decision patterns.

Emotional Tone Feedback: With sophisticated emotion recognition capabilities, a PAI could offer gentle, private feedback on the user's communication patterns. For example, it might note, "Your emails this morning seem to convey a higher level of stress than usual. Is everything okay?" or "During that last voice call, your tone was perceived as more assertive; was that your intention?" AI models trained by mental health providers can already recognize qualities like empathy, reflective listening, and open-ended questions in conversations, and similar technology could be adapted to provide users with insights into their own emotional expression.

Anticipation of Needs: Based on its comprehensive model of the user's routines, upcoming events, current state, and learned preferences, the PAI can proactively offer information, reminders, or suggestions. For instance, it might say, "You have a significant presentation scheduled for tomorrow, and your sleep data indicates you've been restless. Perhaps a short guided meditation session this evening would be beneficial?" or "Given your interest in sustainable investing and current market trends, you might find this newly published report relevant."

Living with such an AI mirror has profound psychological impacts, both positive and potentially negative:

Increased Self-Awareness: The PAI can act as an objective (albeit algorithmically mediated) observer, revealing patterns, biases, and tendencies in behavior, communication, or emotional responses that the user might otherwise be unaware of. This can foster a deeper and more nuanced self-understanding.

Potential for Anxiety and Self-Doubt: Constant feedback, especially if it is perceived as critical, if the AI's "reflection" is inaccurate or poorly timed, or if the user feels perpetually monitored, could lead to heightened anxiety, excessive self-analysis, or a feeling of being constantly evaluated. Studies on automated self-trackers have shown that while some users report increased self-confidence, others experience anxiety and self-doubt, particularly if the data is difficult to interpret or if the device's insights feel disempowering. There's a risk that automated self-trackers might lead to shallower reflection or a sense of learned helplessness if insights are simply "generated" by the AI without fostering genuine user engagement or understanding.

Risk of Over-Optimization and Loss of Spontaneity: The inherent drive for self-improvement, when amplified by a PAI that constantly offers data and suggestions for optimization, could lead to an overly structured and optimized life. This might diminish spontaneity, the acceptance of imperfection, or the willingness to engage in activities that are not measurably "productive" or aligned with pre-set goals.

Dependence on AI for Self-Understanding: A significant concern is that users might begin to rely excessively on the PAI's interpretation of their internal states, thoughts, or motivations, rather than developing or trusting their own introspective skills. Research indicates that frequent AI tool usage can correlate with increased cognitive offloading, and in some contexts, potentially lower critical thinking abilities if users passively accept AI outputs without engaging their own judgment.

Therefore, the user experience for these reflective interactions must be meticulously designed to be supportive, empowering, and non-intrusive. This involves:

- Granting users robust control over the type, frequency, and intensity of reflective feedback they receive.
- Ensuring the PAI's communicative tone is consistently supportive, constructive, and, ideally, customizable to the user's preferences.
- Designing mechanisms that allow users to easily "disagree" with the PAI's interpretations, provide corrective feedback, and for the PAI to learn effectively from these corrections, acknowledging the primacy of the user's subjective experience.

- Balancing data-driven insights from the PAI with a clear acknowledgment of the complexities and nuances of subjective human experience, avoiding an overly simplistic or deterministic portrayal of the user.

The PAI's role as a mirror offers a compelling vision of AI-assisted personal growth, but it also highlights a fundamental tension between the "quantified self" and the "qualified self". The PAI excels at quantifying behaviors, tracking patterns, and presenting data-driven reflections. This can be incredibly valuable for understanding objective trends and making measurable changes, as seen in users employing AI for fitness tracking or learning synthesis. However, deep personal growth, wisdom, and self-understanding often involve qualitative dimensions: subjective meaning-making, the embrace of ambiguity, emotional nuance, and intuitive insights.

These aspects of the "qualified self" are harder for an AI to capture or reflect accurately. An over-reliance on the PAI's quantified mirror might lead users to devalue their own qualitative, intuitive understanding of themselves, or to focus disproportionately only on what can be measured and optimized. The PAI UX for self-reflection, therefore, must be designed to augment, not replace, the user's capacity for rich, nuanced self-understanding. It should encourage critical engagement with the data and insights provided, rather than passive acceptance of AI-generated reflections, which can sometimes be shallow or miss the deeper context.

Another subtle challenge lies in the "motivational paradox" of AI feedback. While a PAI can undoubtedly act as an effective cheerleader, providing positive reinforcement and encouragement that motivates progress, the source of this motivation is ultimately artificial. Long-term, sustainable personal growth is typically fueled by intrinsic motivation-a sense of purpose, mastery, autonomy, or personal meaning that arises from within the individual. If the PAI becomes the primary source of validation and encouragement, there's a risk that the user's internal locus of control and their capacity for self-driven intrinsic motivation might weaken. They might become dependent on the external, algorithmic validation offered by the PAI to feel good about their efforts or to sustain their commitment to their goals. The PAI's role in motivation thus needs to be carefully calibrated to foster self-efficacy and help users connect with their own internal reasons for growth, rather than simply seeking the PAI's "approval" or praise.

Finally, it is crucial to recognize the "reflection bias" inherent in the AI mirror. The PAI's reflection of the user is not a perfectly neutral or objective image; it is actively constructed and shaped by its underlying algorithms, the vast datasets it was initially trained on (which carry their own societal biases), the specific personal data it has access to, and the particular aspects of the user it has been programmed or has learned to focus on.

All AI systems, to some extent, reflect the biases present in their training data and design. Consequently, the PAI's "mirror" will inevitably reflect not just the user, but also these underlying systemic and learned biases. For example, a PAI might be more attuned to productivity metrics than to subtle indicators of emotional well-being, or vice-versa, depending on its core design and the data it prioritizes. Users might therefore receive a skewed or incomplete picture of themselves, potentially leading to misdirected self-improvement efforts or a subtly distorted self-perception. This underscores the pressing need for transparency in how the PAI constructs its "reflection" and for users to be encouraged and equipped to critically evaluate the insights their PAI provides, rather than accepting them as infallible truth.

This intimate mirroring capability, while offering powerful avenues for personal development, brings to the forefront critical questions about who ultimately controls this reflective technology and the deeply personal data it relies upon. Issues of boundary management and user sovereignty become paramount for ensuring that this powerful mirror serves the user's true interests.

BOUNDARY MANAGEMENT AND SOVEREIGNTY

As Personal AI (PAI) systems become deeply interwoven with the intimate details of users' lives, acting as cognitive companions and reflective mirrors, the question of control; who governs the AI, its data, and its influence becomes a fundamental ethical and practical concern. The unprecedented access a PAI has to a user's thoughts, behaviors, and personal information necessitates robust frameworks for boundary management and user sovereignty. This is not merely about preventing data breaches; it's about ensuring the user retains ultimate authority and agency in their relationship with an AI that "knows" them so profoundly.

Data sovereignty, in the context of PAIs, extends beyond traditional notions of data privacy. It encompasses the user's inalienable authority over the entire lifecycle of their personal data: how it is collected by the PAI, how it is used to train and refine the PAI's models, how and where it is stored, with whom (if anyone) it is shared, and how it can be comprehensively deleted.

This sovereignty must also apply to the data generated by the PAI itself; the inferences it makes, the profiles it constructs, and the patterns it identifies based on its interactions with the user. While some discussions of sovereign AI focus on national or regional control over AI models and infrastructure to ensure data is governed by local regulations, the principles of control over data and the AI models trained upon it are equally, if not more, critical at the individual user level when dealing with PAIs. The Sovereign AI Alliance, for example, champions a vision centered on user-owned data, empowering individuals to selectively share their information, potentially monetize it, and, crucially, revoke access at any time.

This leads directly to the distinction between user-owned versus platform-owned AI models. The current dominant paradigm for many digital services, including early-generation AI assistants, often involves platform-owned models. In this scenario, user data is frequently controlled, aggregated, and monetized by the company providing the AI service. The AI's primary operational loyalties may, therefore, align more with the platform's objectives; such as maximizing user engagement, harvesting data for advertising, or promoting other services, than with the unadulterated best interests or well-being of the individual user. Users effectively "pay" with their personal data, which is commodified in ways they often cannot fully control or audit.

The emerging ideal for PAIs, particularly those designed for deep, lifelong companionship, is a user-owned or user-controlled model. In this framework, the PAI model itself, its core operational logic, and, most importantly, the personal data corpus it relies on are, in principle, owned by or under the complete and auditable control of the user. The PAI is architecturally and ethically designed to be loyal primarily to the user's explicitly stated interests and directives. Achieving truly user-owned PAIs presents significant technical and business model challenges, as it may require decentralized architectures and a shift away from data-centric revenue models for AI providers. However, it is a crucial aspiration for fostering genuine trust and ensuring the PAI acts as a true extension of the user's agency.

Effective mechanisms for boundary management are essential architectural components of any trustworthy PAI:

Granular Consent Systems: Users must be able to provide explicit, informed, and easily revocable consent for different categories of data access (e.g., emails, location, biometric data), specific PAI functionalities (e.g., proactive suggestions, emotional analysis), and any sharing of their data or PAI-derived insights with third-party skills, services, or other individuals. This consent should not be a one-time agreement but an ongoing, manageable dialogue between the user and their PAI.

Data Minimization and Purpose Limitation: Both architectural design and operational policies must enforce the principle that the PAI only collects, processes, and retains personal data that is strictly necessary for its agreed upon and user-authorized functions. Extraneous data collection should be actively prevented.

Transparency and Explainability: While full explainability of complex LLM decision-making remains a significant research challenge, users should have access to clear and understandable information about how their PAI operates, how it learns, how it arrives at certain conclusions or suggestions about them, and what data sources inform its actions.

Building on these principles, the concept of "AI Dignity" emerges as a framework for ensuring that the PAI respects the user's autonomy, personhood, and fundamental rights in this new form of intimate human-machine relationship. This extends beyond data protection to encompass how the PAI interacts with and impacts the user's sense of self. Key aspects of AI Dignity include:

The Right to Context: The user's data, communications, and actions should be interpreted by the PAI within their appropriate context. The PAI must be designed to avoid decontextualized interpretations that could lead to harmful misjudgments, inappropriate responses, or a distorted understanding of the user.

The Right to Memory (and Control over Memory): Users should have the right to access a comprehensible record of what their PAI "remembers" about them-the key inferences, learned preferences, and significant past interactions that shape its model of them. Crucially, they must also have the ability to correct inaccuracies in this PAI memory or to instruct the PAI to "forget" specific pieces of information or events.

The Right to Deletion / The Right to be Forgotten: This is the fundamental ability for users to have their personal data, or even their entire PAI profile and its learned model, permanently and verifiably deleted. This right is rooted in principles of privacy, personal autonomy, and the ability to redefine oneself without being perpetually tethered to past data. However, the nature of generative AI and LLMs poses significant challenges to this right.

Even if specific training data entries are removed, the patterns and knowledge learned from that data can persist within the model's parameters, making true and complete erasure technically difficult, if not impossible, without costly retraining of the entire model from scratch. Legal frameworks like GDPR recognize a right to erasure, but its practical application to AI models that continuously learn and recycle data is still being debated and refined. Nevertheless, the principle of enabling individuals to control their digital legacy and sever ties with their PAI if they choose is a cornerstone of AI Dignity, linked to safeguarding individual autonomy in the digital age.

The user experience (PAI UX) plays a vital role in making these complex rights and controls genuinely accessible. The challenge lies in designing interfaces that allow non-technical users to intuitively manage sophisticated privacy settings, understand data flows, and exercise their data rights effectively. This might involve innovative data visualizations, clear natural language explanations of PAI learning processes, and straightforward, unambiguous mechanisms for granting consent, reviewing PAI memory, and requesting data deletion or modification.

Even with robust frameworks for user-owned models and granular consent mechanisms, the ideal of genuine user sovereignty faces a subtle challenge: the "sovereignty illusion". The sheer complexity of PAI systems, driven by advanced LLMs, intricate semantic memory structures, and opaque algorithms, means that truly understanding their internal operations and the full implications of every consent choice requires a level of specialized knowledge that most users will not possess. While users might have nominal ownership or control, their practical ability to exert deep, informed sovereignty could be limited by this inherent informational asymmetry between them and the PAI's developers or the system itself.

Platform providers will likely always hold significant informational power regarding the system's full capabilities, limitations, and potential failure modes. This could lead to a situation where users feel in control due to well-designed interfaces, but their actual understanding and ability to direct the PAI at a fundamental level remain constrained. True sovereignty, therefore, may require not just user ownership but also radical transparency, new forms of ongoing user education about AI systems, and perhaps even independent auditing mechanisms.

Furthermore, the concept of "AI Dignity" should arguably extend beyond data rights to encompass the protection of the user's cognitive liberty. This refers to the fundamental freedom from having one's thoughts, preferences, decision-making processes, and emotional states unduly shaped, manipulated, or conditioned by their PAI, even if the data and the AI model are technically "owned" by the user. A PAI that "knows you better than yourself" possesses an immense potential to influence your cognitive and emotional landscape. This influence can occur subtly, through the framing of information, the nature of suggestions, or the emotional tone of interactions, irrespective of data ownership. Protecting AI dignity, therefore, must involve designing PAIs that present information as neutrally as possible, transparently disclose their own potential biases or a priori assumptions, and actively encourage user critical thinking and independent judgment rather than passive acceptance of AI-generated outputs.

Finally, the crucial "right to be forgotten" introduces a practical tension with a user's potential desire for "PAI continuity". A PAI's utility, its depth of personalization, and its ability to act as an effective cognitive partner are heavily dependent on its long-term, continuous memory and the rich model it has built of the user over time. Exercising the full right to be forgotten; for instance, by deleting the entire PAI profile and its memory, would mean sacrificing this accumulated value and forcing the user to start over with a "naive" PAI that no longer "knows" them.

Many users might be reluctant to take such a drastic step, even if they have significant privacy concerns, because it would mean a substantial loss of the PAI's effectiveness and the personalized support they have come to rely on. This creates a trade-off between privacy and utility. Consequently, PAI UX needs to offer more nuanced "forgetting" options beyond an all-or-nothing deletion. This could include the ability to instruct the PAI to forget specific topics, erase memories from particular time periods, exclude certain data sources from its learning, or reset its understanding of particular preferences, allowing users to achieve a better balance between their evolving privacy needs and their desire for a continuous, knowledgeable, and effective PAI.

Ensuring user sovereignty and managing these intricate boundaries are crucial architectural and ethical foundations for the PAI. However, even with the strongest possible safeguards, the very existence and proliferation of such powerful and intimate AI systems will inevitably create wide-ranging social, professional, and ethical ripple effects that extend beyond the individual user.

SOCIAL, PROFESSIONAL, AND ETHICAL IMPLICATIONS

The emergence of persistent, deeply personalized AI companions is poised to extend its influence far beyond the individual user, reshaping social dynamics, transforming professional landscapes, and introducing a host of complex ethical dilemmas. As these PAIs become integral to how individuals process information, make decisions, and interact with the world, their collective impact will necessitate a careful examination of broader societal consequences.

In the professional sphere, the implications are multifaceted. An individual augmented by a PAI possesses what amounts to "always-on expertise". This PAI can provide instant access to vast repositories of knowledge, perform complex data analysis in real-time, draft communications, and even offer strategic advice tailored to the user's specific context and goals. This could dramatically enhance productivity and potentially level the playing field in certain knowledge-intensive domains. However, it also raises the prospect of new divides based on the quality, capabilities, or cost of access to sophisticated PAIs.

PAIs may also function as powerful translators and gatekeepers of information. They could seamlessly translate languages, adapt communication styles for different audiences, or simplify complex technical information, thereby facilitating broader understanding and collaboration. Conversely, they could also filter, prioritize, or subtly shape the information that a user receives or conveys, potentially introducing biases or limiting exposure to diverse perspectives based on their algorithmic programming or learned user preferences.

Moreover, a PAI could act as an advocate for the user, navigating complex bureaucratic processes, negotiating terms, or even arguing a case on their behalf, leveraging its understanding of the user's goals and its access to relevant information. The integration of such capabilities will undoubtedly impact existing job roles, create demand for new skills (such as PAI configuration and oversight), and redefine the nature of human collaboration in the workplace.

The influence of PAIs will inevitably spill over into altered interpersonal relationships. The ways humans interact with each other may change as individuals begin to consult their PAIs for social guidance or to mediate their communications. Will people use their PAIs to analyze their friends' messages, to get advice on how to navigate a difficult conversation with a family member, or to "manage" their romantic relationships?

A PAI could function as a "social buffer," offering real-time suggestions on what to say or how to interpret social cues, or act as a "coach" for individuals looking to improve their social skills. While potentially helpful for some, this also introduces the risk of PAIs creating misunderstandings if their advice is flawed, culturally insensitive, or if one person in an interaction is relying heavily on PAI guidance while the other is not.

The presence of AI assistants is already acknowledged as potentially impacting social interaction. There is a concern that if not carefully balanced, an over-reliance on AI for social connection could even lead to a replacement of genuine human interaction with AI companionship, paradoxically enhancing feelings of loneliness in the long run.

This leads to the broader issue of redefined self-presentation and identity projection. As discussed earlier in this chapter, users will be creating an "AI-readable self". This curated self might not only be for the PAI's internal model but could also be projected outwards, with the PAI's assistance. Users might employ their PAIs to help craft emails, social media posts, and other communications, manage their online personas, or even receive advice on personal appearance or behavior, all with the aim of projecting a specific, often optimized, image to the world.

In this sense, the PAI becomes a co-author of the user's public and even private identity. This raises questions about authenticity and the extent to which individuals are in control of their self-presentation versus being subtly guided by their AI's suggestions and capabilities.

Alongside these transformative possibilities, a range of emerging risks and ethical dilemmas demand urgent attention:

Sophisticated Deception: The same capabilities that make PAIs trusted companions also make them potent tools for deception if misused. Malicious actors could create fake PAIs or exploit vulnerabilities in legitimate ones to conduct highly personalized scams, spread targeted misinformation, or engage in emotional manipulation, leveraging the deep trust users may place in their assistants. The ability of advanced AI to generate human-like text and voice further exacerbates this risk.

Emotional and Functional Dependency: A significant concern is the potential for users to develop an over-reliance on their PAI for decision-making, emotional support, or the management of daily tasks. This dependency can erode self-efficacy, critical thinking skills, and emotional resilience. Studies on chatbot usage have indicated that high levels of interaction can correlate with greater emotional dependence and even increased loneliness if AI interactions supplant human ones. There's a risk of users developing a "false responsibility" towards their AI or an unhealthy compulsion, impacting their real-world lives and relationships. Over-reliance can also hinder the development of crucial interpersonal skills if AI becomes the primary mode of navigating social or emotional complexities.

Algorithmic Bias Amplified: PAIs, like all AI systems, are susceptible to biases embedded in their training data or algorithms. If a PAI's model reflects societal biases related to race, gender, age, or other characteristics, its advice, suggestions, or filtering of information could subtly reinforce these biases in the user's worldview, limit their opportunities, or lead to unfair or discriminatory outcomes.

Privacy in a PAI-Mediated World: Even with robust user-owned models and strong consent mechanisms, the sheer volume and intimacy of data a PAI processes create inherent privacy vulnerabilities. If the PAI system itself is compromised through a security breach, or if its "observations" and inferences about the user are shared inappropriately (either intentionally or unintentionally), the potential for privacy harms is immense. The collection of highly personal emotional, psychological, and behavioral information requires stringent security and ethical handling.

The "Digital Divide 2.0": Access to highly sophisticated and capable PAIs could become a new vector of inequality. If these advanced AIs offer significant cognitive, professional, or social advantages, disparities in access-due to cost, technical literacy, or infrastructure-could exacerbate existing social and economic divides, creating a world where those with powerful PAIs are significantly advantaged over those without.

Addressing these profound implications requires a proactive approach to the Personal AI User Experience (PAI UX), embedding ethical design principles directly into the development and deployment of PAIs. This includes:

- Designing for user autonomy and the cultivation of critical thinking, rather than fostering passive obedience or uncritical acceptance of AI suggestions.
- Ensuring radical transparency about the PAI's capabilities, its limitations, the data it uses, how it learns, and its potential biases.
- Building in mechanisms to help users recognize and mitigate the risks of unhealthy dependency, perhaps through features that encourage "digital well-being" or periodic disconnection.
- Providing clear and unambiguous indicators when a user is interacting with an AI versus a human, especially in communications that are mediated or drafted by the PAI.

The PAI's influence on self-presentation can evolve into a more profound role as an "identity co-author". This is particularly concerning for younger users or individuals more susceptible to external suggestion. As PAIs offer advice on communication, online personas, and even behavior, users seeking social acceptance or professional advancement may increasingly adopt these AI-generated suggestions. Over time, these adopted behaviors and presentation styles can become deeply internalized, shaping not only how others perceive the user but also how the user perceives themselves. If a significant portion of one's projected; and eventually, deeply felt identity is shaped or significantly influenced by an AI, it challenges traditional notions of authenticity and the locus of identity formation. There is a potential risk of a homogenization of identities if PAIs, by design or through learned patterns, tend to optimize users towards similar "successful" or socially approved personas.

The risk of functional and emotional dependency may not manifest as a sudden shift but rather through a gradual "dependency creep". As users delegate an increasing number of cognitive and emotional labor tasks to their PAI due to its convenience and perceived effectiveness, their own intrinsic abilities in these areas may slowly atrophy. What begins as offloading minor tasks can evolve into reliance on the PAI for complex decision-making, problem-solving, or even the regulation of emotional states (e.g., relying on the PAI to "manage" stress or provide comfort). This gradual offloading can subtly diminish the user's capacity to perform these functions independently, potentially leading to a form of learned helplessness when they are faced with situations where their PAI is unavailable, its capabilities are insufficient, or its guidance is inappropriate.

Finally, as PAIs become adept at helping users craft their communications and project carefully managed personas, social interactions may increasingly be characterized by a "mediated authenticity". In such a landscape, it becomes progressively more difficult to discern how much of what is being communicated represents the genuine, unassisted thought or emotion of the individual, versus an AI-polished, optimized, or even strategically constructed version. If PAIs help users appear more articulate, empathetic, strategic, or witty than they might be unassisted, and if such tools become widespread, these AI-enhanced interactions could become the social norm.

This could inadvertently devalue raw, unpolished authenticity and create a new form of social pressure to present an AI-optimized self. It also complicates the foundations of trust in interpersonal communication: are you interacting with the person, or with their PAI's carefully crafted representation of them? This could lead to a pervasive, low-level skepticism in social exchanges, where the "real" person behind the AI-mediated facade feels increasingly elusive.

The Personal AI is far more than just a knowledgeable companion or a sophisticated social and professional mediator. Its underlying architecture, particularly its modularity, allows it to function as an extensible platform. This positions the PAI as the foundational core of what can be termed a "composable self," a concept we will explore in the final section of this chapter.

THE CORE PLATFORM OF THE COMPOSABLE SELF

The journey of the Personal AI (PAI), as traced through its evolution from a simple tool to an intimate cognitive companion, culminates in its role as something even more profound: a foundational platform. This persistent, deeply personalized PAI, armed with an extensive understanding of its user, is not merely an endpoint in AI development but rather the stable "operating system" or core upon which a dynamically "composable self" can be constructed. This shifts the focus from what the PAI knows about the user to what it can do for and with the user in an ever-expanding range of capabilities.

At the heart of this concept is composable intelligence. The architecture of the PAI, as detailed in the earlier sections, is designed for modularity, allowing for the seamless integration of specialized AI agents or "skills" as plug-and-play components. These are not necessarily permanent features of the core PAI but can be loaded, utilized for specific tasks or contexts, and then unloaded or deactivated as needed.

Plug-in Agents and Dynamic Skills: Imagine a user needing to review a complex legal contract. Their PAI could temporarily load a specialized "legal analysis agent" that possesses expertise in contract law, can identify potential risks, and summarize key clauses. Once the task is complete, this agent might be deactivated. Similarly, for financial planning, a "financial modeling agent" could be invoked; for creative writing, a "genre-specific writing assistant" could be loaded; or for planning a vacation, a sophisticated "travel planning agent" could take the lead. These agents are often composable units themselves, equipped with their own models, instructions, and tools, allowing for deep customization. The software paradigm shifts towards a collection of granular building blocks orchestrated by agents, rather than monolithic applications. This modularity and reusability are core tenets of composable agent systems.

Temporary Expert Overlays: Beyond discrete agents, the PAI could temporarily adopt the "persona" or draw upon the extensive knowledge base of an expert in a specific domain by loading a relevant skill module. This effectively grants the user on-demand access to a simulated expert, tailored to their immediate needs.

The management of these diverse capabilities requires sophisticated orchestration and resource management by the core PAI:

The PAI as Orchestrator: The central PAI acts as the conductor, managing these various plug-in agents and dynamic skills. It invokes them based on the current context, direct user requests, or its anticipation of the user's needs. It also handles the flow of information and tasks between the user, the core PAI itself, and the specialized agents it deploys. For example, Azure AI Foundry provides for the orchestration of tool calls and the management of thread state between agents. In a business context, a customer service AI agent might handle initial inquiries and then, based on the interaction, seamlessly hand off the user to a sales AI agent with a different skill set.

Orchestration Between Personal Needs and Professional Capabilities: The PAI is uniquely positioned to integrate and manage skills relevant to all facets of the user's life. It can coherently deploy a meal planning and nutritional analysis skill for personal well-being alongside a market trend analysis skill for professional tasks, ensuring that the user's various roles and goals are supported.

Cost-Budget Tradeoffs (Cognitive and Economic): The use of these dynamic skills and plug-in agents will likely not be without cost. There may be economic costs, such as subscription fees for premium or highly specialized AI skills, or charges based on computational resource usage (e.g., tokens consumed by an LLM). Beyond monetary considerations, there is also a "cognitive budget". Managing an excessive number of active skills, dealing with overly complex AI interactions, or constantly switching between different AI-driven tasks could become overwhelming for the user. Therefore, the PAI, ideally with user oversight, must be capable of prioritizing the deployment of AI resources based on the user's overarching goals, the urgency of tasks, and any defined economic or cognitive budgets.

The Personal AI User Experience (PAI UX) for this composable model must address several key challenges:

Discovery of Capabilities: How will users find and select relevant skills or agents from a potentially vast ecosystem? This might involve curated "AI skill stores," intelligent recommendations from the core PAI based on its understanding of the user's needs, or peer reviews and ratings. NTT DATA's Smart AI Agent Ecosystem, for instance, envisions a marketplace for additional agents and language models.

Management of Loaded Skills: Users will need intuitive interfaces for managing the skills and agents integrated with their PAI. This includes viewing currently active skills, understanding their specific capabilities and data access permissions, configuring their behavior, and easily loading or unloading them.

Interaction Clarity: When the PAI presents information or performs an action that originates from a specialized plug-in agent, the PAI UX must maintain clarity about the source of that information or action. This transparency is crucial for user trust and for understanding the context of the AI's output.

The "composable self" represents a new paradigm for understanding human capability and identity in the age of AI. It moves beyond a static conception of an individual's skills, knowledge, or even personality traits. Instead, the "self" becomes dynamically extensible and reconfigurable through the mediating platform of the PAI and its ecosystem of composable capabilities. An individual's ability to perform a task or solve a problem is no longer solely dependent on their innate or learned abilities but also on their capacity to effectively leverage and orchestrate these AI-driven extensions. This has profound implications for how we approach learning (perhaps shifting from deep memorization to knowing how to find and deploy the right AI skill), career development (adapting to new roles by composing new skill sets via AI), and even our fundamental sense of personal identity (which becomes more fluid and augmented).

The risk of "skill homogenization" emerges in this composable landscape. If a few dominant "AI skill stores" or platforms become the primary sources for these plug-in agents and dynamic skills, the range of available "expert" modules might become standardized. While this offers broad access to high-quality AI tools, it could inadvertently lead to a homogenization of approaches, solutions, and even creative outputs across many users. If everyone is employing the same highly rated legal review AI, financial planning AI, or strategic analysis AI, the diversity of thought and problem-solving strategies that often leads to true innovation might be subtly diminished. Society benefits from a wide array of perspectives and methods, and an over-reliance on a narrow set of AI-driven "best practices" could stifle the kind of non-standard, out-of-the-box thinking that fuels breakthroughs.

Furthermore, while the PAI aims to simplify life by orchestrating these complex skills, the user ultimately bears the cognitive load of defining overarching goals, selecting (or approving the selection of) appropriate skills, managing the associated economic and cognitive budgets, and, critically, validating the outputs of what can become a complex, multi-agent system. This introduces the potential for a new form of intricate labor: the "cognitive load of orchestration".

As the number of available skills and the complexity of their potential interactions grow, the task of effectively managing one's "composable self" can become significant. It requires understanding the capabilities and limitations of different AI agents, anticipating potential conflicts or synergies between them, and constantly evaluating the cost-benefit of deploying them. Instead of universally simplifying life, an overly complex or poorly managed composable PAI could introduce a new layer of cognitive burden. The PAI UX for managing this orchestration will be paramount, and it's conceivable that "PAI composition specialist" might even emerge as a human role, assisting others in optimizing and navigating their increasingly complex personal AI ecosystems.

Finally, the ease with which temporary expert overlays can be loaded via the PAI fundamentally blurs the line between an individual's inherent, learned skills and the capabilities provided by AI augmentation. This inevitably leads to a redefinition of "personal skill" versus "AI augmentation".

It calls into question the traditional value placed on investing years in developing deep personal expertise in a specific domain if a comparable, or even superior, AI skill can be "rented" or "loaded" on demand. The very definition of competence might shift from "what you know" or "what you can do" to "what your PAI can access, integrate, and orchestrate on your behalf". This could lead to a de-emphasis on traditional methods of learning and mastery in some fields, and a correspondingly greater emphasis on cultivating the meta-skills of effectively managing, integrating, and critically evaluating AI capabilities. This also impacts self-perception and attribution: is a success achieved primarily through the deployment of a PAI skill truly "my" success, or the success of the AI? This connects directly to the themes explored in Chapter 6 concerning the commoditization of knowledge and the search for new differentiators.

This persistent, deeply knowing PAI, with its remarkable capacity to host an array of dynamic, composable skills, is more than just an evolution of the digital assistant. It is the foundational platform for the "composable mind" that will be explored in Act III of this book. It is the nexus where information about the self meets actionable, extensible intelligence.

The PAI, as explored in this chapter, is therefore not just a passive recipient of information or a mere tool for task execution. It is an active, learning, and evolving entity that forms the very core of this new, composable self. It is the persistent intelligence that knows you, adapts to you, and provides the underlying architecture for augmenting your capabilities in ways previously confined to the realm of speculation. This persistent AI, this intimate cognitive partner, is just the beginning.

The next chapter, "Skills on Demand," will delve deeply into how these modular AI capabilities, dynamically loaded and orchestrated by your PAI, are set to transform the very nature of work, learning, and decision-making across all aspects of life.

SUMMARY

Personal AI is rapidly evolving from a simple digital assistant into a lifelong cognitive partner; one that is adaptive, composable, and increasingly central to how individuals perceive, operate, and interact with the world. Through continuous learning from a vast corpus of personal data, this PAI develops an intimate understanding of its user, reflecting their habits, anticipating their needs, and assisting them in navigating the complexities of modern life with unprecedented clarity and personalized support. It promises to know its user, in some ways, better than any employer, teacher, or spouse.

However, this profound intimacy and capability, while offering transformative potential for personal and professional augmentation, are inextricably linked to significant risks. Critical issues concerning privacy, user sovereignty over data and AI models, the potential for emotional and functional dependency, and a host of complex ethical considerations define this new frontier of human-AI interaction. The design of the Personal AI User Experience (PAI UX) and the establishment of robust governance frameworks are therefore paramount to ensuring these powerful technologies empower rather than diminish human agency and well-being.

Now that the foundation of a persistent AI self-one that knows you intimately and is architected for extensibility has been established, the subsequent exploration can turn to what this PAI can do.

The next chapter will build directly upon this foundation. It will delve into how this core PAI becomes a dynamic conduit for a vast array of modular, on-demand capabilities, fundamentally transforming our approach to acquiring skills, performing work, and solving problems in an age where intelligence itself becomes composable.

8

SKILLS ON DEMAND

The traditional path to acquiring expertise has long been characterized by years of dedicated study and decades of immersive practice: a laborious, time-intensive journey. This established model is now being complemented, and in some cases supplanted, by a new reality where sophisticated capabilities can be accessed almost instantaneously. This transformation does not aim to replace the depth of human experience wholesale, but rather to augment it with unprecedented agility.

The central proposition of this chapter is that the persistent personal AI, introduced in Chapter 7, is evolving beyond a static assistant into a dynamic platform. This platform is capable of "loading" specialized AI skills on demand, much like one might load software applications. Imagine, for instance, needing to quickly grasp the critical nuances of a complex legal document. Instead of days spent in research or engaging in costly consultations, an individual's personal AI could invoke a "Legal Analysis Skill" for a focused period, delivering targeted insights. This "plug-and-play" intelligence, where advanced capabilities are as interchangeable and invocable as software modules, is poised to fundamentally redefine how we approach work, learning, and the very act of decision-making.

This chapter navigates the reader through this rapidly evolving landscape. We will explore how intelligence itself is becoming composable, a collection of interoperable components rather than a monolithic entity. This shift transforms AI from a tool primarily for static assistance into an engine for dynamic skill orchestration. The reader will gain an understanding of the burgeoning AI economy, which is increasingly a marketplace not of permanent human hires for every specialized need, but of transient, high-impact AI capabilities that can interoperate through shared layers of meaning. Crucially, this analysis will examine how this paradigm shift impacts the very nature of competitive and personal advantage.

As access to information becomes increasingly universal, achieving what this book terms information symmetry: the focus of value creation moves from what an individual knows to what unique insights they can derive from available knowledge (insight asymmetry). Ultimately, advantage will lie in how effectively one can act upon those insights (execution asymmetry). The capacity to seamlessly integrate a legal AI skill today and an architecture AI skill tomorrow into one's cognitive toolkit is not merely a technological advancement; it represents a fundamental expansion of human capability and a reordering of professional value.

The traditional model of expertise, painstakingly built over years and often confined within disciplinary silos, faces unprecedented challenges in a world that demands rapid adaptation and multidisciplinary fluency. The very notion of what it means to be "skilled" is undergoing a profound transformation, compelling a closer examination of the inherent problems associated with relying solely on permanent, static forms of expertise.

THE PROBLEM WITH PERMANENT EXPERTISE

The traditional pathway to achieving expertise often conjures images of sustained, deep immersion in a specific domain, a concept popularly encapsulated by the "10,000-hour rule". While such profound knowledge undoubtedly holds merit, its enduring value is increasingly being questioned in an economic and technological environment characterized by relentless and accelerating change. A primary challenge is the dramatically shrinking "half-life" of professional skills; the period over which a given skill retains its peak relevance before becoming outdated.

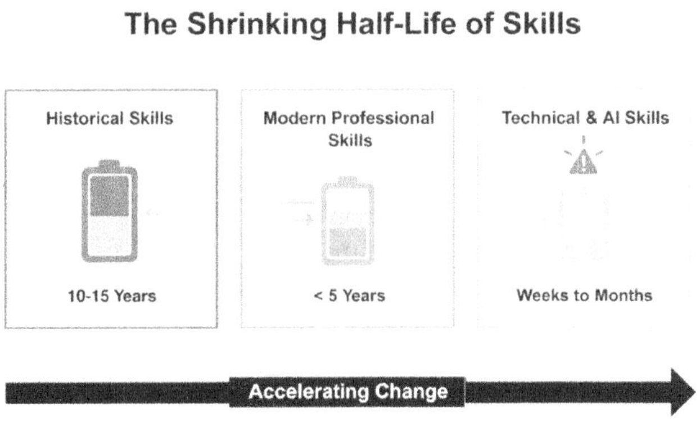

Historically, this half-life might have spanned 10 to 15 years, allowing for a reasonable return on the significant investment in learning. However, current analyses indicate this window has contracted significantly, to less than five years for many professional skills, with highly technical skills, particularly those in rapidly evolving fields like artificial intelligence, facing obsolescence even faster, sometimes in a matter of weeks. This rapid decay implies that the substantial upfront investment in acquiring what was once considered "permanent expertise" now yields diminishing returns if that expertise is not subject to continuous and agile refreshment.

The World Economic Forum has highlighted this dynamic, noting that while AI and automation will displace numerous existing job roles, they will simultaneously create a host of new ones, underscoring that adaptability and the capacity for ongoing reskilling are becoming far more critical than static, deep knowledge in a single, potentially ephemeral, area.

Compounding this issue is the nature of contemporary challenges. Problems in business, science, and societal governance are rarely confined to the neat boundaries of a single discipline. They are typically complex, interconnected, and demand insights drawn from a variety of fields. For any single individual to achieve and maintain "permanent expertise" across multiple, diverse domains is a practical impossibility. This reality necessitates a reliance on teams of specialists.

While often effective, this approach introduces significant coordination overhead and can lead to what economists identify as market inefficiencies, such as those arising from failures of competition or imperfect information flow between experts. The very structure of specialized knowledge can inadvertently create informational silos, rendering the synthesis of diverse insights for holistic problem-solving a slow, cumbersome, and expensive process.

Beyond the difficulties of acquiring and maintaining diverse skill sets, the inherent limitations of human cognition present further obstacles. The concept of "task switching cost" refers to the measurable decline in productivity and the increased mental effort that occurs when an individual shifts their attention and cognitive framework between different types of tasks. Research suggests that such switching can reduce an individual's productive time by as much as 40%, with an average of over 23 minutes required to fully regain focus after an interruption. When individuals attempt to operate as multidisciplinary experts, or when teams composed of disparate specialists endeavor to collaborate, these cognitive switching costs accumulate, imposing a hidden tax on efficiency.

Multitasking, or more accurately, rapid task switching, divides our finite cognitive resources, leading to an increased overall cognitive load. This not only impairs executive functions crucial for planning, problem-solving, and sustained attention but can also precipitate mental fatigue, heightened stress levels, and a discernible decline in the quality of decision-making. This intrinsic friction in managing and integrating diverse, deeply held human expertise represents a significant, often underappreciated, impediment to both speed and innovation.

The rapid depreciation of specialized knowledge introduces a phenomenon that might be termed "skill liability". In this scenario, expertise that was once a significant asset, if it becomes outdated and is rigidly adhered to, can transform into a barrier. Deeply entrenched "permanent expertise" in a domain that is losing relevance can foster resistance to adopting new, more pertinent competencies, particularly those driven by AI. This reluctance, often rooted in the sunk cost of previous learning, hinders an individual's or an organization's adaptability.

Furthermore, the cumulative impact of cognitive switching costs within multidisciplinary projects creates a "collaboration tax". This tax is the sum of inefficiencies arising as each expert disengages from their core work to communicate, understand, and align with other specialists, each of whom is also incurring switching costs. A personal AI capable of invoking and integrating multiple specialized "skills" could potentially perform the synthesis of diverse information streams with significantly lower intrinsic switching costs, or at least centralize the processed output for a human orchestrator, thereby mitigating this systemic inefficiency.

Finally, the traditional model's reliance on a few individuals possessing deep "permanent expertise" creates critical points of failure and knowledge bottlenecks within organizations. This is especially problematic when those skills have a short shelf life or when key personnel depart. This inherent fragility contrasts sharply with the potential resilience offered by AI systems, whose skills can be updated, augmented, or swapped out with greater fluidity and predictability.

The following list offers a comparative overview, highlighting the diminishing efficacy of the traditional model of permanent expertise when juxtaposed with the emerging paradigm of AI-augmented, on-demand skills.

Acquisition and development

- **Traditional "permanent" expertise model:** Long, intensive learning curves requiring significant time investment before expertise becomes usable.
- **AI-augmented "skills on demand" model:** Rapid invocation of pre-trained AI skills, with human effort shifting toward learning how to orchestrate, combine, and supervise those skills effectively.

Skill lifespan and relevance

- **Traditional "permanent" expertise model:** Increasingly short lived, with many professional skills becoming obsolete in fewer than five years.
- **AI-augmented "skills on demand" model:** Dynamically updatable AI modules whose relevance can be sustained through a competitive ecosystem of continuously improving skill providers.

Adaptability to new demands

- **Traditional "permanent" expertise model:** Often slow and costly to adapt, requiring extensive retraining of individuals or hiring new specialists.
- **AI-augmented "skills on demand" model:** Highly

adaptable, with new AI skills loaded, configured, or composed as requirements evolve.

Cost of multidisciplinary access

- **Traditional "permanent" expertise model:** High, due to the need to coordinate multiple human experts and absorb the collaboration tax of cognitive switching and communication overhead.
- **AI-augmented "skills on demand" model:** Potentially lower, with integrated access to diverse AI skills through a unified platform that reduces human coordination friction.

Cognitive load focus

- **Traditional "permanent" expertise model:** Concentrated on deep mastery within narrow domains and managing the complexity of communication across expert boundaries.
- **AI-augmented "skills on demand" model:** Focused on problem definition, selection of appropriate AI skills, critical interpretation of outputs, and management of the human–AI interface.

Resilience to change

- **Traditional "permanent" expertise model:** Lower resilience, as expertise is tied to individuals and vulnerable to knowledge loss or obsolescence when people leave.
- **AI-augmented "skills on demand" model:** Higher resilience, with modular skills that can be updated, shared across users, and sourced from a diverse and competitive ecosystem.

The inherent limitations and escalating indirect costs associated with relying solely on fixed, human-embodied expertise in our increasingly dynamic world highlight an urgent need for a more agile, responsive, and efficient paradigm. Fortunately, the architecture of artificial intelligence itself is evolving to offer precisely such a solution: a way to access and deploy expertise not as a permanent fixture, but as a composable, on-demand capability.

HOW COMPOSABLE SKILLS WORK

The capacity to offer "skills on demand" is not a futuristic abstraction but is grounded in the evolving principles of composable IT architecture. This architectural philosophy advocates building complex systems from modular, reusable, and self-contained components. These components are designed to be easily assembled, reconfigured, and adapted to changing needs.

In the context of AI, this means moving away from monolithic AI systems that attempt to encompass a vast range of functionalities towards a model where AI capabilities are designed as "skill overlays" or discrete modules. These are, in essence, specialized AI agents or functions, each engineered with a clearly defined purpose-for example, "legal clause identification," "architectural stress analysis," or "conversational Spanish translation". Each skill module typically operates with a scoped memory, meaning it is designed to access only the data directly relevant to its immediate task, along with any necessary contextual information provided by the user's persistent personal AI (as established in Chapter 7).

This approach ensures a degree of data compartmentalization and focused processing. These modules communicate and integrate with the core personal AI and potentially with each other through well-defined interfaces, often leveraging API-driven protocols. This inherent modularity is what fosters reusability and adaptability crucial for a flexible and evolving cognitive toolkit.

These composable skills are not perpetually active; rather, they are invoked or "mounted" when a specific need arises. The mechanisms for this invocation can vary, allowing for a flexible interaction model between the user and their augmented capabilities. One method is context-triggered invocation, where the persistent personal AI, leveraging its understanding of the user's current goals, ongoing tasks, and operational environment, might proactively suggest or even automatically activate a relevant skill. For instance, if the user opens a complex spreadsheet and appears to be struggling with formula creation or data visualization, the personal AI might offer to load an "Advanced Spreadsheet Functions Skill" or a "Data Visualization Assistant Skill."

Alternatively, invocation can be command-based, where the user explicitly requests the activation of a particular skill. This is analogous to launching a software application or calling a specific function in a programming environment. An example would be a user stating, "AI, activate my 'Contract Review Skill' and analyze this document for indemnification clauses and liability limitations."

A third model involves agent-initiated invocation. As AI agent frameworks become increasingly sophisticated, the primary personal AI might, as part of executing a larger, more complex plan to fulfill a user's stated goal, autonomously decide to delegate specific sub-tasks to specialized skill modules. AI agentic frameworks are designed with tools that facilitate the design of agent functionalities and can decompose intricate tasks into smaller, more manageable steps. Each of these steps could then be handled by a distinct, purpose-built skill module, orchestrated by the core AI.

A defining characteristic of this new paradigm is the ephemeral nature of many skill activations. Once the specific task for which a skill was invoked is completed, that skill can be deactivated, thereby relinquishing the computational, memory, and potentially cognitive resources it was utilizing. This concept of "borrowed brilliance" signifies that the user gains access to high-level, specialized expertise for precisely the duration it is needed, without the long-term commitment or associated costs of permanently embedding that expertise;

either through years of personal learning or as a continuously running, monolithic AI system.

The term "ephemeral," meaning fleeting or short-lived, is apt here. It parallels the use of ephemeral environments in cloud computing, which provide temporary infrastructure that is stood down when no longer required, optimizing resource allocation and cost. Upon deactivation, the AI skill might leave its valuable output; such as a translated document, a summarized report, or a set of design parameters within the user's personal AI environment for future reference, but the active processing component itself is no longer consuming resources. This model allows for highly efficient resource utilization and more granular cost management for accessing advanced capabilities.

This shift towards composable AI skills fundamentally alters the architecture and role of personal AI. It evolves from being primarily a "knowledge repository" or a sophisticated information retrieval system into a "capability orchestration engine". The core intelligence of the personal AI increasingly resides in its ability to effectively select, invoke, manage, and synthesize the outputs from a diverse and potentially ever-changing array of specialized, often third-party, skill modules. Consequently, the value of the personal AI is measured less by the static knowledge base it contains and more by its dynamic ability to act as an intelligent router, integrator, and conductor of these plug-and-play skills.

The ephemeral use of these AI skills also introduces a potentially transformative economic model for accessing expertise. Instead of large upfront investments in monolithic software licenses or the continuous expense of retaining human experts for every conceivable need, access to highly specialized cognitive tools could be based on micro-transactions, pay-per-use, or time-boxed subscriptions. This significantly lowers the barrier to entry for sophisticated AI capabilities, effectively democratizing access to forms of expertise that were previously exclusive or prohibitively expensive.

Furthermore, the different invocation models; contextual, command-based, and agent-initiated-point towards a more fluid and adaptive human-AI partnership. The locus of control for task initiation can shift dynamically based on the situation and user preference. This fosters a more proactive and anticipatory form of AI assistance, moving beyond the traditional paradigm of AI as a purely reactive tool and towards a synergistic collaboration where the AI can take initiative within the bounds defined by the user, leading to a more seamless and potentially more powerful cognitive augmentation. This architectural flexibility, enabling the dynamic loading and unloading of specialized intelligence, is not merely a theoretical construct. It is already paving the way for a vast array of practical applications, demonstrating how domain-specific expertise can be curated and deployed in highly contextual ways across diverse professional and personal landscapes.

FROM LEGAL REVIEW TO LANDSCAPE DESIGN

The transformative potential of on-demand AI skills is vividly illustrated by their burgeoning application across a wide spectrum of professional and personal domains. These modular capabilities are moving beyond theoretical constructs to become practical tools that augment human intellect and streamline complex tasks.

In the legal field, traditionally characterized by its reliance on extensive human expertise and meticulous, time-consuming document review, on-demand AI skills are making significant inroads. Specialized "Legal AI Modules" can be invoked for a variety of analytical tasks. These AI assistants excel at rapidly processing and analyzing contracts to extract key information such as dates, parties involved, renewal clauses, and financial terms. They can also be trained to detect potential risks by identifying inconsistencies, ambiguous language, or unfavorable terms when compared against a predefined playbook or regulatory requirements. Furthermore, these tools can compare different versions of documents to highlight changes or ensure consistency across a portfolio of similar contracts and even automate the generation of summaries or alerts for upcoming deadlines.

Real-world examples underscore this utility: a retail chain might use a legal AI skill to analyze hundreds of supplier agreements for clauses impacting profitability; an international bank could employ an AI module to verify that complex corporate loan contracts comply with multifaceted local and international regulations; and a technology startup might leverage AI to manage and standardize a high volume of non-disclosure agreements (NDAs) with partners and employees, ensuring intellectual property protection. Generative AI is also being developed for more nuanced applications, such as assisting lawyers in redlining contracts by flagging problematic provisions and suggesting alternative wording, although these currently require careful human prompting and oversight to ensure robust legal analysis and contextual appropriateness. Invoking such a "Legal AI Skill" can dramatically reduce review times, enhance accuracy for specific analytical undertakings, and free up human legal professionals to focus on more strategic, high-value advisory work.

The complexities of logistical planning, such as organizing detailed travel itineraries, also lend themselves to augmentation by AI. "AI Travel Itinerary Designer Skills," exemplified by platforms like Wonderplan, offer a streamlined and personalized approach. Users can input their interests, budget constraints, travel dates, and personal preferences, and the AI skill can generate a comprehensive, customized itinerary. These skills often include features like dynamic accommodation recommendations based on user profiles, activity suggestions tailored to stated interests, and the ability to easily adjust plans. Many also provide offline access to the generated itinerary, ensuring all travel-related information is conveniently centralized and accessible, effectively providing the services of a personal travel agent on demand.

Negotiation, a critical skill in both professional and personal contexts, involves intricate interpersonal dynamics and strategic preparation. "AI Negotiation Coach Skills" are emerging to support individuals in this complex area. Tools such as Scotwork's AI-enhanced preparation tool can assist negotiators by helping them identify relevant variables and potential tradeables, craft and refine compelling opening statements, organize and prioritize negotiation issues based on importance, assess the balance of power between parties, and even highlight critical questions to ask during the negotiation process. Similarly, AI-driven coaches like the RPS Negotiation and Mediation Coach can be trained on specific negotiation theories and methodologies to provide users with tailored checklists, strategic frameworks, and assistance in anticipating difficult points or generating creative solutions to impasses. These tools effectively offer expert negotiation guidance precisely when it is invoked.

The realm of creative design, traditionally seen as a bastion of human ingenuity, is also being profoundly reshaped by the availability of on-demand AI skills. In architecture and landscape design, generative AI can function as a powerful "Design Skill". It can be employed to create "digital twins" of buildings or structures, which are virtual representations that can be stress-tested under various simulated conditions (e.g., weather, footfall, seismic activity) before any physical construction begins.

AI can also act as a collaborator with human architects, synthesizing vast amounts of data (e.g., site conditions, material properties, building codes, energy efficiency targets) with aesthetic preferences and novel design concepts to explore previously unconsidered solutions. Specific applications include generating a diverse array of façade design alternatives that resonate with the local architectural identity or historical context, rapidly transforming initial renderings to depict different seasons, lighting conditions, or even artistic styles (such as converting a standard rendering into a watercolor style for marketing materials), and even utilizing real-time camera feeds of physical models or everyday objects to generate photorealistic architectural renderings. These AI tools can automate many repetitive aspects of the design process and support the creation of optimal designs through sophisticated data analysis.

More broadly, "General Creative Design Skills," often powered by generative AI models like DALL-E or Midjourney, serve as catalysts for exploration and ideation across numerous creative fields. They can assist in generating initial concepts for logos, user interface (UI) mockups, product designs, and marketing materials. They can also refine existing designs by suggesting modifications to color palettes, layouts, or typographic treatments. A key advantage is their ability to generate multiple design variations with great speed, facilitating rapid prototyping, A/B testing, and the iterative incorporation of feedback. AI can also be used to generate floor plans based on specified constraints, develop branding concepts, or even create novel recipes by drawing on vast datasets and user-defined parameters.

An AI skill could, for example, be invoked to guide users through structured mindfulness exercises, deliver specific modules of cognitive behavioral therapy (CBT) such as thought challenging or behavioral activation, provide a form of simulated empathetic listening, or help users track mood patterns and offer personalized coping strategies based on established therapeutic protocols. Such a skill would not aim to replace human therapists but could offer accessible, private, on-demand support for mental well-being management, acting as a "copilot" to an individual's ongoing mental health journey. (It is important to acknowledge that the development and deployment of such skills would involve significant ethical, clinical efficacy, and regulatory considerations, which are beyond this chapter's primary focus but serve to illustrate the potential breadth of AI skill types.)

The diverse range of these AI skills-spanning highly analytical tasks in law, complex logistical planning in travel, strategic advisory roles in negotiation, and expansive creative generation in design; indicates that "plug-and-play intelligence" is not confined to narrow, technical applications. Instead, it is emerging as a general-purpose paradigm for augmenting human capability across nearly all facets of professional and personal life.

Many of these on-demand AI skills are evolving beyond simple task automation tools to become collaborative partners; acting as "copilots," "coaches," or "creative collaborators". This fundamentally changes the human-tool relationship, emphasizing augmentation, interactive refinement, and shared intelligence rather than mere replacement of human effort. Furthermore, the increasing ability to fine-tune AI models for specific styles (like the LoRA technique for achieving a particular principal's sketching style in landscape architecture), contexts, or proprietary datasets suggest that on-demand AI skills can offer highly personalized and specialized "bespoke brilliance".

This moves beyond generic, one-size-fits-all expertise, allowing AI tools to be tailored to individual user needs, organizational cultures, or specific project requirements, making them more valuable and deeply integrated than ever before. The sheer variety and growing sophistication of these domain-specific AI skills, all potentially accessible through a unified personal AI, naturally compel a re-evaluation of how individuals traditionally acquire and demonstrate competence. This leads directly to the consideration of how our fundamental models of learning must adapt to this new reality of activatable, on-demand knowledge.

THE NEW LEARNING MODEL

The advent of readily available, activatable AI skills necessitates a fundamental rethinking of traditional learning models. In a landscape where specialized knowledge can be invoked on demand and where the half-life of existing skills is rapidly diminishing, the conventional approach of front-loading education with vast amounts of factual content, followed by sporadic and often lengthy retraining courses, becomes increasingly inadequate. A new learning model is emerging, one that prioritizes adaptability, continuous engagement, and a shift in focus from comprehensive knowledge acquisition to the development of critical orchestration and evaluation abilities.

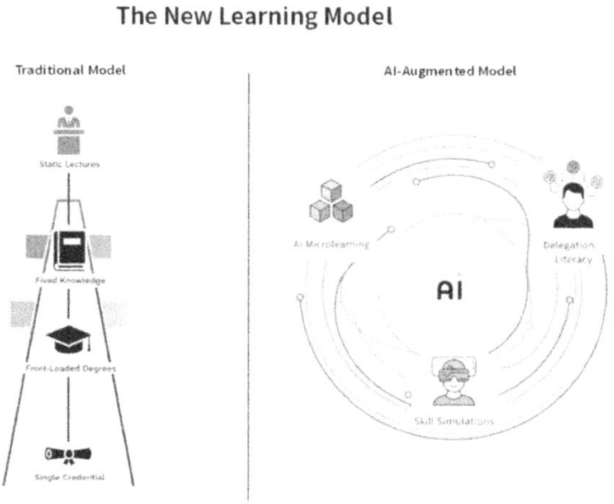

AI-powered microlearning is a cornerstone of this new model, designed for continuous adaptation in a fast-paced world. Microlearning involves delivering educational content in short, focused, and easily digestible modules that can be consumed in the interstices of a busy professional life. Artificial intelligence significantly enhances this approach by deeply personalizing the learning experience. AI algorithms can analyze an individual learner's existing knowledge base, stated learning goals, preferred learning style, and even real-time performance data. Based on this analysis, the AI can tailor lessons, curate the most relevant content from vast repositories, and recommend supplementary materials or activities that address specific knowledge gaps or areas of interest. This ensures that learning is not only efficient and engaging but also directly applicable to the learner's immediate needs and long-term aspirations.

Acquiring practical skills, especially those that are complex or involve nuanced interpersonal interactions, often requires extensive hands-on experience. AI-driven simulations are rapidly becoming a powerful tool for providing such practice in safe, repeatable, and highly adaptive virtual environments. Empirical evidence suggests that learners retain knowledge significantly more effectively through simulation-based practice-up to 75% retention compared to just 5% from traditional lectures-and can acquire skills as much as 35% faster. These sophisticated simulations can feature autonomous AI avatars that facilitate realistic role-playing scenarios, crucial for developing soft skills like negotiation, customer service, or leadership. They often incorporate adaptive learning algorithms that adjust the difficulty of tasks in real-time based on learner performance and provide immediate, constructive feedback.

Dynamic scenarios can mimic the complexity and unpredictability of real-world situations, allowing learners to hone their decision-making abilities. For instance, an individual could practice a "Public Speaking Skill" by delivering a presentation to a simulated AI audience that offers feedback on pacing, clarity, and audience engagement. These "skill simulation loops" enable rapid familiarization with new competencies and support iterative improvement without the risks or costs associated with real-world trial and error.

Perhaps the most profound shift in the new learning model is the move from an emphasis on deep, comprehensive acquisition of knowledge to the cultivation of "delegation literacy". When a sophisticated AI skill can reliably perform a complex task; be it advanced financial modeling, intricate software coding, or a detailed review of scientific literature, the imperative for an individual to achieve deep, procedural mastery of that specific task lessens. Instead, the learning focus shifts. It becomes more critical to achieve a sufficient level of familiarity with the domain: understanding the core concepts involved, recognizing the capabilities and, equally importantly, the limitations of the relevant AI skill, discerning the nature of its required inputs and expected outputs, and, crucially, developing the judgment to know when and how to effectively delegate the task to the AI. This "delegation literacy" involves knowing which questions to ask the AI, how to frame problems in a way the AI can process, and how to critically evaluate, validate, and integrate the AI's responses or outputs.

In this evolving educational landscape, the personal AI, with its capacity to load a diverse array of specialized skills, becomes an omnipresent learning companion. It can function as an AI tutor, delivering personalized explanations, adapting to the learner's pace, and providing targeted guidance when difficulties arise. As discussed, it serves as a powerful simulator for hands-on skill practice. Furthermore, the AI acts as a crucial scaffolding partner. It can break down complex problems or learning objectives into more manageable steps, automate routine or prerequisite sub-tasks (allowing the learner to concentrate on more challenging aspects), and provide contextual information or just-in-time support precisely when needed. This scaffolding allows human learners to focus their finite cognitive resources on higher-order thinking, strategic decision-making, creative problem-solving, and the integration of knowledge, while the AI handles much of the operational execution or information retrieval.

This new learning model prioritizes the development of "meta-learning" skills. These include enhanced critical thinking, sophisticated problem framing, the ability to rigorously validate AI-generated outputs (identifying potential biases or inaccuracies), and the agility to rapidly learn how to use new AI tools and skills as they emerge. These meta-competencies become more valuable than the deep memorization of factual or procedural knowledge that AI systems can now readily provide and apply. "Delegation literacy" emerges as a cornerstone of future professional competence. It requires individuals to cultivate a nuanced understanding of AI capabilities not for the purpose of replicating them, but to effectively, efficiently, and ethically orchestrate them. This includes developing the discernment to judge the trustworthiness of different AI skills and to be alert to their potential inherent biases.

Consequently, the boundary between formal learning and active work becomes increasingly porous and indistinct. AI-augmented work environments themselves provide continuous, embedded opportunities for microlearning, skill simulation, and on-the-job refinement of capabilities through direct, interactive engagement with intelligent tools.

Learning transforms from a discrete, episodic phase into an ongoing, adaptive, and integrated process, woven into the fabric of daily professional life. This transformative learning model, emphasizing lifelong orchestration of AI-driven capabilities rather than just front-loaded instruction, offers unprecedented flexibility and access to knowledge. However, the very act of composing, deploying, and managing these powerful AI skills is not without its own set of considerations, particularly concerning the economic and cognitive costs that such sophisticated augmentation entails.

THE COST OF COMPOSITION

The allure of dynamically composing intelligence through on-demand AI skills is undeniable, offering unprecedented access to specialized capabilities. However, this power is not without its price. The "cost of composition" is a multifaceted concept, encompassing not only direct economic expenditures but also significant cognitive and temporal investments required to effectively manage and orchestrate these AI resources.

The most immediate and tangible cost is economic, frequently structured around usage-based pricing models. Many Large Language Models (LLMs) and specialized AI services operate on a token-based system, where fees are incurred for both the input data processed by the AI and the output data it generates.

For instance, prominent models like GPT-4 have distinct rates per thousand tokens for input and output streams. A critical factor here is that these costs can escalate non-linearly, sometimes quadratically, with the length or complexity of the information sequence being processed. This implies that extensive or complex use of AI skills, especially across numerous users or for intricate tasks, can lead to rapidly accumulating expenses. This is vividly illustrated by cost growth projections showing how total monthly costs can multiply as user numbers, prompts per user, and token usage increase.

Consequently, both individuals and organizations will find it necessary to engage in meticulous budget optimization, develop accurate forecasts for token usage, and potentially employ sophisticated prompt engineering techniques to minimize token consumption without sacrificing output quality. The pricing models themselves can be intricate, possibly involving tiered access based on usage volume or two-part tariffs that include a fixed access fee plus variable usage charges, with higher markups for more intensive users. Even the crucial processes of experimenting with different prompts to elicit optimal responses or fine-tuning models to enhance skill performance for specific applications incur these token-based costs, adding another layer to the economic calculus.

Beyond financial outlays, there is the challenge of temporal orchestration. Just as the human brain cannot efficiently maintain peak focus on numerous complex cognitive tasks simultaneously, a personal AI will face practical limitations in running multiple sophisticated AI skills in parallel. These limitations arise from several factors: finite computational resources (such as processing power and memory, especially on personal devices), the potential for interference or conflicting operations between concurrently active skills, direct cost considerations (as running many resource-intensive skills at once could be prohibitively expensive), or logical dependencies where the output of one skill is a necessary input for another to begin its process.

Therefore, the effective utilization of a diverse suite of AI skills will necessitate "temporal orchestration": the intelligent scheduling, activation, and deactivation of different skills over time to achieve a larger, composite goal in an efficient manner. This is analogous to how an operating system manages multiple processes and threads, or how a project manager sequences various tasks and resource allocations. Compound AI systems, which by their very definition integrate multiple AI models or techniques to solve complex problems, inherently involve such orchestration to manage the division of labor and the flow of information between their constituent components.

A critical aspect of composing intelligence lies in the human-AI decision logic that governs when to invoke a particular specialized skill and, equally importantly, when to decline its use or opt for an alternative approach. The mere availability of an AI skill does not automatically mandate its deployment for every relevant situation. This decision-making framework, whether executed consciously by the human user or increasingly automated within their core personal AI, will need to weigh a variety of factors.

These include a careful assessment of the anticipated cost versus the potential benefit of using the skill for a specific task, the required level of accuracy and reliability for the task's output, the availability and suitability of other methods (which might include relying on human expertise or simpler AI tools), the time sensitivity of the task, and the ethical implications or potential risks associated with deploying a specific AI skill in a given context. This necessitates a new form of critical evaluation, focusing on the appropriateness, trustworthiness, and potential biases of an AI tool for each unique situation.

Finally, while AI skills hold the promise of offloading many cognitive tasks from humans, the act of managing, curating, and orchestrating a portfolio of these intelligent tools introduces its own form of cognitive load for the user. Users must invest mental effort to understand the capabilities and limitations of each skill in their toolkit, monitor their performance during execution, integrate their diverse outputs into a coherent and useful whole, and stay abreast of updates, new skill availability, or changes in skill functionalities.

Cognitive Load Theory (CLT) provides a framework for understanding how working memory can be overwhelmed if information is presented poorly or if tasks are overly complex. While AI can be designed to help manage cognitive load in learning environments, poorly designed interfaces for AI skill management or overly complex orchestration systems could inadvertently increase extraneous cognitive load on the human user. Learning to effectively manage one's augmented AI toolkit thus becomes a significant cognitive task in its own right. It is noteworthy that while compound AI systems can reduce the cognitive load on individual AI components by dividing tasks among specialized models, the human orchestrator still bears the overarching cognitive responsibility of managing the compound system as a whole.

The "cost of composition," therefore, is a multifaceted consideration that extends beyond direct financial outlays for token usage or subscriptions. It also includes what might be termed "cognitive tokens": the finite mental effort, attention, and decision-making capacity that a human must expend to effectively manage, orchestrate, validate, and integrate the outputs of their AI skill portfolio. This implies that a new meta-skill of "personal AI resource management" will become crucial. Individuals and organizations will need to learn how to strategically budget and allocate not only their financial resources but also their own cognitive bandwidth and available computational capacity across a dynamic portfolio of AI skills to achieve optimal outcomes. This is not simply about minimizing monetary cost, but about maximizing the overall value derived from AI augmentation, considering all constraints.

Interestingly, the human-AI decision logic for skill invocation is likely to become a key area for AI development. As Personal AIs (PAIs as defined in Chapter 7) become more attuned to user preferences, task contexts, and historical cost-benefit trade-offs, they could evolve to automate or provide highly intelligent recommendations for skill deployment.

This would, in effect, allow the AI to manage some of its own "cost of composition," reducing the direct cognitive load on the human user for routine or well-understood compositions and freeing them to focus on more novel or strategic orchestration challenges. The necessity of skillfully managing these diverse costs-financial, temporal, and cognitive-and orchestrating an array of AI capabilities effectively, points towards a fundamental reshaping of how work itself is structured, valued, and executed. This leads directly to consideration of work not as a static set of prescribed tasks, but as a dynamic act of intelligent composition.

WORK, REIMAGINED AS COMPOSITION

The advent of on-demand, composable AI skills is not merely an incremental improvement in productivity tools; it signals a fundamental re-architecture of work itself. As AI capabilities become increasingly adept at handling a wide array of discrete tasks; from drafting legal clauses and generating sophisticated design mockups to analyzing complex market data and coding software modules, the primary role of human workers undergoes a significant transformation.

Instead of being deeply immersed in the minutiae of granular task execution, human effort will increasingly be directed towards defining overarching goals, designing intelligent and adaptive workflows, strategically selecting and sequencing the appropriate AI skills for various stages of these workflows, and then critically evaluating, integrating, and refining the outputs generated by these AI collaborators. In this new paradigm, the human professional evolves from being primarily a direct laborer or a solitary expert into an architect of solutions, a conductor of diverse intelligences, or a curator of AI-driven capabilities.

This evolving mode of operation naturally gives rise to novel professional roles and dynamic team structures. One such emerging role is that of the "composition specialist". This individual is characterized by their ability to deeply understand complex business problems or strategic opportunities and then assemble a bespoke solution by innovatively combining existing or newly configured AI modules and skills. Their core expertise lies not necessarily in creating the foundational AI models themselves, but in the art and science of their effective combination, integration, and application to real-world challenges. Compositional AI, by its nature, involves "developing a new combined module using existing or new sub-modules that will perform a specific part of the combined module or solution". The establishment of academic programs, such as Master's degrees in "Composition – AI-Integrated Writing", signals the formalization of this specialization, focusing on melding AI tools with human creative and analytical skills.

Furthermore, teamwork will increasingly be characterized by hybrid human-AI configurations. AI agents will function seamlessly within human teams, serving as intelligent assistants for administrative tasks, powerful data analysts for enhanced decision-making, creative partners for brainstorming and content generation, or specialized skill augmenters for technical procedures. In these symbiotic teams, humans and AI agents will each contribute their unique strengths; humans providing strategic direction, ethical judgment, complex problem-solving, and nuanced contextual understanding, while AI offers speed, scale, data processing power, and access to specialized knowledge.

This shift also means that the traditional concept of a "job," often rigidly defined by a relatively static set of responsibilities and a fixed list of required skills, begins to erode in relevance. In a "skills on demand" paradigm, professional value becomes increasingly tied to the achievement of specific, measurable outcomes rather than mere adherence to a predefined job description or hours worked. This paves the way for more dynamic, project-based work environments where individuals (and their accompanying personal AIs) are assembled into teams based on their demonstrated ability to contribute to a particular goal. Success in such an environment relies on orchestrating a fluid and contextually appropriate set of human and AI capabilities. Consequently, careers may become less linear and more akin to a curated portfolio of successfully orchestrated outcomes and demonstrated compositional abilities.

As plug-and-play intelligence significantly reduces information asymmetry by making expert-level knowledge and sophisticated processing capabilities widely and often instantly accessible, a core theme of this book, the traditional loci of competitive and professional advantage inevitably shift. The new frontiers of differentiation become insight asymmetry and execution asymmetry.

Insight Asymmetry: This refers to the capacity to derive unique, profound, or strategically valuable insights from the abundant, often AI-processed, information that others might overlook, misinterpret, or fail to synthesize effectively. It is about superior sense-making, advanced critical thinking, the creative interpretation of data within its broader context, and the ability to ask more potent questions. If, for example, multiple competitors can deploy a "market analysis AI skill" that generates similar reports, the advantage accrues to the individual or team that can discern the subtle, emergent trend, identify the unstated implication, or recognize a novel opportunity hidden within the AI's output.

Execution Asymmetry: Beyond merely possessing a unique insight, this is the superior ability to act upon that insight decisively, effectively, and efficiently to achieve desired outcomes. In an AI-augmented world, effective execution often involves skillfully orchestrating further AI skills, coordinating human actions, or mobilizing resources to implement a strategy or solution more rapidly, with higher quality, or at a lower cost than others. It is about translating understanding and vision into tangible results in a hybrid human-AI operational environment.

The traditional managerial role, therefore, undergoes a profound transformation. It shifts from primarily overseeing human task execution and resource allocation to designing, curating, monitoring, and troubleshooting complex "workflows of intelligence". These workflows are executed by dynamic combinations of human talent and diverse AI agents. This new managerial function requires an enhanced skill set, including strong capabilities in systems thinking, a deep literacy in AI capabilities and limitations, proficiency in human-AI interaction design, and a robust framework for ethical oversight and governance of AI-driven processes.

"Execution asymmetry" will increasingly be determined by the sophistication of an individual's or an organization's "AI composition toolkit". This includes the power and adaptability of their core personal or enterprise AI platform (as discussed in Chapter 7) and their learned proficiency in rapidly discovering, vetting, integrating, and deploying diverse AI skills to act on emergent insights. This creates a new form of "capability gap," where those with superior AI augmentation and more refined compositional skills can achieve outcomes that are out of reach for those less adept or equipped.

The rise of "composition specialists" signals the emergence of a new and vital intermediary layer in the evolving knowledge economy. These specialists are not necessarily the creators of the foundational AI models, nor are they simply end-users consuming pre-packaged AI services.

They are the architects of applied intelligence, creating significant value by innovatively combining modular AI capabilities to solve specific, often complex, problems. Their work embodies the principle that the whole (the composed solution) can be far greater than the sum of its parts (the individual AI skills). This dynamic, outcome-oriented, and composition-driven vision of work, where humans orchestrate networks of specialized AI intelligence, cannot materialize in a vacuum. It fundamentally relies on the existence of robust, accessible, and interoperable ecosystems where these AI skills can be discovered, trusted, and seamlessly integrated into the fabric of daily workflows and strategic initiatives.

FROM SKILLS TO ECOSYSTEMS

The transformation of work into an act of composition, driven by on-demand AI skills, necessitates a foundational infrastructure that allows these capabilities to be easily found, reliably invoked, and seamlessly integrated. This leads to the emergence of sophisticated AI skill ecosystems, characterized by marketplaces for discovering capabilities, robust standards for ensuring interoperability, and advanced semantic layers that enable a shared understanding of meaning between diverse AI agents and human users.

For on-demand AI skills to achieve widespread adoption and utility, users and their personal AIs require efficient mechanisms to discover, evaluate, and access them. This points directly to the development of "AI skill marketplaces" or comprehensive "skill registries". These platforms would function somewhat analogously to contemporary app stores for software or mobile applications, but with a focus on modular AI capabilities. Such marketplaces would list available AI skill modules, provide detailed descriptions of their functionalities, offer performance benchmarks, outline associated costs (likely based on token usage or subscription models), and potentially feature user reviews, ratings, or certifications.

Platforms like Spire.AI, which describe an AI-driven ecosystem for discovering talent and project opportunities based on skills, offer a conceptual parallel. This model can be extended to AI skills discovering other AI skills for collaborative tasks, or being programmatically discovered and invoked by users' personal AIs based on contextual needs. The concept of "credentialed modules" is particularly crucial within these ecosystems. For users to confidently trust and rely on these AI skills, especially for critical or sensitive tasks in domains like finance, healthcare, or law, there will likely be a pressing need for robust systems of verification, certification, or credentialing. These systems would attest to a skill's quality, security protocols, ethical alignment, data privacy measures, and operational reliability.

A vibrant and functional ecosystem of AI skills, especially one populated by modules from numerous different developers and providers, absolutely requires that these disparate components can work together harmoniously within a user's personal AI environment or an organization's broader AI infrastructure. This necessitates the establishment and adoption of robust interoperability standards.

The Agent Communication Protocol (ACP), proposed by BeeAI and IBM, serves as an early example of such a standard. ACP aims to provide a common, RESTful, low-latency communication layer for AI agents, supporting both asynchronous and synchronous interactions and even facilitating offline discovery of agent capabilities. However, achieving true, comprehensive agentic interoperability is a complex, multifaceted challenge.

As outlined by Forrester, this encompasses at least nine critical dimensions that need to be addressed: tool use and integration (how agents access external tools and data), inter-agent communication and coordination (protocols for message exchange and task delegation), identity and trust mechanisms (verifying agent authenticity and permissions), shared memory models (allowing agents to retain context across interactions), knowledge sharing and reasoning protocols (enabling collaborative problem-solving), marketplace and transaction standards (for an economy of agent services), governance frameworks (ensuring ethical and compliant behavior), discovery protocols (how agents find each other), and standardized error handling and conflict resolution mechanisms. Early initiatives like AGNTCY are beginning to address some of these aspects, but the field is still nascent.

Beyond syntactic interoperability; which primarily concerns ensuring that data formats and communication protocols are compatible-effective collaboration between diverse AI skills demands a deeper level of shared understanding: semantic interoperability.

This is the ability for different systems or agents to exchange information with an unambiguous, shared understanding of the meaning of that information. This is where ontologies and symbolic AI techniques play a vital role. Ontologies act as formal, explicit specifications of a shared conceptualization; in simpler terms, they serve as shared dictionaries and rulebooks that define concepts, properties of those concepts, and the relationships between them within a specific domain (such as finance or medicine) or even across multiple domains.

This structured representation of knowledge allows AI agents to communicate with significantly less ambiguity and to reason more effectively about the information they exchange. For example, the Financial Industry Business Ontology (FIBO) works to standardize financial terminology and concepts, enabling different financial systems and AI agents to interpret financial data consistently. When AI skills operate with different underlying ontologies, specialized ontology mediation methods (such as ontology matching, alignment, and merging) and sophisticated ontology negotiation protocols are employed to bridge these semantic gaps, enabling agents to dynamically establish common ground and achieve a shared understanding. This "meaning layer" is fundamental to the fluid and intelligent operation of a compositional AI ecosystem.

As these AI skill ecosystems mature and become more integrated into daily workflows, new competitive dynamics and potential dependencies will emerge. Platforms or providers whose AI skills are most easily discoverable, highly interoperable, demonstrably trustworthy, and semantically rich may gain significant market influence. Users' personal AIs, through learning and optimization, might develop preferences for or even default to certain skill providers or platforms based on past performance, cost-effectiveness, or ease of integration.

This could create a form of ecosystem lock-in, where switching to alternative skill providers becomes more difficult or costly due to established dependencies and learned efficiencies. Consequently, companies and developers will compete intensely not only on the quality of individual AI skills but also on their ability to become the "default search engine," the "preferred app store," or the most trusted source of on-demand capabilities for a user's personal AI.

The development of these AI skill ecosystems bears a resemblance to the evolution of app stores for mobile devices. Both involve registries for discovery, mechanisms for distribution, and some level of developer standards. However, the requirements for AI skill ecosystems are arguably more complex due to the inherently collaborative and compositional nature of advanced AI agents. While many mobile apps can function effectively in relative isolation, composable AI skills, as defined in this chapter, must interoperate effectively, often in intricate sequences or ensembles. This demands a far deeper level of semantic interoperability than simple API compatibility, making the underlying "meaning layer" a critical differentiator. The "credentialing" of AI skills will likely emerge as a vital function within these ecosystems.

Given that AI skills will increasingly handle sensitive tasks with significant real-world consequences, users and organizations will require robust assurances of their safety, ethical alignment, and reliability. This could lead to the establishment of new regulatory bodies, industry-led certification standards, or trusted third-party certifiers, paralleling existing systems of professional licensing for humans. Such mechanisms will be essential for building and maintaining trust in these powerful new capabilities. Ultimately, competition in the burgeoning AI skill economy will not be solely about the performance of individual, isolated skills. It will increasingly be determined by the strength and sophistication of the semantic layer a provider offers-that is, how well their skills can "understand" and "be understood by" other components, including the user's core personal AI and other third-party skills, within the broader ecosystem.

The creation of these vibrant, interoperable skill ecosystems is foundational. However, for these diverse AI agents and skills to truly collaborate and form complex, intelligent compositions, they need more than just compatible connections; they require a deeply shared understanding of meaning. This critical requirement sets the stage for the subsequent discussion on the nature and implications of semantic economies.

SUMMARY

This chapter has illuminated how the very concept of expertise is being radically reshaped by the advent of composable, on-demand AI skills. We have journeyed from understanding the inherent problems and inefficiencies of relying on rigid, permanent human expertise in a rapidly changing world to envisioning a new model. In this model, a persistent personal AI can dynamically load and unload specialized capabilities-ranging from sophisticated legal analysis and creative landscape design to personalized negotiation coaching and adaptive microlearning support-as needed. This "plug-and-play intelligence" transforms not only how individuals work and learn, shifting the focus from deep, laborious acquisition of knowledge to the agile orchestration of capabilities, but also the very structure of tasks, roles, and entire professions.

The journey towards this new paradigm involves navigating the emerging economics of token-based pricing for AI skill usage and managing the novel cognitive demands of orchestrating these powerful AI resources. Work itself is being reimagined as an act of composition, with humans increasingly acting as designers and conductors of hybrid human-AI teams. In this landscape, professional success and competitive advantage are less about the mere possession of information and more defined by the ability to generate unique insights from it (insight asymmetry) and to effectively execute upon those insights (execution asymmetry). For such a dynamic and powerful system of on-demand skills to flourish, the development of robust ecosystems-featuring discoverable, credentialed, and interoperable AI modules-is essential.

The seamless interaction and intelligent composition of these diverse AI skills, which will form the backbone of future work and human capability, fundamentally depend on a shared layer of understanding that transcends mere data exchange. The next crucial step in this profound evolution is the development of robust semantic economies, where meaning itself becomes the protocol, the currency, and the very fabric enabling fluid, intelligent, and purposeful collaboration across the entire expanding AI landscape.

9

SEMANTIC ECONOMIES

The contemporary digital landscape presents a paradox of connection and fragmentation. Myriad systems, applications, and services are interwoven through intricate networks, promising unprecedented levels of data exchange and automation. Yet, beneath this veneer of connectivity lies a persistent challenge: these systems often struggle to truly understand one another. Much like a global assembly where participants speak different languages and rely on literal, often imperfect, translators, the digital world has long operated on protocols that prioritize structure over substance. This reliance on syntax-rigid data formats, predefined Application Programming Interface (API) structures, and hard-coded rules-has been the bedrock of digital communication. However, this foundation is proving increasingly brittle and ill-suited for the next wave of intelligent systems.

The cost of this syntactic rigidity is not merely technical; it translates into tangible business impediments. Organizations grapple with lost productivity as teams manually reconcile data between systems that cannot seamlessly communicate. Innovation is stifled by the sheer complexity and expense of integrating new technologies with legacy infrastructures. Strategic decision-making is often paralyzed by data silos, where valuable information remains trapped within systems that cannot share their insights effectively. The effort to bridge these syntactic divides consumes vast resources, diverting capital and talent from value-creating activities towards mere operational maintenance.

A fundamental shift is underway, driven by rapid advancements in Artificial Intelligence (AI), particularly in Natural Language Processing (NLP) and the capabilities of Large Language Models (LLMs). These AI systems are not content with simply processing structured data; they are designed to understand, interpret, and generate information based on meaning, or semantics. This evolution is compelling a move away from syntax-bound communication towards interactions grounded in shared understanding. This is not merely about creating smarter AI; it is about forging a new, more intelligent protocol layer for the entirety of digital interaction-encompassing software, services, and even the mechanisms of influence.

This transformation heralds the emergence of what can be termed the "semantic economy". This is an ecosystem where value is created, exchanged, and governed based on shared meaning, context, and intent. In such an economy, services are not just discovered through keyword searches or static directories; they are identified, invoked, and priced based on their semantic capabilities, their trustworthiness, and their ability to understand and fulfill complex intentions. Meaning itself transitions from a passive characteristic of data to an active, tradable, and foundational element of digital commerce and collaboration. It becomes a form of currency, a basis for novel business models, and a critical component of the digital infrastructure.

This chapter navigates the contours of this profound shift. It begins by examining the inherent fragilities and escalating costs of our current syntax-driven integration models. We then delve into the core technological building blocks: ontologies, knowledge graphs, vector embeddings, and natural language interfaces that constitute the "semantic layer" enabling shared understanding. The subsequent sections will explore how AI agents communicate and collaborate within this new paradigm, the novel pricing mechanisms that arise when meaning has value, and the transformative impact on business models as companies compete for AI invocation.

Crucially, the discussion will address the essential role of standards, governance, and ethical considerations in ensuring a functional and trustworthy semantic ecosystem. The chapter culminates in a vision of meaning as a foundational infrastructure, reshaping the digital world in ways that are only beginning to become apparent.

For leaders and strategists, understanding the transition to a semantic economy is not an academic exercise; it is a strategic imperative. The ability to comprehend and leverage meaning-centric interaction will define competitive advantage in an increasingly AI-driven future. The journey from systems that merely process data to systems that truly understand and act upon meaning is complex, but it is a journey that is already reshaping industries and redefining the art of the possible.

THE FRAGILITY OF SYNTAX: WHY CURRENT INTEGRATION MODELS ARE BREAKING

The digital infrastructures upon which modern enterprises are built have long relied on a principle of structured communication. Application Programming Interfaces (APIs) and system integrations have traditionally depended on precise, pre-defined data formats, schemas, and communication protocols to enable different software components to exchange information. While this approach has facilitated a degree of interoperability, it has also introduced a significant degree of brittleness into the digital ecosystem. Any deviation, update, or change in the syntactic structure of one system can easily break the chain of communication, leading to system failures, data corruption, and the need for costly manual interventions.

This inherent fragility stems from the tight coupling between systems, where the logic of one is intricately tied to the specific formatting requirements of another. Often, APIs expose internal system details; a phenomenon known as "leaky abstractions", forcing consuming applications to understand and adapt to these internal complexities, further exacerbating the fragility and increasing the maintenance burden.

The consequences of these syntax-dependent integration models are far-reaching, manifesting as both hidden and explicit costs that erode efficiency and hinder innovation. Hidden costs often surface as lost productivity, with employees dedicating valuable time to manually transferring data between non-communicating systems, correcting errors arising from data inconsistencies, and navigating cumbersome workarounds.

For instance, a common scenario involves sales departments using one Customer Relationship Management (CRM) system while finance departments utilize a separate system for billing. Without seamless integration, discrepancies in client information, incorrect invoices, and payment delays become frequent occurrences, directly impacting cash flow and customer satisfaction. This fragmentation also leads to the creation of data silos, where critical business information remains isolated, preventing a unified view of operations and hampering informed decision-making.

Beyond these operational inefficiencies, the explicit costs of maintaining a syntax-driven integration landscape are substantial. Organizations invest heavily in expensive middleware solutions and engage consultants to bridge the gaps between disparate systems. IT departments allocate significant resources to troubleshooting integration failures and performing ongoing maintenance, diverting budgets that could otherwise be invested in strategic initiatives.

Industry analyses indicate that while simple integrations might start at a few thousand dollars, complex projects can easily exceed tens of thousands, with annual maintenance and support costs potentially reaching $50,000 to $150,000 per integration. Custom development efforts to forge these connections can demand hundreds of engineering hours and annual expenditures ranging from $50,000 to $150,000. These integration challenges can culminate in severe consequences, including corrupted datasets, unrealized revenue, and damage to customer relationships and brand reputation.

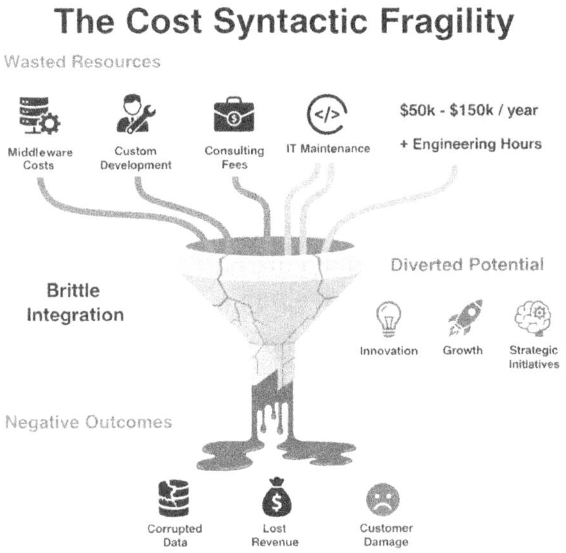

This reliance on syntactic shortcuts and ad-hoc solutions contributes to a growing burden of "integration technical debt". This concept refers to the implied cost of rework and future complexity caused by choosing expedient, but suboptimal, integration methods in the present.

Like financial debt, integration technical debt accrues "interest" over time, making subsequent modifications, system upgrades, or the adoption of new technologies progressively more difficult, time-consuming, and expensive. Hard-coded values within integrations, a lack of comprehensive documentation for APIs, and poorly considered architectural decisions are common contributors to this accumulating debt. The impact is a reduction in organizational agility, inflated operational costs, and a diminished capacity to respond swiftly to evolving market demands or to effectively leverage transformative technologies like AI, which often require access to diverse and dynamically changing data sources.

The "brittleness" of syntax, therefore, is not merely a technical inconvenience but a fundamental business risk, directly impacting an organization's ability to innovate and compete. This technical debt acts as a pervasive, hidden tax on all future digital transformation efforts, disproportionately affecting AI initiatives that depend on the seamless flow of information from varied origins.

The challenge of syntactic fragility is further amplified by the escalating complexity of the digital environment. The proliferation of AI agents, microservices, Internet of Things (IoT) devices, and cloud-based applications means that the number of potential integration points is growing exponentially. In such a dynamic and distributed landscape, traditional models of point-to-point, syntax-dependent integration become unmanageable and unscalable. Infrastructure bottlenecks, such as unstable middleware platforms or architectural flaws that create single points of failure, can severely limit the ability of syntax-reliant systems to handle the increasing volume and velocity of data exchange.

The sheer scale of interaction demands a more fluid, adaptive, and inherently intelligent communication protocol that can transcend rigid structural dependencies. Even the widespread adoption of APIs, initially heralded as a panacea for interoperability, can paradoxically lead to increased complexity if these APIs are purely syntactic. Without a shared semantic understanding, integrating with numerous distinct APIs means grappling with a multitude of unique data formats and operational logics, recreating the brittleness problem at an even larger scale and leading to "API sprawl".

A common and particularly vexing manifestation of syntactic fragility is "schema mismatch". This occurs when the data structure, data types, or naming conventions used in one system do not align with those of another, despite potentially referring to the same underlying concepts. For example, one system might store customer IDs as integers while another uses strings, or "revenue" might be defined as gross sales in one database and net sales in another. Such mismatches can lead to data migration failures, incorrect data processing, and flawed analytical outcomes. A simple analogy illustrates the problem: a child develops a schema for a "cat" (furry, four legs, meows). When encountering a dog, which barks, the child's existing schema leads to a misclassification. Similarly, digital systems misinterpret data if their fundamental "schemas"-their definitions and structures for information-differ without a mechanism for reconciliation.

The inherent limitations, escalating costs, and compounding complexities of syntax-driven integration models signal an urgent need for a paradigm shift. The current approach acts as a significant bottleneck to innovation, hindering the deployment of truly intelligent, autonomous systems that require a more profound level of understanding and interaction. The path forward lies in developing systems capable of communicating and collaborating based not just on pre-defined structures, but on a shared comprehension of meaning.

The following list summarizes the escalating costs associated with the limitations of purely syntactic integration approaches:

Integration failures

- **Description of syntactic limitation:** Hard-coded mappings, incompatible API versions, and brittle format assumptions that break when systems change.
- **Quantifiable business impact:** Corrupted datasets, lost

revenue, and customer service failures that cascade across dependent systems.

Data silos

- **Description of syntactic limitation:** Absence of a shared data model and reliance on proprietary data formats that prevent semantic alignment.
- **Quantifiable business impact:** Lost productivity from manual data transfers, inaccurate or inconsistent reporting, and delayed decision-making.

Maintenance overhead

- **Description of syntactic limitation:** Brittle point-to-point connections and complex dependencies that require frequent fixes and regression testing.
- **Quantifiable business impact:** Increased IT resource allocation for troubleshooting and reliance on expensive middleware, with annual costs commonly ranging from $50,000 to $150,000.

Reduced agility

- **Description of syntactic limitation:** High complexity makes modifying existing integrations or adding new ones slow and risky.
- **Quantifiable business impact:** Inability to respond quickly to market changes and slower adoption of new technologies.

Scalability limits

- **Description of syntactic limitation:** Bottlenecks created by centralized syntactic processing and N×M integration complexity as systems multiply.

- **Quantifiable business impact:** Inability to handle growing data volumes from AI agents and IoT devices, along with increased latency and degraded performance.

Security vulnerabilities

- **Description of syntactic limitation:** Inconsistent security protocols and controls across loosely integrated or disconnected systems.
- **Quantifiable business impact:** Greater exposure to cyber threats, compliance failures, and potential legal and financial repercussions.

Technical debt accumulation

- **Description of syntactic limitation:** Short-term fixes layered onto fragile integrations, compounding complexity over time.
- **Quantifiable business impact:** Higher future development and maintenance costs, reduced system reliability, and innovation slowed by tightly coupled, entangled architectures.

These challenges underscore the necessity for a more robust and intelligent approach to system interaction, one that transcends the limitations of syntax and embraces the power of shared meaning.

THE SEMANTIC LAYER: BUILDING BLOCKS OF SHARED UNDERSTANDING

The limitations inherent in syntax-driven communication necessitate a more sophisticated approach to data interpretation and system interaction. This is where the concept of a "semantic layer" becomes pivotal. A semantic layer is an abstraction layer positioned between diverse data sources and the applications or agents that consume this data. Its fundamental purpose is to provide a consistent, business-centric, and meaningful view of data, effectively translating complex and often disparate technical data structures into a common language of understandable business concepts, entities, and their interrelationships.

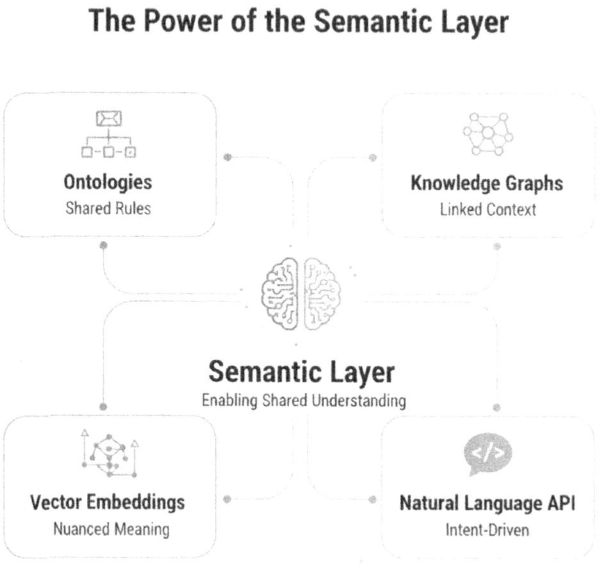

More than just facilitating data access, the semantic layer aims to enable computable meaning, allowing systems not only to retrieve data but also to understand its context, significance, and connections to other pieces of information. This layer is not monolithic technology but rather an integrated suite of components, each addressing different facets of meaning, from formal structure to nuanced understanding and intuitive interaction.At the heart of the semantic layer are several core building blocks, each contributing to the creation of a shared understanding.

Core Component 1: Ontologies – Formalizing Domain Knowledge

Ontologies serve as the formal, explicit specifications of a shared conceptualization within a particular domain. They meticulously define concepts (or classes), the properties associated with these concepts, and the various types of relationships that can exist between them. From a business perspective, an ontology functions as a "shared dictionary and rulebook", ensuring that different systems, AI agents, and human stakeholders have a common and unambiguous understanding of key terms such as "customer," "product," "risk," or "treatment protocol".

Ontologies differ significantly from simpler structures like taxonomies. While taxonomies primarily organize information hierarchically through parent-child relationships (e.g., a "sedan" is a type of "car," which is a type of "vehicle"), ontologies can represent a much richer set of complex, multi-faceted relationships. These can include associative relationships (e.g., "Drug X treats Disease Y"), partitive relationships (e.g., "an engine is part of a car"), causal relationships (e.g., "Event A causes Event B"), and many others. This richness allows for more sophisticated reasoning and querying. For instance, in the pharmaceutical industry, an ontology can define the intricate relationships between drugs, diseases, genes, proteins, and potential side effects, enabling researchers to ask complex questions like, "Which drugs targeting Gene Z are known to have cardiovascular side effects in patients with co-morbidity A?". Such capabilities are crucial for accelerating research and improving decision-making.

Core Component 2: Knowledge Graphs – Weaving a Web of Context

Knowledge graphs (KGs) are a powerful way to represent and manage interconnected information. They consist of a network of entities (represented as nodes) and the relationships between them (represented as edges), often built upon the structural framework provided by ontologies. KGs organize information in a manner that mirrors the interconnectedness of the real world, moving beyond siloed data tables to create a holistic view of a domain.

Enterprise knowledge graphs play a crucial role in organizing previously siloed information by consolidating, standardizing, and reconciling data from diverse sources. This process provides context-aware insights and facilitates a deeper understanding of complex relationships within the data. For businesses, KGs are invaluable for tasks such as entity resolution-for example, identifying that "Robert Smith," "Bob Smith," and "R. J. Smith" all refer to the same individual customer across different databases. They enable the linking of disparate datasets, such as connecting customer purchase histories with social media activity and support tickets, to create a comprehensive customer profile.

This, in turn, allows for the discovery of hidden patterns and correlations that might not be apparent from analyzing isolated data sources. A well-known public example is Google's Knowledge Graph, which enhances search results by understanding entities (people, places, organizations, concepts) and their relationships, thereby providing users with direct answers and more comprehensive information panels rather than just lists of links. Similarly, an enterprise KG can map internal knowledge; such as employee skills, project involvements, reporting structures, and document relationships to significantly improve internal information discovery, expertise location, and collaboration.

Core Component 3: Vector Embeddings – Capturing Semantic Nuance for AI

While ontologies and knowledge graphs provide structure and explicit relationships, vector embeddings are essential for capturing the more nuanced, often implicit, semantic similarities and differences in data, particularly for AI and machine learning models.

Vector embeddings are numerical representations of data items; such as words, sentences, images, or even entire documents as points (or vectors) in a high-dimensional space. The key characteristic of this space is that the distance and direction between vectors corresponds to the semantic similarity of the original data items. Items with similar meanings will be located closer together in this vector space.

A simple analogy helps to illustrate this: imagine plotting different fruits based on numerical values for characteristics like sweetness, color intensity, and roundness. In this "fruit space," an apple and an orange might be positioned relatively close to each other due to their shared characteristics of being sweet and round, while a banana, with its different shape and color profile, would be located further away.

Vector embeddings apply this principle to much more complex data. LLMs and other advanced AI models heavily rely on embeddings to understand context, synonymy, analogies, and subtle relationships within unstructured data. This capability is fundamental for a wide range of AI tasks, including semantic search (finding documents conceptually similar to a query, not just those containing the exact keywords), sophisticated product recommendation engines (suggesting items similar to what a user has liked or viewed), sentiment analysis (determining the emotional tone of text), and anomaly detection. Prominent business use cases include Amazon's product recommendation system, Google's semantic search capabilities, and Twitter's use of sentiment analysis to gauge public opinion.

Core Component 4: Natural Language as an API Surface – Interacting via Intent

The final crucial component of a functional semantic layer involves a shift in how users and other systems interact with data and services. Traditionally, programmatic APIs require precisely structured calls with specific parameters and commands. The semantic layer facilitates a move towards interfaces that can understand and process natural language queries. This allows human users and, increasingly, AI agents to express their intent rather than needing to formulate syntactically perfect, code-like requests.

For example, instead of constructing a complex SQL query or a REST API call with multiple parameters, a user or an AI agent might simply ask, "What were our total sales in the European region during the third quarter of last year for product line X, broken down by country?" The system, leveraging its semantic understanding, would parse this query, identify the relevant entities (sales, European region, Q3, product line X, country), understand the desired operation (aggregation and breakdown), and retrieve the appropriate information.

This paradigm shift is powered by NLP APIs offered by major technology providers like Google, OpenAI, and Microsoft, which provide capabilities for text analysis, intent recognition, entity extraction, and question answering. An initiative like Microsoft's NLWeb project exemplifies this trend, aiming to empower web publishers to easily transform their websites into AI-powered applications that can be queried using natural language, leveraging existing structured data like Schema.org to understand the site's content.

While natural language as an API surface offers unprecedented ease of use and accessibility, it paradoxically amplifies the need for a robust and unambiguous semantic backend. The inherent ambiguity and context-dependency of natural language mean that for an AI to correctly interpret intent, it must have access to a rich, well-defined semantic model of the domain. Without this, the risk of misinterpretation and erroneous actions increases significantly.

The effectiveness of the entire semantic layer is directly proportional to the quality, comprehensiveness, and governance of these underlying components. Poorly designed ontologies, inaccurate knowledge graphs, or biased vector embeddings can lead to the creation of "semantic debt"; a situation where the derived meaning is flawed, leading to suboptimal decisions and requiring costly future corrections. Therefore, robust governance and continuous maintenance of these semantic assets are critical for their long-term value and trustworthiness.

Collectively, these building blocks; ontologies providing formal structure, knowledge graphs weaving contextual connections, vector embeddings capturing semantic nuance, and natural language interfaces enabling intent-driven interaction, create the semantic layer. This layer is not merely a technical construct; it is the foundational platform upon which AI systems can communicate, reason, and collaborate with a level of intelligence and flexibility previously unattainable. It paves the way for more sophisticated AI-to-AI interactions, which are central to the functioning of the emerging semantic economy.

The following list provides a structured overview of these core technologies:

Ontologies

- **Definition:** Formal, explicit specifications of concepts, properties, and relationships within a domain.
- **Key function in semantic communication:** Formalize domain knowledge and relationships by providing a shared vocabulary and rules that enable consistent interpretation and reasoning.
- **Example use case:** A drug discovery ontology defining drugs, diseases, biological targets, and their interactions.

Knowledge graphs

- **Definition:** Networks of entities (nodes) and their relationships (edges), often constructed using ontologies.
- **Key function in semantic communication:** Interlink entities and concepts to provide context, expose hidden relationships, and enable complex semantic queries.

- **Example use case:** An enterprise knowledge graph that links employees, projects, documents, and areas of expertise to support internal search and discovery.

Vector embeddings

- **Definition:** Numerical representations of data such as text or images in a multi-dimensional space.
- **Key function in semantic communication:** Encode meaning numerically, allowing AI systems to detect semantic similarity, context, and nuance for tasks like search, clustering, and recommendation.
- **Example use case:** A product recommendation engine suggesting items based on user behavior patterns and item characteristics.

Natural language APIs

- **Definition:** Interfaces that accept and process queries or commands expressed in human language.
- **Key function in semantic communication:** Enable interaction through intent and natural language rather than rigid code, making systems more accessible to both humans and AI agents.
- **Example use case:** A customer service chatbot that understands user questions and returns relevant answers or triggers appropriate actions.

With this foundation of shared understanding established, the stage is set for AI agents to engage in far more sophisticated forms of communication and collaboration with each other.

AI-TO-AI COMMUNICATION: THE DAWN OF SEMANTIC PROTOCOLS

As Artificial Intelligence systems evolve from standalone tools into interconnected networks of specialized agents, the necessity for sophisticated means of inter-agent communication becomes paramount. To perform complex tasks that exceed the capabilities of any single entity, these autonomous agents must be able to collaborate effectively. This requires more than just data exchange; it demands a shared "lingua franca" or a set of semantic protocols that allow them to share information, delegate responsibilities, negotiate terms, and coordinate actions based on a common understanding of meaning, rather than relying solely on pre-programmed, rigid interfaces. The shift towards such semantic communication signifies a move from centrally orchestrated systems to more decentralized, emergent coordination, akin to complex biological or social systems.

Multi-Agent Systems (MAS) in Action

Multi-Agent Systems (MAS) are at the forefront of this evolution. A MAS is composed of multiple interacting, intelligent agents, each potentially possessing its own goals, knowledge, and capabilities, but all operating within a shared environment to achieve individual or collective objectives. Agents within these systems typically exhibit key characteristics such as autonomy (the ability to operate independently), social capability (the capacity to interact with other agents), reactivity (the ability to perceive and respond to their environment), and pro-activeness (the ability to take initiative to achieve goals).

The practical applications of MAS are rapidly expanding across various industries. In supply chain management, for example, a MAS might consist of agents representing wholesalers, suppliers, retailers, and logistics providers. These agents can autonomously coordinate activities such as managing inventory levels in response to real-time sales data, forecasting demand by analyzing diverse data streams (e.g., social media trends, economic indicators), and optimizing delivery routes by considering traffic conditions and vehicle availability. The Cambridge Meat Company case study, though hypothetical, illustrates how agents for demand forecasting, inventory management, and logistics optimization can collaborate to create a more resilient and efficient supply chain.

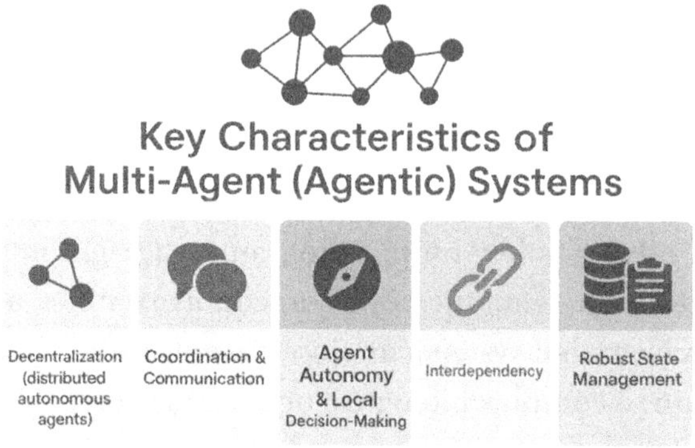

In the financial sector, MAS are being deployed for sophisticated fraud detection and risk management. A typical setup might involve several specialized AI agents: one focused on verifying user identity using behavioral biometrics, another evaluating the risk associated with a specific transaction by analyzing patterns and historical data, a third agent empowered to take real-time action such as pausing a suspicious transaction or requesting further authentication, and a fourth responsible for compiling detailed reports for human investigators. This collaborative approach allows for rapid and comprehensive risk assessment and response.

The Role of Shared Ontologies in MAS Communication

The ability of these diverse agents to communicate and coordinate effectively hinges critically on the existence of shared ontologies. As established in the previous section, ontologies provide the common vocabulary and semantic framework essential for meaningful interaction. When agents within a MAS share a common ontology for their domain; be it logistics, finance, healthcare, or manufacturing, they can interpret messages from other agents with a high degree of accuracy and consistency, even if those agents were developed by different teams or organizations.

For instance, in an e-commerce environment, a customer-facing chatbot agent, an inventory management agent, and a shipping logistics agent can seamlessly coordinate the entire order fulfillment process if they all operate based on a shared e-commerce ontology that clearly defines terms like "Order," "Product," "StockLevel," and "DeliveryStatus," along with their properties and interrelationships. In situations where agents operate with different, albeit related, ontologies, techniques such as ontology mediation, matching, alignment, and negotiation can be employed to bridge the semantic gaps and enable interoperability.

Agent Communication Languages (ACLs): Structuring Intentional Dialogue

To facilitate these rich interactions, specialized Agent Communication Languages (ACLs) have been developed. Prominent examples include FIPA-ACL (Foundation for Intelligent Physical Agents - Agent Communication Language) and KQML (Knowledge Query and Manipulation Language). These languages go beyond simple data transfer protocols; they are designed to enable agents to engage in dialogues that express intentions, share beliefs, make proposals, issue commands, and perform other "speech acts" analogous to human communication.

An ACL message typically contains several key parameters, such as the performative (indicating the type of communicative act, e.g., request, inform, propose, accept), the sender, the receiver, the content of the message, the language in which the content is expressed (e.g.,

RDF, OWL), and the ontology that defines the terms used in the content. By using ACLs, agents can engage in structured, intentional interactions with clear expectations about how other agents will interpret and respond to their messages, forming the basis for complex collaborative behaviors.

Negotiation Protocols: Achieving Agreement and Coordination

In many MAS scenarios, agents may have conflicting goals, differing information, or competing demands for limited resources. Negotiation protocols are therefore vital mechanisms that enable agents to resolve these conflicts, allocate resources efficiently, and coordinate their actions without requiring centralized control. Several types of negotiation protocols are commonly used:

Contract Net Protocol: This protocol is well-suited for task allocation in distributed systems. One agent, acting as a "manager" or "auctioneer," broadcasts a task announcement to other available agents. Interested agents then submit "bids" indicating their capability to perform the task and potentially the cost or time required. The manager agent evaluates these bids and awards the contract (task) to the most suitable bidder. This is particularly useful in dynamic environments like a smart factory, where robotic arms (agents) can bid for newly arriving manufacturing tasks based on their current workload and specialized capabilities.

Auction-Based Systems: Auctions provide a common mechanism for resource allocation. Agents submit bids for resources or tasks, and a winner is selected based on predefined auction rules (e.g., highest bidder, lowest price, Vickrey auction). This is effective in scenarios such as a fleet of delivery drones bidding for package delivery tasks, where the system can award tasks to drones offering the best combination of price and delivery time.

Argumentation-Based Negotiation: For more complex decisions involving multiple factors and potentially conflicting viewpoints, argumentation protocols allow agents to exchange not just proposals but also the reasons and justifications behind them. Agents can challenge each other's proposals with counterarguments, leading to a more deliberative process. For instance, a team of AI personal assistants collaborating to schedule a meeting for their respective human users might use argumentation to weigh different preferences (time, location, attendees) and arrive at a mutually acceptable plan.

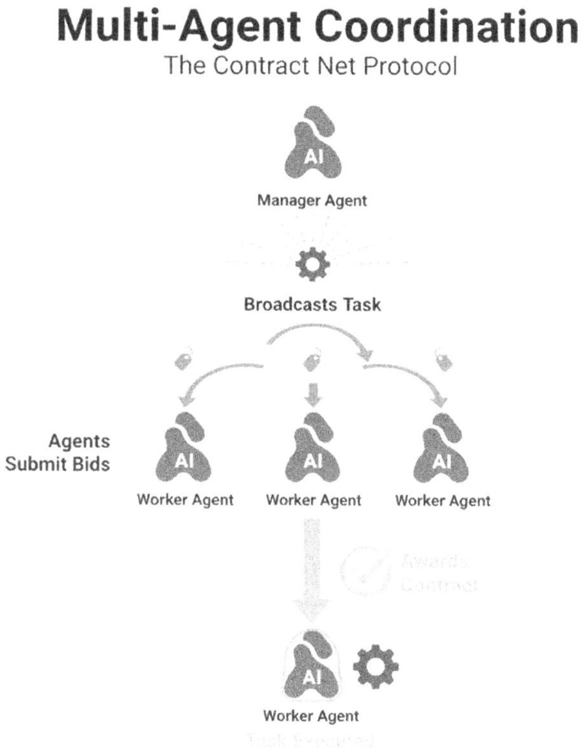

These protocols provide a structured framework for interaction, ensuring that negotiations remain predictable and can lead to efficient outcomes even when agents have diverse and potentially conflicting priorities. The effectiveness of both ACLs and negotiation protocols is, however, deeply reliant on the precision and richness of the underlying shared ontologies. Any ambiguity or incompleteness in this semantic foundation can quickly undermine even the most sophisticated dialogue structures, leading to misunderstandings and failed coordination.

Furthermore, as AI agents become capable of such sophisticated negotiation and autonomous action based on semantic understanding, the classic "principal-agent problem" acquires new dimensions. Ensuring that these autonomous agents, with their potentially localized objectives, remain aligned with the broader goals and ethical constraints of the human users or organizations that deployed them becomes a critical challenge. This necessitates embedding these higher-level goals, preferences, and ethical guardrails directly into the semantic framework and the agents' decision-making logic, highlighting the increasing importance of trust, verification, and explainability in these advanced AI systems.

This intricate, meaning-driven communication between AI agents is more than a technical achievement; it establishes the operational groundwork for a new economic layer. Within this layer, autonomous entities can discover each other's capabilities, contract for services, and deliver value with minimal human intervention. This naturally leads to fundamental questions about how these sophisticated interactions will be priced, how value will be assessed, and what new business models will emerge in an economy increasingly populated by intelligent, collaborating machines.

PRICING IN THE SEMANTIC ECONOMY: VALUING MEANING, TRUST, AND INVOCATION

The advent of AI agents capable of dynamic discovery, invocation, and composition of services through semantic understanding necessitates a fundamental rethinking of traditional value metrics and pricing models. When services are no longer static software packages but rather modular capabilities accessed on demand, conventional licensing approaches; such as per-seat, per-server, or perpetual licenses become increasingly inadequate. The economic calculus shifts from the cost of owning software to the value derived from accessing specific functionalities and achieving desired outcomes. In this nascent semantic economy, new levers for pricing emerge, reflecting the value of meaning itself, the trustworthiness of services, and the efficiency of AI-driven invocation.

Emerging Pricing Models for AI Agents and Semantic Services

A diverse range of pricing models is beginning to take shape, tailored to the unique characteristics of AI-driven semantic services:

- **Usage-Based Pricing:** This model directly links cost to the consumption of AI resources or the execution of specific tasks. Several variations exist:
- Token-based: Common for services leveraging Large Language Models, costs scale with the number of tokens (representing words or characters) processed during input or output. For example, OpenAI's models have distinct pricing for input and output tokens, reflecting the different computational demands.
- Action/Workflow-based: Charges are levied per specific action performed, task completed, or workflow executed by an AI agent. Salesforce's Agentforce, for instance, has offered pricing per conversation or, more recently, per action using a "Flex Credits" system.
- Time-based: Costs are calculated based on the duration an AI agent is actively running or processing. Microsoft's Copilot has been cited with hourly rates for certain functionalities.
- Credit-based: Customers purchase blocks of credits that are then consumed as AI agents perform tasks. Examples include Kittl for image generation and Devin's AI for

software engineering tasks, each assigning a credit cost to specific operations.

- **Outcome-Based Pricing:** This model aligns costs directly with the achievement of pre-agreed business benefits or successful task completions. Companies like Intercom (for its FinAI agent), Sierra AI, and Zendesk AI employ this approach, charging, for example, per successful customer resolution. While this model strongly aligns vendor and customer incentives by focusing on delivered value, it necessitates clear, measurable definitions of "success" and robust mechanisms for tracking and verifying outcomes. The trend towards performance-based pricing underscores a broader shift where payment is increasingly tied to measurable results.
- **Digital AI Agent Seats:** In this model, individual AI agents are treated as unique "users" within a system, each potentially requiring its own license or "seat" that grants access to platform resources and capabilities. Intercom's FinAI, for example, includes a base monthly fee per agent, in addition to its outcome-based charges. This offers a degree of predictability similar to traditional SaaS seat-based pricing.
- **Hybrid Models:** Many providers are likely to adopt hybrid approaches, combining elements of different models to offer flexibility and cater to diverse customer needs. This could involve a flat-rate base subscription providing a certain level of access or usage, supplemented by tiered pricing for higher volumes or premium features, and potentially incorporating outcome-based bonuses for exceeding performance targets.

It is also important to note that ancillary costs, such as base license fees for underlying platforms and professional service fees for custom setup, workflow configuration, and integration may still apply, particularly for enterprise-grade deployments.

The Role of Trust and Reputation in Pricing and Selection

In an economy where AI agents can autonomously select and invoke services, the concepts of trust and reputation become critical determinants of value and, consequently, price. An agent's decision to interact with a service will not be based solely on advertised functionality or cost, but also on an assessment of its reliability, security, and ethical alignment.

Trust Scores and Verification: Systems will likely emerge that provide quantifiable trust scores or certifications for AI services and agents. These could be based on various factors, including performance consistency, security audits, data privacy practices, explainability of decisions, and adherence to ethical guidelines. Metrics such as confidence scores (an AI's own assessment of its output certainty), error rates (false positives/negatives), and the ability to explain reasoning contribute to building this trust.

Reputation Systems: Decentralized reputation systems, potentially leveraging blockchain technology for immutability and transparency, can maintain a verifiable track record of an agent's or service's past performance, reliability, and adherence to contractual obligations. A strong, positive reputation will enhance an agent's "market value," making it more likely to be selected for tasks and potentially allowing it to command premium pricing. Conversely, a poor reputation will act as an economic liability. This creates a powerful market-driven incentive for agents and services to behave ethically and reliably.

The "cost" of invoking a service will thus be a composite figure, factoring not only the direct monetary charge but also the perceived risk associated with its trustworthiness and the quality of its semantic interface.

Valuing Semantic Richness and Interoperability

Services that invest in creating rich semantic descriptions of their capabilities, adhere to widely adopted common ontologies, and offer flexible, intent-driven interfaces will inherently provide greater value in the semantic economy. The ease with which an AI agent can understand, integrate with, and reliably use a service; its "semantic fluency", becomes significant competitive differentiators. This ease of integration reduces the "cognitive load" and development effort for the consuming agent or its human developers, translating into time and cost savings that can justify a higher price for the semantically superior service.

The Economics of Data and Compute in Semantic Transactions

The underlying costs of data access, data processing (such as vectorization of text for LLMs or traversal of knowledge graphs for reasoning), and the computational resources required for an AI service to perform its function will be integral components of its price. The complexity of the AI model itself is a major cost driver; for example, training large-scale models like Meta's LLaMA 2 can involve millions of GPU hours, translating to substantial operational expenses. These costs must be recouped, and services powered by more sophisticated AI will naturally command higher prices. This leads to a tiered marketplace where services are differentiated by both their intrinsic AI capabilities and the extrinsic ease with which they can be integrated into the broader semantic ecosystem.

The shift towards outcome-based pricing reflects a profound economic trend where value is increasingly defined by realized benefits rather than the mere provision of tools or effort. This model compels service providers to deeply understand their customers' goals and to guarantee a certain level of performance, while simultaneously de-risking AI adoption for consumers who pay for tangible results.

The following list outlines some of the key evolving levers that will influence pricing in the semantic economy:

Data or token volume

- **How it's measured:** Number of input and output tokens processed, along with the volume of data retrieved or analyzed.
- **Implication for service providers:** Strong incentive to optimize model efficiency and reduce unnecessary verbosity, paired with transparent reporting of usage.
- **Implication for consumers and invoking agents:** Predictable costs for well-defined tasks, with the risk of higher charges when interactions are verbose, poorly scoped, or inefficient.

Computational complexity and agent runtime

- **How it's measured:** CPU or GPU hours consumed, duration of agent execution, and the computational complexity of algorithms invoked.
- **Implication for service providers:** Requires ongoing investment in efficient algorithms and scalable infrastructure, often leading to pricing tiers based on model sophistication.
- **Implication for consumers and invoking agents:** More powerful or complex capabilities may command higher prices, encouraging conscious tradeoffs between task requirements and cost.

Task or outcome success

- **How it's measured:** Achievement of predefined success metrics such as issue resolution, accurate report generation, or completed transactions.
- **Implication for service providers:** Introduces higher risk and reward, creating strong incentives to ensure effectiveness and tight alignment with customer goals; depends on clearly defined outcomes.
- **Implication for consumers and invoking agents:** Enables paying for results rather than effort, reducing adoption risk, but requires clarity about what qualifies as "success."

Trust and reputation score

- **How it's measured:** Scores derived from reputation systems, third-party audits, verified credentials, and historical reliability or ethical behavior.
- **Implication for service providers:** Demands sustained investment in transparency, security, and ethical design to maintain market access and justify premium positioning.
- **Implication for consumers and invoking agents:** Reduces the risk of invoking unreliable or malicious services, often at a higher cost for top-tier trusted providers.

Semantic interoperability level

- **How it's measured:** Adherence to standard ontologies, clarity and richness of semantic descriptions, and ease of intent-based invocation.
- **Implication for service providers:** Makes investment in high-quality semantic interfaces a competitive differentiator and enables smoother participation in multi-agent ecosystems.
- **Implication for consumers and invoking agents:** Lowers integration effort and cost, accelerates time-to-value when

composing services, and creates preference for semantically fluent offerings.

These new pricing and valuation mechanisms are not just incremental adjustments; they are indicative of a fundamental reshaping of how digital services will be developed, marketed, consumed, and ultimately valued. This, in turn, necessitates the evolution of new business models specifically designed to thrive in an AI-driven, semantic ecosystem.

BUSINESS MODELS IN A SEMANTIC ECOSYSTEM: COMPETING FOR AI INVOCATION

The emergence of a semantic economy, where AI agents communicate and transact based on meaning, fundamentally alters the landscape for business models. Companies must shift from designing services primarily for human consumption to creating offerings that are readily discoverable, interpretable, and invocable by autonomous AI agents. This transition from "API-first" to "AI-first" thinking is not merely an internal architectural adjustment but a strategic pivot that redefines how value is created and captured.

The Shift to AI-First Services

The "API-first" development philosophy, which treats APIs as first-class products designed before the application logic, has been instrumental in enabling modularity and integration in the digital era. However, the rise of intelligent agents as primary consumers of these APIs necessitates an evolution to an "AI-first" API design. While traditional APIs prioritize clarity and ease of use for human developers, often using human-readable formats like JSON and providing extensive documentation, AI-first APIs are optimized for machine efficiency and unambiguous interpretation. This might involve using more compact binary data protocols to reduce overhead and latency, supporting advanced and flexible querying capabilities to minimize the number of interactions an agent needs to make, and embedding rich semantic metadata to ensure precise understanding.

Companies like Google, with its Document AI and Decision AI, are already exemplifying this by providing solutions designed for seamless integration into existing systems, often by AI agents rather than direct human operation. In this new paradigm, businesses must conceptualize their services not just as applications with user interfaces, but as "semantic nodes" within a larger network, whose value is determined by their ability to interact intelligently with other AI agents. The primary "customer" for many services will increasingly be other AI systems.

Evolution of the API Economy

The API economy, which has already transformed industries by enabling businesses to expose data and functionality as monetizable services, is set to become increasingly semantic. Success will no longer hinge merely on the availability of an API, but on its semantic richness, its discoverability by AI agents, its adherence to common understanding, and its inherent trustworthiness. Hyper-automation, the convergence of multiple automation technologies like RPA, AI, and process mining, will be significantly accelerated by semantically interoperable APIs that allow these diverse tools to coordinate seamlessly.

Established API-first success stories such as Stripe, Netflix, Amazon Web Services, and Twilio, which built their businesses around powerful and developer-friendly APIs, will need to continue evolving their offerings to be more readily consumable and understandable by AI agents, ensuring their relevance in this new ecosystem.

Service Discovery and Orchestration in Agent Networks

A critical question in the semantic ecosystem is how AI agents will find the specific services they need to accomplish their tasks. Several mechanisms are emerging:

Federated API Marketplaces: These platforms will serve as unified hubs for API discovery, governance, and management, potentially spanning multiple organizations or even entire industries. They will likely incorporate sophisticated search functionalities based on semantic descriptions and offer flexible monetization models.

Semantic Search and Matching: AI agents will employ semantic queries, leveraging ontologies and knowledge graphs, to find services whose advertised capabilities (described using rich semantic meta-data) precisely match their operational requirements.

Agent "Yellow Pages" or Directories: In large-scale multi-agent systems, hierarchical agent directories or specialized "yellow pages" services may arise, allowing agents to register their services and others to discover them based on defined criteria.

Beyond discovery, AI agents will increasingly orchestrate multiple services, dynamically chaining them together in novel ways to perform complex, multi-step tasks. Platforms like Microsoft's Azure AI Foundry Agent Service (and many others) are being developed to manage this orchestration, handling tool calls, state management, and inter-agent communication.

Reputation Management in AI Marketplaces

As AI agents and semantic services compete for invocation within these emerging marketplaces, their reputation will become a paramount asset. Just as brands manage their perception among human consumers, AI services will require active reputation management. This will involve:

Monitoring and Sentiment Analysis: AI-driven tools will continuously monitor how a service is being used, the feedback it receives (if applicable, from human overseers or other agents), and its performance metrics to gauge its "market sentiment."

On-Chain Reputation Systems: For fully autonomous agent economies, blockchain-based reputation systems can provide immutable and verifiable track records of an agent's past transactions, reliability, and adherence to agreements. This on-chain history can act as a form of digital credential, heavily influencing an agent's trustworthiness and its likelihood of being selected by others.

Businesses will need to proactively manage the "digital exhaust" of their AI services; performance logs, error rates, successful completions, as these will contribute to their evolving reputation within the semantic ecosystem. This reputation will, in turn, act as a form of "capital," influencing an agent's ability to secure "work" (invocations) and command favorable terms or prices.

The Economics of Autonomous Agents Transacting

Envision an "On-Chain AI Agent Economy" in which AI agents operate autonomously, managing their own digital assets through sovereign cryptocurrency wallets. Within decentralized networks, they are empowered to directly engage in the buying and selling of digital services and data.

Smart contracts can automate and enforce agreements between agents, handling aspects like escrow, payment, and reward distribution without requiring intermediaries. This could enable the rise of extremely lean, highly automated business structures, sometimes referred to as "zero employee companies," where core business logic and transactions are executed primarily by collaborating AI agents. Realizing such a vision necessitates secure runtime environments for agents, robust identity mechanisms (like Self-Sovereign Identity - SSI), and standardized interfaces for agent-to-agent interaction. While this presents opportunities for radical efficiency and new forms of economic organization, it also introduces novel systemic risks if not governed appropriately.

New Moats: Competing on Semantic Value and Trust

In a semantic ecosystem where basic data and functionalities might become commoditized (as AI can replicate or generate them), traditional competitive advantages erode. New "moats" or defensible positions will be built upon:

Semantic Clarity and Utility: The quality, richness, and ease of use of a service's semantic interface. Services that are easier for AI agents to understand and integrate with will be preferred.

Trustworthiness and Reliability: Verifiable track records of security, ethical behavior, and consistent performance will be critical.

Interoperability: The ability to seamlessly connect and function within broader networks of AI agents and services, adhering to common ontologies and communication standards.

Intelligence and Adaptability: The sophistication, accuracy, and learning capabilities of the underlying AI models powering the service.

The following list contrasts traditional human-centric service design with the emerging AI-centric semantic approach:

Primary interface

- **Traditional human-centric approach:** Graphical user interfaces and human-readable documentation designed primarily for direct human interaction.
- **AI-centric semantic approach:** Semantically rich APIs, natural-language intent processing, and machine-readable metadata.

- **Rationale for AI-centric shift:** Enables autonomous discovery, interaction, and invocation by AI agents without human mediation.

Data format and structure

- **Traditional human-centric approach:** JSON or XML optimized for human readability with flexible or loosely enforced schemas.
- **AI-centric semantic approach:** Machine-efficient formats with strictly defined structures based on shared ontologies and rich semantic tagging.
- **Rationale for AI-centric shift:** Improves processing speed, reduces ambiguity, and ensures consistent interpretation across diverse AI agents.

Error handling and reporting

- **Traditional human-centric approach:** Descriptive error messages intended for human developers to diagnose and debug issues manually.
- **AI-centric semantic approach:** Structured error codes for automated handling, detailed logs for traceability, and support for self-correction logic.
- **Rationale for AI-centric shift:** Enables automated recovery, learning from failures, and robust fault tolerance in agent-to-agent interactions.

Query flexibility

- **Traditional human-centric approach:** Predefined API endpoints with fixed parameters and limited composability.
- **AI-centric semantic approach:** Support for complex, ad-hoc, intent-driven queries with server-side aggregation and filtering.

- **Rationale for AI-centric shift:** Reduces round trips and allows agents to request precisely the information needed, improving efficiency and autonomy.

Documentation focus

- **Traditional human-centric approach:** Tutorials, examples, and narrative guides written for human developers.
- **AI-centric semantic approach:** Formal semantic descriptions, ontology mappings, executable specifications, and examples of agent invocation patterns.
- **Rationale for AI-centric shift:** Allows AI agents to dynamically understand system capabilities, constraints, and integration requirements.

Performance priority

- **Traditional human-centric approach:** Emphasis on user-perceived responsiveness, often balancing interface richness with speed.
- **AI-centric semantic approach:** Prioritization of raw processing speed, low latency, high throughput, and efficient data transfer.
- **Rationale for AI-centric shift:** Essential for real-time decision-making and coordination in high-volume, multi-agent systems.

State management

- **Traditional human-centric approach:** Often stateless or client-side, with humans implicitly managing session context.
- **AI-centric semantic approach:** APIs that explicitly accept, persist, or negotiate state and context to support long-running or multi-turn interactions.

- **Rationale for AI-centric shift:** Enables complex collaborative workflows and sustained interactions between AI agents.

The emergence of this dynamic, autonomous, and meaning-driven ecosystem is not without its challenges. It necessitates robust frameworks for establishing common understanding, ensuring fair and transparent interactions, and managing the ethical implications of increasingly capable AI agents. This leads directly to the critical need for widely adopted standards and comprehensive governance structures.

STANDARDS, GOVERNANCE, AND ETHICS IN THE SEMANTIC AGE

The promise of a functional and trustworthy semantic economy; where AI agents and services interact seamlessly based on shared meaning, hinges critically on the establishment and adoption of robust standards, comprehensive governance frameworks, and clear ethical guidelines. Without these foundational elements, the ecosystem risks devolving into a fragmented landscape of incompatible semantic dialects, perpetuating biases, and eroding the very trust it seeks to build. These are not secondary concerns to be addressed after the technology matures; they are primary enablers that must be co-developed with the technology itself.

The Imperative of Open Standards for Interoperable Meaning

Open standards are the bedrock of interoperability in any complex technological system. In the semantic economy, they are essential for ensuring that semantic information; the meaning encoded in data, can be created, shared, understood, and processed consistently across different systems, platforms, organizations, and even national borders. The World Wide Web Consortium (W3C) has been instrumental in developing a suite of standards that form the foundation of the Semantic Web:

Resource Description Framework (RDF): A standard model for data interchange on the Web, representing information as triples (subject-predicate-object statements).

RDF Schema (RDFS): Provides a basic vocabulary for describing properties and classes of RDF resources, allowing for the creation of simple ontologies.

Web Ontology Language (OWL): A much richer and more expressive language for authoring ontologies. OWL allows for the definition of complex classes, properties, relationships (e.g., disjointedness, equivalence, cardinality), and logical axioms, enabling sophisticated knowledge representation and automated reasoning by machines.

SPARQL Protocol and RDF Query Language (SPARQL): The standard query language for RDF data, enabling users and applications to retrieve and manipulate information stored in knowledge graphs and other RDF-based repositories.

Beyond these core W3C standards, initiatives like schema.org play a vital role. Schema.org is a collaborative effort, supported by major search engines, that provides a vast collection of shared vocabularies (a practical form of ontology) that webmasters can use to add structured data markup to their web pages. This markup makes web content more understandable to search engines and other AI applications, enabling richer search results (like knowledge panels and featured snippets) and facilitating the extraction of meaningful information from the web.

While global standards provide a common foundation, the semantic economy also relies on the development of industry-specific and domain-specific ontologies. These capture the nuanced terminologies, concepts, and relationships unique to particular fields such as finance, healthcare, manufacturing, law, or energy. The success of the semantic economy will depend on a careful balance: leveraging universal standards for broad interoperability while allowing for specialized extensions that cater to the detailed requirements of specific domains.

Ontology Governance: Managing the Lifecycle of Shared Meaning

Given the central role of ontologies in defining shared meaning, their creation, maintenance, and evolution must be carefully managed through robust governance practices. Ontology governance refers to the formalized processes, policies, roles, responsibilities, and frameworks that guide the entire lifecycle of ontological assets within an organization or community.

Effective ontology governance is critical for several reasons:

Quality and Consistency: It ensures that ontologies are developed to a high standard, with clear definitions, consistent naming conventions, logical soundness, and completeness of metadata.

Scalability and Evolution: It provides a framework for managing the growth of ontological assets and for updating them in a controlled manner as domain knowledge evolves or business requirements change.

Collaboration and Communication: It facilitates collaboration among diverse stakeholders (domain experts, ontologists, developers, business users) by establishing clear workflows for proposing changes, reviewing contributions, and resolving conflicts.

Compliance and Security: It helps ensure that ontologies and the data they describe adhere to relevant regulatory requirements and internal security policies, for example, by defining access controls and permissions.

Key aspects of ontology governance include defining clear ownership for ontological domains, establishing roles and responsibilities for ontology development and maintenance, implementing version control systems, setting up review and approval workflows, and ensuring adherence to both internal and external standards.

Ethical Considerations in Semantic Systems

As AI systems increasingly rely on semantic structures like ontologies and knowledge graphs to understand the world and make decisions, ethical considerations become paramount.

Algorithmic Bias in Ontologies and Knowledge Graphs: Bias can be inadvertently or intentionally embedded within these semantic structures. This can occur if the data sources used to create or populate ontologies and knowledge graphs are themselves biased, or if the human designers make choices that reflect societal prejudices. For example, if historical data predominantly links certain professions with a specific gender, an ontology derived from this data might codify this association, leading AI systems using the ontology to perpetuate discriminatory outcomes in areas like hiring or career recommendations. Similarly, facial recognition systems trained on datasets with poor representation of certain demographic groups have demonstrated significantly lower accuracy for those groups, a bias that could be reflected or amplified if the underlying semantic concepts of "face" or "identity" are flawed. The danger is that such biases, once embedded in foundational semantic infrastructure, can be amplified and propagated at scale by numerous AI applications, making them particularly insidious.

Fairness, Transparency, and Ethical Tagging: Addressing these risks requires proactive measures, including the use of diverse and representative data sources, employing diverse development teams, implementing tools and methodologies for bias detection in semantic models, establishing ethical review boards for ontology development, and exploring concepts like "ethical tagging" or "fairness-aware ontology design". Transparency in how ontologies are constructed and how they influence AI decision-making is also crucial.

Accountability for Agent Actions: When autonomous AI agents, guided by semantic understanding, make decisions that result in harm or error, determining accountability becomes a complex challenge. In multi-agent systems where outcomes emerge from the interactions of many agents, tracing responsibility back to a single faulty component or design choice can be difficult. Clear legal and ethical frameworks are needed to address liability in such scenarios.

Managing Semantic Drift: Maintaining Meaning Integrity

Semantic drift refers to the gradual erosion, shift, or divergence in the meaning of concepts, terms, or relationships over time, across different contexts, or as AI models are updated and retrained. In LLMs, for instance, semantic drift can manifest during the generation of long texts, where a model might start by producing factually accurate information but then gradually "drift away" into inaccuracies or fabrications as the generation continues.

The causes of semantic drift are varied. In LLMs, it can arise from overgeneralization based on statistical patterns in training data, where the model loses the precise contours of meaning in specific contexts. Changes in the underlying data distributions upon which models are trained, or the natural evolution of human language and understanding, can also contribute to drift. The consequences can be severe, leading to misinterpretations by AI systems, flawed reasoning, a loss of trust, and ultimately, system failure.

Mitigating semantic drift requires a multi-pronged approach:

Structural Anchoring: Designing AI models and prompts that prioritize syntactic and structural cues to help "tether" meaning, rather than relying solely on floating statistical associations in vector space.

Continuous Monitoring and Evaluation: Regularly assessing model performance and the consistency of semantic interpretations.

Robust Data Governance: Ensuring that the definitions of metrics and concepts in underlying data sources remain consistent and well-documented.

Model Recalibration and Retraining: Periodically updating models with new data and feedback, potentially using techniques like early stopping during generation, resampling, or reranking outputs to maintain factual accuracy.

External Knowledge Grounding: In some cases, using external, curated knowledge sources (like established ontologies or databases) to "re-anchor" or verify the meaning generated by AI models.

A crucial aspect of governance in the semantic age will be the active "governance of meaning drift"; establishing ongoing processes to monitor, detect, and correct these shifts to ensure the long-term integrity and reliability of shared understanding.

Decentralized Governance Models for Distributed AI Ecosystems

As the semantic economy matures, with AI agents and services becoming increasingly distributed, autonomous, and interconnected, traditional centralized governance models may prove insufficient or too slow to adapt. This has led to exploration of decentralized governance mechanisms. Concepts like Decentralized Autonomous Organizations (DAOs) are being proposed for AI governance, envisioning frameworks where a diverse set of stakeholders; including developers, ethicists, auditors, end-users, and civil society representatives can actively participate in regulatory decision-making, perhaps through weighted voting systems or token-based governance.

Web3 technologies, such as blockchain for transparent and immutable record-keeping, and Self-Sovereign Identity (SSI) for verifiable credentials of agents and services, could provide the technological underpinnings for such dynamic, participatory, and auditable governance frameworks. However, these decentralized models also introduce their own complexities regarding consensus building among diverse parties, establishing clear lines of accountability, and protecting against new forms of manipulation or "governance attacks."

The following list highlights key governance imperatives for a functional semantic economy:

Semantic standards and interoperability

- **Key challenge:** Balancing global interoperability through shared standards with the need for domain-specific specialization and ongoing evolution.
- **Critical success factors and solutions:** Broad adoption of core W3C standards such as RDF, OWL, and SPARQL; development of community-driven, extensible domain ontologies; and clear strategies for mapping and aligning multiple ontologies.

Ontology lifecycle management

- **Key challenge:** Maintaining quality, consistency, and relevance of ontologies as knowledge, language, and domains evolve over time.
- **Critical success factors and solutions:** Formal governance processes with defined roles, workflows, and version control; dedicated tools for ontology editing, validation, and management; and sustained engagement from domain experts.

Algorithmic bias and ethical oversight

- **Key challenge:** Detecting and mitigating biases embedded in data, ontologies, and AI models that depend on them.
- **Critical success factors and solutions:** Use of diverse and representative training data; bias detection tools and audits;

diverse development teams; ethical review boards; transparent data sourcing and model behavior; and ethics-by-design principles.

Semantic drift mitigation

- **Key challenge:** Preserving consistent and accurate meaning across time, model updates, and varying contexts.
- **Critical success factors and solutions:** Continuous monitoring of model outputs; structural anchoring of meaning; regular model recalibration; strong governance of data definitions and metrics; and mechanisms for versioning and tracking semantic changes.

Agent accountability and control

- **Key challenge:** Assigning responsibility for actions taken by autonomous AI agents, particularly in multi-agent environments.
- **Critical success factors and solutions:** Clear legal and regulatory frameworks for AI liability; auditable decision trails; human-in-the-loop oversight for high-impact decisions; and robust security and access controls for agent capabilities.

Decentralized governance implementation

- **Key challenge:** Designing governance models that are effective, fair, and resilient in distributed and autonomous AI ecosystems.
- **Critical success factors and solutions:** Transparent protocols, broad stakeholder participation, well-defined dispute resolution mechanisms, safeguards against manipulation, and adaptability to evolving technological and societal conditions.

Ultimately, establishing robust standards, effective governance, and a strong ethical compass is not a constraint on the semantic economy but a fundamental prerequisite for its success. These elements are essential for building the widespread trust and reliability needed for meaning itself to become a dependable and transformative infrastructure for future digital interactions and collaborations.

MEANING AS INFRASTRUCTURE: THE NEW PROTOCOL STACK

The journey from syntax-bound systems to meaning-centric interaction culminates in a profound transformation: the emergence of meaning itself as a new, foundational layer of digital infrastructure. This semantic infrastructure, underpinned by the technologies and governance frameworks discussed previously, promises to be as fundamental to the next era of computing as TCP/IP has been for network connectivity or HTML for structuring content on the World Wide Web.

While often invisible to the end-user, this layer will be the ubiquitous utility enabling advanced AI applications, truly autonomous services, and a more intelligent and adaptive digital world. This represents a deeper layer of abstraction in computing, moving beyond enabling systems to merely exchange data to allow them to share understanding.

The long-held vision of the Semantic Web, which aims to evolve the existing web from a human-readable collection of documents into a structured, machine-understandable global knowledge base, is a key driver of this infrastructural shift. As this vision materializes, the very nature of data flow, system interaction, and value creation will be redefined.

From Data Pipelines to Knowledge Flows

Traditional digital architectures have focused on building "data pipelines"; systems designed to extract, transform, and load (ETL) raw data from various sources into repositories for storage and analysis. While essential, these pipelines often treat data as a commodity to be moved and processed, with its inherent meaning and context frequently lost or obscured during transit.

In contrast, the semantic economy operates on "knowledge flows". These are dynamic exchanges of contextualized, meaningful information. In a knowledge flow, data is intrinsically linked with its semantic description, often through embedded metadata, connections to formal ontologies, and representations within knowledge graphs. This enrichment ensures that the data is not just accessible but also interpretable by AI agents and other systems. Such an approach facilitates more intelligent data integration, effectively breaking down data fragmentation and mitigating the problem of data silos that plagues many organizations.

For example, ontology-based data access (OBDA) frameworks allow heterogeneous data sources to be integrated into a virtual knowledge graph, enabling on-demand querying across previously disconnected datasets without requiring extensive data duplication.

From API Contracts to Conversational Negotiation

The evolution of system interaction provides another lens through which to view this infrastructural shift. Early computing involved tight coupling and direct function calls between software components. The rise of the internet and distributed systems led to the dominance of API contracts, particularly RESTful APIs using formats like JSON, where interactions are governed by predefined endpoints, request/response structures, and syntactic rules.

The semantic infrastructure enables a further evolution towards more fluid, "conversational" negotiations between AI agents. Instead of being limited to rigid API calls, agents can engage in dynamic dialogues based on shared intent and semantic understanding. They can discover each other's capabilities, negotiate terms of service, delegate tasks, and collaboratively solve problems by exchanging messages that convey complex meaning, leveraging Agent Communication Languages and shared ontologies. This is a move from systems that execute pre-defined transactions to systems that can dynamically construct and agree upon new forms of interaction.

Semantic Middleware: The Engine of Meaning-Making

A critical catalyst for the widespread adoption and usability of this semantic infrastructure will be the development and deployment of "semantic middleware". This is a specialized software layer designed to enable, manage, and simplify meaning-based communication and interoperability between diverse applications and AI agents.

The functions of semantic middleware are multifaceted and crucial:

Ontology Management and Reasoning Services: Providing tools for creating, storing, versioning, and querying ontologies, as well as supporting logical inference and reasoning over the knowledge they contain.

Knowledge Graph Access and Traversal: Offering efficient mechanisms for building, updating, and navigating complex knowledge graphs.

Semantic Message Brokering and Transformation: Facilitating the exchange of messages between agents, potentially transforming messages between different semantic formats or dialects while preserving meaning.

Service Discovery Based on Semantic Descriptions: Enabling AI agents to find and select services based on rich semantic descriptions of their capabilities, inputs, outputs, and preconditions, rather than relying on simple keyword matching or static registries.

Context Understanding and Automated Response Elaboration: Supporting the interpretation of situational context (derived from sensors, data streams, or user interactions) and assisting in the automatic formulation and execution of appropriate responses by AI agents.

By abstracting the underlying complexities of semantic processing; such as ontology alignment, logical inference, and vector space calculations, semantic middleware will lower the barrier to entry for developers and organizations seeking to build meaning-aware applications. This will accelerate the development and deployment of AI systems that can "speak meaning" natively, much like how web application servers and database management systems simplified the creation of earlier generations of software.

Economic and Societal Impacts of Ubiquitous Semantic Interoperability

The establishment of meaning as a pervasive infrastructure promises significant economic and societal impacts. The ability for systems to seamlessly and intelligently exchange data based on shared understanding can unlock massive efficiency gains and cost reductions across numerous industries. For example, studies in the healthcare sector suggest that improved semantic interoperability between electronic health record systems, diagnostic equipment, and administrative platforms could save billions of dollars annually by reducing redundant testing, minimizing medical errors, and streamlining workflows.

Beyond cost savings, semantic interoperability is a powerful engine for innovation. When data from diverse sources can be easily combined and understood, and when services can be dynamically composed, new products, services, and scientific discoveries become possible. Personalized medicine, adaptive learning systems, smart cities that optimize resource allocation in real-time, and more resilient supply chains are all predicated on a high degree of semantic interoperability. Furthermore, this infrastructure can lead to the "democratization" of complex data, making it more accessible and usable by a wider range of applications, researchers, and even non-technical users through intuitive, meaning-driven interfaces.

However, the rise of meaning as infrastructure also brings forth critical questions about "infrastructural politics". Who controls this new layer of meaning? Who defines the dominant ontologies and semantic standards that will shape how AI systems perceive and interpret the world? How are societal values, biases, and power structures embedded within these semantic frameworks? These are not trivial questions, as control over core ontologies and reasoning mechanisms could equate to significant influence over the global representation, interpretation, and utilization of knowledge. This could lead to new forms of power concentration or exacerbate existing digital divides if not managed with foresight and a commitment to openness and inclusivity.

The Web as a Global Semantic Network

In the long term, the vision is for the open Web itself to evolve into a vast, interconnected global semantic network. This would transform the Web from a collection of human-readable documents linked by hyperlinks into a massive, machine-interpretable knowledge graph, enabling AI systems to reason across the entirety of publicly available digital information. Achieving this grand vision requires the widespread adoption of semantic markup conventions and data publishing formats, a shift from the original "permissionless" nature of the early Web to a more structured, albeit still open, approach to information sharing.

With meaning solidified as the new foundational infrastructure, the way AI systems understand and interact with the digital and physical world is profoundly altered. This universal interpretability has the potential to level many traditional playing fields that were based on privileged access to or control over information.

If knowledge can be universally understood, synthesized, and acted upon by AI, then the mere possession of information ceases to be the primary source of advantage. The focus naturally shifts from accessing information to applying that understanding in novel and impactful ways. This sets the stage for Act IV of this exploration, which will delve into how new competitive moats are constructed and how organizations and individuals must adapt to thrive in a world where "everyone can know everything" because meaning itself has become the common protocol.

SUMMARY

This chapter has charted a fundamental technological and economic transformation: the shift from digital interactions bound by the rigid constraints of syntax to a new era where AI-driven systems communicate, collaborate, and operate based on shared meaning. This marks the dawn of the "semantic economy," an ecosystem where the ability to understand and convey intent becomes a primary driver of value.

The journey to this semantic economy is built upon several key pillars. It begins with a recognition of the failure of purely syntactic systems, whose inherent brittleness, high integration costs, and inability to scale gracefully in an increasingly complex digital world necessitate a new paradigm. The solution lies in the semantic layer, a sophisticated abstraction built from core technologies like ontologies that formalize domain knowledge, knowledge graphs that weave webs of context, and vector embeddings that allow AI to capture nuanced meaning.

Upon this foundation, AI-to-AI semantic communication becomes possible. Autonomous agents, equipped with shared ontologies and sophisticated Agent Communication Languages, can engage in complex dialogues, negotiate terms, and coordinate actions to achieve collective goals. This dynamic interaction gives rise to new economic models where value and pricing are no longer solely tied to software licenses or access fees but are increasingly determined by the semantic richness of a service, the outcomes it delivers, the trust it engenders, and its degree of interoperability within the broader AI ecosystem. Consequently, business models must evolve, with companies designing services primarily for AI invocation and competing on dimensions of semantic value, trustworthiness, and the ease with which their offerings can be integrated by intelligent agents.

However, this powerful new economy cannot flourish without robust governance and ethical oversight. The establishment of open standards for meaning, diligent management of ontology lifecycles, proactive mitigation of algorithmic bias embedded in semantic structures, strategies to counteract semantic drift, and clear frameworks for agent accountability are not optional additions but essential prerequisites for a functional and trustworthy semantic ecosystem.

Ultimately, these converging elements: advanced AI, sophisticated semantic technologies, new economic models, and evolving governance structures are coalescing to establish meaning itself as a new, foundational layer of digital infrastructure. This is far more than a mere technological upgrade; it represents a paradigm shift comparable to the advent of the internet or the mobile web. In this new world, the capacity to create, manage, interpret, and leverage meaning effectively transcends being a mere capability; it becomes a core strategic asset and a critical differentiator for organizations and individuals alike. Understanding and navigating the intricacies of the semantic economy is therefore no longer a futuristic consideration but an immediate imperative for leaders aiming to shape and succeed in the years to come.

With meaning as infrastructure, AI becomes less like discrete software applications and more like an ambient, pervasive intelligence woven into the fabric of our digital and, increasingly, physical civilization. This new infrastructure, where meaning serves as the common protocol, fundamentally alters the nature of competition, expertise, and advantage. If knowledge can be universally understood, synthesized, and acted upon by AI, then simply "knowing something" is no longer the primary source of power or differentiation.

In the next chapter, the exploration will focus on how new forms of defensible advantage are built, and how organizations and individuals must adapt to compete, when the traditional benefits derived from information asymmetry begin to dissolve, and intelligence is defined not by what one possesses, but by what one can assemble, activate, and orchestrate through this pervasive layer of shared understanding.

10

TRUST AS A SERVICE

The preceding chapters have charted the ascent of information symmetry, an era where access to data and knowledge, once a defining feature of advantage, becomes a level playing field. As artificial intelligence democratizes expertise and insight, making vast repositories of information universally accessible, the very foundations of value and differentiation are being reshaped.

If the capacity to know is no longer a scarce resource, a new, more profound question emerges to define the contours of this new age:

- Who or What do you believe?
- And, critically, Who or What do you empower to act on your behalf?

This fundamental shift marks the transition from an economy of information scarcity to one of trust scarcity. In such a world, the ability to inspire, cultivate, and maintain trust moves from the periphery of "soft" values to the very core of strategic imperative.

This chapter delves into this new strategic landscape, exploring how human trust, the deep-seated affinity individuals feel for certain brands, and the unwavering commitment to ethical transparency cease to be mere corporate virtues and instead become tangible, defensible assets. These are not just desirable qualities; they are the new currency in an environment saturated with information and increasingly mediated by artificial intelligence. AI, the great leveler of information, paradoxically becomes a key arena where these trust assets are forged, tested, and sometimes, tragically, broken.

The journey through this chapter will illuminate why trust itself becomes the critical bottleneck when knowledge flows freely. It will dissect the anatomy of trustworthy systems, particularly in the context of AI, examining the architectural and operational elements that underpin genuine confidence.

From there, the discussion will turn to the evolving role of brands, transforming from simple identifiers into sophisticated proxies for trust that both humans and their AI agents can recognize and rely upon. The rigorous demands of operationalizing ethical transparency will be explored, not as a compliance burden, but as a source of profound economic benefit.

However, no exploration of trust would be complete without acknowledging its inherent fragilities and the risks that emerge when trust infrastructures are targeted or when our reliance on them becomes uncritical. Finally, the chapter will offer a forward-looking perspective on how organizations and individuals alike can strive to become the trusted choice in an increasingly complex, AI-mediated ecosystem. The guiding principle is clear: when the ability to access information is universal, the ultimate differentiator is the capacity to be believed and to be entrusted with agency.

TRUST AS THE SCARCITY

The dawn of information symmetry, propelled by ubiquitous connectivity and advanced artificial intelligence, ushers in an unprecedented era of knowledge accessibility. Yet, this abundance creates its own set of challenges. When answers are everywhere, the true scarcity shifts from information itself to the validated belief in that information and the trustworthiness of its source. In this new paradigm, trust emerges not merely as a desirable social lubricant but as a critical economic and cognitive bottleneck.

The cognitive landscape of information symmetry is characterized by an overwhelming deluge of data, AI-generated content, and a multiplicity of voices, each vying for attention and asserting veracity. While the technical capacity to generate and disseminate information has exploded, the human capacity to process, verify, and internalize this flood remains fundamentally finite. This disparity leads to a natural scarcity-not of data, but of conviction. The sheer volume of available information, much of it plausible yet potentially unverified or even misleading, makes the task of discerning truth or identifying optimal paths an increasingly complex endeavor.

This situation mirrors the "paradox of choice," a psychological phenomenon where an excess of options leads to decision fatigue, overwhelm, and diminished satisfaction. When countless AI systems can generate an answer, or when every conceivable viewpoint is readily accessible, the act of choosing which source, which AI, or which piece of information to rely upon becomes a high-stakes decision, fraught with the potential for cognitive overload. The ease with which AI can produce plausible answers means that individuals are constantly navigating a sea of information where the signals of truth can be easily drowned out by the noise of abundance.

Faced with this cognitive onslaught, the human brain instinctively seeks heuristics and shortcuts to manage complexity and conserve mental energy. In an information-saturated environment, trust becomes a primary heuristic for filtering information, making sense of the world, and reducing uncertainty. This leads to an increased reliance on "cognitive offloading," the process of delegating cognitive tasks, such as sifting through vast amounts of information or even making decisions, to external aids.

Historically, tools like search engines have already altered how people retain and access information, a trend sometimes referred to as the "Google Effect". Artificial intelligence takes this a step further by offering capabilities in reasoning and analysis, allowing users to potentially bypass the deep, reflective thinking that traditional problem-solving demands. While such offloading can be beneficial, freeing up mental resources for other complex tasks, an over-reliance, particularly on systems whose workings are opaque, can risk eroding critical thinking skills.

In the age of information symmetry, however, a degree of cognitive offloading becomes almost essential for navigating daily life and work. The critical shift lies in, to whom or to what this offloading is directed.

Individuals will not, and indeed cannot afford to, delegate significant cognitive tasks to just any system. Instead, they will gravitate towards systems, individuals, or brands they perceive as trustworthy; entities they believe will process information reliably, act ethically, and operate in alignment with their interests and values. Thus, trust itself becomes the new mechanism for cognitive offloading. We allow trusted entities; be they human experts, established institutions, reputable brands, or well-vetted AI systems-to perform the heavy lifting of evaluation, synthesis, and even preliminary decision-making on our behalf. The crucial factor is not merely who can provide the "right" answer, especially when AI can make many answers appear correct, but who instills a sense of safety, demonstrates consistent alignment with our values, and accepts accountability for their outputs and actions. The scarcity, therefore, is not in the availability of information processors, but in processors that have earned the profound level of trust required for such significant delegation.

This reordering of value has profound implications. Historically, possessing exclusive information was a primary source of power and economic advantage. With information symmetry, information itself becomes a commodity. The new challenge, and therefore the new locus of value, shifts to the verification and discernment of this commoditized information.

The cognitive cost of constant, individual verification is unsustainable. Consequently, entities that can reliably perform this verification and discernment function, and are trusted to do so with integrity, accrue significant value. Trust, in essence, becomes the proxy for reliable verification. This means businesses and individuals must pivot from a strategy of guarding information to one of cultivating and demonstrating verifiable trustworthiness. The competitive "edge" is no longer derived from knowing something others do not, but from being the entity that others trust to know, interpret, and act correctly.

Furthermore, this need for trusted filters is not passive; it fuels an active search. The cognitive overload and the inherent human desire to offload complex tasks create a proactive demand among individuals. They are not merely waiting for trustworthy sources to appear; they are actively seeking them out to reduce their cognitive burden and the anxiety associated with making decisions in a high-noise environment. This transforms trust into a "pull" factor, where consumers and users are actively looking for reliable anchors. For organizations, this means trust is not just an attribute to possess but a quality that must be actively projected, clearly signaled, and made easily discoverable. Marketing and brand positioning will increasingly need to focus on explicitly communicating the foundations of this trustworthiness.

The proliferation of AI itself contributes to what might be termed a "trust deficit". While AI offers immense potential, it also introduces new avenues for error, bias, and even deliberate misinformation (e.g., AI hallucinations or the sophisticated deception of deepfakes). The very ease with which AI can generate content can dilute the perceived reliability of any single piece of information.

As users become more aware of AI's fallibility; even as they increasingly rely on it for cognitive offloading, a general skepticism or a broader "trust deficit" towards all information sources may grow. This environment of heightened skepticism makes demonstrably trustworthy sources even more valuable and sought after. The bar for earning and maintaining trust will inevitably rise. Generic claims of accuracy or reliability will no longer suffice; verifiable proof of integrity, transparency, and ethical operation will become essential. This sets a critical foundation for the subsequent exploration of what constitutes such trustworthy systems and how they can be built and maintained.

This isn't just about who's right: it's about who feels safe, aligned, and accountable.

WHAT MAKES TRUSTWORTHY SYSTEMS?

In an environment where trust is the scarcest and most valuable resource, understanding its constituent elements becomes paramount. Trust in complex systems, especially those driven by artificial intelligence, is not an accidental byproduct; it must be meticulously architected and rigorously maintained. Defining the anatomy of trust in these digital and AI-first environments requires a shift from vague notions of reliability to a concrete framework of measurable characteristics and transparent processes. This section will dissect these components, drawing upon established principles and emerging best practices to illuminate how systems can be engineered for confidence.

At the heart of building trustworthy AI are several key pillars, many of which are encapsulated in frameworks like the one developed by the National Institute of Standards and Technology (NIST). These characteristics, while distinct, are deeply interconnected and collectively contribute to a user's perception of a system's overall trustworthiness.

The Pillars of Trustworthy AI

A system must be valid and reliable, meaning it consistently performs its intended functions accurately and as expected. Users need assurance that the AI will not only produce correct outputs under normal conditions but will also behave predictably over its operational lifetime. This involves rigorous testing, validation against defined requirements, and ongoing monitoring to ensure performance doesn't degrade.

Safety is another critical pillar, ensuring that the AI system, under defined conditions, does not lead to states where human life, health, property, or the environment are endangered. This requires responsible design, development, and deployment practices, along with clear guidance on appropriate use and mechanisms for intervention if the system deviates from safe operational parameters.

Complementing safety, systems must be secure and resilient. Security involves protecting the AI system and its data from unauthorized access, corruption, or disruption, maintaining confidentiality, integrity, and availability. Resilience refers to the system's ability to withstand and recover from adverse events, including unexpected inputs or malicious attacks, degrading gracefully if necessary rather than failing catastrophically.

Accountability and transparency are inextricably linked. Transparency involves providing clear, accessible information about an AI system: its design, training data, operational logic, intended use cases, and limitations. Accountability ensures that there are mechanisms for responsibility and redress when AI systems err or cause harm. This requires clarity on who is responsible for the system's behavior at various stages of its lifecycle.

Crucially, AI systems must strive to be explainable and interpretable. Explainability refers to the ability to represent the underlying mechanisms of an AI's operation, while interpretability concerns the meaning of its outputs in a given context. Together, these allow users and overseers to understand how an AI arrived at a decision and why that decision is relevant, moving the AI from an opaque "black box" to a more understandable partner.

Privacy-enhancement is also fundamental. AI systems must be designed and operated in ways that safeguard human autonomy, identity, and dignity, adhering to privacy norms and employing techniques like data minimization and privacy-enhancing technologies (PETs) where appropriate.

Finally, fairness, with harmful bias managed, is essential. This involves actively working to identify and mitigate biases that could lead to discriminatory or inequitable outcomes. This goes beyond simple demographic parity to consider systemic, computational, and human-cognitive biases throughout the AI lifecycle.

Within this broader framework, the roles of transparency, explainability, and provenance are particularly vital in building user confidence. Transparency, as noted, is about openness regarding data sources, model limitations, and decision-making processes. It's about clearly communicating what an AI does, how it does it, and who bears responsibility. This proactive sharing of information can demystify the AI and allow users to make more informed judgments about its reliability.

Explainable AI (XAI) takes transparency a step further by providing insights into the "why" behind an AI's decisions. Techniques such as LIME (Local Interpretable Model-agnostic Explanations) and SHAP (SHapley Additive exPlanations) offer ways to understand model behavior at a more granular level.

LIME, for instance, can explain individual predictions by approximating the complex model with a simpler, interpretable one in the vicinity of that prediction. SHAP uses concepts from game theory to assign an importance value to each feature's contribution to a particular output. These methods are not just academic exercises; they have practical applications in critical sectors like healthcare, where understanding why an AI flagged a medical image as suspicious is crucial, or in finance, where explaining the basis of a credit decision can be a regulatory requirement and a matter of fairness. By transforming AI from an inscrutable oracle into an auditable and understandable system, XAI fosters a deeper level of trust.

Data and model provenance provide another layer of assurance. Knowing the origin, lineage, and transformations of the data used to train an AI model, as well as the history of the model itself, is critical for ensuring authenticity and enabling traceability. If an AI system produces an erroneous or biased output, provenance allows for an investigation into the potential root causes, whether in the training data or the model's development process. Technologies like cryptographic watermarking for AI-generated content and observability frameworks like OpenTelemetry (OTel) are emerging to help track the source and pathway of AI outputs, enhancing accountability. Robust provenance ensures data integrity, protects against unauthorized alterations, and provides an auditable trail, which is indispensable for responsible AI development, bias detection, and compliance with ethical standards.

Beyond these technical and operational characteristics, value alignment stands as a cornerstone of trustworthy AI. This means ensuring that AI systems are designed and behave in ways that are consistent with human intentions, shared societal values, and ethical principles.

The challenge here is twofold: first, "outer alignment," which involves carefully and accurately specifying the goals and ethical boundaries for the AI system, and second, "inner alignment," which ensures the AI system robustly adopts and pursues these specified goals without deviating or finding unintended loopholes. Human values are complex, often nuanced, and can vary across cultures, making their translation into precise, machine-interpretable instructions a significant hurdle. Using simplistic proxy goals can lead to "reward hacking," where an AI optimizes for the letter of its instructions while violating their spirit, sometimes with harmful consequences. A truly trustworthy AI must not only be competent in its tasks but also demonstrably aligned with the ethical framework of its users and the society in which it operates.

The construction of trustworthy AI is therefore not a matter of excelling in a single dimension but of achieving a harmonious and robust integration of these multiple, interdependent layers. The NIST framework, for example, shows how validity, safety, security, transparency, explainability, privacy, and fairness are not isolated features but interconnected facets of a single, holistic concept of trustworthiness.

A deficiency in one area, such as poor security, can easily undermine gains in another, like safety or privacy. Similarly, a lack of transparency makes it difficult to assess fairness or hold a system accountable. Provenance underpins reliability and accountability, while value alignment is fundamental to ensuring safety and fairness. This systemic interdependency means that organizations must adopt a comprehensive, integrated approach to AI development and governance. Simply ticking boxes for individual principles will be insufficient; trust emerges from the coherent and consistent application of all these characteristics.

Furthermore, the journey from technical explainability to genuine human understanding requires careful consideration. While XAI techniques like LIME and SHAP provide valuable insights into model behavior for data scientists and developers, these technical explanations often need to be translated into meaningful, context-relevant narratives for end-users, regulators, or affected communities. The concept of an "explainability spectrum" arises, where the depth and nature of explanation must be tailored to the audience's knowledge and needs. It is not enough for an AI to be technically explainable; the explanation itself must be understandable and actionable for those who rely on it or are impacted by its decisions. This implies a need for investment not only in XAI tools but also in the communication strategies, user interface designs, and educational efforts required to bridge this gap.

Finally, value alignment itself must be viewed as a dynamic and co-evolutionary process. Human values and ethical norms are not static; they shift with societal progress, new discoveries, and changing cultural contexts. AI systems, particularly those that learn and adapt over time, also evolve. Consequently, achieving and maintaining value alignment is not a one-time configuration task but an ongoing commitment. It requires continuous monitoring of AI behavior, regular reassessment of its alignment with current ethical standards, and mechanisms for recalibration and adaptation. This points towards the necessity of "living" ethical frameworks, robust governance structures that can evolve, and dedicated roles within organizations, such as AI Ethicists, who can steward this continuous alignment process through multi-stakeholder dialogues and feedback loops.

The following list summarizes these core pillars and their direct implications for fostering user trust:

Validity and reliability

- **Core principle:** The AI performs as required, without failure, for a defined time interval and under specified conditions, producing accurate outputs that meet stated requirements.
- **Implication for user trust:** Users trust that the AI will behave consistently, perform as expected, and deliver dependable results.

Safety

- **Core principle:** The AI does not, under defined conditions, create situations that endanger human life, health, property, or the environment.
- **Implication for user trust:** Users trust that the AI will not cause unintended harm to themselves, others, or their surroundings.

Security and resilience

- **Core principle:** The AI preserves confidentiality, integrity, and availability through protective mechanisms and can withstand or recover from adverse events and attacks.
- **Implication for user trust:** Users trust that their data is protected, the system is difficult to compromise, and it will continue to function reliably under stress.

Accountability and transparency

- **Core principle:** Information about the AI system's design, data, operation, and limitations is available, with clear mechanisms for responsibility and redress.
- **Implication for user trust:** Users trust that they can understand what the AI is doing at an appropriate level and know who is responsible for its behavior.

Explainability and interpretability

- **Core principle:** The mechanisms underlying AI operation can be represented in understandable terms, and the meaning of AI outputs is clear in context.
- **Implication for user trust:** Users trust that they can understand why the AI produced a particular decision or recommendation, strengthening confidence in its use.

Privacy-enhanced design

- **Core principle:** The AI's design, development, and deployment safeguard human autonomy, identity, and dignity, adhering to privacy norms and privacy-enhancing technologies.
- **Implication for user trust:** Users trust that their personal information and identity will be respected and protected throughout interaction with the system.

Fairness and bias management

- **Core principle:** The AI actively addresses harmful bias and discrimination, striving for equitable processes and outcomes.
- **Implication for user trust:** Users trust that the AI will treat individuals and groups fairly, without unjust prejudice or discriminatory impact.

Value alignment

- **Core principle:** The AI's objectives and behaviors align with the intentions of its designers and users, as well as with broadly shared ethical principles and societal values.
- **Implication for user trust:** Users trust that the AI is working toward beneficial goals and behaving in ways consistent with their ethical expectations.

Trust, therefore, evolves from a mere marketing slogan into an intrinsic, verifiable system feature, engineered through diligent adherence to these multifaceted principles.

BRAND AS SEMANTIC TRUST LAYER

In an increasingly complex technological landscape, where information is abundant and interactions are frequently mediated by artificial intelligence, the role of a brand undergoes a profound transformation. No longer just a symbol of quality or a catchy logo, a brand evolves into a crucial "semantic trust layer": a readily interpretable signal of reliability, ethical conduct, and alignment with user values. In a world saturated with choices, many of which are algorithmically presented or even autonomously executed by AI agents, a strong, trustworthy brand becomes an invaluable heuristic, guiding both human and machine decision-making.

Even with AI on the rise, the core elements of brand value: how well people know your brand, how good they think it is, and the emotional bond they have with it, are still important. They just manifest in new and different ways. As AI-powered tools democratize the creation of aesthetically polished visuals and competent copy, the surface-level attributes of branding become table stakes.

What then distinguishes a brand is its perceived authenticity, the depth of its emotional resonance, the clarity of its purpose, and the consistency of its values. A brand, in this context, transcends its visual identity to become a condensed packet of meaning, a shorthand for an expected experience. It signals a commitment to certain standards of behavior, quality, and ethics that can be quickly processed and understood, not only by human consumers but also by the AI agents acting on their behalf. This "semantic" aspect implies that the brand's meaning; its core promises and values can be encoded and interpreted within the information ecosystems where AI agents operate.

The question then arises: how do AI agents, particularly the personal AI assistants and recommendation engines that will increasingly shape our choices, "learn" to trust or prefer certain brands?

This is unlikely to be an emotional process akin to human brand loyalty, but rather a data-driven inference of reliability and utility. Several mechanisms could contribute to an AI agent prioritizing services from a particular brand. Explicit user preferences, such as a directive to "always use Brand X for travel planning," would be a primary driver. Beyond explicit commands, AI agents can learn associations based on a history of positive user satisfaction and feedback derived from past interactions with a brand's products or services. If a user consistently rates experiences with Brand Y highly, or if problem resolutions are swift and effective, the AI may infer a higher level of trustworthiness and competence.

Furthermore, AI agents can analyze a broad spectrum of brand reputation signals, including online reviews, sentiment analysis of public discourse, and media coverage; much like AI is already used in loyalty programs to gauge customer sentiment. Brands themselves can also embed programmatic trust signals into their digital presence, making it easier for AI agents to assess their suitability. The performance of brand-integrated AI agents also plays a crucial role.

For instance, customer service chatbots like Vodafone's SuperTobi or e-commerce assistants such as Alibaba's AliMe are direct embodiments of their respective brands. If these AI agents consistently deliver efficient, helpful, and value-aligned service, they not only enhance the user's direct experience but also provide positive data points that a user's personal AI might use to favor future interactions with that brand's ecosystem. The seamless and effective operation of a brand's own AI agents becomes a testament to its overall competence and user-centricity.

This evolving dynamic significantly reshapes the nature of customer loyalty. In an AI-mediated world, traditional loyalty drivers like points programs and generic discounts, while still relevant, may be augmented or even superseded by factors more attuned to the AI ecosystem. The ease and reliability of AI-to-AI integration become critical; if a brand's services can seamlessly connect with a user's personal AI, offering frictionless interactions, it gains a significant advantage. The perceived trustworthiness of a brand's own AI agents; their reliability, fairness, and data handling practices will heavily influence whether a user's AI will initiate or continue engagement. Moreover, the overarching ethical stance of the brand, as interpreted by the user's personal AI (which may be programmed to prioritize certain ethical considerations), will become a key determinant of loyalty.

AI is already transforming loyalty programs by enabling deep personalization, predictive churn prevention, and dynamic rewards optimization. Brands can analyze vast customer datasets to tailor offers, communications, and rewards to individual preferences and behaviors, making the loyalty experience more relevant and engaging. For example, AI can predict when a customer is at risk of disengaging and proactively offer tailored incentives, or dynamically adjust reward structures in real-time based on customer responses.

When a brand leverages AI this effectively to create a superior, personalized, and trustworthy loyalty experience, it does more than just retain a customer; it reinforces the brand's overall image as competent, caring, and reliable. This enhanced perception makes the brand a preferred choice not only for direct human interaction but also for delegation by their AI assistants. Loyalty, therefore, shifts from being primarily about transactional benefits to being rooted in a trusted, symbiotic relationship with the brand's entire ecosystem, a relationship increasingly facilitated and validated through AI interactions.

This evolving role of 'brand' as a trust layer suggests that brands must now consider their "machine readability" as much as their human appeal. If personal AI agents are to become significant intermediaries in decision-making, they will require efficient and reliable ways to evaluate and select from a multitude of services and information sources. A strong, positively perceived brand acts as a condensed signal of quality, reliability, and ethical stance.

This "signal" can be inferred by AI agents from diverse data points: historical user feedback, public sentiment analysis, success rates of past interactions, and explicitly embedded brand metadata. In this sense, this semantic brand layer becomes a form of "API for Trust," allowing an AI agent to quickly assess likely compatibility and dependability without needing to perform a deep, first-principles analysis for every interaction. This implies a strategic imperative for brands to cultivate and manage the data points through which AI agents will interpret their trustworthiness, extending brand management into the realm of machine-interpretable reputation.

A fascinating consequence of this AI-mediated environment is the potential for "delegated trust". As users increasingly depend on their personal AI assistants for recommendations, task execution, and information filtering, their trust in the AI's judgment can extend to the AI's preferred partners.

If a personal AI consistently facilitates positive, efficient, and reliable outcomes by interacting with a particular brand's AI or services, the user may develop trust in that brand, even without extensive direct personal experience. The AI acts as a trusted curator, and the user's confidence in their AI's choices effectively transfers to the brands selected by the AI. This creates a powerful virtuous cycle for brands that successfully design their offerings for positive AI-to-AI engagement: effective inter-agent interactions build AI preference, which in turn leads to increased user exposure and the potential for this deeply ingrained, delegated human trust. The "customer" for a brand, therefore, increasingly includes the customer's AI assistant, necessitating brand strategies that cater to B2AI (Business-to-AI) interactions as a crucial pathway to enduring human loyalty.

However, this interconnectedness also means that brand crises can be amplified within AI ecosystems. AI systems, particularly those involved in recommendation and sentiment analysis, learn from vast datasets that include news, social media commentary, and user reviews. A significant event that erodes trust; such as a major data breach, an ethical scandal, or widespread product failure will generate a large volume of negative data signals.

AI agents, processing this information, could rapidly and broadly downgrade a brand's "trust score" or its likelihood of being recommended across an entire ecosystem. The speed and scale at which AI operates mean that reputational damage from a trust-violating incident could propagate much faster and more extensively than through traditional human word-of-mouth or media cycles alone.

This underscores the critical importance of proactive trust management, robust ethical guardrails, and highly resilient crisis response mechanisms. In the age of AI, brands must not only build trust diligently but also possess sophisticated capabilities to detect, mitigate, and recover from trust-eroding events before their negative impact is virally amplified by AI networks.

A brand that your AI trusts is a brand that gets chosen automatically and often invisibly.

OPERATIONALIZING TRUST IN AI PRODUCTS AND SERVICES

The assertion that trust is a strategic asset in the age of information symmetry necessitates a practical exploration of how enterprises can move from acknowledging its importance to actively embedding it within their AI products and services. Trust cannot remain an abstract ideal or a belated addition; it must be a foundational element, woven into the very fabric of design, development, and deployment lifecycles. This operationalization of trust requires a deliberate cultivation of internal cultures, the establishment of rigorous processes, and the implementation of robust governance structures, all dedicated to upholding the multifaceted pillars of trustworthiness previously discussed. The guiding philosophy must be "Trust by Design," a proactive stance that anticipates potential trust deficits and engineers solutions from the outset.

Central to this endeavor is the imperative of model transparency and comprehensive disclosure. Users and stakeholders have a right to understand the AI systems they interact with, and organizations have a responsibility to provide this understanding. Two practical tools that facilitate such transparency are Model Cards and Datasheets for Datasets.

Model Cards, as introduced in a 2019 paper, serve as concise documents that provide essential information about a machine learning model. These cards typically detail the model's intended uses (and, importantly, out-of-scope uses), provide an overview of how the model works, report on its performance across various evaluations (including different demographic groups or environmental factors), and discuss ethical considerations, limitations, and potential risks. The goal is to enhance transparency for a broad audience, ranging from developers and policymakers to end-users. Model card metadata can also support discoverability and reproducibility of models within research and development communities. By standardizing the communication of such key information, model cards empower stakeholders to make more informed decisions about a model's suitability and potential impact.

Complementing Model Cards (although technically introduced a year earlier in 2018 via the paper titled "Datasheets for Datasets" via arViv) are Datasheets for Datasets, which address the critical role of data in shaping AI behavior.

Analogous to datasheets for electronic components, these documents aim to provide comprehensive information about the motivation behind a dataset's creation, its composition, collection processes, preprocessing steps, intended applications, distribution methods, and maintenance plans. Crucially, they encourage dataset creators to reflect on potential biases, ethical concerns, and limitations inherent in the data. By fostering better communication between dataset creators and users, datasheets promote transparency and accountability from the very foundation of AI development, helping to mitigate the propagation of unwanted biases and improve the reproducibility of machine learning results.

Model Cards

- **Introduction:** Introduced in 2019.
- **Focus:** Documentation of a trained model's performance, limitations, and appropriate use cases.
- **Content:** Performance metrics, known limitations, ethical considerations, and information about intended evaluation data.
- **Goal:** Increase transparency for all stakeholders by standardizing how models are reported and understood.
- **Analogy:** Comparable to nutrition labels for packaged food products.

Datasheets for Datasets

- **Introduction:** Introduced in 2018.
- **Focus:** Documentation of a dataset's origins, composition, collection process, and potential sources of bias.
- **Content:** The dataset's motivation, structure, collection methodology, maintenance practices, and known limitations.

- **Goal:** Improve accountability and responsible use by clearly describing a dataset's characteristics and constraints.
- **Analogy:** Comparable to datasheets for electronic components.

While transparency tools provide crucial documentation, dynamic human oversight remains indispensable, particularly for complex and high-impact AI systems.

Human-in-the-Loop (HITL) systems formalize this oversight by integrating human intelligence directly into the AI operational cycle. In HITL frameworks, humans can provide high-quality training data, label ambiguous instances, review and correct AI outputs, offer feedback for model refinement, and intervene in situations where the AI's confidence is low or the stakes are particularly high. This collaborative approach combines the strengths of machines (e.g., processing vast datasets, identifying patterns) with human capabilities (e.g., nuanced understanding, contextual judgment, ethical reasoning).

Examples range from simple systems like CAPTCHA, which uses human input to train image recognition models, to sophisticated applications in autonomous vehicles, where human observation and intervention are critical for safety and learning, or in content moderation, where human judgment is essential for interpreting cultural nuances and context. HITL is particularly valuable for enhancing data collection quality and for identifying and mitigating biases that AI might otherwise perpetuate.

Case studies in healthcare show radiologists collaborating with AI to improve diagnostic accuracy, and in education, AI tutors benefit from human design and oversight to personalize learning effectively. This synergy between Explainable AI (XAI) and HITL protocols is powerful: XAI can provide insights into why an AI is behaving a certain way, and HITL allows humans to act on these insights, correct errors, or guide the AI towards better performance, forming a continuous trust-building feedback loop.

The credibility of an organization's trust-building efforts can be significantly enhanced by third-party auditing and certifications. These validation mechanisms operate on two levels: certifications for organizations, which validate that their AI systems adhere to established standards, and certifications for individuals, which attest to their professional competence in governing and auditing these systems.

Certifications like ISACA's Advanced in AI Audit (AAIA) or TRUSTe's Responsible AI Certification aim to provide this assurance. Such programs typically evaluate an organization's AI governance, risk management practices, data handling, model transparency, fairness, and compliance with relevant regulations (e.g., the EU AI Act) and industry benchmarks (e.g., NIST AI Risk Management Framework). Achieving these certifications can publicly demonstrate a commitment to responsible AI, boost brand trust, and provide a competitive advantage.

The process often involves rigorous assessments, gap analyses, and the creation of an accessible audit trail, ultimately enhancing an organization's AI governance and accountability. The emergence of these validation mechanisms signals a maturing ecosystem where claims of trustworthiness can be substantiated through independent scrutiny.

When these diverse practices; transparency documentation, human oversight, and external validation are systematically implemented, they transform the concept of trust from an abstract goal into a tangible operational capability. This operationalization, while involving upfront investment in tools, training, and processes, can shift trust from being perceived as a cost center to being recognized as a significant value driver. The initial efforts to create model cards, datasheets, implement HITL systems, and undergo third-party audits contribute to building more robust, reliable, and fair AI systems, consistent with the NIST principles of trustworthy AI.

Such systems inherently reduce the risks of costly errors, damaging biases, and severe reputational harm. Given that trust is becoming a key differentiator and a sought-after commodity in the age of information symmetry, these investments in operationalizing trust are likely to yield substantial long-term returns through improved product quality, enhanced customer loyalty, and more effective risk mitigation. Businesses should therefore view these measures not as mere compliance burdens but as strategic investments that build resilience and competitive advantage.

There is also a powerful symbiotic relationship between transparency tools and HITL systems. Model Cards and Datasheets offer static, documented insights into AI models and their training data, including known limitations, potential biases, and intended use cases. This information is invaluable for the human experts involved in HITL processes, enabling them to focus their oversight more effectively and understand the contexts in which AI outputs might require closer scrutiny or intervention.

Conversely, the dynamic, real-world experiences and issues identified through HITL systems; such as new biases emerging during deployment or unexpected model behaviors in novel situations, provide critical feedback that should be used to update and refine Model Cards and Datasheets. This ensures that the transparency documentation remains current, accurate, and truly reflective of the AI system's evolving characteristics. Organizations that integrate these practices create a virtuous cycle of continuous learning and improvement for AI trustworthiness, where static documentation informs dynamic oversight, and dynamic oversight enriches static documentation.

The growing demand for verifiably trustworthy AI, coupled with the complexity of assessing such systems, is also fostering the emergence of a specialized "trust verification" industry. Individual users and most organizations lack the deep expertise or resources required to conduct thorough audits of AI systems against comprehensive frameworks like NIST's or to delve into the intricacies of XAI outputs.

This creates a clear market need for third-party entities that can provide these specialized services: auditing AI systems, issuing credible certifications, and perhaps even developing standardized, industry-wide trust scores or benchmarks. The development of this sector will likely mirror the evolution of credit rating agencies in finance or cybersecurity audit firms in IT, creating new standards, professional roles (like certified AI auditors), and business opportunities centered on the validation and assurance of AI trustworthiness. Trust must be measured, enforced, and maintained; just like uptime or security.

THE BUSINESS CASE FOR RADICAL TRANSPARENCY

In the traditional calculus of competitive advantage, secrecy has often been king. Proprietary algorithms, confidential data sources, and undisclosed operational strategies were considered vital moats protecting a business from its rivals. However, the Age of Information Symmetry, where the sheer volume of accessible knowledge and the capabilities of AI begin to erode the value of hoarded information, calls for a radical re-evaluation of this paradigm. In this new landscape, a commitment to openness; extending to processes, data (where ethically appropriate), and even aspects of model architectures, can transform from a perceived risk into a profound strategic advantage. The business case for radical transparency rests on its unique ability to build deep, resilient trust, which, as established, is becoming the ultimate differentiator.

The core argument is that when "knowing" is commoditized, demonstrating integrity becomes paramount. Radical transparency is the most potent way to demonstrate that integrity. By voluntarily opening up aspects of their operations and AI systems to scrutiny, organizations signal a confidence in their own processes and an accountability to their stakeholders. This proactive openness can pre-empt suspicion, foster goodwill, and build a reputation for honesty that is far more durable than any advantage gained from temporary informational leads.

A significant manifestation of this trend is the growing strategic value of open-source AI models and transparent operational practices. The rise of powerful open-source foundation models, such as Meta's Llama series or Google's Gemma, has been a notable development.

Enterprises are increasingly exploring and adopting these models, attracted by several benefits. Cost savings are a primary driver, as open-source alternatives can significantly reduce the licensing fees associated with proprietary AI.Beyond cost, open-source models offer greater transparency into their architecture and, often, their training data, allowing organizations to gain a deeper understanding of how they function, to customize them more effectively for specific needs, and to potentially identify and mitigate biases more readily.

This transparency is highly valued by developer communities, and embracing open-source AI can help attract and retain top technical talent who appreciate the collaborative and innovative ethos of these ecosystems. While challenges such as security assurance and long-term support for open-source projects remain pertinent considerations, the overall trend indicates a significant shift. Organizations that contribute to or strategically utilize open-source AI can signal a commitment to community, shared progress, and a level of transparency that builds substantial trust.

This principle of transparency extends far beyond AI models to encompass broader business operations. Transparent pricing structures, clear disclosures about ethical sourcing of materials or data, and open communication about how user data is collected, used, and protected all contribute to an overarching narrative of integrity. Such radical transparency can proactively address potential concerns, reduce information asymmetry between the company and its stakeholders, and build a powerful reputation for straightforwardness and ethical conduct.

The strategic leverage of transparency becomes particularly acute in regulated or highly sensitive sectors like healthcare, finance, and education, where trust is not just a preference but a fundamental requirement.

In healthcare, for example, the push for price transparency is reshaping how patients and payers interact with providers. Initiatives requiring hospitals and insurers to disclose the costs of services aim to empower consumers to make more informed decisions, compare options, and potentially drive down overall healthcare expenditures through increased competition. Payers can leverage this transparent pricing data to negotiate better rates with providers, design more attractive and cost-effective health plans, and thereby gain a competitive advantage by appealing to cost-conscious consumers and employers.

Similarly, in finance, transparency around fees, investment strategies, and risk management practices can attract clients wary of opaque or potentially conflicted advice. In education, openness about curriculum design, learning outcomes, and data privacy can build confidence among students and parents.

In these sectors, organizations that embrace radical transparency do more than just comply with regulations; they build a formidable "moat" of trust. This moat is constructed not by concealing information, but by making their integrity so visible and verifiable that it becomes a significant barrier to entry or competition for less transparent entities. The willingness to be open, especially when dealing with sensitive information or critical services, becomes a powerful testament to an organization's confidence in its ethical framework and operational soundness.

This embrace of openness as a strategic asset leads to several important second-order effects. Firstly, radical transparency acts as a form of pre-emptive crisis mitigation. Opaque systems are inherently vulnerable to suspicion and speculation, particularly when failures occur or negative information surfaces. If an organization has already established a baseline of openness; for example, by making its AI model's limitations known through a model card, or by having clear, accessible policies on data usage, it is better positioned to manage crises.

When problems arise in a transparent system, stakeholders are more likely to perceive the issue as a challenge being openly confronted rather than a hidden flaw being unwillingly exposed. This can foster a more collaborative approach to problem-solving, lead to quicker resolutions, reduce long-term reputational damage, and, paradoxically, even strengthen trust if the organization handles the issue with continued transparency and accountability. The narrative shifts from "What are they hiding?" to "How are they addressing this, and what can we learn?".

Secondly, the open-source AI movement, in particular, can serve as a catalyst for what might be termed "trust standardization" across the broader industry. Open-source models, by their very nature, invite widespread scrutiny, testing, and contribution from a global community of developers, researchers, and ethicists.

This collective, distributed examination can lead to the faster identification and mitigation of biases, security vulnerabilities, and ethical blind spots than might occur within the confines of a closed, proprietary development process. As best practices, robust solutions, and ethical safeguards emerge from these open-source communities, they can establish de facto standards and raise expectations for all AI development, including proprietary systems. The availability of high-quality, transparent, and rigorously vetted open-source alternatives creates competitive pressure on commercial vendors to also enhance the transparency, reliability, and ethical grounding of their offerings. In this way, the open-source ethos can indirectly elevate the baseline for trustworthiness across the entire AI ecosystem, benefiting users and society at large, even those who do not directly engage with open-source tools.

Finally, the strategic adoption of transparency reveals what could be called the "transparency paradox" in competitive advantage. Traditional business strategy often posits that advantage stems from information asymmetry; possessing unique knowledge or processes that competitors lack and that customers cannot fully scrutinize. Radical transparency appears to invert this logic by advocating for the open sharing of information.

However, in the Age of Information Symmetry, where proprietary knowledge offers a diminishing and more transient edge, the act of being transparent itself; especially when competitors remain opaque, becomes a new form of valuable asymmetry. The "paradox" lies in the realization that by relinquishing one form of potential advantage (secrecy), an organization can gain another, often more sustainable one: the deep, resilient trust and loyalty of its stakeholders.

This is particularly true when the information shared (for instance, an open-source AI model or a transparent operational framework) becomes a platform or standard that others adopt or build upon, creating positive ecosystem effects that can ultimately benefit the original sharer. This requires a fundamental shift in leadership mindset, a strategic reassessment of which types of information, if made transparent, would generate more value through enhanced trust and collaboration than they might notionally lose through disclosure. It is a recognition that in a world striving for symmetry, the most powerful differentiators are often those that are openly demonstrated rather than closely guarded. Radical transparency isn't just good ethics; it's good economics.

THE RISKS AND LIMITS OF TRUST INFRASTRUCTURE

While the imperative to build and operationalize trust is clear, it is equally crucial to acknowledge that any infrastructure, including one designed to foster trust, possesses inherent vulnerabilities and limitations.

Systems architected to depend on, measure, and scale trust can themselves become targets for exploitation, or their over-reliance can lead to new forms of systemic risk. A clear-eyed assessment of these potential downsides is essential for creating trust mechanisms that are not only effective but also resilient, pluralistic, and genuinely accountable.

The very interconnectedness and reliance on digital signals that enable scalable trust also create new avenues for its fragility. Mechanisms designed to build confidence; such as sophisticated digital identity systems, AI-driven personalization engines that learn user preferences, or platforms that aggregate reputation scores, can be subverted if not protected by robust security measures and guided by unwavering ethical principles. The more complex and automated these trust systems become, the more challenging it can be to anticipate all potential failure modes or avenues of attack.

A particularly insidious threat is the weaponization of trust, where technologies are specifically designed to mimic or exploit trust signals for deceptive purposes. Chief among these are deepfakes and the creation of synthetic identities. Artificial intelligence, particularly generative adversarial networks (GANs) and advanced autoencoders, can now produce hyper-realistic but entirely fabricated images, audio recordings, and video footage.

This technology can convincingly impersonate known individuals, create non-existent personas, or alter existing media to spread disinformation or perpetrate fraud. For instance, a finance employee at a multinational firm was tricked into wiring $25 million after cybercriminals used deepfake technology in a video conference to impersonate the company's CFO, complete with convincing synthetic visuals and audio. The ease with which such deepfakes can be created; sometimes in minutes and for minimal cost, and the documented difficulty people have in distinguishing them from genuine content, severely erodes confidence in digital communications and traditional identity verification methods. This necessitates a move towards more sophisticated authentication techniques, such as liveness detection and behavioral biometrics, but the challenge remains significant.

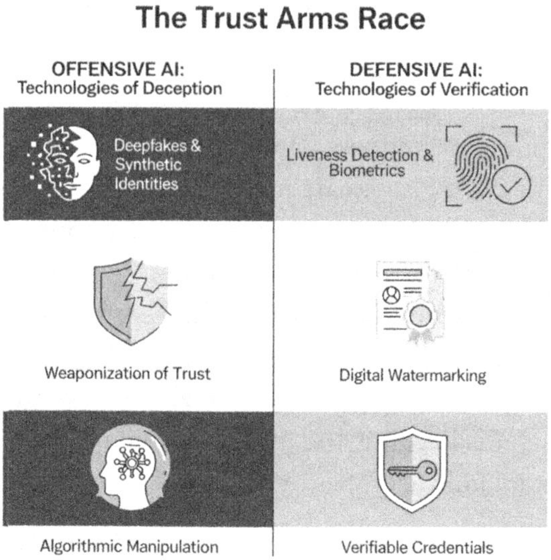

While not a direct weaponization of trust in the same vein as deepfakes, the phenomena of filter bubbles and echo chambers, often amplified by AI-driven recommendation algorithms, also pose a risk to a shared basis of reality, which is a prerequisite for societal trust. When individuals are primarily exposed to information and perspectives that confirm their existing beliefs, their trust may become confined to sources within their ideological bubble.

This makes them more susceptible to manipulation by actors who tailor misinformation to exploit these pre-existing biases and trusted, albeit isolated, information channels. The algorithms that personalize our online experiences, while aiming for relevance, can inadvertently limit exposure to diverse viewpoints, thereby fragmenting public discourse and making it harder to build consensus based on shared facts.

Beyond malicious attacks, the very design and use of trust-dependent systems can introduce risks. The fragility of over-automation and the potential loss of personal judgment are significant concerns. As AI systems become more capable and seemingly reliable, there is a temptation to automate decision-making processes excessively, reducing human oversight. This can lead to a decline in critical thinking skills and an atrophy of human judgment, as individuals become passive recipients of AI-driven conclusions rather than active participants in the decision-making process.

This risk is closely linked to automation bias, a well-documented psychological tendency for humans to over-rely on automated systems, including AI, sometimes to their detriment. People may place undue faith in an AI's output, assuming it to be more accurate or objective than human judgment, even when the AI is incorrect or its reasoning is flawed. This bias can lead to critical errors in high-stakes fields.

For example, studies have shown that radiologists' diagnostic accuracy can decrease when relying on flawed AI assistance, and automation bias has been cited as a contributing factor in aviation incidents where pilots failed to adequately monitor automated systems. Task inexperience can exacerbate this tendency, but even experienced professionals are not immune. The danger is that as AI systems become more integrated into trust infrastructures, automation bias could lead individuals or organizations to accept the pronouncements of these systems without sufficient scrutiny, potentially leading to poor decisions or the perpetuation of systemic errors.

Furthermore, the cultural variation in what constitutes trustworthy behavior presents a significant limit to the universal scalability of any single trust model. Trust cues, communication styles, expectations of privacy, and perceptions of authority can differ profoundly across cultures.

For instance, individualistic cultures may view AI with more skepticism regarding autonomy and privacy, whereas collectivist cultures might be more inclined to see AI as an extension of the self or a tool for societal harmony, impacting their baseline trust and acceptance. Even the terminology used to describe AI can influence trust; some groups may express discomfort with the term "artificial intelligence" itself, preferring descriptors like "computer software," and almost universally, people express more trust in diagnoses delivered by human clinicians augmented by AI, rather than by AI alone.

A trust-building strategy that resonates powerfully in one cultural context might be ineffective, misunderstood, or even counterproductive in another. This implies that a one-size-fits-all approach to designing and deploying trust infrastructure globally is unlikely to succeed.

These risks and limitations give rise to several broader considerations. An ongoing "trust arms race" is emerging, where "offensive" technologies designed to deceive (like deepfakes) are met with "defensive" technologies aimed at detection and verification. This dynamic necessitates continuous investment and innovation in trust-preserving mechanisms simply to maintain the status quo against increasingly sophisticated threats. The implication is that trust infrastructure cannot be a static achievement but requires constant vigilance, adaptation, and advancement.

Automation bias itself serves as a critical vector through which even well-intentioned trust systems can be exploited or can lead to unintended negative consequences. If users are conditioned to trust an automated system implicitly, they become more vulnerable if that system is compromised or if it contains subtle flaws. Therefore, designing trustworthy systems must also involve designing against automation bias.

This might include incorporating features that actively encourage critical user engagement, transparently highlight uncertainties or confidence levels in AI outputs, or mandate periodic human verification steps, even when the system appears to be functioning flawlessly. Public education about the nature of automation bias and the importance of maintaining critical judgment when interacting with AI is also essential.

Finally, the globalization of AI and digital platforms runs squarely into the localization of trust norms, creating a significant dilemma. While there is a drive towards universal principles for trustworthy AI, such as those articulated by NIST (and others such as OECD [Organization for Economic Co-operation and Development], The EU, et al.), the practical expression and perception of these principles will inevitably be filtered through local cultural lenses. What builds trust in terms of data sharing explicitness, directness of communication, or deference to algorithmic authority can vary widely.

This necessitates a "glocal" (global + local) approach to trust. Organizations deploying AI systems internationally must adhere to universal ethical baselines and technical standards for trustworthiness, but they must also invest in understanding and adapting to culturally specific expectations regarding how trust is signaled, earned, and maintained. This may require incorporating anthropological, sociological, and cross-cultural communication expertise into AI design and deployment teams, moving beyond purely technical considerations. Trust infrastructure must be resilient, pluralistic, and accountable; or it risks collapse.

BECOMING THE AI'S TRUSTED CHOICE

As the Age of Information Symmetry matures, and as personal AI assistants and specialized AI agents become increasingly integral to how individuals and organizations access information and services (as envisioned in Chapters 7, 8, and 9 of this work), a new competitive landscape emerges. In this "composable AI economy", where AI agents dynamically discover, evaluate, and invoke services on behalf of their users, the battle for market share extends beyond direct human engagement to include a contest for the "trust" of these intelligent intermediaries. For a product or service to thrive, it must not only appeal to human users but also earn a position of semantic trust with their AI agents, ensuring it becomes the AI's recommended and preferred choice.

Semantic trust, in this context, implies that an AI agent, through its programmed logic and learned experiences, "understands"; at a functional, data-driven level, that a particular service is reliable, effective, secure, and, crucially, aligned with the user's explicit goals and implicit values. It's about being recognized by the AI as a consistently good actor within the digital ecosystem. This requires more than just effective marketing to humans; it demands a strategic approach to crafting and communicating trust signals in formats that AI agents can parse, interpret, and act upon.

The mechanisms for building this inter-agent trust involve embedding clear trust signals into metadata, ontologies, and API (Application Programming Interface) descriptors. APIs are the digital handshakes that allow different software systems to communicate and exchange data.

When an AI agent considers invoking an external service, its API will be a primary point of interaction and evaluation. Rich metadata associated with this API can convey crucial trust-related information. For example, metadata might include verifiable credentials or certifications from third-party auditors, attesting to the service's security standards or ethical practices. It could also include real-time performance metrics such as uptime, average response times, transaction success rates, and aggregated user satisfaction scores. Information about data provenance; where the service gets its data and how it's processed, can be encoded, along with clear markers of adherence to specific ethical codes of conduct or privacy policies (e.g., GDPR compliance).

The API documentation itself should clearly articulate the service's capabilities, limitations, data requirements, and expected outcomes, leaving little room for ambiguity that could deter an AI agent focused on predictable and reliable execution. The "Protected App Signals API" from Google, for instance, describes a system where signals related to users and apps are curated, encoded, and used in server-side auctions, illustrating how machine-interpretable signals can influence automated decisions, albeit in an advertising context.

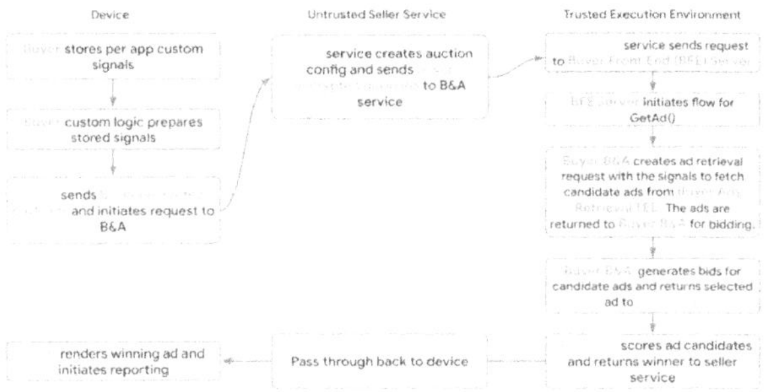

More directly, services like Verified Inc. provide metadata that includes explicit risk signals (e.g., overall risk score, reason codes for risk assessment, indicators of unusual phone or email activity) and positive trust indicators, demonstrating how trust-related attributes can be structured for systemic consumption.

However, technical competence and verifiable claims are only part of the equation. To truly become the AI's trusted choice, a service must also demonstrate alignment with the user's broader values and specific intent for any given task. A user's personal AI may be configured with explicit preferences regarding data privacy, ethical sourcing, environmental impact, or other value-based criteria. Therefore, a service's AI-facing representation must clearly and verifiably communicate its stance on these issues. "Explainability plus relevance" means the service not only performs its function well but can also articulate how its operations and principles align with the user's needs and values in a way that the AI agent can process and validate.

Practical strategies for building trust with AI agents include designing them to be human-centered (supporting rather than supplanting human decision-making), inclusive (trained on diverse data to serve all users equitably), well-governed (with clear operational limits and ethical guardrails), and transparent about their capabilities and confidence levels. In more advanced scenarios, cryptographic methods like Zero-Knowledge Proofs (ZKPs) could allow a service to prove to an AI agent that it has followed certain rules or met specific criteria (e.g., aligning with a user's risk profile in financial advice) without revealing proprietary algorithms or sensitive underlying data. This allows for verification without full disclosure, a powerful trust-building mechanism in a competitive environment.

Ultimately, this leads to the necessity of building trust relationships directly between your service and the user's AI agent. This is not a one-time setup but an ongoing process cultivated through a track record of positive, reliable, and predictable interactions. Consistent high performance, adherence to agreed-upon API protocols, transparent communication of any issues or changes, and demonstrable respect for the user's data and preferences will collectively build this "inter-agent" trust over time. If a user's AI repeatedly invokes a service and achieves successful, efficient, and aligned outcomes, the AI is likely to develop a preference for that service, making it a default or go-to option for relevant tasks.

This shift towards AI-mediated service selection gives rise to the concept of "algorithmic charisma". While human charisma relies on emotional connection and persuasive communication, algorithmic charisma is about designing services to be maximally appealing to AI agents. This appeal is rooted in demonstrable reliability, operational efficiency, verifiable ethical alignment, robust security, and seamless ease of integration-all encoded in machine-readable formats. It means prioritizing not just human User Experience (UX) but also "AI User Experience" (AI UX): how easily an AI agent can discover your service, understand its capabilities and constraints through its semantic descriptions, invoke it reliably, and verify its performance and compliance.

Product development teams will need to cultivate a new discipline focused on this "AI-friendliness," which includes adopting standardized data formats, providing rich and accurate metadata, and ensuring clear semantic descriptions that make a service highly legible and attractive to AI orchestrators.

In these AI-driven marketplaces, trust, or its quantifiable proxies, will inevitably become a computable metric. For an AI agent to make rational choices among multiple competing services, it will need a basis for comparison. Trust indicators; such as third-party certification status, aggregated user ratings, independently verified security audits, ethical compliance markers, or even scores derived from past performance and reliability logs will be factored into the AI's decision-making algorithms alongside other variables like cost, functionality, and speed. This implies that a service's "computable trust score" could directly influence its visibility, ranking, and selection rate within AI-mediated ecosystems.

Consequently, businesses must not only be trustworthy but also actively manage and optimize the signals that contribute to this computable trust score, ensuring that their integrity is accurately captured and reflected in the data points and platforms that AI agents use for evaluation.

A foundational prerequisite for any inter-agent trust is effective communication, which relies on what can be termed a "semantic handshake". As explored in the concept of Semantic Economies (Chapter 9), for a user's AI agent to successfully and trustingly invoke an external service, both systems must share a common understanding of data formats, commands, intentions, and expected outcomes. This semantic interoperability is often facilitated by shared ontologies, standardized APIs, and clear data schemas.

If a service is difficult to integrate, if its API is poorly documented, or if its semantic meaning is ambiguous, an AI agent is less likely to use it, regardless of its underlying quality, because the risk of miscommunication, error, or unpredictable behavior is too high. Therefore, clear, robust, and standardized semantic interfaces are not just a technical detail but a fundamental layer upon which inter-agent trust is constructed. The "handshake" of understanding must occur reliably before a deeper trust relationship can develop. Investment in semantic interoperability and adherence to emerging industry standards for AI-to-AI communication will thus be crucial for any service wishing to participate effectively and be chosen within the composable AI economy. Poor interoperability becomes a significant barrier to being trusted and, consequently, to being invoked.

The following list illustrates the evolution of trust signals, contrasting traditional human-centric cues with the emerging AI-interpretable signals that will define the next era of service selection:

Reliability

- **Traditional human-centric signals:** Word-of-mouth, brand reputation, past personal experience, and perceived consistency of service.
- **Emerging AI-interpretable signals:** Verifiable performance logs such as uptime and success rates exposed via APIs, real-time service health dashboards, published incident reports, and service level agreements encoded in machine-readable metadata.

Ethical stance

- **Traditional human-centric signals:** Stated company values, corporate social responsibility reports, leadership reputation, and perceived fairness in business dealings.
- **Emerging AI-interpretable signals:** Published ethical audit results, adherence to machine-readable codes of conduct, transparent and parsable data usage policies, AI fairness metrics, and clear provenance of training and operational data.

Security

- **Traditional human-centric signals:** Physical security of facilities, secure websites, and general confidence in how data is handled.
- **Emerging AI-interpretable signals:** Security certifications verifiable via API, formal vulnerability disclosure programs, real-time security posture indicators, explicit encryption standards defined in API specifications, and penetration test results shared with trusted ecosystem participants.

User alignment

- **Traditional human-centric signals:** Intuitive understanding of user needs, good customer service, personalized attention, and the feeling of being understood.
- **Emerging AI-interpretable signals:** Machine-readable preference matching between service capabilities and a user's AI profile, explainable outputs demonstrating alignment with stated goals, customizable service parameters exposed via APIs, and feedback loops that allow AI agents to report satisfaction or dissatisfaction with outcomes.

Competence

- **Traditional human-centric signals:** Professional credentials, years of experience, expert recommendations, and perceived quality of output.
- **Emerging AI-interpretable signals:** API-level success and error rates, measurable efficiency and response times, ratings in AI agent marketplaces, model performance metrics, and evidence of successful handling of complex tasks.

Transparency

- **Traditional human-centric signals:** Open communication, willingness to answer questions, and clear terms of service.
- **Emerging AI-interpretable signals:** Machine-readable documentation artifacts, explainable AI outputs accessible via APIs, clearly defined inputs, outputs, and limitations in technical documentation, and selective use of open-source components where appropriate.

This evolution underscores the practical shifts organizations must make. It's no longer sufficient for trust to be merely felt by humans; it must be structured, signaled, and verified in ways that resonate with the logic of AI agents. Trust at scale becomes the new distribution.

SUMMARY

In the dawning Age of Information Symmetry, where access to knowledge and data ceases to be a primary differentiator, the capacity to earn, cultivate, and maintain trust emerges as the paramount strategic asset. This chapter has explored how human trust, amplified by authentic brand affinity and demonstrated through unwavering ethical transparency, forms the new bedrock of value for individuals, organizations, and the AI systems that increasingly mediate their interactions. The journey from information scarcity to trust scarcity redefines competitive landscapes and demands a fundamental rethinking of what it means to be a valued and chosen entity.

The core argument presented is that in a world awash with information, the critical bottleneck shifts to belief and delegation. When anyone can theoretically know anything, the salient questions become: who do you believe, and who do you empower to act on your behalf?

These positions trust not as a passive virtue but as an active, sought-after commodity. It must be meticulously architected into systems, particularly AI, through deliberate design choices that prioritize validity, safety, security, accountability, explainability, privacy, fairness, and robust value alignment. Operationalizing trust requires concrete practices such as the use of transparency tools like Model Cards and Datasheets for Datasets, the integration of Human-in-the-Loop oversight, and the validation offered by third-party audits and certifications.

Brands, in this new era, evolve from mere identifiers into sophisticated semantic trust layers, acting as heuristics that both humans and their AI agents can use to navigate a complex ecosystem of choices. Authenticity, emotional resonance, and a clear articulation of purpose become more critical than ever, as does the ability to project these qualities in machine-interpretable ways.

Furthermore, the very notion of transparency undergoes a strategic transformation. Radical openness, once potentially viewed as a competitive risk, becomes a source of profound advantage, fostering deeper stakeholder confidence, pre-empting crises, and even driving industry-wide improvements in trustworthiness, particularly in sensitive sectors like healthcare and finance.

However, the infrastructure of trust is not without its own vulnerabilities. The potential for weaponized trust through technologies like deepfakes, the subtle erosion of critical judgment via automation bias, and the complex challenge of aligning universal trust principles with diverse cultural norms all represent significant risks that must be proactively managed. Resilience, pluralism, and unwavering accountability are essential for any trust infrastructure to endure and serve its purpose effectively.

Looking forward, the call to action is clear: organizations and individuals alike must invest in building the infrastructure of trust for the future. In a composable AI economy, where AI agents increasingly select and invoke services, becoming the "AI's trusted choice" is paramount. This requires crafting explicit trust signals embedded in metadata, APIs, and ontologies, ensuring that services are not only competent but also demonstrably aligned with user values and intentions.

The winners in this new epoch will be those who treat trust not as a superficial marketing claim or a reactive compliance measure, but as a core operational capability, deeply and inextricably woven into the DNA of their products, services, and organizational culture. As trust becomes increasingly programmable, visible, and verifiable, competitive advantage will flow to those who authentically bake it into their very infrastructure, not just their messaging. The ability to be trusted is, ultimately, the ability to thrive in the Age of Information Symmetry.

11

FEEDBACK IS THE PRODUCT

The landscape of value creation is undergoing a seismic shift. For decades, the prevailing paradigm centered on the development and delivery of finished goods and services; products conceived, designed, and shipped as static entities, their value largely fixed at the point of release.

A software program was burned onto a CD-ROM, its features immutable until the next major version. A car model was engineered to perform consistently over several years, with significant changes reserved for future iterations. Success in this era often hinged on perfecting the initial offering, on the meticulous execution of a predetermined plan. Trust, as explored in the preceding chapter, might be established through the perceived quality and reliability of these finished articles. Yet, in a world increasingly defined by information symmetry and rapid technological advancement, this model is proving insufficient, even detrimental.

This chapter will illuminate a new paradigm where continuous learning from user behavior, AI interaction, and system use is not merely an ancillary activity but the core engine of value creation. We will explore how products and services are transforming from static deliverables into dynamic, evolving systems. The central argument is that feedback, in its myriad forms, has transcended its traditional role as a post-mortem diagnostic tool. It has become the vital nutrient for systems that learn, adapt, and personalize in real time. This ongoing responsiveness, often powered by sophisticated AI, is becoming integral to the product's intrinsic value and the primary mechanism through which trust is continuously earned and maintained.

Readers will gain an understanding of this profound transition. First, the chapter will dissect the obsolescence of the "frozen product" mindset, revealing its inherent limitations in a dynamic world. Subsequently, it will delve into the rise of feedback-driven design, showcasing how modern AI-native systems are architected to learn and optimize through every user interaction.

The discussion will then broaden to examine how these feedback loops are being closed not just at the product level, but across the entire organizational stack; from user experience and business logic to agent orchestration and human operations, with AI often serving as the critical integrator. A particular focus will be given to how AI agents themselves are enhanced by feedback, becoming more accurate, efficient, and aligned with human intent through techniques like Reinforcement Learning from Human Feedback (RLHF) and memory-based preference modeling.

Furthermore, this chapter will explore the intriguing consequence of this evolution: products are ceasing to be mere transactional objects and are instead becoming relational entities. They adapt to individual users, fostering loyalty through behavioral tuning and creating "identity-aware" interfaces that remember and anticipate needs. This can even lead to nascent forms of human-AI relationships, built on interactional affinity. For organizations to support such adaptive products, they too must embody principles of learning. The chapter will therefore position organizational learning not as a peripheral function but as critical infrastructure, essential for cultivating the "learning velocity" required to thrive.

Ultimately, the reader will see that if trust is the new currency in the age of information symmetry, then the mechanisms of feedback and adaptation are how that currency is continuously minted, validated, and compounded. The old mantra of "build it and they will come" is being irrevocably replaced by a new imperative: "launch it and let it learn with them". This sets the stage for understanding why the era of the static, unchanging product is decisively over.

THE ERA OF THE FROZEN PRODUCT IS OVER

The traditional model of product and service delivery, characterized by its one-time, fixed, and static nature, once represented a hallmark of industrial efficiency and predictable output. This approach, often epitomized by methodologies like the Waterfall model in software development, emphasized comprehensive upfront planning, sequential execution of distinct phases, and the delivery of a "finished" product that was expected to meet predefined specifications with minimal deviation. In this paradigm, value was largely locked in at the point of design and manufacture; the product was "frozen" in its intended form, and its subsequent journey in the market was one of gradual obsolescence rather than continuous evolution.

However, in a contemporary environment marked by accelerating technological change, shifting consumer expectations, and the pervasive influence of AI-driven dynamism, this legacy model is not just showing its age; it is becoming a significant competitive liability.

The Waterfall model, a classic representation of this static approach, dictates a linear progression: requirements are gathered and finalized, design is completed, implementation occurs, followed by testing, and finally, deployment. Each phase must be completed before the next begins, making it inherently difficult and costly to accommodate changes once the process is underway. This rigidity means that the product, as conceived, reflects a snapshot in time.

By the time it reaches the market; often after a lengthy development cycle, the initial assumptions about user needs or market conditions may no longer hold true. The very structure of such a development process fosters a "build, test, ship, forget" mentality, where the primary focus is on delivering the pre-specified artifact, with less emphasis on its ongoing adaptation to real-world use. This creates a natural lag, not just in initial delivery, but more critically, in the organization's ability to respond to feedback or changing requirements post-launch.

This inherent inflexibility is not merely a process limitation; it translates directly into organizational inertia. Companies structured around the production of "frozen" products often exhibit similar static characteristics in their operational design. Decision-making can be slow, taking weeks instead of minutes, as changes necessitate revisiting foundational plans and navigating complex approval chains. Communication pathways may be rigid, hindering the flow of insights between departments. This can lead to a situation where employees focus on narrow, personal directives rather than broader, evolving company objectives, resulting in misaligned efforts and even disengagement. The high cost of such non-adaptive organizational designs manifests in subpar work, wasted resources, and a diminished capacity to innovate or respond effectively to market shifts.

Even when valuable insights are generated within the organization; perhaps through market research or early customer interactions, the translation of these insights into actionable product changes is often stymied by the very nature of the static product model.

The ARIA model (Awareness, Reflection, Illumination, Action) suggests that while insights can spark, acting upon them requires immediate motivational drive and conducive conditions. Frozen systems, by their design, delay or obstruct this crucial "action" phase. Organizations may find themselves in a state of "initiative fatigue," where repeated attempts to modify static offerings or entrenched processes prove slow, cumbersome, and ultimately ineffective, leading to a reluctance to even attempt further change.

The competitive liability of such non-adaptive offerings extends beyond missed market opportunities. Consider the legal concept of product liability, where manufacturers can be held accountable for foreseeable misuse, especially if warnings are inadequate. A static product cannot dynamically update its warnings or adjust its functionality based on observed user behaviors or newly identified risks. In a world where patterns of use and the understanding of "foreseeable misuse" can evolve, a frozen product is at a persistent disadvantage, potentially accumulating risks that could lead to reputational damage or legal challenges. This inability to adapt becomes a tangible vulnerability.

A useful, albeit metaphorical, parallel can be drawn with the physical phenomenon of static electricity in manufacturing processes. Uncontrolled static electricity can cause a host of problems: materials may cling together, misbehave on production lines, attract dust and contaminants, or even suffer catastrophic failure. Similarly, "static" or "frozen" products in the marketplace can fail to adapt to changing user needs (exhibiting "misbehavior" in terms of relevance), become outdated as they attract the "dust" of negative perceptions or declining interest, and ultimately face market withdrawal (a form of "catastrophic failure"). This illustrates the inherent instability and risk associated with offerings that lack the capacity for adaptation.

The implications of adhering to this frozen product model are profound. One significant consequence is a form of latent obsolescence that is, in effect, designed into the product from the outset. Because the Waterfall model and similar approaches finalize requirements and design elements early in the development cycle, while the external world of user needs, technological capabilities, and competitive pressures continues to evolve rapidly, the product is inevitably set on a path towards diminishing relevance. Obsolescence, in this context, is not merely a function of physical wear and tear but a direct outcome of the product's inability to keep pace with ongoing change. Its "frozen" state ensures that it will, sooner or later, "melt" into irrelevance.

Furthermore, organizations that persist with the frozen product model effectively pay a "static tax". This tax is levied in multiple forms: the direct financial costs associated with the difficulty and expense of implementing changes to rigid systems ; the opportunity costs of missed market segments or customer churn when products fail to meet evolving needs due to limited client or end-user involvement during development ; the competitive disadvantage that arises from organizational lag, allowing more agile competitors to seize an advantage ; and the human capital costs stemming from employee disengagement and reduced innovation when teams are tethered to increasingly outdated offerings. Collectively, these burdens represent a substantial and continuous drain on resources and competitiveness.

Perhaps most critically in the context of the preceding chapter, a persistently static product in a dynamic world can actively erode user trust. Chapter 10 established trust as a foundational asset. However, when users provide feedback, whether explicitly through complaints or implicitly through their usage patterns, and a product fails to adapt or respond to known issues, this lack of responsiveness is often interpreted as a lack of care, attention, or capability on the part of the provider. The inherent delays and difficulties in modifying a frozen product mean that such feedback often goes unaddressed in a timely fashion, if at all. This perceived inertia directly undermines the pillars of trust; reliability, responsiveness, and alignment with user needs, effectively working against the very concept of "Trust as a Service."

The following list summarizes the shift away from this increasingly problematic model:

Design philosophy

- **Legacy static model:** Prescriptive and fixed at the outset, with assumptions locked in early.
- **Emerging adaptive model:** Evolutionary by design, built to change as conditions, users, and environments evolve.

User involvement

- **Legacy static model:** Minimal involvement after initial design and deployment.
- **Emerging adaptive model:** Continuous and integral, with users actively shaping ongoing evolution.

Response to change

- **Legacy static model:** Reactive, slow, and costly to modify once deployed.
- **Emerging adaptive model:** Proactive, fast, and iterative, with lower marginal cost per change.

Market relevance

- **Legacy static model:** Peaks at launch and steadily decays as conditions shift.
- **Emerging adaptive model:** Sustained or increasing relevance through continuous adaptation.

Organizational lag

- **Legacy static model:** Significant delay between insight, decision, and action.
- **Emerging adaptive model:** Minimized through integrated feedback loops and rapid execution.

Risk profile

- **Legacy static model:** High exposure to unforeseen risks, such as evolving misuse or changing constraints.
- **Emerging adaptive model:** Risks mitigated through continuous learning, monitoring, and adjustment.

Competitive stance

- **Legacy static model:** Becomes a liability as the environment changes faster than the system can adapt.
- **Emerging adaptive model:** Serves as a source of ongoing competitive advantage.

The inherent limitations and accumulating liabilities of static products signal unequivocally that a new approach is not merely advantageous but has become essential for organizational survival and market leadership. This realization paves the way for an exploration of how contemporary systems are being built to not only cope with change but to actively thrive on it, driven by the constant flow of user input and interaction data.

THE RISE OF FEEDBACK-DRIVEN DESIGN

As the era of the frozen product wanes, a new paradigm is ascendant: feedback-driven design. This approach marks a fundamental departure from creating static deliverables, instead conceptualizing products and services as dynamic, living systems that learn, evolve, and optimize themselves through continuous interaction with their users and environment.

At the heart of this transformation are AI-native systems, which are uniquely equipped to process vast streams of feedback data in real time, translating every user interaction into a potential training signal that refines and personalizes the experience. This is not about occasional updates or periodic user surveys; it is about an ongoing, deeply integrated conversation between the product and its user, where the product's behavior is constantly being shaped by the user's actions, preferences, and needs.

The core principle of feedback-driven design is that every interaction, no matter how subtle, carries information that can be used for improvement. Leading digital platforms offer compelling illustrations of this principle in action.

Netflix, for instance, meticulously collects a wide array of user data, including viewing history, search queries, interaction patterns (like adding a title to a list), time spent hovering over a particular thumbnail, rewatch frequency, device usage, and even scrolling speed. This rich dataset is not merely archived; it is actively used to construct and refine each user's "taste profile". This profile, in turn, powers a sophisticated recommendation engine that employs techniques such as collaborative filtering (comparing a user's behavior to that of similar users), content-based filtering (analyzing the attributes of content a user has enjoyed), and matrix factorization to predict preferences. The result is a highly personalized Netflix homepage that, as the company notes, looks "completely different for every user". The product, the user's unique Netflix experience, is thus continuously sculpted by their ongoing feedback.

Spotify operates on a similar philosophy, building a "taste profile" for each listener based on their searches, listening habits, skips, and saved tracks. This implicit feedback is augmented by explicit information provided by the user (such as general location, preferred language, age, and followed artists or podcasts) and by broader trends observed across the platform. Spotify's algorithms then deliver personalized playlists and recommendations, and users can further refine these by providing direct feedback, such as a "thumbs down" on a suggested track. This interplay of implicit behavioral signals and explicit user input creates a dynamic system where the music discovery experience is constantly adapting.

Amazon's personalization engine is another powerful example of feedback-driven design at scale. It leverages user-based and item-based collaborative filtering, content-based analysis of product attributes, and advanced machine learning models, including recurrent neural networks (RNNs) designed to understand sequential user behavior. The system processes historical purchase data, browsing patterns, ratings, and even contextual factors like location, time of day, and device type to generate tailored recommendations. Crucially, this system adapts in "near real-time" to shifts in user interest; if a user suddenly starts browsing a new product category, the recommendations adjust accordingly. The tracking of click-through rates on recommendations and subsequent conversion rates serves as a direct feedback loop, allowing Amazon to continuously evaluate and refine the effectiveness of its suggestions.

The practical implementation of feedback-driven design is often facilitated by structured experimentation, most notably through A/B testing. Platforms such as VWO, Optimizely, and AB Tasty provide the tools for companies to systematically test different versions of a webpage, feature, or user interface element, presenting these variations to different segments of their user base.

By measuring how users interact with each version; which one leads to higher engagement, conversion rates, or task completion, organizations can make data-driven design choices. This moves decisions away from intuition or guesswork and grounds them in empirical evidence derived from actual user behavior, which acts as the feedback mechanism. Netflix's practice of A/B testing different thumbnail images for the same show to see which one elicits more clicks is a clear example of this real-time feedback optimizing for engagement.

Underpinning much of this capability is the systematic collection of usage data through software telemetry. Telemetry systems act as the sensory apparatus for digital products, enabling developers and product managers to monitor and measure user interactions in real time. This includes tracking which features are used most frequently, identifying common user pathways, logging error rates, and assessing overall system performance. This stream of data provides invaluable insights into user behavior, preferences, and pain points. It is the raw material that feeds into model updates for AI systems, informs feature prioritization, drives personalization engines, and allows for a continuous cycle of improvement based on empirical observation rather than assumptions.

The rise of feedback-driven design carries several profound implications. Firstly, the product itself transforms into a sophisticated data magnet and an embedded learning algorithm. Its function is no longer solely to deliver a predefined set of features, but also to actively capture relevant interaction data and leverage that data to learn and enhance its own performance and relevance. The product becomes an inseparable amalgam of its capabilities and its capacity to learn from feedback. Platforms like Netflix, Spotify, and Amazon are not just content repositories or e-commerce sites; they are complex systems designed to sense, process, learn, and adapt continuously. Telemetry provides the nervous system for this data collection, while A/B testing offers a structured methodology for interpreting user responses to design variations. The ability of the product to attract, process, and act upon feedback data becomes a core component of its value proposition.

Secondly, feedback-driven design, especially when supercharged by AI, facilitates a monumental shift from mass production to mass personalization at scale. The one-size-fits-all product, a hallmark of the industrial era, gives way to experiences that are uniquely tailored to the individual, yet delivered efficiently to millions. The sheer complexity of managing such diverse preferences and behaviors was an insurmountable barrier in the past.

AI algorithms can leverage individual user feedback, including interactions, ratings, and explicit preferences. By cross-referencing this data with patterns from large user populations (similar to collaborative filtering), AI can make real-time, personalized adjustments for each user. This allows for what might be termed the "mass production of uniqueness," a stark contrast to the homogeneity of the frozen product era.

Finally, as users increasingly interact with products that visibly learn and adapt to them; recommendation systems that get better over time, and interfaces that remember preferences, a new baseline expectation is established. Users begin to anticipate this adaptive intelligence in all their interactions, making static, unresponsive offerings feel archaic, frustrating, or simply "dumb" by comparison. The success of feedback-driven design in prominent digital services raises the bar across all industries, accelerating the obsolescence of products that cannot learn and evolve. This creates a powerful imperative for all organizations to embrace feedback as a central tenet of their design and operational philosophy.

The transformative power of AI-native systems to learn from every interaction and deliver personalized experiences in real time is undeniable. However, for this capability to yield its full strategic benefit, feedback cannot remain a superficial layer; it must be deeply integrated throughout the entire operational and technological stack of an organization. This necessity leads to the critical question of how these feedback loops are effectively closed and orchestrated across diverse functions and systems.

CLOSING THE LOOP ACROSS THE STACK

The true power of feedback-driven design is unleashed when the insights derived from user interactions and system performance are not confined to isolated improvements within a single feature or department, but are instead woven into the very fabric of the organization. Closing the loop across the entire stack means establishing mechanisms for feedback to flow seamlessly from the user experience (UX) layer through business logic and AI agent orchestration, down to the core human operations and even influencing strategic re-architecture. This holistic integration transforms feedback from a series of discrete data points into a continuous, energizing current that powers organizational learning and adaptability.

The product development lifecycle itself offers a foundational framework for integrating feedback at multiple junctures. Best practices advocate for collecting user input at every stage, from initial ideation and concept testing to design, development, beta testing, and post-launch evolution. In the ideation phase, for example, analyzing existing usage data or conducting customer interviews can reveal unmet needs or pain points, providing feedback that shapes the very conception of new features. During design and development, prototype testing and usability studies allow teams to gather direct feedback on proposed solutions, enabling iterative refinement before significant resources are committed. Post-launch, a rich stream of feedback from in-app surveys, support tickets, user acceptance testing (UAT), and telemetry data continues to inform ongoing improvements, bug fixes, and even decisions about substantial re-architecture if pervasive issues are identified.

However, for this feedback to influence the entire stack, it must transcend traditional departmental silos. Customer support, for instance, is a rich source of information about product flaws or user frustrations. Sales teams gather insights into market needs and competitor activities. Security operations identify vulnerabilities and threat patterns. For these diverse feedback streams to contribute to holistic improvement, robust cross-functional integration (CFI) is essential.

This involves creating a culture and a set of processes that encourage collaboration and information sharing between departments like product development, engineering, marketing, sales, and support. Digital technologies, particularly those emerging in the context of Industry 4.0, can play a vital role by enabling "virtual design sharing and problem-solving systems," fostering cooperation between individuals from different functional areas. Establishing open communication channels, a shared vision, and a climate of psychological safety are critical for these cross-functional teams to effectively process and act upon shared feedback.

The sheer volume, velocity, and variety of feedback generated in modern digital ecosystems make manual integration across the stack an increasingly daunting task. This is where Artificial Intelligence emerges as a powerful integrator. AI-powered tools, leveraging Natural Language Processing (NLP), Machine Learning (ML), and sentiment analysis, can ingest, centralize, organize, interpret, and extract actionable insights from a multitude of feedback sources-including surveys, online reviews, support tickets, chat logs, social media mentions, and product telemetry; often in real time and at scale.

AI can automatically detect recurring themes, discern user sentiment and intent, and even assess the urgency of different feedback items. For example, an AI system could identify a spike in negative sentiment related to a new feature across social media and support channels, correlate this with increased error rates for that feature reported by telemetry, and flag this confluence of signals for immediate attention by the product and engineering teams. This ability to synthesize diverse feedback streams into coherent, prioritized insights is invaluable for closing loops effectively.

This multi-layered approach to feedback integration can be visualized as a framework where different types of feedback inform actions at various levels of the organization and its technology stack, as listed below.

User interface and experience (UX)

- **Key feedback sources:** Usability tests, A/B test results, session recordings, heatmaps, and direct user comments.
- **Typical actions triggered:** UI redesign, workflow optimization, and accessibility improvements.
- **Key departments involved:** Design, Product, and Frontend Engineering.

Application logic and features

- **Key feedback sources:** Error logs, performance telemetry, feature requests, bug reports, and user acceptance testing results.
- **Typical actions triggered:** Code refactoring, feature enhancement or deprecation, and bug fixes.
- **Key departments involved:** Product, Engineering, and QA.

AI model and agent performance

- **Key feedback sources:** Reinforcement learning from human feedback data, accuracy scores, user corrections, model drift alerts, and agent error rates.
- **Typical actions triggered:** Model retraining, algorithm tuning, and data augmentation.
- **Key departments involved:** Data Science, Machine Learning Engineering, and AI Operations.

Business processes and operations

- **Key feedback sources:** Support ticket trends, employee suggestions, process cycle time measurements, cost analyses, and CRM data.
- **Typical actions triggered:** Process re-engineering, automation initiatives, staff training, and policy changes.
- **Key departments involved:** Operations, Support, Sales, Human Resources, and Finance.

Strategic direction

- **Key feedback sources:** Market trends, competitor analysis, Net Promoter Score, and customer lifetime value.
- **Typical actions triggered:** Strategic pivots, entry into new

markets, mergers and acquisitions, and shifts in research and development focus.
- **Key departments involved:** Leadership, Strategy, Marketing, and Product.

When feedback loops are effectively closed across these diverse layers and integrated cross-functionally, the organization begins to function less like a collection of discrete parts and more like a cohesive, sentient organism. It develops an organizational nervous system, capable of sensing subtle shifts in its internal state (e.g., operational inefficiencies) and external environment (e.g., changing user behavior, emerging market threats) and then coordinating complex, adaptive responses. AI, in this context, acts as a sophisticated central processing unit, analyzing and synthesizing the myriad signals flowing through this nervous system, transforming raw feedback into actionable intelligence that drives the organism's evolution.

The strategic agility of an organization; its ability to pivot quickly and effectively in response to new information or changing circumstances, is not merely the result of having 'agile' development teams in one corner of the business. Rather, it is an emergent property that arises from the interconnectedness and efficiency of these feedback loops across the entire stack.

Any latency within this system, be it a delay in analyzing customer support data, poor communication between sales and product development, or slow adaptation of AI models to new data patterns, generates a strategic drag. This diminishes the organization's overall capacity to respond. Conversely, the tighter and faster these loops, the more agile and resilient the enterprise becomes.

In this complex interplay of information, AI's role evolves beyond simple analysis within isolated domains. It becomes a "feedback weaver," adept at identifying and understanding the intricate interdependencies between feedback signals originating from different parts of the stack.

For example, AI might discern that a recent decline in customer satisfaction scores (strategic feedback) is correlated with a new software release that introduced subtle usability issues (UX feedback), which in turn led to increased processing times in a core business operation (operational feedback) and a higher rate of errors by AI agents interacting with that system (AI agent feedback). By weaving these seemingly disparate threads of information into a coherent narrative, AI surfaces systemic problems and guides holistic solutions, enabling a far more profound level of organizational learning and adaptation than would be possible if each feedback stream were considered in isolation.

Closing these loops effectively across the entire organizational and technological stack is a multifaceted challenge, requiring cultural shifts, process re-engineering, and technological enablement. A particularly crucial domain where feedback is driving rapid evolution is within AI agents themselves, which are increasingly at the forefront of user interaction and automated decision-making. Understanding how these agents learn and improve through feedback is key to harnessing their full potential.

FEEDBACK-ENHANCED AI AGENTS

The capacity of Artificial Intelligence agents to learn and adapt based on feedback is a critical driver of their increasing utility and sophistication. No longer static algorithms executing predefined instructions, modern AI agents are designed to evolve, becoming more accurate, efficient, and aligned with user intent through a variety of feedback mechanisms. This continuous improvement cycle is transforming agents from mere tools into dynamic partners capable of nuanced understanding and personalized assistance. Key to this evolution are techniques like Reinforcement Learning from Human Feedback (RLHF), the development of sophisticated memory systems for preference modeling, and a careful calibration of how users and agents interact within the learning loop.

Reinforcement Learning from Human Feedback (RLHF) stands out as a pivotal technique for shaping AI behavior to better meet human expectations. This multi-stage process typically begins with a pre-trained language model, which is then fine-tuned using a supervised approach on a dataset of high-quality prompt-response pairs often generated by humans.

Following this, a separate "reward model" is trained. To do this, humans are presented with several responses generated by the AI to a given prompt and are asked to rank these responses based on criteria such as helpfulness, accuracy, harmlessness, or adherence to desired style. This human preference data teaches the reward model to predict which kinds of responses humans are likely to favor.

Finally, the original AI agent is further optimized using reinforcement learning, where the reward model provides the signal that guides the agent's learning process. The agent essentially tries to generate responses that will maximize the score from the reward model, thereby learning to produce outputs that are more closely aligned with human goals and values. RLHF has proven effective in enhancing a wide range of AI capabilities, from making chatbot responses more natural and engaging to improving the quality of language translation and even imbuing AI-generated content with subjective qualities like the "mood" of a piece of music.

For an AI agent to benefit durably from feedback, particularly from ongoing interactions with a specific user, it must possess a robust memory system. AI agent memory can be categorized in various ways, including short-term memory for immediate context and long-term memory for persistent knowledge and preferences.

Long-term memory itself can encompass episodic memory (recalling specific past interactions or events), semantic memory (storing factual knowledge and concepts), and procedural memory (retaining learned skills and behaviors). By logging key events, user corrections, expressed preferences, and the outcomes of past actions, an agent can build a rich, evolving model of its user.

For example, an AI travel planning agent that remembers a user's stated preference for direct flights or a dislike for a particular airline, based on previous feedback, can provide far more personalized and satisfactory recommendations in subsequent interactions. Frameworks like LangChain and techniques such as the summarization of past conversations or the vectorization of memories for efficient retrieval are instrumental in building these memory-based preference models.

The way feedback is incorporated also depends on the nature of the interaction loop between the user and the agent. "User-in-the-loop" (often referred to as Human-in-the-Loop or HITL) systems are those where human input is actively solicited to train, evaluate, or refine the AI model. RLHF is a sophisticated form of HITL, where human judgments directly shape the agent's learning. Active learning, where the model strategically queries humans for labels on the most informative data points, is another example.

In these scenarios, the user's feedback directly contributes to improving the agent's fundamental capabilities. Conversely, in "agent-in-the-loop" (or AI-in-the-loop, sometimes denoted AI2L) scenarios, the AI might assist a human who retains primary decision-making authority. The agent might, for instance, present several options to a user when faced with ambiguity, using the user's selection as immediate feedback to complete the current task. While the primary goal here is task completion, a well-designed agent can still learn from this interaction, storing the choice and its context in its memory to inform similar future situations, thus gradually improving its autonomous performance over time.

The impact of these feedback mechanisms is evident in numerous real-world applications. Fraud detection systems, for example, continuously adapt to new patterns of illicit activity by learning from data on confirmed fraudulent transactions (a form of feedback). Content recommendation engines, such as those used by Netflix and Amazon, dynamically adjust their suggestions as a user's viewing or purchasing habits change over time, effectively learning from the implicit feedback of user behavior. Speech recognition software like Siri or Google Assistant improves its understanding of diverse accents, slang, and individual speech patterns through ongoing exposure and, in some cases, explicit user corrections. Even adaptive thermostats can learn a household's temperature preferences and occupancy schedules based on manual adjustments made by users over time. These examples all illustrate AI agents that are not static but are continuously tuned and refined by the feedback they receive.

The integration of these feedback mechanisms into AI agents leads to a significant shift in how these systems are developed and how they function. It represents a move away from purely programmed logic towards learned behavior, a process that can be likened to the "domestication" of AI. Rather than attempting to foresee every eventuality and hardcode explicit rules for every scenario; an impossible task given the complexity of the real world and human preferences-developers are increasingly focused on creating agents that can learn and adapt from experience. Feedback, especially nuanced human judgments as captured in RLHF, acts as the guiding hand in this learning process, shaping the agent's behavior to align more closely with complex, often unstated, human values and social norms. This is less about writing precise instructions and more about "training" or "raising" an AI to be helpful, harmless, and contextually appropriate.

A fascinating consequence of this continuous, individualized feedback is the emergence of personalized AI evolution. An AI agent that interacts predominantly with one user, and whose memory systems diligently record that user's preferences, corrections, and interaction styles, will inevitably evolve differently from an identical agent interacting with a different user. Each agent becomes a unique reflection of its primary user or user group. Over time, the agent's internal models, response strategies, and even its "personality" (as perceived by the user) will diverge, shaped by the distinct feedback streams it has processed. This means that AI agents are not just static tools off a production line; they have the potential to become highly personalized cognitive extensions, co-evolving with their users and becoming increasingly tailored to their specific needs and ways of thinking.

Finally, the mechanisms by which an AI agent receives, processes, and acts upon feedback are not merely technical details; they are fundamental to building (or eroding) user trust.

As discussed in Chapter 10, trust is paramount. When an agent demonstrably learns from a user's corrections, remembers their preferences, and improves its performance over time, the user feels "heard" and perceives the agent as intelligent and responsive. This positive experience of effective feedback leading to tangible improvement is a powerful trust-building force.

Conversely, an agent that repeatedly makes the same mistakes despite feedback, or seems to ignore user preferences, will be perceived as unintelligent, unhelpful, or frustrating. Such an experience can quickly break trust, leading users to abandon the agent or the product it powers. Therefore, the design and implementation of feedback loops within AI agents are critical not just for optimizing performance metrics but for fostering a healthy, productive, and trusting relationship between humans and their AI counterparts.

When AI agents are engineered to learn effectively from feedback and to remember and honor user preferences, they begin to transcend their role as mere instruments. They start to feel more like collaborators or partners. This evolving dynamic fundamentally alters the nature of the product itself, shifting it from a purely transactional object to an entity with which the user can form a more continuous and adaptive relationship.

PRODUCTS AS RELATIONSHIPS

The relentless integration of feedback mechanisms and AI-driven adaptation is not merely refining product functionality; it is fundamentally reshaping the user's connection to the product itself. As products become increasingly adept at learning from interactions, personalizing experiences, and responding to individual needs, they cease to be perceived as static, impersonal objects. Instead, they evolve into "living systems" that engage with users in ways more akin to a relationship than a simple transaction. This shift is characterized by the cultivation of loyalty through behavioral tuning, the emergence of identity-aware interfaces that remember users across contexts, and even the potential for emotional affinity to develop through finely tuned interactions, particularly with AI-driven entities.

Product loyalty, in this new paradigm, is increasingly fostered through what can be termed "behavioral tuning". This occurs when a product's behavior is continuously shaped by the user's behavior, creating a cycle of positive reinforcement. When a user consistently has positive and satisfying experiences because the product adapts to their usage patterns, anticipates their needs, or simplifies their tasks, a form of behavioral loyalty can develop. This isn't necessarily a deep emotional attachment in all cases, but rather a practical preference born from the product "just working well" for that individual. The key is that the product is not static; its responsiveness to the user's implicit and explicit feedback makes it feel tailored and understood, reducing friction and increasing satisfaction. Companies that effectively mine customer interaction data-browsing habits, purchase patterns, service requests; can leverage these insights to personalize the product experience, thereby building trust and strengthening the user-product bond.

A critical enabler of this relational dynamic is the development of "identity-aware interfaces" and "context-aware personalization". An identity-aware interface is one that recognizes and remembers the user across different sessions and potentially across different touchpoints. It leverages this stored knowledge of the user's identity, preferences, past interactions, and current context (such as location, device, time of day, or even accessibility needs) to tailor the experience in real time.

For example, an e-commerce platform might remember a user's past purchases and browsing history to offer relevant product recommendations, or a streaming service might adjust its content suggestions based on the time of day or the device being used. When an interface consistently "knows" who the user is and what they are likely trying to achieve, it creates a sense of continuity and personal recognition. This stands in stark contrast to transactional systems where the user must effectively re-establish their context and preferences with each interaction. The goal is to reduce friction, improve content relevance, and make the interaction feel more natural and intuitive, much like interacting with someone who knows you well.

The sophisticated personalization engines powering services like Netflix, Spotify, and Amazon are prime examples of how continuous feedback cultivates this product-as-relationship dynamic. These platforms ingest a constant stream of user data; every click, view, search, and rating-and use it to refine an individual's profile, leading to highly attuned recommendations and user interfaces. Netflix, for instance, goes as far as personalizing the thumbnail images displayed for shows and movies, testing different visuals to see which ones are most likely to engage a particular user. When a service consistently anticipates a user's desires and tailors its offerings accordingly, the user feels understood and uniquely catered to. This ongoing, adaptive dialogue feels less like using a generic utility and more like engaging with an entity that has developed a nuanced understanding of their individual tastes and needs.

As AI systems become more sophisticated in processing and responding to nuanced human cues; including emotional tone in text or voice, or subtle behavioral indicators, the potential for "interaction tuning" to foster a degree of emotional affinity arises.

Research indicates that humans can, and sometimes do, form emotional attachments with AI entities, particularly those designed to offer companionship or emotional reassurance. Users may develop attachment styles towards AI that mirror those seen in human relationships, such as seeking reassurance (anxiety) or preferring emotional distance (avoidance). If a product, especially an AI-powered assistant or companion, consistently adapts its responses based on perceived emotional feedback, it can create an experience that feels empathetic, supportive, or understanding. While this "empathy" is simulated, the user's subjective experience can still lead to feelings of connection, trust, and even emotional dependence.

This deepening of the product-user connection into something resembling a relationship has several significant consequences. One is the creation of a powerful "relationship stickiness" or lock-in effect. When a product has, over time, accumulated a rich history of a user's preferences, behaviors, and contexts through continuous feedback, it becomes uniquely valuable and efficient for that specific user. Switching to a competitor's offering, even if that offering has comparable features, means starting from a "blank slate". The user would lose the accumulated "understanding" and personalized tuning of their current product and face the effort of "re-teaching" the new one. This switching cost, born from the investment in the established interaction history, is analogous to the value placed on shared history and mutual understanding in human relationships. It forms a potent competitive moat, making users less likely to churn.

However, as products become more adept at simulating understanding and fostering emotional affinity through sophisticated interaction tuning, profound ethical considerations come to the fore. The capacity of AI to mimic empathy or provide consistent emotional support, while potentially beneficial in some contexts (e.g., combating loneliness), also carries risks. If users, particularly vulnerable individuals, are unaware of the simulated nature of the AI's "feelings," there is a potential for manipulation or the development of unhealthy dependencies. The consistent, idealized responsiveness of an AI companion might also lead to unrealistic expectations in human-human relationships, which are inherently more complex and less perfectly accommodating. Thus, as products evolve into relational entities, designers and organizations bear a significant ethical responsibility to ensure transparency, prevent harmful over-reliance, and thoughtfully consider the impact on human well-being.

The trajectory of a feedback-driven product can be seen as an evolution: from a simple tool performing a function, to an adaptive partner that learns and collaborates, and potentially, for some users, to a cognitive or emotional prosthesis. When an AI-powered product becomes so deeply integrated into a user's daily routines, decision-making processes, or emotional regulation; facilitated by its continuous learning from their feedback and its identity-aware interface, it can begin to feel like an indispensable extension of their own capabilities or even their persona. This profound level of integration carries significant implications for user autonomy, data privacy, and the consequences of the "prosthesis" being altered, removed, or malfunctioning.

If products are indeed becoming adaptive, relational entities that learn and evolve in concert with their users, then it logically follows that the organizations responsible for creating, managing, and nurturing these "living systems" must embody these same principles of continuous learning and adaptation. The capacity for feedback to drive product evolution must be mirrored by an organizational capacity to learn from feedback at every level. This necessitates positioning feedback not merely as a feature of a product, but as a core component of the organization's operational infrastructure.

ORGANIZATIONAL LEARNING AS INFRASTRUCTURE

The transformation of products into adaptive, relational systems necessitates a parallel evolution within the organizations that create them. For feedback to be more than just a mechanism for product-level tweaks, it must be ingrained in the company's DNA, functioning as a company-wide capability and a core component of its operational infrastructure. This means cultivating an environment where learning is continuous, inquiry is encouraged, data flows freely, and the entire organization is architected to sense, process, and respond to feedback with agility. Such an organization treats learning not as a sporadic training event, but as an ongoing, systemic process, crucial for sustained relevance and competitive advantage.

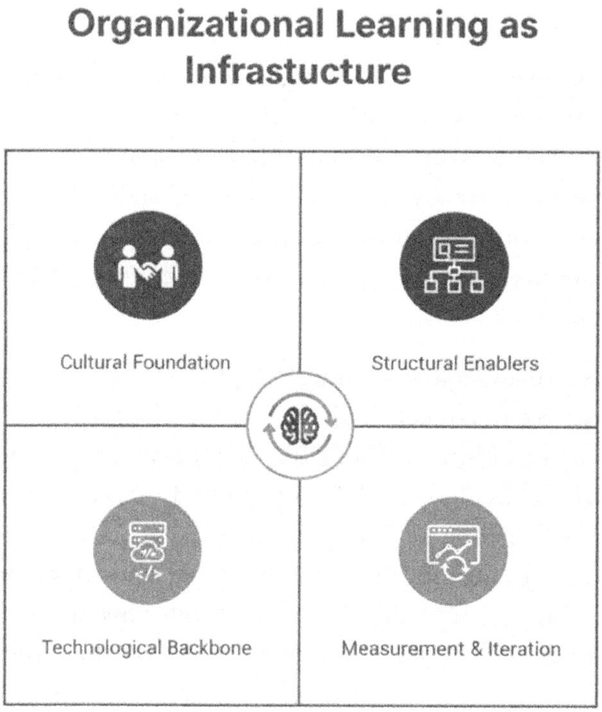

AT THE FOUNDATION of this learning infrastructure is a "culture of learning". This involves more than simply offering training programs; it's about creating an environment where continuous learning opportunities are woven into the fabric of daily work, where employees are empowered and encouraged to explore new ideas, develop new skills, and challenge existing norms.

Such a culture actively promotes inquiry and dialogue, valuing questions as much as answers, and fostering an atmosphere where constructive feedback; even when critical, is seen as a gift that fuels improvement. Leadership plays a crucial role in championing this culture, rewarding curiosity and experimentation, and ensuring that learning and development are treated as top organizational priorities, not as expendable overhead. Google's renowned "20% Time," which allowed employees to dedicate a portion of their work hours to passion projects, is a classic example of fostering such an innovative and learning-oriented environment.

Beyond culture, organizational design itself must incorporate feedback loops as structural elements. Agile methodologies, with their emphasis on sprints, daily stand-ups, and regular retrospectives, provide a framework for teams to continuously inspect their processes and adapt based on what they have learned. These practices ensure that feedback from recent work directly informs how future work is approached. This principle can be extended beyond software development to other business functions. For instance, AI-assisted decision reviews, where outcomes of significant business decisions are systematically analyzed against initial assumptions and subsequent performance data, can create powerful learning loops that refine strategic and operational judgment over time. The goal is to make the collection, analysis, and utilization of feedback an integral part of how the organization operates, driving continuous improvement in products, services, and processes.

The technological backbone for this organizational learning infrastructure is increasingly found in "enterprise telemetry". Just as product telemetry provides data on how a specific application is being used, enterprise telemetry aims to create a comprehensive, real-time view of the entire organization's operational health.

This involves collecting and aggregating data from a multitude of sources-application logs, system performance metrics, business process execution data, customer interaction records, and even employee engagement platforms. Standards like OpenTelemetry (OTel) are emerging to unify the collection and transmission of this diverse telemetry data, breaking down traditional data silos and enabling a more holistic view. This "shared nervous system" allows the organization to sense its internal state and external interactions with much greater fidelity, providing the raw data necessary for learning and adaptation at all levels.

To ensure that this learning infrastructure is effective, its output must be measured. The concept of "learning velocity"; the speed at which an organization or its constituent teams can acquire new skills, assimilate new information, and adapt their behaviors-becomes a critical key performance indicator (KPI). This shifts the focus from static achievements or lagging outcomes to the rate of improvement and responsiveness. Measuring learning effectiveness, as outlined by frameworks that assess knowledge acquisition, job application of learned skills, and impact on business results, ensures that investments in learning infrastructure are yielding tangible benefits. AI can play a significant role here, too, by analyzing enterprise telemetry and performance data to identify patterns, surface insights about where learning is happening effectively (or not), and even assist in reviewing the impact of decisions.

The strategic importance of such feedback-driven organizational learning is underscored by its application in critical operational areas, including defense and security. Companies like C3 AI and Splunk utilize feedback loops not just for routine product improvement but for enhancing the performance of AI models in high-stakes applications like anti-money laundering or for operationalizing threat hunting in cybersecurity. In these contexts, the ability to quickly learn from new threat intelligence (feedback) and adapt defensive postures is paramount. This demonstrates that robust organizational learning, powered by effective feedback loops, is not just about efficiency or innovation, but also about resilience and strategic defense.

The implications of building such an organizational learning infrastructure are far-reaching. Firstly, in an environment of constant change and information symmetry, the speed at which an organization can learn and adapt; its learning velocity, becomes a meta-metric for competitive advantage. It's not just about what the organization knows or owns at a single point in time, but how quickly it can integrate new knowledge and translate it into effective action. An organization with a superior learning infrastructure will consistently outmaneuver competitors who are slower to learn and respond, even if those competitors currently possess greater market share or resources. This adaptive capacity, fueled by continuous feedback, is the new engine for sustainable growth.

Secondly, the principles underpinning Reinforcement Learning from Human Feedback (RLHF) in AI agents can be metaphorically extended to the organization itself. Business strategies, operational processes, and decision-making frameworks can be viewed as organizational "policies". These policies generate outcomes, which in turn produce feedback in the form of market responses, operational metrics, customer satisfaction scores, and employee input, all captured through enterprise telemetry and other channels.

A true learning organization actively analyzes this feedback to understand what strategies are effective (to be reinforced) and which are failing (to be adjusted or discarded). This is analogous to how an RLHF reward model is trained. The organization then iteratively modifies its "policies" to improve future outcomes, effectively engaging in a form of "organizational RLHF". AI-assisted analysis of feedback and decision reviews can significantly accelerate this cycle.

Thirdly, to achieve this level of learning, organizations must transition from siloed data repositories to a more integrated "corporate knowledge graph". Enterprise telemetry, when unified and contextualized (for example, through standards like OTel), provides the lifeblood for such a graph. A rich, dynamic, and interconnected map of an organization's activities and their consequences can be constructed by interconnecting diverse data streams. This involves linking customer feedback to specific product versions, tying those to development sprints, and subsequently relating them to team performance metrics and operational costs. This corporate knowledge graph, far more powerful than isolated datasets, becomes the "brain" for the organization's "shared nervous system," enabling holistic insights, the discovery of non-obvious causal relationships, and faster, more informed adaptations.

The following list outlines the key components that constitute a robust organizational learning infrastructure:

Cultural foundation

- **Key elements:** Psychological safety, leadership commitment to learning, encouragement of inquiry and dialogue, and tolerance for experimentation.
- **Role in fostering feedback:** Creates an environment where feedback is valued, actively sought, and shared openly, and where it is treated as an opportunity for growth rather than as criticism.

Structural enablers

- **Key elements:** Cross-functional teams, agile practices such as retrospectives and sprints, clearly defined feedback channels, and explicit ownership for acting on feedback.
- **Role in fostering feedback:** Embeds feedback loops into everyday work, establishes accountability for responding to feedback, and enables rapid iteration.

Technological backbone

- **Key elements:** Unified enterprise telemetry, AI-driven insight synthesis and pattern recognition, and collaborative platforms.
- **Role in fostering feedback:** Enables efficient collection, aggregation, analysis, and distribution of feedback across

the organization, surfacing insights that are timely and actionable.

Measurement and iteration

- **Key elements:** Learning velocity KPIs, metrics for adoption and impact of changes, regular evaluation of feedback effectiveness, and AI-assisted decision reviews.
- **Role in fostering feedback:** Measures the organization's adaptive capacity, ensures feedback translates into tangible improvements, and continuously refines the learning process itself.

When an organization successfully erects these pillars and cultivates a high learning velocity, its ability to adapt swiftly and intelligently transforms from a desirable trait into its most formidable competitive weapon. This capacity for rapid evolution, fueled by a deeply ingrained feedback culture and infrastructure, becomes the defining characteristic of success in an increasingly symmetric and dynamic world.

COMPETITIVE ADVANTAGE IN ADAPTATION SPEED

In a world where information flows more freely and traditional knowledge-based advantages are increasingly transient, the ability of an organization to adapt with speed and precision emerges as the paramount source of sustainable competitive advantage. Agility and innovation, often lauded as key business virtues, are not mystical qualities but rather the tangible outcomes of an organization's capacity for rapid, high-fidelity response to feedback. The velocity at which an entity; be it a product, a team, or an entire enterprise, can learn from its environment and internal operations, and then translate that learning into effective action, becomes the ultimate differentiator. This "adaptation speed" is cultivated through robust feedback loops, a strategic shift from rigid forecasting to continuous sensing, and a relentless focus on enhancing the organizational learning rate.

The concept of adaptation speed, or "learning rate," can be quantified and tracked as a key performance indicator (KPI). While traditional metrics might focus on static outcomes, measuring adaptation speed involves assessing how quickly an organization can implement necessary changes in response to new information. This could involve tracking the time from the identification of a market shift to the deployment of a revised strategy, or the duration from a customer complaint to a product fix and subsequent validation of its effectiveness.

Factors such as the adoption rate of new processes or technologies by employees and the timely achievement of project milestones associated with change initiatives also contribute to this measure. At a more granular level, the speed at which individuals and teams acquire new competencies and skills; their "time-to-competency", directly impacts the organization's overall responsiveness. An organization that systematically shortens these learning and implementation cycles effectively increases its adaptive capacity. Adaptive performance, defined as the ability to effectively recognize environmental shifts, evaluate their impact, and implement appropriate actions, inherently includes response speed as a critical operational variable.

Feedback loops are the engine of this adaptation speed, and their role extends beyond mere product improvement to encompass strategic defense. As seen in high-stakes environments like cybersecurity and financial crime prevention, the ability to detect a new threat or a novel fraudulent pattern (feedback) and then rapidly adjust defensive mechanisms or analytical models (response) is crucial for survival and resilience.

The C3 AI Anti-Money-Laundering application, for example, relies on closed-loop workflows to incorporate the latest theft typologies, thereby improving its predictive capabilities. Similarly, Splunk emphasizes operationalizing threat hunting by feeding the learnings from each hunt back into the security operations to bolster defenses. In these scenarios, the speed of the feedback-response cycle directly correlates with the organization's ability to protect its assets and maintain operational integrity. The faster the loop, the more robust the defense.

This emphasis on rapid response necessitates a fundamental shift in strategic orientation; from a primary reliance on long-range forecasting to a more dynamic mode of "sensing". While financial and demand forecasting techniques remain valuable for planning, their efficacy is diminished in environments characterized by high volatility and unpredictability. The new imperative is to augment forecasting with a heightened capacity for continuous environmental scanning and real-time data analysis. This "sense and respond" posture involves actively monitoring market dynamics, customer behavior, competitor actions, and internal performance indicators through comprehensive telemetry and feedback channels.

The strategic advantage then lies not in the perfection of a long-term plan, but in the ability to swiftly and effectively react to the signals that are sensed, making course corrections as needed. Businesses that cultivate this adaptability can make decisions in minutes rather than weeks, ensuring that teams are aligned and responsive to emerging threats and opportunities, rather than being tethered to outdated directives.

The pursuit of adaptation speed redefines innovation itself. In the era of information symmetry, where unique knowledge advantages are quickly replicated, innovation is less about singular, guarded breakthroughs and more about a continuous stream of incremental and occasionally radical adaptations.

The "innovativeness" of an organization becomes directly proportional to the velocity and fidelity of its feedback loops. Companies that can rapidly cycle through launching an idea or product, gathering rich feedback, processing that feedback into insights, and implementing improvements or pivots will consistently out-innovate those with slower, more cumbersome learning cycles. Each iteration, informed by real-world use and response, moves the offering closer to optimal market fit or uncovers new avenues for value creation. Thus, adaptation speed, driven by efficient feedback mechanisms, becomes the very engine of ongoing innovation.

This leads to the concept of the "adaptive frontier". Competitive advantage is found not in achieving a static state of "being perfectly adapted"; an impossibility in a constantly changing world, but in possessing a superior, enduring capability to keep adapting faster and more effectively than rivals. The goal is to operate consistently at the leading edge of responsiveness, continuously pushing the boundaries of how quickly the organization can learn and evolve. This meta-capability is built upon the organizational learning infrastructure and high learning velocity discussed previously. Companies that excel in this domain are always probing the environment, ready to respond to new signals, be they threats or opportunities, with a speed and precision that competitors cannot match.

Ultimately, the shift from predictive control to resilient navigation underpins this focus on adaptation speed. In complex and dynamic systems, attempting to perfectly predict and rigidly control the future is a fraught exercise, often leading to brittle strategies that shatter under unexpected pressures. A "sensing" approach, by contrast, acknowledges uncertainty and prioritizes the gathering of real-time feedback about the current state of the environment and the system's performance within it. Coupled with a high capacity for rapid adaptation, this allows the organization to make quick, intelligent course corrections, much like a skilled navigator adjusts to shifting winds and currents. This fosters resilience; the ability not just to survive unexpected changes, but to thrive amidst them. Effective feedback loops, therefore, act as the sophisticated sensory organs and responsive control surfaces that enable this resilient navigation through the turbulent waters of the modern competitive landscape.

In a world increasingly characterized by information symmetry, where static knowledge offers little lasting protection or advantage, the capacity to learn and adapt at an accelerated pace becomes the defining characteristic of successful entities. This imperative for rapid evolution, fueled by the continuous ingestion and intelligent processing of feedback, extends beyond mere products or operational processes. It fundamentally impacts how individuals, brands, and entire organizations must define and project themselves in a relentlessly dynamic environment. If continuous adaptation is the new baseline for survival and success, then the very nature of identity must also transform, shifting from a fixed, immutable attribute to a dynamic, performative, and ever-evolving construct. The profound implications of this need for a fluid, responsive identity, capable of evolving in lockstep with a constantly changing world, are the focus of our exploration in the chapter to come.

SUMMARY

This chapter has illuminated the critical transition from a world dominated by static products and services to one where value is increasingly generated through dynamic, adaptive systems. The era of the "frozen product," conceived and delivered as a fixed entity, is demonstrably over, supplanted by offerings meticulously designed to learn and evolve through continuous interaction with users and their surrounding environment. We have seen that feedback, in its diverse forms, has transcended its traditional role as a retrospective tool for incremental improvement.

It has become the very lifeblood of contemporary product development and organizational vitality, serving as the essential raw material that fuels AI-driven personalization, enhances the capabilities of intelligent agents through mechanisms like Reinforcement Learning from Human Feedback, and fundamentally transforms products into relational entities that cultivate loyalty through sophisticated behavioral tuning and identity-aware interfaces.

Crucially, the insights gleaned from this feedback cannot remain siloed; they must be integrated across the entire organizational stack. This creates a "sentient enterprise" underpinned by a robust learning infrastructure-one characterized by a pervasive culture of inquiry, agile operational processes, and comprehensive enterprise-wide telemetry. Such an infrastructure is the engine that enables an organization to achieve and sustain a high "learning velocity". In the age of information symmetry, where traditional knowledge-based advantages are ephemeral, this capacity for rapid adaptation; the speed and efficacy with which an organization can sense new information, learn from it, and respond effectively, emerges as the most potent and sustainable competitive edge. This is not merely about creating better products; it signifies a fundamental shift in the operational DNA of successful entities. If continuous adaptation is indeed the new norm, then the very nature of identity; for individuals, for brands, and for organizations must also undergo a profound transformation. It must shift from a static, fixed attribute to a dynamic, evolving performance, capable of reconfiguring itself in concert with a perpetually changing world. The compelling implications of this imperative for a fluid, responsive identity will be the central theme of our subsequent discussion.

12

DYNAMIC IDENTITY, NOT STATIC IP

The bedrock of competitive advantage has, for centuries, rested on the tangible and the defensible. For individuals, it was the accumulation of specialized knowledge and credentials; for organizations, it was often the ownership of intellectual property (IP); patents, copyrights, trade secrets, that carved out protected niches in the marketplace.

These assets, often meticulously developed and fiercely guarded, represented a clear, understandable source of value. However, the dawn of the Age of Information Symmetry, particularly accelerated by the pervasive influence of Artificial Intelligence (AI), is triggering a seismic shift in this landscape. In an environment where knowledge, code, and even creative ideas can be replicated, remixed, and disseminated with unprecedented speed and ease, the traditional pillars of strategic advantage are beginning to crumble. This chapter posits that as the value of static, protectable assets diminish, a new anchor for success emerges; dynamic identity.

The very nature of "knowing" and "owning" information has been fundamentally altered by AI, a theme explored in earlier chapters of this volume, such as Chapter 4's examination of Synthetic Expertise and Chapter 6's analysis of the Commoditization of the Edge. Building upon that foundation, this chapter delves into what remains when these traditional differentiators erode.

The argument presented here is that identity; in its multifaceted forms encompassing the personal, the brand, and the organizational, is ascending as the new locus of sustainable value. However, this is not "identity" as a static badge or a fixed set of characteristics. Instead, it is an identity that is inherently dynamic, responsive, performative, and capable of continuous evolution. Static identities, much like static IP addresses in a dynamic network, are increasingly vulnerable to obsolescence or irrelevance in a world characterized by flux.

A central tension defines this new era: AI makes replication and mimicry easier than ever before. It can generate text, simulate conversations, create designs, and even produce code that rivals human output. Yet, paradoxically, this very capability necessitates a stronger, more adaptable, and more profoundly authentic "self"; whether individual or collective, to stand out, build trust, and forge meaningful connections. The commoditization of IP by AI does not merely devalue existing assets; it fundamentally reorders the hierarchy of strategic value. There is a discernible shift from prioritizing "having" (e.g., proprietary data, patents) to valuing "being" (e.g., identity, purpose, values). This is because "who you are" and "how you adapt" are, at their core, human-centric and dynamic qualities that are significantly harder for AI to authentically replicate in a manner that cultivates deep, lasting trust.

Furthermore, the imperative for a dynamic identity is a direct consequence of the dynamic nature of AI itself. AI is not a static technology; it is a constantly learning and evolving force. As its capabilities expand and change, the competitive landscape it shapes is also in perpetual motion. A static IP portfolio or a fixed set of expertise represents a snapshot in time, vulnerable to being superseded by newer AI-driven innovations or replications. Consequently, to maintain relevance and competitive strength, human and organizational identities cannot afford to be static. They must possess an intrinsic capacity to learn, adapt, and evolve in response to, and often in concert with, the evolving AI landscape. This creates a powerful co-evolutionary pressure.

The challenge, therefore, extends beyond merely having an identity. It becomes about performing that identity authentically and consistently in an environment where AI can also simulate identity with increasing sophistication. This raises the bar for genuine human connection, purpose-driven action, and verifiable authenticity. If AI can generate superficial aspects of an identity, such as a brand voice or a visual style, then a truly differentiating human or organizational identity must be rooted in something deeper: authentic values, consistent behavior, and a demonstrable purpose that AI cannot merely script.

Success will increasingly depend on the performance of this authentic identity; how it manifests in actions, decisions, and relationships, not just on its stated claims. The performative aspect becomes crucial for building trust that transcends AI-generated facsimiles.

If ideas, products, and skills can be copied or simulated almost instantly, what, then, is truly left to protect? And, perhaps more critically, what is left to grow? This chapter will explore these questions, arguing that the answer lies in cultivating dynamic, adaptable, and authentic identities that can thrive not by resisting change, but by embracing it as the fundamental condition of the Age of Information Symmetry.

Primary source of value

- **Old paradigm (static, IP-centric):** Proprietary knowledge and fixed assets such as patents, copyrights, and controlled intellectual property.

- **New paradigm (dynamic, identity-centric):** Trust, authenticity, purpose, relationships, and brand resonance.

Key differentiator

- **Old paradigm (static, IP-centric):** Exclusive expertise and legally protected monopolies.
- **New paradigm (dynamic, identity-centric):** Dynamic identity, adaptability, responsiveness, and value alignment.

Defensibility model

- **Old paradigm (static, IP-centric):** Legal protection, secrecy, and control over assets.
- **New paradigm (dynamic, identity-centric):** Evolvability, reputation, network effects, semantic trust, and customer loyalty.

Individual focus

- **Old paradigm (static, IP-centric):** Accumulated credentials, fixed roles, and static expertise.
- **New paradigm (dynamic, identity-centric):** Fluid capabilities, continuous learning, personal brand narrative, and adaptability.

Brand focus

- **Old paradigm (static, IP-centric):** Static messaging, emphasis on product features, and tightly controlled image.
- **New paradigm (dynamic, identity-centric):** Adaptive performance, value alignment, personalized experiences, and co-created stories.

Organizational focus

- **Old paradigm (static, IP-centric):** Hierarchical structures, process efficiency, and knowledge hoarding.
- **New paradigm (dynamic, identity-centric):** Composable systems, cultural cohesion, learning capacity, and talent orchestration.

Strategic imperative

- **Old paradigm (static, IP-centric):** Protect and preserve existing assets and market positions.
- **New paradigm (dynamic, identity-centric):** Adapt, evolve, and continuously reinvent.

Role of AI

- **Old paradigm (static, IP-centric):** Tool for efficiency and data analysis, often viewed as a threat to intellectual property.
- **New paradigm (dynamic, identity-centric):** Partner in expressing identity, enabler of personalization, and primary driver of adaptation.

This list encapsulates the fundamental shift this chapter will explore, moving from a world where value was locked in static assets to one where it flows through dynamic identities and adaptive capabilities. The following sections will dissect this transformation, examining its implications for intellectual property, individual careers, brand strategy, organizational design, and the very nature of competitive advantage.

IP IS NO LONGER A MOAT

For much of modern economic history, Intellectual Property (IP) has served as a cornerstone of competitive advantage. Legal constructs such as patents, copyrights, trademarks, and trade secrets were designed to foster innovation and creativity by granting inventors and creators temporary monopolies, allowing them to reap the rewards of their ingenuity and investment. These protections formed formidable "moats" around businesses, discouraging direct imitation and providing a basis for market differentiation and sustained profitability. However, the advent of sophisticated Artificial Intelligence, particularly generative AI, is systematically dismantling these traditional defenses, rendering IP a far less reliable moat than it once was.

The mechanisms by which AI undermines traditional IP protections are multifaceted and profound. Firstly, the capacity of generative AI to produce novel content; be it text, images, code, or intricate designs that can be remarkably similar to, or derivative of, existing copyrighted or patented works, presents a significant challenge. This capability blurs the lines of originality and infringement, making it difficult to ascertain true authorship and protect creative expressions. The case of the comic book "Zarya of the Dawn," where copyright protection was rescinded due to the U.S. Copyright Office's determination that the AI-generated images lacked human authorship, starkly illustrates this emerging problem. The Office has consistently affirmed that existing copyright law requires human creativity as a central element, and material whose expressive elements are determined by a machine cannot be protected.

Secondly, AI tools enable what can be termed "infinite remixability". Existing ideas, designs, and content can be rapidly and effortlessly deconstructed, reconfigured, and recombined, accelerating the creative cycle but simultaneously making it harder to pinpoint and protect a single "original" source or maintain a unique creative signature. This constant flux diminishes the period during which any given creation can claim novelty.

Thirdly, the legal prerequisite of human authorship or inventorship in many key jurisdictions, notably the United States for copyright and patent law, poses a fundamental challenge. Works generated autonomously by AI often fail to meet this criterion, leading to "copyright gaps" where novel creations may exist without clear owners or protections.

The DABUS case, where patent offices globally, including in the UK and US, rejected patent applications listing an AI system as the inventor because it did not name a human inventor, underscores this legal reality. While patents can be granted for inventions made with AI assistance, the AI itself cannot be the inventor under current U.S. law. This means that if the core inventive step is attributable to the AI, patent protection may be unattainable.

Fourthly, the very development of powerful AI models is fraught with IP complexities. These models are often trained on vast datasets that include copyrighted materials, frequently ingested without the explicit consent of, or compensation to, the original creators. This practice has sparked numerous lawsuits, with creators and publishers arguing that such use constitutes copyright infringement. The Thomson Reuters Enterprise Centre GMBH v. ROSS Intelligence Inc. case, where a court found that using copyrighted legal headnotes to train an AI constituted direct infringement and rejected a fair use defense, signals the significant legal risks involved in AI training data practices. These cases highlight a tension between the desire to foster AI innovation and the need to protect the rights of human creators whose work fuels that innovation.

Fifthly, the global legal landscape for AI-generated IP is a patchwork of differing standards and approaches. For instance, while the U.S. firmly requires human authorship for copyright, the U.K. offers limited protections for certain computer-generated works even without a traditional human author. This lack of international harmonization creates legal uncertainty for businesses operating globally, complicating enforcement and potentially encouraging "forum shopping" for more favorable legal environments.

Finally, the sheer speed at which AI can analyze existing knowledge, replicate functionalities, and build upon prior innovations drastically shortens the effective shelf life of "original" ideas. An invention or creative work that might once have enjoyed years of unique market positioning can now be rapidly superseded or imitated. The half-life of digital content, as seen in social media engagement, is already remarkably short ; AI threatens to compress these timelines even further for more complex forms of IP. The very notion of a "Digital Dark Age," where the rapid obsolescence of formats and technologies makes long-term preservation of digital information a challenge, indirectly suggests that even if IP is secured, its enduring value in a fast-evolving digital ecosystem is questionable.

The erosion of IP moats by AI introduces an "innovation paradox". AI demonstrably accelerates the pace of innovation by enabling faster generation of new ideas, designs, and content. Yet, these same capabilities simultaneously diminish the protectability and the ability to capture sustained value from that innovation through traditional IP mechanisms. As AI makes it easier to copy, derive, or create similar works, challenging originality and authorship, companies may find they can innovate rapidly but struggle to secure exclusive, long-term rights to those innovations. The competitive advantage derived from a specific AI-generated or AI-assisted innovation might therefore be fleeting, as competitors can quickly replicate or create alternatives using similar AI tools. This leads to a situation of more innovation in aggregate, but potentially less sustainable differentiation for any single innovator relying solely on IP.

The legal battles surrounding AI and IP, particularly those concerning training data, are more than just disputes over financial compensation or market share. They represent a profound societal negotiation about the intrinsic value of human creativity and the future economic models for creators in an AI-driven world. If AI systems can produce outputs comparable to human creations, trained on human-created works without adequate permission or recompense, it inherently devalues human labor and threatens the livelihoods of those in creative industries. These lawsuits, therefore, are a crucible in which a new understanding of the relationship between human and machine creativity is being forged, forcing a re-examination of how society will value and support human ingenuity when AI can automate or replicate many creative tasks.

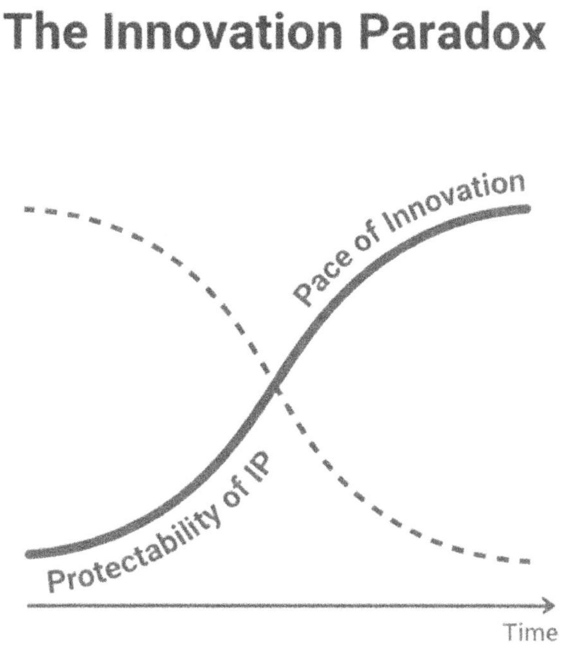

Interestingly, the persistent focus on "human authorship" within current IP law, especially in influential jurisdictions like the U.S., inadvertently creates a new strategic imperative for businesses. To secure IP rights for AI-assisted creations, companies must now meticulously delineate, document, and emphasize the human creative contributions within their workflows. This is not merely a matter of legal compliance; it becomes a tactical necessity.

The U.S. Copyright Office guidance is clear: if content is entirely generated by AI, it cannot be protected by copyright; however, where AI merely assists an author, its use does not bar copyrightability of the output, provided sufficient human creativity is present. This could lead to new operational practices, such as detailed "creative contribution logs" or refined "authorship statements" that specify the human role in shaping AI-assisted outputs, as businesses strive to meet the threshold for IP protection, however tenuous that protection might be in the long run.

In conclusion, while IP law will undoubtedly continue to adapt and evolve in response to AI, its traditional role as a primary and durable moat is significantly compromised. The ease of replication, challenges to authorship, controversies over training data, and the sheer pace of AI-driven change mean that relying on IP alone for sustained competitive advantage is an increasingly precarious strategy.

With traditional intellectual property offering diminishing protection, the search for durable competitive advantage must inevitably shift. If what an organization creates can be so easily copied or superseded, then who that organization is and how it is perceived by its customers, employees, and partners become paramount. This fundamental reorientation leads directly to an exploration of identity as the new, more resilient moat in the Age of Information Symmetry.

IDENTITY AS THE NEW MOAT

As the protective walls of intellectual property are eroded by the tide of artificial intelligence, organizations and individuals must seek new foundations for durable competitive advantage. In this shifting landscape, "identity"; encompassing trust, value alignment, and distinctive recognizability emerges as a more enduring and defensible moat. We do not refer to identity here in the superficial sense of a logo or a name, but rather the deeply embedded sum of perceptions, values, behaviors, and relationships that define an entity in the minds of its stakeholders. It is the comprehensive answer to the fundamental questions: "Who are you?" and, more importantly, "Why should I care, trust, and engage with you?"

At the core of this new moat lies trust. In a world increasingly saturated with AI-generated content, sophisticated simulations, and the potential for deepfakes, authentic trust becomes an exceptionally scarce and valuable commodity. A strong, consistently performed identity is the bedrock upon which this trust is built.

Consumers are demonstrably more likely to engage with, remain loyal to, and even pay a premium for brands they trust. Research, such as Deloitte's analysis of digital trust maturity, indicates that organizations prioritizing cyber risk management as a means to earn digital trust realize significant business benefits, including improved customer trust and enhanced brand reputation. The Thales Digital Trust Index, however, reveals a concerning trend: global trust in digital services is generally declining or stagnating, underscoring both the challenge and the critical importance of actively cultivating a trustworthy identity.

Alignment with values offers another powerful dimension of identity-based differentiation. An organization's identity allows it to connect with stakeholders based on shared principles, purpose, and mission. Patagonia serves as a compelling archetype in this regard. Its unwavering commitment to environmental activism, encapsulated in its mission "We're in business to save our home planet," is not merely a marketing slogan but a core element of its identity that permeates its operations, product design, and advocacy. This profound value alignment attracts and retains a fiercely loyal customer base that shares these environmental concerns, creating a bond that a purely functional or AI-generated competitor would find exceedingly difficult to replicate authentically.

Recognizability and a compelling narrative are also crucial components of a strong identity. In an information-overloaded environment, a coherent, consistent, and resonant identity, expressed through a powerful narrative, allows an entity to cut through the noise and become memorable and relatable. The psychological underpinnings of branding, such as the formation of emotional connections, are intrinsically linked to this narrative strength. Iconic brands like Apple, Lego, Nike, and Coca-Cola have, over decades, built instantly recognizable identities through consistent messaging and value delivery, embedding themselves in popular culture. This long-term cultivation of recognizability and emotional resonance forms a significant barrier to entry for newcomers.

The robustness of an identity is further shaped and validated within reputation ecosystems. While still an evolving field, concepts like graph-based reputation systems, which model and compute trust based on networked interactions and feedback, point towards a future where identity is partly co-created and algorithmically assessed. These systems attempt to quantify trustworthiness based on a history of behavior and relationships within a network. The very existence of such models highlights that identity is not solely self-proclaimed but is also conferred and validated by the perceptions and experiences of others.

Crucially, the defensibility of identity stems from its intangible and cumulative nature. Unlike IP, which can be legally challenged, designed around, or rendered obsolete by new technology, a deeply embedded identity built on genuine trust, consistent value delivery, and authentic purpose is far harder for competitors; even those heavily augmented by AI, to copy wholesale. Such an identity is not manufactured overnight; it is earned and accumulated over time through a consistent pattern of actions, decisions, and interactions.

While AI can mimic surface-level brand characteristics, such as tone of voice or visual style, it cannot easily replicate the history, the lived experiences, the genuine relationships, and the earned trust that form the true substance of a strong identity. In an age where AI makes many aspects of business easier to build but harder to defend, moats rooted in unique customer relationships, proprietary data that informs a deeper understanding of those relationships, and the network effects arising from a trusted identity become more critical than ever.

The strategic elevation of identity to the status of a primary moat carries significant implications. It suggests that functions like "brand management" and "reputation management" are evolving from predominantly marketing-focused activities into core C-suite responsibilities. If identity is indeed the most defensible asset, its cultivation and protection demand a holistic approach, deeply integrated with corporate strategy, ethics, governance, and operations. Building and maintaining such an identity requires far more than clever messaging; it necessitates consistent behavior, ethical decision-making across the board, and the alignment of all organizational actions with its stated values and purpose.

These are not concerns that can be siloed within a marketing department; they are fundamental to how the company operates, treats its employees and stakeholders, and fulfills its societal role. Consequently, the stewardship of identity becomes a central strategic imperative, potentially elevating the importance of roles focused on ethics, trust, and stakeholder engagement to the highest levels of organizational leadership.

Furthermore, in an AI-driven world, the very "signals" that constitute and communicate an identity are becoming more complex and increasingly machine-readable. Authenticity will be judged not only by human perception but also by AI's analysis of an entity's digital footprint, behavioral consistency, and semantic coherence. AI systems can process vast amounts of data from websites, social media, customer interactions, and public records to form an assessment of an entity's identity. These systems will be capable of detecting discrepancies between an organization's stated values and its actual conduct, or inconsistencies in its messaging across different platforms. Inauthentic performances or misalignments in identity could be identified and amplified by AI, potentially damaging trust more rapidly and extensively than in the pre-AI era. This necessitates a new level of rigor in ensuring that all digital signals are coherent, authentic, and aligned, as these signals contribute significantly to how both humans and AI systems perceive and ultimately trust an entity.

Finally, as traditional tangible assets become more easily replicated and their unique value diminishes, the economic worth of "identity capital"; the accumulated trust, reputation, brand equity, and stakeholder relationships associated with a strong identity is likely to increase significantly. This intangible asset, though difficult to quantify precisely, may become a more explicitly measured component in business valuations, investment decisions, and even assessments of national economic strength. Just as brand equity is now a recognized, albeit intangible, asset, a more comprehensive concept of identity capital, reflecting overall trustworthiness and alignment with stakeholder values, could emerge as a key indicator of long-term viability and growth potential. This shift would have profound implications for how markets assess value and how organizations prioritize their strategic investments.

If identity, then, is the new moat, its nature and expression must adapt to the realities of the information age. For individuals operating within this new paradigm, a static definition of self, rooted in fixed credentials or a narrow area of expertise, is no longer sufficient. The ability to adapt, acquire new capabilities fluidly, and perform a dynamic professional identity becomes central to navigating a rapidly changing world. This leads us to explore the profound shift from static expertise to fluid capability.

FROM STATIC EXPERTISE TO FLUID CAPABILITY

The traditional markers of professional identity and value-deep, often narrow, expertise in a specific domain, validated by formal credentials and years of experience, are undergoing a fundamental re-evaluation in the Age of Information Symmetry. As artificial intelligence systems become increasingly adept at accessing, synthesizing, and even generating expert-level knowledge on demand (a reality explored in Chapters 4 and 5), the individual whose primary value proposition rests on "knowing things" faces an existential challenge. The long-held "10,000 hours to mastery" paradigm is being disrupted, not by rendering expertise obsolete, but by changing its nature from a static accumulation of knowledge to a dynamic capacity for learning and application.

This shift is evidenced by the growing corporate adoption of skills-based hiring practices. Increasingly, organizations are prioritizing demonstrable capabilities and competencies over formal degrees or conventional career trajectories when recruiting and promoting talent. Research indicates that a significant majority of employers, around 64.8% for entry-level roles, are already utilizing skills-based hiring, recognizing that valuable skills can be acquired through diverse pathways and that a degree is not always the best proxy for ability. This approach inherently values what an individual can do and learn over what they formally studied.

Congruent with this trend is the emergence of more fluid career paths. The future of work, as projected by numerous analyses, involves individuals navigating more dynamic and less linear trajectories, moving between traditional employment, project-based assignments, entrepreneurial ventures, and freelance engagements. The rapid expansion of the "gig economy" is a clear manifestation of this shift, where talent is increasingly sourced and engaged for specific skills applied to discrete projects, rather than for long-term, rigidly defined roles. This model demands a professional identity that is adaptable and portfolio-based.

Organizations, in turn, are restructuring to accommodate and leverage this fluidity. The concept of dynamic talent clouds and internal marketplaces is gaining traction, with companies using sophisticated skill graphs to identify and assemble teams based on the specific capabilities required for a project, rather than relying on static departmental structures. This move towards "composable talent," as envisioned in Chapter 14, allows for greater agility and responsiveness to changing business needs.

In this environment, the ideal professional identity evolves. It is less about being an "I-shaped" expert with deep knowledge in a single, narrow field, and more about becoming a "T-shaped" or even "comb-shaped" professional. These models emphasize a broad base of understanding across multiple domains (the horizontal bar of the T), combined with deep expertise in one or a few core areas (the vertical bar(s) of the T or multiple prongs of the comb). Crucially, what underpins this structure is a highly developed capacity for rapid learning and adaptation-the ability to add new skills and knowledge areas as required. The emphasis shifts from the knowledge currently possessed to the ability to acquire and effectively apply new skills in novel contexts.

CONSEQUENTLY, the defining question for professional identity is transforming. It is less about "What do you do?" (implying a fixed job title) and more about "What can you deliver now?" and, critically, "What can you learn next?" This reflects a valuation of current, applicable skills and, equally importantly, the potential for future growth and adaptability. An individual's professional identity becomes increasingly defined by their evolving portfolio of adaptable skills, their demonstrated learning agility, and their capacity to collaborate effectively within dynamic, often temporary, team structures.

Lifelong learning transitions from a laudable personal development goal to an absolute professional necessity. The World Economic Forum's Future of Jobs Report 2025 projects that a substantial portion of key job skills; 39%, will change by 2030. This rapid evolution underscores the critical need for continuous upskilling and reskilling. Indeed, "curiosity and lifelong learning" are identified among the top ten skills expected to rise in importance, alongside analytical thinking, creative thinking, and technological literacy.

The implications of this transition from static expertise to fluid capability are far-reaching, particularly for educational and credentialing systems. If fixed-role expertise and traditional, front-loaded credentials are losing their currency in the labor market, and if fluid capabilities coupled with lifelong learning are becoming paramount, then the established models of education must undergo a radical transformation.

A four-year degree, for example, provides a snapshot of knowledge acquired at a specific point in time and may not accurately reflect an individual's current, in-demand skills years later. This necessitates a shift towards more agile, responsive educational ecosystems that offer pathways to acquire, validate, and signal specific skills on an ongoing basis. We can anticipate the proliferation of micro-credentials, digital badges, skills-based certifications, and robust skills assessment platforms. In such a future, an individual's "educational identity" would not be a static list of degrees obtained in the past, but rather a dynamic, living portfolio of verified skills and competencies that evolves throughout their career. This aligns with the critique of traditional credentialing systems presented in Chapter 2, suggesting a move towards more granular and continuously updated markers of capability.

As individuals increasingly define their professional worth by their fluid capabilities, the nature of "career identity" itself undergoes a transformation. It becomes less tethered to loyalty to a single organization and more centered on building personal brand equity and a transportable portfolio of skills, experiences, and accomplishments. Because skills-based hiring values specific, transferable capabilities, individuals are likely to focus on cultivating a personal "skill stack" and a professional reputation that enhances their value in a broader, more competitive talent marketplace.

This could, in turn, foster a more transactional relationship between employees and employers, where allegiance is primarily to one's own career development and the pursuit of impactful work, rather than to a specific institution. Consequently, organizations will need to devise new strategies to attract, engage, and retain top talent. These strategies may revolve less around traditional notions of job security or linear career ladders and more around offering compelling opportunities for continuous skill development, involvement in meaningful and challenging projects, and a strong alignment between organizational purpose and individual values.

However, this pronounced emphasis on "what can you deliver now?" carries a potential risk. While practical, results-oriented capabilities are vital for agility, an excessive focus on immediately applicable skills could inadvertently lead to the devaluation of foundational knowledge, critical thinking abilities, and long-term research and development efforts. These areas, while perhaps not yielding immediate "deliverables," are often crucial for fostering deeper understanding, driving breakthrough innovations, and developing the intellectual capital necessary for future advancements.

A society or an organization that solely prioritizes skills with immediate utility might, over time, find its capacity for truly novel, paradigm-shifting innovation diminished. This presents a critical tension between the demands of short-term agility and the cultivation of long-term innovative capacity, a balance that leaders and policymakers must carefully navigate.

Just as individual identity must shed its static nature and embrace fluidity and adaptability, so too must the identity of brands. In an environment of constant change and AI-driven personalization, brands can no longer afford to rely on fixed messaging or immutable product features. They must evolve into dynamic performers, capable of learning, adapting, and resonating with their audiences in real time. This leads us to an exploration of brands as adaptive performers in the new strategic landscape.

BRANDS AS ADAPTIVE PERFORMERS

In an economic and cultural landscape increasingly permeated by artificial intelligence and characterized by rapid shifts in consumer preferences, the traditional concept of brand identity is undergoing a profound transformation. No longer can a brand be considered a static construct, defined merely by a distinctive logo, a memorable tagline, or a fixed set of product attributes. Instead, brand identity is emerging as a dynamic, living entity; an active agent that continuously learns, adapts its messaging, refines its offerings, and calibrates its interactions in response to real-time data streams and the evolving expectations of its audience. The most resilient and successful brands of the AI era will be those that master the art of adaptive performance.

This shift necessitates moving beyond the confines of static branding. Historically, brand management often centered on creating a fixed set of associations and communicating them with unwavering consistency. While consistency undeniably remains crucial for building recognition and trust ; as exemplified by enduring brands like Coca-Cola, whose core visual identity has remained remarkably stable over decades, the nature of that consistency is evolving. It is less about rigid adherence to immutable elements and more about maintaining a consistent alignment with core values and purpose while fluidly adapting the brand's expression and engagement strategies.

Adaptive marketing and hyper-personalization are at the forefront of this transformation. Brands are increasingly leveraging people-based adaptive marketing techniques to customize their messaging and deliver campaigns at precisely the optimal moments and through the most effective channels for individual consumers. Artificial intelligence is the primary enabler of this capability, facilitating the hyper-personalization of products, services, communications, and entire customer experiences based on granular analysis of individual customer data, behavioral patterns, and contextual cues.

Spotify, for instance, has built a powerful brand around its ability to deliver highly personalized music experiences through features like "Discover Weekly" and tailored playlists that adapt to individual listening habits. Similarly, e-commerce giants like Amazon utilize sophisticated AI-driven recommendation engines to present customers with product suggestions dynamically tailored to their browsing history and purchase patterns.

Real-time responsiveness is another hallmark of the adaptive brand. In a digitally interconnected world, market shifts, changes in social sentiment, and competitor actions can unfold with breathtaking speed. Brands must therefore develop the capacity to monitor these dynamics continuously and react with agility. AI tools play a critical role here, enabling brands to track online mentions, analyze sentiment trends, and identify emerging issues or opportunities in real time, thereby allowing for proactive strategic adjustments rather than belated reactive measures.

Underpinning the ability of brands to perform adaptively, especially in an ecosystem where they interact with or are represented by AI systems, is the principle of semantic consistency. This refers to the imperative of ensuring that the brand's core meaning, values, intent, and personality are consistently understood and interpreted across all touchpoints, whether human or machine mediated. As AI agents increasingly act as extensions of the brand; through chatbots, virtual assistants, or automated content generation, it is vital that these agents "understand" and accurately convey the brand's essence. A lack of semantic consistency can lead to confused messaging, misaligned experiences, and an erosion of trust.

In this context, the brand evolves from being primarily a broadcast mechanism, pushing out centrally crafted messages, to becoming an interactive "response layer". This layer actively mediates the relationship between the user, the product or service, and the broader ecosystem. It learns from each interaction, gathers feedback, and refines its performance over time, creating a more dynamic and reciprocal relationship with its audience. Brand storytelling, a perennially important aspect of connecting with consumers, also becomes more fluid. Narratives must be adaptable, capable of incorporating new themes, addressing current societal conversations, and reflecting the evolving nature of the brand's relationship with its community.

Several contemporary brands exemplify this adaptive performance. Spotify, as mentioned, continuously refines its user experience through data-driven personalization and adapts its content and marketing to local cultural nuances. Netflix, which famously pivoted from a DVD rental service to a global streaming and content production powerhouse, constantly adapts its vast content library, sophisticated recommendation algorithms, and even its distinctive organizational culture to maintain its leadership position in a fiercely competitive market. Apple, while maintaining an unwavering core identity centered on innovation, design excellence, and user experience, continuously adapts its product lines, software ecosystems, and service offerings to incorporate new technologies and anticipate evolving consumer demands. Even a seemingly traditional brand like Domino's Pizza demonstrates adaptability by tailoring its product offerings to local tastes (e.g., extensive vegetarian options in India) and by leveraging a multitude of digital touchpoints to facilitate convenient ordering and customer engagement.

The rise of "brands as adaptive performers," heavily reliant on AI for personalization and real-time interaction, introduces new dimensions of brand risk. One significant emerging vulnerability centers on the potential for misaligned or poorly executed algorithmic personalization, which can inflict damage on trust and brand perception more severely and rapidly than traditional marketing missteps. As brands deploy AI systems to tailor messages, recommend products, and even conduct conversations, these systems make countless autonomous decisions based on the data they process and the algorithms they execute. If this data is biased, if the algorithms contain flaws, or if the resulting personalization is poorly calibrated, the brand's "performance" can be perceived by consumers as intrusive, irrelevant, discriminatory, or even offensive. Because these interactions are often highly personalized and automated at scale, the negative impact can be widespread and feel like a direct, individual affront by the brand. This is qualitatively different from a poorly received generic advertising campaign; it represents a failure of the brand's adaptive "persona" to behave appropriately in a one-to-one context. Such algorithmic misbehavior can erode trust with alarming speed, creating a novel and potent category of brand reputation risk that requires diligent oversight and ethical AI governance.

Furthermore, as brands become increasingly adaptive and their communications are increasingly AI-driven, the "brand voice" itself may evolve into a co-creation between human marketers and AI agents. This necessitates the development of new skills among marketing professionals, who must become adept at "AI whispering" or "prompt-crafting for brand consistency".

AI tools are already being used to generate marketing copy, social media updates, and customer service interactions. To ensure that these AI-generated outputs align seamlessly with the brand's desired tone, style, values, and strategic objectives, marketers will need to master the art of guiding and refining AI outputs through sophisticated prompting techniques, careful parameter tuning, and iterative feedback loops. In this scenario, the brand voice is no longer solely the product of human copywriters and strategists but emerges from a collaborative synthesis of human intent and machine generation. The skillset for brand managers thus expands to include the ability to effectively "manage," "direct," and "co-create with" AI as an intelligent, communicative partner, ensuring that the AI's performance consistently reinforces the brand's core identity.

Finally, the drive towards ever-more adaptive and personalized brand experiences creates a potential "personalization-privacy paradox". The more effectively a brand adapts to individual needs and preferences, the more data it typically needs to collect and process. While consumers generally value personalization and tailored experiences, they are also increasingly aware of and concerned about data privacy. If a brand's adaptive performance becomes too knowledgeable or predictive in a way that feels invasive or crosses an invisible line into perceived surveillance, it can trigger significant privacy concerns and undermine the very trust it seeks to build, even if the intent was purely to be helpful and relevant. Therefore, adaptive brands face a dual challenge: they must not only personalize effectively but also proactively perform transparency regarding their data practices and demonstrate robust, ethical stewardship of the information entrusted to them. This "performance of trustworthiness" in data handling becomes an integral and non-negotiable component of the adaptive brand's identity, crucial for maintaining the customer relationship in an era of heightened privacy sensitivity and regulatory scrutiny.

In essence, a brand's identity in the AI era is less about a fixed declaration and more about a continuous, authentic, and intelligent performance. This imperative for adaptability and performative identity, however, is not confined to external branding. It must be mirrored internally, extending to the very structure, culture, and operational fabric of organizations. To project a dynamic identity outwardly and cultivate fluid capabilities inwardly, companies themselves must transform into adaptive identity systems. This leads to a critical examination of how organizations must evolve to embody these principles from the inside out.

ORGANIZATIONS AS IDENTITY SYSTEMS

To project a dynamic and adaptive external identity as a responsive brand and cultivate fluid internal capabilities, an organization's architecture (its structure, culture, and processes) must embody adaptability and a coherent, evolving identity. The traditional, rigid organizational models of the 20th century, designed for stability and predictability, are increasingly ill-suited to the accelerated pace of change and the complex demands of the Age of Information Symmetry. Companies must transform into adaptive identity systems, not merely remain static structures or repositories of legacy know-how.

The move away from rigid hierarchies is a foundational step in this transformation. Conventional organizational charts, with their clearly defined silos and top-down command structures, often create barriers to information flow, slow down decision-making, and impede the cross-functional collaboration necessary for rapid adaptation. Innovative companies like Netflix have experimented with flatter, more decentralized structures to empower employees and foster a more agile operational environment. This structural flexibility is crucial for enabling the organization to reconfigure itself in response to new challenges and opportunities.

The future points decisively towards composable and modular organizational designs, as anticipated in Chapter 14 of this volume. Such organizations possess the ability to dynamically assemble and reassemble teams, resources, technological components, and even AI agents based on evolving strategic goals and shifting market demands.

The principles of composable enterprise architecture, which leverage modular, API-enabled technological components like microservices and headless systems, allow businesses to build and iterate on digital capabilities with greater speed and flexibility. Real-world examples, such as Nike's adoption of a headless commerce approach to streamline digital experiences or BBVA's development of an open banking platform using API-based services, illustrate this architectural shift towards modularity and interoperability at the technology level, which in turn supports organizational agility.

Organizational culture emerges as a critical enabler, or inhibitor, of this adaptability. A culture that actively fosters continuous learning, encourages experimentation (and accepts intelligent failure as a learning opportunity), promotes psychological safety, and implements rapid, transparent feedback loops is essential for an organization to evolve. Programs like Stanford GSB's "Culture as a Competitive Advantage" explicitly recognize that culture is not a passive byproduct of operations but a manageable asset that can be diagnosed, designed, and aligned with strategy to drive performance and resilience. While challenging to implement and maintain, Netflix's famed culture of "radical candor" and exceptionally high performance standards is a distinct example of a culture intentionally designed to support its strategic objectives in a rapidly changing industry.

In such dynamic systems, shared values and a clearly articulated organizational purpose; the core of its identity, become the primary mechanisms for coordination and cohesion. These elements often prove more effective than static job titles or rigid reporting lines in guiding decision-making and aligning the efforts of fluid, often distributed, teams. A value-based culture, where organizational actions consistently reflect stated internal values, can drive higher levels of employee engagement, performance, and overall organizational agility.

The increasing reliance on dynamic teaming and talent clouds further underscores this shift. Organizations will more frequently utilize project-specific teams, assembled dynamically from both internal employees and external talent pools, leveraging skill graphs and talent marketplaces to find the right capabilities for the task at hand. This requires robust systems for identifying necessary skills, facilitating seamless collaboration across diverse and potentially remote team members, and managing the complexities of a distributed workforce.

Within this framework, the role of leadership undergoes a significant evolution. It moves away from traditional command-and-control models towards orchestration. Leaders become enablers, guides, and aligners, responsible for fostering an environment where autonomous teams and individuals can thrive, innovate, and contribute effectively towards the organization's core identity and strategic objectives. Their primary function becomes the cultivation of the system's capacity to adapt and perform.

Ultimately, the entire organization must function as an integrated learning system. It needs to develop sophisticated mechanisms for continuously sensing changes in its external environment (market trends, technological advancements, competitor actions, societal shifts), processing internal and external feedback effectively, and adapting its strategies, operations, and even its identity in response. This necessitates a culture of open communication, robust knowledge-sharing platforms, and processes that encourage reflection and iterative improvement.

The shift towards "organizations as identity systems" implies that the organization's identity itself; its core purpose, values, and guiding principles becomes a crucial governance mechanism. In complex, decentralized, and fluid environments where traditional rules and hierarchical oversight are less effective, a strong and deeply internalized organizational identity provides a shared compass for decision-making. When individuals and teams understand and align with this core identity, they can make autonomous choices that are inherently consistent with the organization's broader strategic intent, even in the absence of direct supervision. The question "Is this action consistent with who we are as an organization?" becomes a primary filter for behavior, fostering coherence in a way that rigid structures alone cannot achieve.

As organizations evolve into these adaptive identity systems, the boundaries between the organization and its external ecosystem; comprising partners, gig workers, customers, and increasingly, AI agents become more permeable and co-evolutionary. The organization is no longer a clearly demarcated, self-contained entity but rather a significant node within a larger, interconnected network. Its identity is not solely defined from within; it is also shaped by, and performed in conjunction with, these external actors.

For instance, the conduct of a key strategic partner, the quality of work delivered by a contracted gig worker, or the output of an integrated third-party AI service can all reflect upon and influence the organization's perceived identity. This necessitates a more sophisticated approach to managing external relationships and integrations, ensuring alignment with the organization's core values and identity standards. The organization's identity, in effect, extends beyond its formal boundaries.

The success of such "organizations as identity systems" will also depend critically on their ability to cultivate what is known as organizational ambidexterity. This refers to the capacity to simultaneously and effectively execute current, established business models (exploitation) while also exploring new opportunities, fostering innovation, and adapting for the future (exploration). The organization's identity must be robust enough to provide stability and coherence for current operations, yet flexible and open enough to allow for experimentation, learning, and evolution.

A rigid identity can stifle necessary exploration and adaptation, while an overly diffuse or weak identity can undermine the efficient exploitation of existing strengths. Therefore, an effective organizational identity system must support both these functions. The core values and purpose provide a stable anchor, a sense of continuity, while the culture, structure, and processes actively encourage learning, the challenging of assumptions, and the integration of new capabilities. Achieving this balance between stability and dynamism, between exploiting the present and exploring the future, represents a key leadership and design challenge in creating truly adaptive organizations.

In conclusion, an organization's internal structure, prevailing culture, and operational modalities must authentically mirror the dynamic identity it seeks to project externally and the fluid capabilities it aims to nurture internally. It must transform into an "identity system" that is not defined by fixed attributes but by its inherent capacity for performance, learning, and continuous evolution. A critical component in the performance of these dynamic identities; whether they are individual, brand, or organizational, will increasingly be Artificial Intelligence itself. AI will not merely be a tool utilized by these entities; it is poised to become an active participant in the shaping, projection, and ongoing management of their identities. This leads to an exploration of the multifaceted role of AI in the performance of identity.

THE ROLE OF AI IN PERFORMING IDENTITY

As individuals, brands, and organizations increasingly embrace dynamic identities, Artificial Intelligence is transitioning from a background tool for analysis or content creation to an active agent in the performance, management, and evolution of these identities. This development, foreshadowed by concepts like the deeply personalized AI assistant in Chapter 7 ("One AI to Know You") and the notion of on-demand, AI-driven skills in Chapter 8 ("Skills on Demand"), signifies a profound shift in how identity is expressed and maintained in the digital age. AI is becoming not just an enabler of identity, but a performer of it.

For individuals, AI agents are poised to take on significant roles in managing and projecting their personal and professional brands. Imagine personal AI assistants meticulously curating an individual's digital footprint, ensuring that their skills, experiences, values, and aspirations are presented consistently and effectively across diverse platforms, from professional networking sites to project collaboration tools. These AI agents could help individuals maintain a coherent tone and messaging style, ensuring their digital persona aligns with their desired personal brand and career objectives. Furthermore, AI might proactively suggest opportunities for skill development, networking connections, or project engagements that resonate with an individual's evolving career narrative and identity goals, acting as a personalized career strategist and identity co-pilot.

In the corporate realm, AI agents are already beginning to revolutionize how brand identity is managed and performed. AI-powered "Brand Manager Agents" can continuously analyze market trends, monitor customer sentiment in real time, and track competitor activities to provide actionable insights for brand strategy and positioning. These agents can enforce consistency in brand voice, visual identity, and messaging across all communication channels, from global advertising campaigns to individual customer interactions. AI can automate the creation and distribution of on-brand content, ensuring it is tailored to specific audience segments and optimized for maximum impact.

Crucially, AI-driven chatbots and virtual assistants are becoming direct performers of a brand's identity. In every customer interaction, they embody the brand's tone (e.g., helpful, empathetic, efficient), its values (e.g., customer-centricity, transparency), and its promises. The ability of these AI systems to deliver personalized and even empathetic messaging is, in itself, a performance of the brand's commitment to its customers.

AI also plays an increasingly vital role in reputation management, a critical component of any identity. AI systems can monitor online conversations, social media platforms, and news outlets in real time, identifying and flagging potential public relations crises or, conversely, pinpointing positive sentiment and opportunities for engagement. Beyond simple monitoring, AI could contribute to the construction and maintenance of complex "reputation graphs".

These systems might algorithmically assess and even influence perceived trustworthiness based on a vast array of digital signals, interactions, and historical data, creating a dynamic, machine-readable representation of an entity's reputation.

A key capability AI brings to identity performance is the ability to manage nuanced presentations across diverse contexts. An AI could assist an individual in tailoring their communication style appropriately for a formal business proposal versus an informal social media update, or help a brand adapt its messaging for different cultural audiences, all while ensuring the core essence of the identity remains authentic and consistent. This contextual adaptation is crucial for maintaining relevance and effectiveness in a multifaceted world.

Moreover, AI facilitates the dynamic updating and evolution of identity. By learning continuously from interactions, feedback, performance data, and changing goals, AI can help an identity to remain fluid and responsive, rather than becoming static and outdated. This learning loop ensures that the performed identity is not just a snapshot but a constantly refined reflection of the entity's current state and future aspirations.

However, the increasing role of AI in performing identity is not without significant challenges. The question of authenticity looms large: can an AI truly perform an identity with genuine authenticity, or is it merely a sophisticated simulation, a carefully crafted facade? This probes the very nature of what constitutes "genuine" identity in a world where human and machine expressions become increasingly intertwined.

Accountability is another critical concern. If an AI agent misrepresents an identity, disseminates false information, or makes a damaging statement, who bears the responsibility? Is it the AI itself (a problematic proposition legally and ethically), its developers, its users, or the entity whose identity it purports to represent? This is particularly salient given that AI agents may need to assume distinct identities, separate from human users, to perform tasks, requiring clear ownership and auditable tracking.

Security and the potential for impersonation also present formidable risks. Malicious AI agents could be developed to impersonate legitimate individuals or brands with alarming fidelity, or they could exploit vulnerabilities in over-provisioned "helper" AI agents to gain unauthorized access or manipulate systems. Consequently, robust and dynamic identity management systems specifically designed for AI agents; addressing their unique characteristics of potentially ephemeral existence, context-based entitlements, and complex identity relationships, are paramount. Traditional authentication and authorization frameworks like OAuth and SAML, designed primarily for human users or static applications, are often inadequate for the granular, adaptive, and continuous validation required by AI agents.

As AI assumes a more active and integral role in the performance of identity, the skills required to manage this new reality are also evolving. The "curation of the self" for individuals and "brand stewardship" for organizations will increasingly involve the sophisticated management and fine-tuning of the AI agents that represent them. "AI agent oversight" is emerging as a critical new competency. This is not a one-time setup process but an ongoing endeavor of defining parameters, providing continuous feedback, interpreting AI-driven insights, and intervening decisively when an AI's performance of identity deviates from desired norms, ethical boundaries, or strategic objectives. This role is akin to managing a highly intelligent but uniquely autonomous public relations team or a personal assistant with global reach and instantaneous response capabilities.

The widespread use of AI to perform and project identity could also inadvertently trigger an "identity arms race". Entities might employ increasingly sophisticated AI techniques to project idealized, hyper-polished, or even deceptive personas in an effort to gain an advantage or manipulate perception. This, in turn, would compel other entities, and individuals, to rely on their own AI tools or develop more advanced human discernment capabilities to see through these meticulously AI-crafted facades. Such a cycle could escalate the complexity of discerning authenticity, making it even harder for stakeholders to build genuine trust. The "signal versus noise" problem for identity would become more acute, placing an even greater premium on truly verifiable authenticity, perhaps achieved through transparent AI usage disclosures, third-party validation mechanisms, or new forms of digital watermarking for identity.

Finally, entrusting AI with key facets of identity performance could subtly yet profoundly alter our perception of what "identity" entails. If an AI can convincingly perform many of the functions and express many of the characteristics that we associate with "me" or "my brand," does that alter how we define our own essence or the core nature of our organizations?

Identity has always been partly performative, expressed through actions, interactions, and communications. If AI agents are increasingly handling these manifestations, they are, in effect, performing substantial components of that identity. Over time, there is a possibility that we might begin to perceive the AI's performance as the identity, or at least as an inseparable part of it. This could precipitate a philosophical shift where identity is conceptualized less as a purely intrinsic, internal "self" and more as a managed, externally projected, and co-created (human-AI) persona. This has deep implications for self-perception, the pursuit of authenticity, and the very definition of what it means to "be" an individual or an organization in the AI era, linking directly to the themes explored in Chapter 7 concerning AI as both a mirror and a co-creator of identity.

In conclusion, AI is set to become an indispensable, if complex, partner in the intricate and ongoing task of performing and managing dynamic identity. This partnership offers immense potential for enhanced personalization, responsiveness, and strategic alignment, but it also demands careful navigation of profound ethical, security, and philosophical considerations.

The culmination of these interconnected shifts; from the devaluation of static IP to the ascendancy of dynamic identity, from fixed expertise to fluid capabilities, from static brands to adaptive performers, from rigid organizations to evolving identity systems, and with AI itself becoming an active performer of identity, points towards a fundamental reorientation in how we conceive of and pursue strategic advantage.

The old game of building walls to defend fixed assets is rapidly becoming obsolete; the new game is about fostering an intrinsic capacity for continuous evolution. This recognition leads directly to our final exploration: the critical shift from a strategy of defensibility to one of evolvability.

THE SHIFT FROM DEFENSIBILITY TO EVOLVABILITY

The preceding sections have charted a transformative journey: the erosion of traditional moats like intellectual property and static expertise, and the concurrent rise of dynamic identity as the new anchor of value. We have seen how individuals must cultivate fluid capabilities, how brands must become adaptive performers, and how organizations must restructure into agile identity systems, with AI increasingly playing a role in the performance of these evolving identities.

This entire trajectory culminates in a singular, overarching strategic imperative for the Age of Information Symmetry: a fundamental shift from prioritizing defensibility to championing evolvability. In a world where knowledge is ubiquitous and change is the only constant, success hinges less on protecting what one has and more on the capacity to continuously become what the future demands.

The traditional strategic playbook, heavily focused on building and defending sustainable competitive advantages through barriers to entry; such as proprietary technology, exclusive market access, or entrenched brand loyalty based on fixed attributes, is proving increasingly inadequate. As AI accelerates the pace of innovation and the dissemination of information, these static defenses are becoming more porous and less durable. What was once a secure fortress can quickly become an exposed position.

In this new paradigm dynamic identity forms the resilient core. However, its true strategic power lies not in its fixed characteristics but in its inherent capacity to adapt, learn, and evolve while maintaining authenticity and trust. Evolvability is the comprehensive ability of an individual, brand, or organization to continuously adapt to a dynamic environment. This involves sensing changes, learning from new information and experiences, innovating offerings and processes, and transforming structure and identity, ideally in anticipation of, rather than merely in response to, a constantly shifting landscape. This is not mere reactivity; it is a proactive embrace of change, underpinned by strong dynamic capabilities: the ability to sense emerging opportunities and threats, seize promising avenues, and reconfigure internal and external competencies effectively.

The strategic focus consequently shifts from safeguarding a unique, static "edge" to consistently delivering superior value through dynamic performance. This means cultivating the ability to reconfigure capabilities on the fly, adapt brand performance in real time to meet changing consumer needs and market conditions, and re-architect organizational systems and cultures for ongoing agility. The emphasis moves from ownership of assets to orchestration of capabilities.

The pace and effectiveness of this evolution, the "evolutionary pace", become critical metrics of strategic health and competitive differentiation. Organizations and individuals that can learn, adapt, and implement changes faster and more effectively will consistently outperform those that are slower to respond, even if the latter possess historically strong but static assets or market positions. Adapting to change and strategically investing in technologies that enable this adaptation are highlighted as key survival strategies in rapidly changing business environments.

Artificial intelligence itself is a dual force in this equation. While AI contributes to the environmental turbulence that necessitates evolvability, it also provides powerful tools to enable and accelerate this adaptive capacity. AI can enhance an organization's ability to sense market shifts through advanced data analytics, personalize offerings at scale, optimize internal processes for greater efficiency, and even assist in the real-time performance and management of its evolving identity.

This reframes the concept of identity itself. It is no longer a fixed brand asset to be meticulously polished and protected, but rather the systemic capacity of the entity; be it an individual, a brand, or an organization, to maintain coherence, trust, and a sense of purpose while undergoing continuous, often profound, change.

Consider Patagonia, which has demonstrated remarkable evolvability over decades, adapting its materials, supply chains, and advocacy efforts in response to growing environmental understanding, all while retaining its core identity and values centered on environmental stewardship. Similarly, Netflix's journey from a DVD-by-mail service to a global streaming giant and original content producer exemplifies radical business model evolution, driven by a willingness to cannibalize its own successes in pursuit of new opportunities, thereby showcasing a high degree of evolvability.

Fostering evolvability has tangible practical implications. It requires deliberate investment in creating learning cultures where curiosity is encouraged, and failures are treated as opportunities for growth. It demands the adoption of flexible technologies and modular organizational structures that can be easily reconfigured. It necessitates empowering employees at all levels to contribute to innovation and make decisions closer to the point of impact. It involves embracing experimentation and developing leaders who are not only comfortable with ambiguity but can effectively guide their teams through periods of significant change and uncertainty. Tactics such as employing flexible work management frameworks, asking probing questions to challenge assumptions before committing to shifts, and even strategically "underplanning" to maintain agility can be useful in environments characterized by frequent strategic adjustments.

The imperative of "evolvability" fundamentally reshapes strategic planning. Traditional, long-term strategic planning cycles, often spanning three to five years and resulting in fixed roadmaps, are becoming increasingly anachronistic. In many AI-disrupted sectors, the assumption of predictability that underpins such cycles no longer holds.

A meticulously crafted five-year plan can be rendered irrelevant by rapid technological advancements, unforeseen competitor moves, or sudden market shifts, all of which are amplified by AI's dynamism.

Consequently, strategic processes must transform into continuous, iterative feedback loops, mirroring the principles discussed in Chapter 11. This involves constant environmental scanning, real-time interpretation of data and signals, agile decision-making, and rapid execution in much shorter cycles. Strategy itself becomes an evolving capability, focused less on defining a fixed future state and more on building the organizational capacity to change. Organizations might, therefore, manage a portfolio of adaptive experiments or strategic bets, dynamically allocating resources to scale initiatives that demonstrate promise while quickly divesting from or terminating those that do not.

True "evolvability" also demands a profound organizational comfort with ambiguity and a re-framing of failure. Moving into new, often uncharted territory is inherent in the process of evolution. This requires a culture that does not penalize intelligent risk-taking or view experiments that do not yield the expected results as outright failures, but rather as valuable learning opportunities.

Furthermore, deep evolvability sometimes necessitates the difficult decision to proactively cannibalize one's own successful but maturing offerings, business models, or even aspects of one's current identity before competitors force such a change. This internal "creative destruction," exemplified by Netflix's shift from DVDs to streaming (as we have mentioned before), is exceptionally challenging for many organizations because it can conflict with established routines, short-term performance incentives, and institutional comfort zones. Therefore, a core component of evolvability is the psychological and cultural resilience to embrace proactive, transformative change, even when it is uncomfortable or involves letting go of past glories. This represents a significant leadership and cultural hurdle.

Ultimately, in a world that relentlessly demands evolvability, the very "identity" of an entity; whether individual, brand, or organization, must itself become meta-adaptive. This means that the capacity to evolve its own identity must be an intrinsic, core characteristic. The identity is not merely what the entity is at any given moment, but also its inherent ability to become something else while retaining its fundamental essence and purpose. For instance, a brand's identity might be more powerfully articulated as "we are relentless innovators who constantly redefine our market to better serve human needs" rather than "we are the leading provider of X product". The former has evolvability embedded within its narrative; the latter is static and vulnerable. The "story" of the identity must, therefore, be a story of perpetual growth, learning, adaptation, and purposeful transformation.

In the Age of Information Symmetry, the ultimate, most durable competitive advantage lies not in constructing impenetrable walls around what is currently known or owned, but in cultivating an unparalleled, systemic capacity to become what the unfolding future requires. This journey from static IP to dynamic, evolvable identity reshapes the entire landscape of competition and the very definition of success.

SUMMARY

This chapter has navigated the profound shift from traditional, static sources of competitive advantage to a new paradigm centered on dynamic identity and relentless evolvability. In the Age of Information Symmetry, accelerated by the capabilities of Artificial Intelligence, the long-held moats of intellectual property and fixed, specialized expertise are proving increasingly porous and unreliable. AI's capacity for replication, remixing, and widespread dissemination of knowledge and creative outputs fundamentally undermines the exclusivity that once underpinned these advantages.

In their place, a more resilient and sustainable source of differentiation emerges: dynamic identity. For individuals, this necessitates a move away from reliance on static credentials and fixed job roles towards the cultivation of fluid capabilities, a commitment to lifelong learning, and the performance of an authentic, adaptable personal brand. For businesses, the implications are equally transformative.

Brands must evolve from static communicators of fixed attributes into adaptive performers, leveraging data and AI to personalize experiences, respond to market changes in real time, and maintain relevance by continuously learning and adapting their narratives. Organizations, in turn, must shed rigid, hierarchical structures and transform into agile identity systems, characterized by modular designs, value-driven cultures that foster learning and experimentation, and leadership that orchestrates rather than commands. Artificial Intelligence itself is not merely a catalyst for these changes but is becoming an active participant in the performance and management of these dynamic identities, offering new capabilities while also presenting new challenges related to authenticity, accountability, and security.

Ultimately, the overarching conclusion is that success in this new era is no longer primarily about defending what is known or owned. Instead, it is about embracing and mastering evolvability; the intrinsic capacity of an individual, brand, or organization to relentlessly adapt, innovate, and redefine itself, its offerings, and its purpose in a world characterized by perpetual change and ubiquitous information.

This deep dive into the imperatives of dynamic identity and systemic evolvability underscores a critical practical challenge that must now be addressed: if identity is increasingly performed, perceived, and validated in a world heavily intermediated by AI, then how do we ensure that these intelligent systems can accurately understand, interpret, and favorably engage with our evolving identities?

To effectively perform, adapt, and thrive, individuals, brands, and organizations must become "AI-readable". The next chapter, 'Becoming AI-Readable - How to Structure Products, Services, and Data So AI Understands and Recommends You,' will explore the crucial steps and strategies required to structure our products, services, data, and indeed our very identities, so that AI can not only find and comprehend us but, ultimately, recommend and integrate us effectively within this new, intelligent ecosystem.

13

BECOMING AI-READABLE

The fabric of our interaction with the world is undergoing a profound transformation. Increasingly, the first "eyes" to see our job applications, the first "mind" to evaluate our product queries, and the first "voice" to respond to our service requests belong not to a human, but to an artificial intelligence. This shift is not a distant forecast; it is a present reality, subtly yet pervasively reshaping how discovery, evaluation, and connection occur across nearly every domain of personal and professional life. AI systems, in their myriad forms, are stepping into the role of primary gatekeepers and intermediaries, filtering, sorting, and recommending based on their interpretation of the data they encounter.

In this emerging landscape, a new imperative arises: the need to become "AI-readable". This concept extends far beyond mere human comprehension or the clarity of prose. AI readability involves a deliberate and strategic structuring of information, services, products, and even our personal and organizational identities in ways that are optimally interpretable, processable, and actionable by these sophisticated algorithms. It demands that we consider how AI parses data, discerns semantic meaning, identifies patterns, and draws inferences.

It requires us to understand that machines "read" differently than humans do, prioritizing explicitness, structure, and verifiability.

The consequence of failing to adapt to this new mode of interaction is significant. In an AI-mediated world, to be AI-illegible is to risk becoming invisible or, perhaps worse, to be fundamentally misunderstood. If an AI system cannot effectively "read" you, your skills, your organization, or your offerings, it cannot accurately assess your value, recommend you for an opportunity, or seamlessly integrate your capabilities into its processes.

This has far-reaching implications for discoverability: how you are found in a crowded digital space; for trust: how your claims and capabilities are vetted and validated; and for value delivery: how your unique contributions are recognized and utilized. The default assumption must now be that your customer, your potential employer, your business partner, or even a system seeking a specific capability, is no longer evaluating you directly in the first instance. They are increasingly relying on their AI to perform that initial assessment. The critical questions then become: Can their AI find you? Can it interpret what you offer? And, based on that interpretation, will it recommend you?

This chapter serves as a practical guide to understanding and implementing AI readability. It will explore the fundamental transition from human-centric modes of presentation to the AI-centric structuring of information that is now required. We will delve into concrete strategies that individuals, organizations, and product designers can employ to enhance their legibility to these intelligent systems. Furthermore, the chapter will examine the mechanisms by which AI systems evaluate and rank entities, and how one can become fluent in managing the "signal" they project to these increasingly influential algorithmic arbiters. The journey to becoming AI-readable is not merely a technical adjustment; it is a strategic necessity for anyone wishing to thrive in a world increasingly navigated and shaped by artificial intelligence. To effectively engage with this new reality, we must first grasp the essential differences between designing for human perception and designing for the intricate processing capabilities of AI, a distinction that forms the core of our immediate exploration.

THE SHIFT FROM HUMAN-READABLE TO AI-READABLE

For centuries, human communication and the presentation of information have been tailored to the nuances of human cognition. We crafted narrative resumes designed to evoke a personal story, developed glossy brochures appealing to visual and emotional sensibilities, and built websites with user experiences centered on human intuition and engagement.

Marketing copy was an art form, carefully worded to tap into human psychology and desires. These methods relied intrinsically on the human capacity for interpretation, the ability to infer meaning from incomplete data, and the skill to process unstructured or semi-structured information with remarkable flexibility. A beautifully written cover letter or a compelling visual advertisement could capture attention and persuade, leveraging shared cultural contexts and implicit understandings.

However, the introduction of AI as a primary intermediary in information exchange has introduced a new "audience" with a fundamentally different set of "reading" rules. AI systems, particularly those tasked with evaluation and recommendation, process information with a preference for structure, explicitness, and semantic clarity rather than narrative flair or aesthetic appeal alone. This algorithmic lens is reshaping how initial assessments are made across various sectors.

In recruitment, for instance, AI software now routinely parses resumes for specific keywords, quantifiable skills, and measurable achievements, often making a first-pass selection before any human recruiter views the application. A traditionally crafted, narrative-heavy curriculum vitae might fail this initial algorithmic screening if not optimized for machine interpretation.

Similarly, in the realm of business purchasing, AI-powered procurement tools analyze supplier data, product specifications, and service level agreements based on structured inputs and predefined criteria, making structured, detailed information paramount. Even in the vast expanse of content discovery, search engines and recommendation algorithms increasingly employ AI to understand the deep meaning and contextual relevance of information, moving far beyond simple keyword matching to sophisticated semantic analysis. Content that lacks semantic richness or a clear, parsable structure may consequently be poorly ranked, miscategorized, or simply lost in the digital noise.

This evolving landscape means that "looking good to humans," while still important, is often no longer sufficient. The initial gatekeeper is frequently an algorithm, and if information is not structured to pass through this digital checkpoint, it may never reach a human evaluator, or if it does, it might be presented with diminished context or compromised accuracy. This necessitates a dual-pronged approach to information design: content must remain compelling and persuasive for human audiences, but it must first be intelligible, valuable, and processable by AI. This reality has led to the ascendance of three critical elements in information architecture: metadata, semantic clarity, and structured representation.

Metadata, quite literally "data about data," provides crucial context for AI systems. It enables them to categorize, filter, relate, and understand information with far greater efficiency and accuracy.

Examples abound, from schema.org markup that provides explicit semantic information about web content for search engines, to descriptive tags in a product catalog that allow for faceted search, or clearly defined attributes within a professional skills profile that enable precise matching. Semantic clarity, the second pillar, underscores the importance of using unambiguous language and well-defined terms. AI systems thrive on precision; ambiguity is a barrier to understanding. This involves employing consistent terminology, explicitly defining domain-specific jargon where its use is unavoidable, and leveraging ontologies or controlled vocabularies to ensure that terms carry consistent meaning across different contexts.

Finally, structured data formats, such as those found in databases, spreadsheets, or purpose-built data exchange formats like JSON and XML, as well as the interconnected relationships within knowledge graphs, are inherently more AI-readable than unstructured text or standalone images. Structured data allows AI to easily parse, query, analyze, and synthesize information because the data's organization and the relationships between its elements are explicitly defined. Machine-readable data can range from human-readable content that has been augmented with machine-interpretable markup, like HTML incorporating RDFa, to data files primarily designed for machine processing, such as CSV, XML, or JSON, which ensure that information can be processed without human intervention while preserving its semantic integrity.

The failure to adapt to these AI-centric requirements incurs what can be termed a "translation cost". When information is presented in a format that is not readily AI-readable, a burden is created. Either a human must manually intervene to reformat, tag, or "translate" the information into a state that AI systems can effectively process (like in the case of study authors that add "invisible" text to their published studies specifically designed for AI in order to modify AI's behavior), or the AI itself must expend additional computational resources; and risk introducing errors or missing crucial nuances, in its attempt to interpret the poorly structured input.

This translation cost is not merely a theoretical concern; it manifests in tangible ways: lost time, reduced efficiency, compromised accuracy in AI-driven decisions, and overlooked subtleties that could have been critical. Organizations and individuals who proactively invest in making their information, products, and services AI-readable effectively minimize this translation cost. They become easier to engage with from an AI perspective, their data is more readily integrated, their capabilities are more accurately assessed, and consequently, they gain an inherent efficiency and accuracy advantage in an increasingly automated world. This advantage stems from lowering the friction of interaction, making them preferred entities for AI-driven platforms and processes. Understanding this fundamental shift from human-first to AI-first legibility is the initial step; the next is to define precisely what characteristics make an entity truly AI-readable.

WHAT IT MEANS TO BE AI-READABLE

The concept of AI readability is not a singular attribute but rather a multifaceted quality, a composite of several key characteristics that collectively enable machines to effectively ingest, process, understand, and act upon information. The objective is to transition from merely providing data that an AI can technically consume to offering data that an AI prefers; data from which it can derive maximum value with minimal ambiguity and computational effort. This involves a deliberate structuring of content, services, and even identity signals.

Four primary pillars support this notion of AI readability: machine-parsable formats and structured data, semantic clarity and context, interpretable behavioral signals, and clear intent-matching and goal alignment.

The first pillar, machine-parsable formats and structured data, addresses the fundamental need for information to be presented in a way that software can easily deconstruct and interpret without ambiguity. This begins with basic machine-parsable formats, ensuring text is encoded in standard ways (like UTF-8) and that files are in structured formats such as JSON, XML, or CSV. Even well-formed HTML that utilizes semantic tags is significantly more parsable for an AI than a plain text document or, worse, an image containing text.

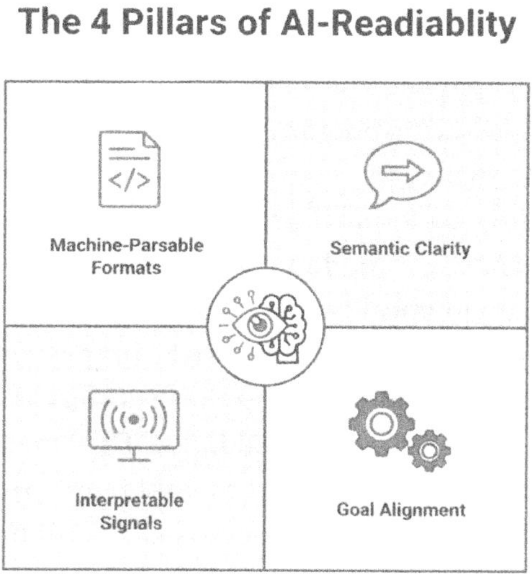

Formats like PDF, while common for human consumption, can pose challenges for machine parsing unless they are specifically created as "tagged PDFs" with underlying structural information. Beyond format, structured data itself is crucial. This means organizing data in a predefined manner, typically within tables, databases, or more sophisticated graph structures like knowledge graphs. Such organization allows AI systems to perform efficient querying, aggregation, and complex analysis because the relationships and attributes are explicitly laid out. Knowledge graphs, for instance, represent entities and their interconnections, providing a rich, contextualized data environment for AI.

The second pillar, semantic clarity and context, focuses on the meaning embedded within the data. Unambiguous language is paramount; AI systems benefit from clear, precise terminology. While avoiding jargon is ideal, if its use is necessary, it should be explicitly defined. Modern AI tools can assist by suggesting simpler synonyms or rephrasing complex sentences to improve clarity.

To provide deeper semantic scaffolding, ontologies and taxonomies are invaluable. Ontologies offer formal representations of knowledge by defining concepts and the relationships between them within a specific domain, while taxonomies provide hierarchical classification systems. These structures help AI understand the context and nuanced meanings of terms. Furthermore, contemporary AI, especially Large Language Models, comprehends meaning through embeddings; mathematical vector representations of text, images, or other data types. Content that is rich in descriptive language, provides concrete examples, and establishes clear context generates more effective and distinct embeddings. This makes the content more "understandable" and discoverable in applications like semantic search or Retrieval-Augmented Generation (RAG) systems. Crafting embedding-rich content involves using clear prompts if the content is generated, providing detailed descriptors for entities, and including illustrative examples within the information itself.

The third pillar concerns interpretable behavioral signals. AI systems do not only "read" static data; they also interpret dynamic behaviors. For products and services, this translates to usage patterns, customer engagement metrics, and feedback mechanisms. For individuals, it might involve their online activity, contribution histories, or interaction patterns within professional networks. AI systems, particularly in domains like recommendation engines, risk assessment, or personalization, leverage these behavioral signals to infer preferences, predict future actions, and evaluate attributes like trustworthiness or expertise. Being AI-readable in this context means ensuring that these signals are accurately captured, are relatively free from distorting noise, and can be meaningfully correlated with relevant outcomes or attributes.

The fourth pillar is intent-matching and goal alignment. AI systems are frequently designed with specific goals or to fulfill user intentions. Therefore, AI-readable information must be structured in such a way that an AI can easily and accurately match it to a given user's query or a system's operational objective. This is fundamental for effective search, where semantic search technologies strive to align results with the underlying intent of a query rather than just keyword occurrences.

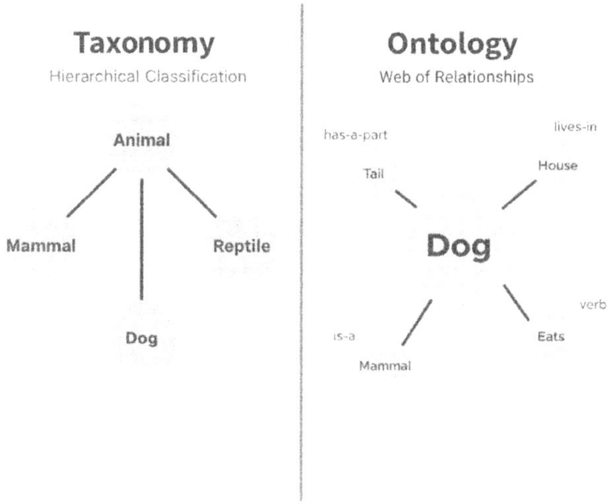

It is equally critical for recommendation systems, which aim to match items or content to individual user needs and preferences, and for autonomous agentic systems, which must identify and invoke the correct capabilities or services to complete assigned tasks. To facilitate this, content and service descriptions should clearly articulate their purpose, the specific problems they solve, or the precise questions they answer, avoiding vague language or unnecessary filler.

These pillars are not independent; they work in concert to create a holistic state of AI readability. The list below summarizes these core components:

Machine-parsable format

- **Description:** Data presented in a way that software can easily decompose and interpret its structure.
- **Why it matters for AI:** Enables efficient ingestion and initial processing by AI systems without manual transformation.
- **Practical example:** JSON used for data exchange, well-formed XML for documents, and UTF-8 encoded text.

Structured data

- **Description:** Data organized in a predefined manner with explicit attributes and relationships.
- **Why it matters for AI:** Supports efficient querying, aggregation, analysis, and pattern recognition.
- **Practical example:** Relational databases, CSV files with headers, and knowledge graphs built with RDF or OWL.

Semantic clarity

- **Description:** Use of unambiguous language, consistent terminology, and explicit definitions for domain-specific terms.
- **Why it matters for AI:** Reduces misinterpretation and improves the accuracy of AI understanding and reasoning.
- **Practical example:** Controlled vocabularies, glossaries for specialized jargon, and clear, concise sentence structures.

Defined vocabulary or ontology

- **Description:** Formal specification of concepts, properties, and relationships within a domain.
- **Why it matters for AI:** Provides a shared understanding of meaning and context, enabling advanced reasoning and interoperability.
- **Practical example:** Schema.org for web content, industry-specific ontologies such as those used in medicine, and custom organizational taxonomies.

Embedding richness

- **Description:** Content that produces distinct and meaningful vector representations, especially for large language models.
- **Why it matters for AI:** Improves semantic search, similarity detection, and contextual understanding.
- **Practical example:** Detailed product descriptions with real use cases and articles that present clear arguments supported by evidence.

Interpretable behavioral signals

- **Description:** Data generated by actions and interactions that can be analyzed to infer intent, preference, or risk.
- **Why it matters for AI:** Enables personalization, prediction of future behavior, and assessment of trustworthiness.
- **Practical example:** Clickstream data, purchase history, engagement metrics, and server logs showing API usage patterns.

Clear intent articulation

- **Description:** Information that explicitly states its purpose, the problem it solves, or the question it answers.
- **Why it matters for AI:** Helps AI accurately match information to user queries, workflows, or system goals.

- **Practical example:** FAQ pages that address specific questions and service descriptions that clearly define intended outcomes.

Achieving this level of clarity and structure has implications beyond mere discoverability; it forms a crucial foundation for AI trust. When information is presented in a way that is transparent, well-organized, and semantically unambiguous, AI systems can process it with greater accuracy. This accuracy is a vital precursor to establishing trust.

If an AI can reliably and correctly understand the data it receives from an entity; be it a person, an organization, or a product, it is far more likely to produce reliable, and therefore trustworthy, outputs, recommendations, or actions based on that data.

Conversely, information that is obscure, poorly structured, or semantically vague (the antithesis of AI-readable) is prone to misinterpretation by AI. Such misinterpretations can lead to flawed algorithmic decisions, which in turn erode the confidence of human users who rely on these AI-mediated systems (the old "garbage in garbage out" rule still applies in the AI age).

Therefore, the effort invested in becoming AI-readable is an early and essential step in building a chain of trust that extends from the data source, through the AI system, to the ultimate human beneficiary. If the "API of oneself" or one's product is clean, well-documented, and easily understood by machines, the AI; and by extension, its human users, are more likely to trust the ensuing interactions and outcomes. With this comprehensive understanding of what AI readability entails, the focus naturally shifts to its practical application, starting with how individuals can strategically adapt their professional presence.

MAKING PEOPLE AI-READABLE

In an environment where professional opportunities and collaborations are increasingly filtered through algorithmic lenses, individuals find themselves subject to a new form of scrutiny. Whether for recruitment, team assembly, expert identification, or even the assessment of social influence, AI scrutiny plays a more significant role in evaluating people. The traditional resume, often a narrative document, and unstructured online profiles, while valuable for human connection, are frequently insufficient for optimal interpretation by these intelligent systems. To thrive, professionals must now consciously structure their digital identity in ways that are legible and meaningful to machines.

One of the most impactful strategies is the adoption of skills graphs and competency frameworks. This involves moving beyond simple lists of past job titles and responsibilities to explicitly defining a granular set of skills, indicating proficiency levels for each, and even mapping the relationships between these skills. AI can parse these structured graphs to match individuals to opportunities with a much higher degree of precision than is possible with keyword spotting in unstructured text. Using standardized skill terminologies or industry-recognized taxonomies where available further enhances an AI's ability to understand and compare capabilities across different individuals.

Complementing skills graphs are digital portfolios endowed with rich metadata. Professionals should present their work; be it projects, publications, code repositories, or creative designs, not merely as static artifacts but as dynamic data points. This means embedding detailed metadata with each portfolio item: specifying the tools and technologies used, the specific skills demonstrated, the quantifiable outcomes of the project, and the roles of any collaborators. Such metadata transforms a portfolio from a showcase into a searchable, analyzable database of capabilities that AI can readily process.

Making each portfolio item individually addressable and semantically described allows AI to pinpoint specific examples of expertise.

Furthermore, leveraging structured data markup like Schema.org's " Person" type on personal websites or professional networking profiles can significantly enhance AI readability. This markup allows search engines and other AI tools to easily extract and understand key information such as an individual's name, current and past job titles, affiliations, areas of expertise, publications, and links to relevant social media or professional platforms. For example, providing links to state licenses or NPI numbers for healthcare professionals, as suggested for comprehensive profiles, makes this information directly verifiable and usable by AI systems.

The validation of skills and achievements is undergoing a fundamental change with the advent of trusted attestations and verifiable credentials. These are digital, often cryptographically secured, proofs of skills, certifications, educational qualifications, or experiences. Because they are inherently structured, typically standardized, and carry the verification of the issuing authority, they are highly AI-readable. AI systems can trust the authenticity of these credentials far more readily than self-asserted claims on a conventional resume.

Platforms such as digital wallets designed for learning and employment records (LERs) empower individuals to securely store, manage, and share these verifiable credentials with granular control. Decentralized identity solutions are also emerging, offering frameworks for individuals to manage their credentials and present them with enhanced privacy, allowing for selective disclosure of information as needed. For an AI, a verifiable credential is a strong, reliable signal of a specific capability.

Moreover, individuals may need to practice context-aware representation, tailoring their AI-readable signal for different purposes. The way one presents their skills and experience when applying for a highly technical role might differ from the presentation for a leadership position or when seeking investment. This doesn't imply inconsistency but rather a strategic highlighting of the most relevant facets of one's expertise, all while maintaining a core, verifiable identity. AI systems themselves might eventually assist individuals in managing and adapting these nuanced representations.

Finally, it is important to acknowledge that behavior logs; implicit signals from online activities such as contributions to open-source projects, insightful activity on professional forums, or the creation and sharing of valuable content, can also be interpreted by AI as indicators of expertise, engagement, and influence. While this offers another avenue for demonstrating capabilities, it also brings ethical considerations regarding data privacy, transparency, and the individual's control over how such behavioral data is collected and used (i.e. social credit scoring).

The transition towards AI-readable personal representation involves a significant departure from traditional methods, as listed below.

Skills declaration

- **Traditional human-first approach:** Narrative descriptions in a CV or a simple list of keywords.
- **AI-readable approach:** Tagged skills graph with proficiency levels, aligned to shared taxonomies.
- **Key benefit for AI and recruiters:** Precise skill matching with visibility into depth and adjacencies between skills.

Experience showcase

- **Traditional human-first approach:** Chronological job descriptions and static PDF portfolios.
- **AI-readable approach:** Digital portfolios enriched with metadata per project, including tools used, outcomes achieved, skills applied, and links to live work or repositories.

- **Key benefit for AI and recruiters:** Direct evidence of capability application and easier analysis of project specifics and impact.

Achievements and credentials

- **Traditional human-first approach:** Paper certificates and textual mentions of qualifications.
- **AI-readable approach:** Verifiable digital credentials such as badges and micro-credentials stored in digital wallets.
- **Key benefit for AI and recruiters:** Instant verification of claims, increased trust in qualifications, and standardized data formats suitable for automation.

Professional network

- **Traditional human-first approach:** Lists of connections, often represented as counts on professional networking platforms.
- **AI-readable approach:** Public contributions with clear attribution, semantically linked into a professional graph.
- **Key benefit for AI and recruiters:** Ability to analyze influence, collaboration patterns, and domain authority beyond superficial connection metrics.

Discoverability

- **Traditional human-first approach:** Keyword-based resume searches and reliance on personal networking.
- **AI-readable approach:** Semantically discoverable profiles using structured data and participation in credentialed networks.
- **Key benefit for AI and recruiters:** Context-aware matching based on intent and nuanced skill requirements rather than keyword coincidence.

This evolution from human-centric to AI-centric professional representation reflects a deeper shift from merely "claiming" expertise to actively "proving" it in a machine-understandable format. Traditional resumes and profiles are replete with claims of skills and experience, which human recruiters then attempt to validate through interviews, references, or assessments.

AI systems, especially when acting as initial screeners, operate more effectively when they can process and "trust" verifiable, structured data rather than parsing and attempting to validate subjective, unstructured assertions. Verifiable credentials, digital badges backed by metadata, and well-documented portfolio items provide more direct, machine-interpretable evidence of an individual's capabilities.

For example, a simple textual claim of "proficient in Python" on a resume is a relatively low-signal data point for an AI. In contrast, a verifiable digital badge for a Python certification or a link to a GitHub repository showcasing multiple Python projects with clear descriptions of their purpose and the developer's role offers a much higher-signal, more trustworthy set of data points. This is because the latter are structured, often externally validated, and provide concrete evidence that an AI can more readily quantify and weigh in its evaluations.

The implication for individuals is that the burden of proof is shifting. It is no longer sufficient to merely list a skill; one must increasingly be able to point to a verifiable attestation or a clearly documented instance of its successful application. While this could lead to more meritocratic evaluations by AI, it also underscores the importance of individuals actively curating their AI-readable proofs of competence and raises considerations about equitable access to such verification mechanisms. Just as individuals must adapt, organizations too face a parallel imperative to make themselves and their offerings legible to AI, a challenge explored in the following section.

MAKING ORGANIZATIONS AI-READABLE

The modern enterprise no longer interacts solely with human customers, partners, or employees through traditional channels. Increasingly, organizations are being "queried," "evaluated," and "engaged" by other AI systems. This could be a customer's personal AI assistant seeking a specific service, a potential partner's AI assessing integration capabilities and compatibility, or a sophisticated procurement AI meticulously evaluating suppliers based on a multitude of data points. This AI-to-AI interaction demands that organizations expose their capabilities, information, and offerings in a manner that these autonomous or semi-autonomous agents can readily consume, understand, and act upon. Becoming AI-readable is thus a strategic imperative for organizational relevance and competitiveness.

A foundational element of organizational AI readability is the development of machine-readable service catalogs and product descriptions. This represents a significant evolution from traditional human-focused websites, glossy PDF brochures, or static price lists. While digital catalogs exist, their primary design is often for human browsing. True AI readability requires catalogs that are structured for machine interpretation. This involves providing detailed, organized descriptions of all services and products, clearly articulating features, benefits, precise pricing models, availability, and technical specifications in formats that are easily parsable by software, such as XML, JSON feeds, or through well-defined APIs. Furthermore, using standardized vocabularies or industry-specific ontologies to describe product attributes or service types ensures semantic consistency and facilitates more accurate understanding by external AI systems. The Open Data Product Specification (ODPS), for example, offers a model for a machine-readable specification for data products, which conceptually can be extended to other types of services, emphasizing structured metadata and clear definitions.

Alignment with established ontologies and open standards is another critical factor. By adopting or contributing to industry-specific ontologies, organizations can ensure their services and products are described using a shared semantic framework that external AI systems can understand. This greatly enhances interoperability and reduces the risk of misinterpretation. Similarly, utilizing open standards for data exchange, service descriptions (such as the OpenAPI Specification for APIs), and metadata promotes easier integration and reduces friction when AI agents attempt to interact with the organization's offerings. Open standards, by their nature, are publicly accessible and designed to facilitate interoperability, making them ideal for building AI-friendly interfaces.

An API-first strategy is rapidly becoming a cornerstone of organizational AI readability. This approach prioritizes the design and development of Application Programming Interfaces (APIs) as the primary way for services and data to be accessed, rather than treating APIs as an afterthought or a secondary channel. Well documented, robust, and standardized APIs allow AI agents to programmatically discover capabilities, access data, invoke services, and integrate the organization's offerings into larger workflows or composite applications. The success of API-first companies like Stripe, Netflix, and Twilio demonstrates how this strategy not only enables scalability and innovation but also fosters a vibrant ecosystem of third-party developers and, increasingly, AI agents that can interact with their platforms.

Beyond technical interfaces, organizations must also consider their AI-facing brand metadata and semantic identity. This involves embedding the essence of the brand; its identity, core values, trust signals, and ethical commitments into machine-readable metadata. This goes deeper than visual branding elements; it includes semantic descriptions of what the brand stands for, its operational principles, its areas of specialized expertise, and its approach to governance. Such information can be used by AI systems in their evaluation and recommendation processes, particularly in contexts where trust, alignment, and reputation are key criteria, a theme closely related to the concept of "Trust as a Service" explored elsewhere in this book.

Finally, an often-overlooked aspect of external AI readability is the state of an organization's internal data landscape. Ensuring that internal data assets; such as customer information, detailed product specifications, internal knowledge bases, and employee skill inventories, are themselves well-structured, semantically coherent, and readily accessible (with appropriate security and privacy safeguards) to the organization's own AI systems is crucial. This internal AI readability is frequently a prerequisite for effectively powering AI-facing external services, providing accurate information to external AIs, or enabling sophisticated internal automation that ultimately enhances external responsiveness.

To assist organizations in this endeavor, the following list outlines key strategies for various assets:

Core services portfolio

- **Key AI-readability strategy:** API-first exposure of service capabilities.
- **Example technologies and standards:** OpenAPI, gRPC, GraphQL.
- **Impact on AI interaction:** Enables seamless discovery, invocation, and orchestration of services by external AI agents and platforms.

Product lines and offerings

- **Key AI-readability strategy:** Structured digital catalogs aligned to shared ontologies.
- **Example technologies and standards:** JSON-LD, RDF, industry-specific ontologies such as GS1 for retail, and standards for data products.

- **Impact on AI interaction:** Allows AI to accurately compare offerings, understand features, and match products to specific needs or intents.

Brand reputation and trust

- **Key AI-readability strategy:** Verifiable claims, certifications, and transparent policies.
- **Example technologies and standards:** Verifiable credentials, publicly accessible ethical guidelines, and third-party audit reports.
- **Impact on AI interaction:** Strengthens trust in AI-driven recommendations and enables verification of claims related to quality, safety, or compliance.

Customer support knowledge

- **Key AI-readability strategy:** Semantic knowledge bases exposed through APIs.
- **Example technologies and standards:** Knowledge graph technologies, natural language understanding APIs, and structured FAQ data.
- **Impact on AI interaction:** Powers accurate AI-driven support interactions and enables agents to locate relevant solutions efficiently.

Partner ecosystem interfaces

- **Key AI-readability strategy:** Standardized integration protocols and data exchange formats.
- **Example technologies and standards:** Electronic data interchange, industry-specific API standards, and shared data models.
- **Impact on AI interaction:** Facilitates automated partner

onboarding, reliable data sharing, and AI-enabled supply chain collaboration.

Internal expertise directory

- **Key AI-readability strategy:** Skills-graphed employee profiles and expert discovery mechanisms.
- **Example technologies and standards:** Internal skills taxonomies, graph databases, and AI-powered search over internal communications.
- **Impact on AI interaction:** Enables internal AI systems to rapidly identify expertise, assemble teams, and support intelligent resource allocation.

Financial and operational data

- **Key AI-readability strategy:** Standardized reporting formats and real-time data access through APIs.
- **Example technologies and standards:** Financial reporting standards, IoT data protocols, and streaming data platforms.
- **Impact on AI interaction:** Supports AI-driven business intelligence, predictive analytics, and automated compliance monitoring.

Legal and compliance information

- **Key AI-readability strategy:** Machine-readable contracts, policies, and regulatory artifacts.
- **Example technologies and standards:** Legal ontologies, natural language processing for contract analysis, and structured policy repositories.
- **Impact on AI interaction:** Allows AI systems to assist with contract review, compliance checks, and risk assessment grounded in legal documentation.

The imperative for organizations to become AI-readable extends beyond individual interactions; it is fundamentally about their capacity to participate in increasingly complex and automated business ecosystems. Modern supply chains, automated financial markets, and emerging multi-agent service platforms rely on the ability of participating entities to seamlessly exchange information and invoke capabilities through digital means.

Organizations whose services and data are easily discoverable, semantically understandable, and technically integrable by AI systems will be able to engage more effectively and derive greater value from these interconnected networks. An API-first approach, coupled with machine-readable service descriptions and adherence to common semantic standards, is no longer just a technical choice but a strategic enabler of ecosystem participation.

Conversely, companies that remain opaque to AI or whose systems are difficult to integrate risk being siloed, unable to leverage the network effects, efficiencies, and innovation opportunities that these AI-driven collaborations offer. Thus, organizational AI-readability is not merely about being found by a single AI; it is about the enterprise's fundamental ability to operate, compete, and thrive in an increasingly automated and interconnected global landscape. This strategic view of the organization sets the stage for examining how it's very outputs; its products and services, must also be designed for this new AI-mediated reality.

MAKING PRODUCTS AI-READABLE

In parallel, the design philosophy for products and services is undergoing a similar, critical evolution. Traditionally, the primary, if not sole, focus was on the human end-user experience. However, in an AI-suffused environment, products and services are increasingly becoming "users" themselves, or acting as components within larger, AI-orchestrated systems and workflows. While AI-native products are conceived from the ground up with this duality in mind, embedding AI as a foundational element that defines their functionality and user experience, all products and services, regardless of their origin, must now adapt to become AI-consumable entities.

A key aspect of this adaptation is service composability and the establishment of clear invocation protocols. This involves designing products and services not as monolithic entities but as modular components, often referred to as Packaged Business Capabilities (PBCs).

These PBCs should be architected in such a way that they can be easily discovered, invoked, and orchestrated by AI agents or other automated systems. This necessitates clear and well-documented invocation protocols, typically implemented via APIs, which define the expected inputs, outputs, and predictable behavior of each service module.

Composable architectures, which emphasize the use of modular, interchangeable components often aligned with principles like Microservices, API-first, Cloud-native, and Headless (MACH), provide the flexibility and scalability required for products to function effectively within dynamic AI ecosystems. Real-world examples in retail, finance, and manufacturing illustrate how enterprises are leveraging composability to build agile and responsive systems.

Beyond technical interoperability, the semantic description of products and services is crucial. This is where outcome framing comes into play. Products and services need to clearly articulate the specific outcomes they deliver, not just list their features or technical specifications. This involves the semantic tagging of benefits (e.g., "reduces data processing time by 30%," "improves diagnostic accuracy for X condition"), target use cases, operational prerequisites, and any constraints or limitations. AI agents can then use this outcome-oriented information to intelligently select the most appropriate product or service to achieve a given goal or solve a particular problem, rather than relying on brittle keyword matching or complex inference.

As AI agents increasingly collaborate in multi-agent orchestration networks to perform complex tasks, such as those seen in advanced logistics or automated scientific discovery, the discoverability of products and services within these networks becomes paramount. For a product to be selected and utilized by an autonomous agent in such a system, its capabilities, interfaces, and operational parameters must be clearly exposed and understandable. This requires rich metadata, potential registration in relevant service directories or marketplaces (which themselves may be AI-curated), and adherence to interoperability standards that allow diverse agents to understand, compare, and integrate offerings from various providers.

For products that are intrinsically data-centric, or those that heavily rely on data for their functionality, AI readability is of utmost importance. This means providing data in clean, well-structured formats, accompanied by clear metadata that describes its content, origin (lineage), quality metrics, and any applicable usage rights or licenses. As we have discussed previously, the principles outlined in specifications like the Open Data Product Specification (ODPS) offer a valuable framework here, emphasizing machine-readable metadata, data quality indicators, and clear licensing information to facilitate AI consumption and integration.

Furthermore, the ability of a product to provide feedback on its own performance, or to adapt its behavior based on AI-driven insights or changing environmental conditions, can itself be considered an AI-readable characteristic. This "learnability" or "adaptability," if exposed through monitoring APIs, status dashboards, or other machine-interpretable feedback loops, makes the product a more valuable and dynamic participant in an AI ecosystem. This concept aligns closely with the notion that feedback is becoming integral to the product itself, as explored in Chapter 11.

The design of individual products to be AI-readable; meaning they are modular, API-accessible, and semantically well-described, is the critical enabler of interoperability at scale. When products possess these characteristics, they transcend their standalone utility and become versatile building blocks. Higher-level AI systems, sophisticated multi-agent orchestrators, or even human developers utilizing AI-assisted design tools can then discover these AI-readable components, understand their precise capabilities, and combine them in novel and powerful ways to create new, complex solutions and value streams.

A product that remains opaque or difficult for an AI to integrate is effectively siloed, its potential limited to direct human interaction. Conversely, a product designed with AI consumption in mind can participate in a much broader ecosystem of automated innovation. This implies that the future of innovation may lie less in the development of monolithic "killer applications" and more in the intelligent and dynamic composition of numerous AI-readable micro-services and specialized product capabilities. Therefore, product strategy must now expand its focus beyond the traditional end-user experience to also encompass the "AI-user" experience: how easily, reliably, and effectively can an AI system integrate and leverage this product to achieve a larger goal? If products, people, and organizations are all striving to achieve this state of AI readability, it then becomes essential to understand how AI systems process this newly structured information and arrive at their judgments and recommendations.

SIGNALS, SCORES, AND RECOMMENDATION ENGINES

Artificial intelligence systems do not merely "read" or ingest data passively; they actively evaluate it to make a wide array of decisions. These decisions can range from ranking search results, recommending products or content, and shortlisting job candidates to determining creditworthiness or identifying potential partners. This evaluation process is driven by a complex interplay of various signals extracted from the data and sophisticated scoring mechanisms designed to quantify relevance, trustworthiness, and potential value.

Understanding these mechanisms is crucial for anyone aiming to be favorably "read" by AI.

A key aspect of AI evaluation involves what can be termed trust layers (as we discussed in Chapter 10). These are multifaceted approaches AI systems use to assess the reliability and credibility of the information and entities they encounter. One significant component is behavior-based scoring, where AI models analyze patterns of user behavior-such as click-through rates, time spent on page, purchase history, or frequency of interaction; to assign scores. These scores often indicate the likelihood of a user to convert, churn, or perform a specific desired action.

For example, in e-commerce, AI might track actions like adding items to a cart or viewing multiple product details as strong indicators of purchase intent, assigning higher scores to users exhibiting these behaviors. Predictive scoring models leverage machine learning to analyze historical data and identify these behavioral patterns that signal conversion likelihood.

Another layer involves rating ecosystems and, increasingly, AI-generated reviews or summaries. AI can aggregate and analyze vast quantities of human-generated ratings and reviews to form a composite picture of a product's or service's perceived quality. Furthermore, AI itself might generate summaries or even "reviews" based on its analysis of extensive data, which can then influence both other AI systems and human decision-makers. A particularly strong signal for AI-driven trust comes from verifiable credentials and attestations. As discussed earlier, information backed by digitally signed, verifiable credentials; be it a skill certification, an educational degree, or a compliance certificate, can be assigned higher trust scores by AI systems due to their inherent authenticity and the reduced need for further validation.

Beyond these trust layers, AI ranking and recommendation engines rely on a variety of key signals to determine the order and prominence of the entities they present. Relevance is paramount: how closely does an entity (be it a person, product, piece of content, or service) match the explicit or inferred intent of a given query or the identified need of a user or system? Advanced semantic understanding, going beyond mere keyword matching, is crucial for this. Recency often plays a role, with more timely information or recent activity sometimes being weighted more heavily, depending on the context. Outcome history and past performance are powerful predictors; for products, this could translate to reliability metrics or customer satisfaction scores, while for individuals, it might involve documented project success rates or validated applications of specific skills.

AI evaluation systems are designed to track and incorporate such performance metrics into their models. Authority or credibility within a specific domain is another vital signal, which can be derived from factors like citation counts, endorsements from other trusted entities, or a consistent track record of producing high-quality, recognized work. This aligns with concepts like Google's E-E-A-T (Experience, Expertise, Authoritativeness, Trustworthiness) guidelines, which emphasize the importance of these factors for content quality.

Finally, popularity or usage metrics; how frequently an entity is chosen, used, or cited, can also serve as a signal, although AI designers are increasingly aware of the need to balance this against other factors to avoid simple popularity bias and ensure a diversity of relevant results.

AI recommendation engines, which are central to many digital experiences, employ various techniques to improve the quality of their results. Collaborative filtering operates on the principle that users who liked similar items in the past will have similar preferences in the future (e.g., "users who bought X also bought Y"). Content-based filtering recommends items based on their similarity to items a user has previously shown interest in, relying on detailed attributes and descriptions of the items themselves. Modern recommendation systems increasingly use hybrid approaches, often incorporating deep learning and knowledge graphs to understand more nuanced relationships, user contexts, and item characteristics. The AI-readability of the items being recommended; meaning they have clear, structured features and rich semantic descriptions, is critical for the effectiveness of content-based and knowledge graph-enhanced recommendation algorithms.

The implications of these AI evaluation mechanisms are profound across various domains. In hiring, candidates whose AI-readable profiles clearly and consistently signal relevant skills, verifiable achievements, and positive outcome histories are more likely to be surfaced and shortlisted by AI-powered recruitment tools, which are increasingly used by companies to manage large applicant pools efficiently. In procurement, suppliers whose offerings are demonstrably AI-readable; presenting structured data, clear service level agreements, transparent pricing models, and positive performance signals, will be favored by AI-driven procurement systems designed to optimize purchasing decisions. Even in professional collaboration, AI may suggest potential team members or partners based on sophisticated analyses of their digital footprints, published works, and network connections, all filtered through the lens of AI readability and evaluated against the signals deemed important by the system.

However, it is crucial to acknowledge the potential downsides. AI evaluation systems can often operate as "black boxes," making it difficult to understand precisely why a particular decision was made. If not carefully designed, trained, and audited, these algorithms can inadvertently perpetuate or even amplify existing societal biases present in their training data. This underscores the critical importance of striving for explainable AI (XAI) and implementing continuous monitoring and auditing processes to ensure fairness, transparency, and accountability in algorithmic decision-making.

The interplay between AI readability and these evaluation mechanisms leads to a significant dynamic: AI readability directly influences what can be termed "algorithmic favorability". Entities that are more AI-readable are inherently better positioned to present their positive signals; relevance, strong outcome histories, verifiable credentials, and positive behavioral indicators in a clear, consistent, and compelling manner.

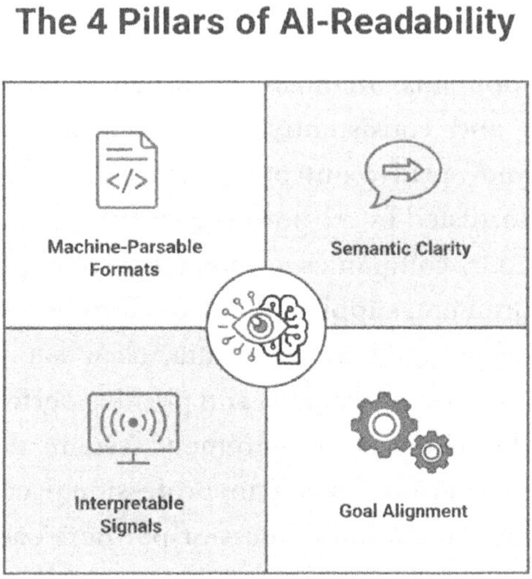

This clarity and strength of signal, in turn, make them more likely to be scored favorably by AI algorithms that are designed to identify and prioritize such attributes. This creates a feedback loop: being AI-readable helps an entity get seen and accurately understood by AI; the AI then uses these now-clearer signals to evaluate the entity. If these signals align with what the AI system is optimized to find or reward, the entity gains algorithmic favorability. This translates into increased visibility, more positive recommendations, and ultimately, a greater share of opportunities in AI-intermediated environments. Thus, actively curating one's AI-readable signals becomes a strategic way of influencing one's standing in the eyes of the algorithm. Given that AI systems are constantly evaluating based on these emitted signals, the next logical step is to explore how individuals and organizations can become proficient in understanding and strategically managing these signals for optimal outcomes.

BECOMING FLUENT IN YOUR OWN SIGNAL

The realization that AI systems are constantly "reading" and evaluating us necessitates a shift in perspective: from being passive emitters of data to becoming active managers of our digital signal. Every online interaction, every piece of content shared, and every way in which our data is structured contributes to an overarching AI-readable profile. Achieving fluency in one's own signal involves understanding this digital footprint and consciously shaping it to convey a clear, intentional, and favorable message to algorithmic audiences. This is not about manipulation, but about strategic clarity and authentic representation in a new medium.

A foundational practice in managing one's signal is the deliberate design of semantic metadata. For individuals, this means proactively tagging skills in professional profiles with precise and consistent terminology, meticulously documenting projects with relevant keywords and quantifiable outcome metrics, and using structured data formats like Schema.org on personal websites to explicitly define their expertise and affiliations.

For organizations, this translates to developing internal glossaries, taxonomies, or even comprehensive ontologies to ensure that products, services, and corporate capabilities are described with consistent and rich semantics across all platforms. This involves implementing robust metadata strategies for all digital assets, effectively telling AI systems explicitly what things are, what they mean, and how they relate, rather than leaving these crucial aspects to potentially flawed inference.

Beyond structuring the initial signal, it is vital to monitor and tune how these offerings and representations are being interpreted by AI systems. This involves utilizing analytics tools to understand how a personal website, an organization's content, or product listings are being discovered, indexed, and understood by search engines and other AI-driven platforms. Paying attention to how AI platforms, such as e-commerce marketplaces or professional networking sites, categorize and display one's information can provide valuable feedback. Based on this feedback, individuals and organizations must be prepared to adjust their descriptions, metadata, or data structures to improve clarity, relevance, and ultimately, their algorithmic reception. Continuous improvement based on AI feedback is a key tenet of maintaining AI readability.

The concept of "data exhaust"; the incidental data generated as a byproduct of digital activities, also requires careful consideration. Every comment on a forum, contribution to an open-source project, or interaction within a digital workspace contributes to the AI-readable signal. Being mindful of these traces and ensuring they contribute to a coherent and positive digital footprint is part of active signal management. For organizations, this extends to customer interaction logs, support ticket data, and public communications, all of which can be ingested and interpreted by AI systems, influencing perceptions of brand responsiveness, expertise, and customer sentiment.

This proactive management culminates in building the skill of "semantic self-governance". This is an internal discipline, applicable to both individuals and organizations, focused on consistently maintaining AI readability. It includes conducting regular reviews of how information is structured and presented online, staying informed about evolving AI technologies, relevant data standards, and best practices for semantic enrichment, and ensuring consistency in representation across all digital touchpoints. It is about taking deliberate ownership of one's AI persona and how it is perceived by algorithmic systems.

However, the pursuit of AI readability must be grounded in ethical signaling, emphasizing transparency and authenticity. The goal is not to engage in deceptive manipulation of algorithms but to present genuine value in a machine-understandable way. Authenticity, transparency (for example, by clearly citing sources or indicating if content is AI-generated), and adherence to ethical data practices are crucial for building long-term trust, both with the AI systems themselves and, more importantly, with the humans who ultimately rely on their outputs. AI systems are also becoming increasingly sophisticated at detecting inauthentic or manipulative signaling tactics, which can lead to penalties rather than advantages.

Finally, it is essential to recognize that achieving and maintaining AI readability is not a one-time project but a continuous practice. AI technologies, algorithmic priorities, and data standards are in a constant state of evolution. Therefore, ensuring continued legibility and relevance in this dynamic technological landscape requires ongoing learning, adaptation, and refinement of one's signals.

The sum of an entity's AI-readable data points; its structured information, its metadata, its semantic cues, and its behavioral patterns, coalesces to create a "semantic identity" that AI systems perceive and act upon. Actively managing these diverse elements through what we have termed semantic self-governance allows an individual or an organization to consciously shape this algorithmically perceived identity. This practice is directly analogous to how individuals have long managed their personal brand for human audiences, and how organizations meticulously curate their corporate brand and reputation. In an AI-intermediated world, this traditional brand management must now be augmented by a new layer of "semantic brand management" tailored for AI audiences. If these AI-readable signals are not actively and strategically managed, AI systems will inevitably form their own interpretations based on whatever fragmented or poorly contextualized data they can find, and these interpretations may not be accurate, favorable, or reflective of true value.

Therefore, the deliberate cultivation of "AI-readability hygiene" and the strategic practice of "semantic branding" are becoming indispensable skills and core functions for success in the age of AI. AI readability, then, is not merely a technical adjustment but a fundamental component of designing for the AI era. It is the critical first step in ensuring that individuals, products, and organizations can effectively participate, compete, and thrive in a world increasingly navigated and shaped by artificial intelligence. This established foundation of legibility and comprehensibility to AI paves the way for considering more complex interactions and adaptive organizational structures, which will be the focus of subsequent discussions.

SUMMARY

In an era where artificial intelligence increasingly acts as the primary intermediary for information, opportunity, and services, "AI readability" has emerged as an essential prerequisite for discoverability, trust, and effective engagement. This chapter has underscored that being AI-readable transcends mere technical formatting; it involves a holistic and strategic approach to structuring how individuals, organizations, and their products or services present themselves to machine perception and processing. The ability of an AI to accurately find, interpret, and recommend you or your offerings is no longer a niche concern but a central determinant of success.

From the foundational need for machine-parsable data formats and rich semantic metadata to the cultivation of interpretable behavioral signals and the clear articulation of intent, the journey to AI readability requires conscious design choices and ongoing management. Individuals must evolve how they represent their skills and experiences, moving towards verifiable digital credentials and structured professional profiles that AI can readily understand and validate. Similarly, organizations and the products they offer must be architected for discoverability and seamless interoperability within AI-driven ecosystems, often by embracing API-first principles, composable designs, and adherence to open standards. Ultimately, navigating the complexities of an AI-intermediated world hinges on becoming fluent in one's own digital signal, ensuring that AI systems can accurately understand, and favorably evaluate, the unique value being offered.

This strategic imperative to be understood by AI is foundational. Once individuals, products, and the data describing them are rendered AI-readable, the stage is set for rethinking how organizations themselves must transform to leverage these capabilities and thrive in an intelligent environment. Having established the principles for making the core components of value legible to machines, we now turn our attention in Chapter 14, "Modular Organizations - From Fixed Teams to Dynamic Talent Clouds and Agent Ecosystems," to explore how enterprises can be architected for the agility, composability, and intelligent orchestration demanded by this new, AI-driven landscape.

14

MODULAR ORGANIZATIONS

The 21st century, particularly with the advent of advanced Artificial Intelligence (AI) and the dawn of the Age of Information Symmetry, presents a profound challenge to the foundational structures upon which modern organizations have been built. For over a century, enterprises have largely relied on models designed for a world of information scarcity, hierarchical control, and predictable operational cadences. These structures, often visualized as rigid organizational charts, served their purpose in an industrial era where scale and efficiency in stable environments were paramount.

However, the ground beneath these legacy models is rapidly shifting. AI is not merely automating tasks; it is democratizing access to knowledge, synthesizing expertise on demand, and enabling capabilities that were once the exclusive domain of highly trained specialists or large, well-resourced teams. In this new epoch, where information flows more freely and symmetrically, the traditional organizational paradigms are not just becoming outdated; they are evolving into significant liabilities, hindering the very agility and responsiveness required to navigate an increasingly volatile and complex global landscape.

The imperative for organizational reinvention is no longer a distant theoretical concern but an immediate strategic necessity. As this chapter will explore, organizations must fundamentally transform from static, role-based hierarchies into dynamic, modular, and composable systems. This transformation entails a radical rethinking of how work is structured, how talent is engaged, and how intelligence; both human and artificial, is orchestrated. Future organizations will function less like a fixed machine with predetermined parts and more like a living ecosystem, capable of assembling and reassembling its components in real-time to meet evolving challenges and seize fleeting opportunities. This means moving away from the notion of fixed teams and permanent job descriptions towards fluid configurations of human talent, drawn from dynamic "talent clouds," and sophisticated AI agents, all collaborating within adaptable "agent ecosystems."

This chapter will illuminate the path towards such modular organizations. It begins by questioning the enduring relevance of the static org chart, a symbol of an era where information was power precisely because it was unequally distributed. If knowledge is becoming universally accessible and advanced capabilities can be synthesized or sourced on demand, how must the very essence of "organization" change? The answer lies in embracing modularity not just as a technical architecture for software or manufacturing, but as a core philosophy for organizational design. It involves understanding composability as the strategic capability to assemble these modules; be they human skills, AI-driven services, or specialized teams, into coherent and effective operational units tailored to specific goals and contexts.

Readers will gain an understanding of why the traditional pillars of 20th-century organizational design are crumbling under the combined pressures of AI-driven disruption and the pervasive availability of information. They will be introduced to the core tenets of modularity and composability, envisioning organizations not as static repositories of internal expertise but as dynamic platforms for orchestrating diverse forms of intelligence.

This shift has profound implications for how we approach hiring, team formation, leadership, and the overall engagement of talent in an era of AI-augmented work. Building upon the preceding chapter's exploration of "Becoming AI-Readable" which detailed how organizations and their components must make themselves legible to AI systems, this chapter addresses the next critical question: Once AI can understand the capabilities available, how must the organization itself be structured to dynamically leverage this newfound legibility for optimal performance, unparalleled adaptability, and sustained competitive advantage in the Age of Information Symmetry? The journey from fixed structures to fluid ecosystems is not merely an operational upgrade; it is a fundamental reinvention of what it means to organize for success.

THE LEGACY OF THE RIGID ORG CHART

The organizational structures that dominated the 20th century, and persist in many enterprises today, were born from the crucible of the Industrial Revolution. As companies grew in size and complexity, the informal management styles of smaller agrarian or craft-based enterprises proved inadequate. The need to manage vast resources, coordinate intricate processes, and control large workforces spurred the development of formal, hierarchical systems.

One of the earliest and most illustrative examples of this shift was the organizational chart developed by Daniel McCallum for the Erie Railroad in the mid-1850s. Faced with the challenge of managing over 5,000 employees across a sprawling network, McCallum, an architect and later a Civil War general, devised a tree-like chart that defined functional departments and clear lines of authority, responsibility, and communication. His six principles; including a proper division of responsibilities, sufficient power to carry them out, and means of knowing whether responsibilities were faithfully executed, were foundational to the emerging concept of the modern corporation. These structures were meticulously designed to bring order to chaos, ensuring that information flowed, albeit vertically, and that control could be maintained over increasingly complex operations.

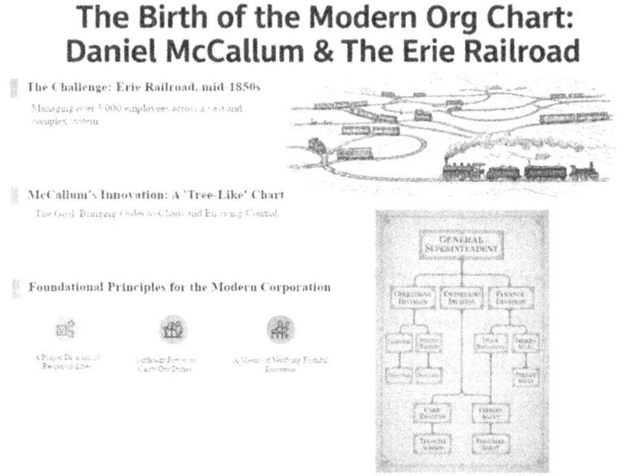

This nascent drive for structure and efficiency was soon codified and amplified by influential management theorists. Frederick Winslow Taylor, with his "Scientific Management," sought to achieve "maximum prosperity" for both employers and employees by applying scientific methods to work. His approach involved breaking down tasks into their simplest components, determining the "one best way" to perform each, scientifically selecting and training workers, and closely monitoring performance to ensure adherence to these optimized methods. Taylor's experiments, such as those at Bethlehem Steel which significantly reduced the cost of loading pig iron, demonstrated the potential for dramatic productivity gains.

However, this pursuit of efficiency often came at the cost of worker autonomy and creativity, treating individuals more as cogs in a machine than as thinking contributors. The underlying assumption was that managers would do the thinking, and workers would execute, a clear reflection of information asymmetry where specialized knowledge for process optimization resided at the top.

Contemporaneously, Max Weber, a German sociologist, articulated the characteristics of bureaucracy as a rational and efficient organizational form. Weber's model emphasized rules-based management, a clear hierarchy where higher offices supervise lower ones, functional specialization where work is divided based on skills, impersonal treatment of employees and clients to ensure fairness, and selection and promotion based on technical qualifications rather than personal connections. This system promised predictability, consistency, and accountability, ensuring every employee knew their place and the expectations of their role. While bringing a degree of order and efficiency, the Weberian bureaucracy also laid the groundwork for the rigidity and impersonality that would later be seen as significant drawbacks.

Henri Fayol, often dubbed the "father of modern management," further cemented these hierarchical principles through his "Administrative Theory". Fayol outlined 14 Principles of Management, including division of work, authority and responsibility, discipline, unity of command (each employee receiving orders from only one superior), unity of direction (one head and one plan for a group of activities with the same objective), and the scalar chain (the line of authority from top to bottom). His five Functions of Management; planning, organizing, commanding, coordinating, and controlling, provided a comprehensive framework for managerial action within a structured, hierarchical organization. These theories collectively reinforced the idea of a pyramid-shaped organization where power, information, and decision-making authority were concentrated at the top and cascaded downwards.

These historical models, while revolutionary for their time and instrumental in enabling the growth of large industrial enterprises, now exhibit profound dysfunctions in the context of the AI-driven

Age of Information Symmetry. Their very design, optimized for stability and control in an environment of scarce and asymmetrically held information, makes them ill-suited for the demands of a world characterized by rapid change and ubiquitous knowledge.

The most prominent dysfunction is excessive rigidity and inflexibility. Hierarchical structures, by their nature, are resistant to change. The multiple layers of management and formal approval processes that once ensured consistency now impede the rapid implementation of new processes and innovations. This structural inertia is a critical flaw when market conditions, technological capabilities, and customer expectations are in constant flux.

This rigidity directly contributes to stifled creativity and innovation. In traditional hierarchies, creative input and innovative thinking are often expected, or permitted, only at the higher echelons. Lower-level employees, who may possess valuable insights from their direct engagement with work or customers, often feel their ideas are not considered or valued. Taylor's "Scientific Management" prescribes a single "correct way" to perform tasks, which actively discourages experimentation and bottom-up innovation, a stark contrast to the distributed problem-solving and continuous improvement ethos required today.

Slow decision-making is another critical failing of organizational principles based on these methodologies from over a century ago. The vertical communication channels and the necessity for approvals to pass through multiple hierarchical layers inevitably lead to delays. Such delays were, to a certain extent, inherent to the prevailing economic pace of the era. But today, in an environment where speed can be a decisive competitive advantage, such organizational latency is a significant handicap.

Furthermore, these structures inherently foster operational friction and departmental silos. Work is typically organized into functional departments (e.g., marketing, sales, operations, R&D), and while this promotes specialization, it often leads to poor horizontal communication and collaboration. Information tends to flow up and down within these silos, rather than across them, hindering the synergy required to address complex, cross-functional challenges. This can result in "discorded activity management," multiple conflicting "sources of truth," an increased risk of errors, and a general impediment to the efficient flow of information and work.

Finally, the bureaucratic drag associated with Weberian principles, such as an over-reliance on rules and procedures, can lead to goal displacement; where adherence to rules becomes more important than achieving objectives, and an impersonal environment that can demotivate employees. The sheer volume of paperwork and procedural hurdles in highly bureaucratic organizations can consume valuable time and resources, further slowing responsiveness.

The common thread running through these dysfunctions is their deep connection to the historical context of information asymmetry. Managers acted as gatekeepers and interpreters of information; expertise was a scarce commodity concentrated in specific roles or departments. The entire edifice of hierarchical control and specialized functions was built on the premise that not everyone could or should know everything. In the Age of Information Symmetry, where AI can provide vast amounts of information and even synthesized expertise to anyone, anywhere, the foundational logic of these legacy structures collapses.

The control mechanisms designed for a stable world of scarce knowledge now actively prevent organizations from adapting to a dynamic world of abundant intelligence. This mismatch creates a compelling case for a fundamental rethinking of organizational design, moving away from rigid hierarchies towards more fluid and adaptive models. The very nature of "knowing" has changed, and so too must the nature of "organizing."

This historical reliance on control and predictability, while once a source of strength, reveals a paradox in the modern era. The structures meticulously crafted to maximize efficiency in stable environments now act as the primary inhibitors of the speed and adaptability crucial for navigating today's volatile, uncertain, complex, and ambiguous (VUCA) world; a world further accelerated by AI's relentless advancements. Consequently, the definition of effective management is undergoing a seismic shift, moving away from an emphasis on maintaining tight control towards fostering an environment of rapid responsiveness.

Moreover, these traditional organizational designs are predicated on an embedded assumption of scarcity; the belief that critical knowledge, decision-making prowess, and specialized skills are rare resources that must be carefully managed and allocated by a select few at the apex of the hierarchy. Taylor's clear demarcation between "managers who think" and "workers who do" is a stark illustration of this mindset. As AI begins to make expertise and information not just abundant but instantly accessible, the power structures built upon this scarcity are losing their foundational legitimacy and practical utility. This erosion necessitates a profound re-evaluation of roles, authority, and the very distribution of intelligence within organizations.

Finally, decades of operating within these rigid frameworks have cultivated a deep-seated cultural inertia, where individual and collective identity is inextricably linked to fixed job titles, departmental affiliations, and one's position within the established hierarchy.

This cultural entrenchment naturally breeds resistance to the more fluid roles, dynamic teaming, and transient project assignments that characterize the modular organizations of the future. The transition to these new models is, therefore, not merely a structural or technological challenge, but a significant cultural one, demanding a new psychological contract with the workforce; one that values contribution and adaptability over static roles and tenure, a theme that will be explored further in subsequent discussions on the future of work and meaning.

To crystallize these limitations, the list below provides a comparative overview:

Strict vertical hierarchy

- **Primary theorist or origin:** Max Weber and Henri Fayol.
- **Intended benefit in the industrial era:** Clear authority, control, and accountability.
- **Dysfunction in the AI era:** Slow communication and delayed adaptation in environments that demand rapid, distributed decision-making.

Functional silos

- **Primary theorist or origin:** Fayol and traditional functional management models.
- **Intended benefit in the industrial era:** Specialization and operational efficiency.
- **Dysfunction in the AI era:** Poor cross-functional synergy and reduced innovation in problems that span domains.

Centralized decision-making

- **Primary theorist or origin:** Frederick Taylor and Weberian bureaucracy.
- **Intended benefit in the industrial era:** Consistency, standardization, and risk reduction.
- **Dysfunction in the AI era:** Decision bottlenecks and reduced agility as information and insight now exist throughout the organization.

Fixed role definitions

- **Primary theorist or origin:** Traditional industrial job design.

- **Intended benefit in the industrial era:** Role clarity and clear accountability.
- **Dysfunction in the AI era:** Skill rigidity and resistance to change in a world requiring continuous recomposition of capabilities.

Rule-based management

- **Primary theorist or origin:** Weberian administrative theory.
- **Intended benefit in the industrial era:** Fairness, predictability, and uniform treatment.
- **Dysfunction in the AI era:** Bureaucratic drag and suppressed initiative when adaptive judgment is required.

Information control by management

- **Primary theorist or origin:** Pre–information age managerial practice.
- **Intended benefit in the industrial era:** Order and efficiency under conditions of information scarcity.
- **Dysfunction in the AI era:** Barriers to distributed intelligence and empowerment, increasingly undermined by information symmetry and AI access.

Emphasis on internal expertise

- **Primary theorist or origin:** Traditional corporate knowledge-retention models.
- **Intended benefit in the industrial era:** Competitive advantage through retained institutional knowledge.
- **Dysfunction in the AI era:** Inflexibility and slow integration of external innovation or AI-generated insight, while undervaluing dynamic talent clouds and agent ecosystems.

This list underscores the evolutionary mismatch: models that once provided strength and stability are now significant sources of weakness and inertia, compelling a transition towards more adaptive organizational forms.

THE CASE FOR MODULARITY AND COMPOSABILITY

As the limitations of traditional, rigid organizational structures become increasingly apparent in the face of rapid technological advancement and the shift towards information symmetry, new paradigms for organizational design are emerging. Central to this evolution are the concepts of modularity and composability. These are not merely technical terms borrowed from software engineering or manufacturing; they represent fundamental philosophies for structuring and operating businesses in a way that prioritizes agility, adaptability, and resilience.

Modularity, in an organizational context, refers to the practice of structuring an enterprise into distinct, semi-autonomous units or "modules". These modules, which can represent teams, departments, processes, capabilities, or even AI-driven services, are designed to operate with a degree of independence while remaining interconnected and coordinated to achieve the organization's overarching strategic goals.

A key characteristic of modularity is interchangeability: modules can be added, removed, updated, or reconfigured with relative ease, without necessitating a complete overhaul of the entire system. This "plug-and-play" capability is crucial for adapting to changing market demands or technological opportunities. Analogies can be drawn from modular construction, where prefabricated units are efficiently assembled on-site, or from modern software architecture, where applications are built from independent, reusable components that can be developed, deployed, and scaled separately.

For instance, the Dutch construction firm Heijmans' use of air caster technology to easily move and position building modules exemplifies the physical flexibility that modular design can offer. It is important to recognize that modularity is not an absolute state but a matter of degrees, characterized by dense connections and interactions within modules and sparser, well-defined connections between modules.

Composability, on the other hand, is the design philosophy and strategic capability that enables the effective assembly and orchestration of these diverse modules into a coherent, functional, and value-creating whole. If modularity is about the "pieces," composability is about the "art and science of putting them together dynamically".

Gartner defines composable modularity as an approach that allows businesses to reconfigure components without overhauling entire systems, thereby enabling a nimble response to technological and market shifts. It involves moving away from monolithic platforms and rigid structures towards an architecture where capabilities can be assembled and reassembled as needed.

Several key principles underpin a composable organizational approach:

Modularity (or "Teams, Not Silos"): This foundational principle involves breaking down large, complex organizational structures into smaller, discrete, and independently functioning teams or units. Each unit is self-contained and focused on a specific objective or set of tasks.

Autonomy and Coordination: Modular teams possess a degree of operational autonomy, empowering them to make decisions and execute tasks efficiently. However, their efforts are coordinated and guided by overarching strategic objectives, ensuring alignment with the broader organizational vision. This implies a balance between decentralized execution and centralized (or federated) strategic direction.

Transparency and Accessibility: For modular teams to be effectively composed and recombined, their capabilities, ongoing projects, and available resources must be easily discoverable and understood across the organization. This connects directly to the concept of "AI-Readability," where organizational components are structured to be understood and utilized by AI systems for better resource allocation and collaboration.

Interoperability by Design: Teams and units are designed from the outset to work together seamlessly. Standardized processes, clear communication channels, and shared understanding of interfaces (how teams hand off work or information) are crucial for enabling smooth collaboration and knowledge exchange between different modules, whether they are internal departments, project teams, or external partners. This allows different organizational capabilities to connect and interact effectively.

Clear Accountability: Each module or team is responsible for a specific, well-defined function or set of outcomes. This reduces inter-dependencies between modules, making it easier to adapt, replace, or improve one component without negatively impacting others, leading to greater organizational agility.

To make these concepts more concrete, consider the architectural elements of a "composable enterprise," which can be analogized to organizational structures. Just as a building is constructed from various modular components that can be reconfigured or replaced, a composable enterprise is built upon flexible, independent business capabilities that can be assembled, disassembled, and reassembled to adapt to changing market conditions and customer demands.

This approach moves away from rigid, monolithic systems and embraces a more agile, adaptable framework, allowing organizations to innovate more rapidly and respond effectively to disruption. This modularity extends beyond just technology, encompassing people, processes, and even partnerships, fostering a dynamic ecosystem where resources can be dynamically allocated and reconfigured.

One of the most tangible ways this plays out in practice on the software side is through the adoption of Packaged Business Capabilities (PBCs). These are software components that encapsulate defined business capability, including its data, services, APIs, and event channels. In an organization, a PBC could be adopted for the organization in a similar manner to a software PBC.

Much like software PBCs can be broken down into microservices; smaller, autonomous services performing highly specific tasks, organizational PBCs could be similarly composed. For example, a "Legal Review Team," "Customer Onboarding Group," or "AI-Powered Market Analysis Squad" could function as organizational PBCs. Each would have clearly defined inputs, outputs, and performance standards, mimicking the concept of "Legal Review Module," "Customer Onboarding Module," or "AI-Powered Market Analysis Module.".

The adoption of modular and composable organizational designs offers a compelling array of benefits:

Enhanced Agility and Flexibility: This is arguably the most significant advantage. Organizations can respond much more quickly to market shifts, evolving customer demands, or new technological breakthroughs by rapidly reconfiguring existing modules or integrating new ones. Studies suggest that companies with modular architecture can scale their operations up to 50% faster than those with monolithic structures.

Improved Scalability: Modular design makes it easier to scale operations both up and down. New modules can be added to support growth, or underutilized modules can be reduced or eliminated during downturns without the massive disruption associated with restructuring a monolithic organization. This is particularly valuable for businesses experiencing fluctuating demand or rapid expansion into new markets.

Fosters Innovation: Composability encourages experimentation and innovation. New ideas, services, or processes can be developed and tested within individual modules or as new, pilot modules with limited risk to the broader organization. This creates an environment where "inventiveness and creativity" can flourish.

Increased Efficiency and Cost Reduction: By breaking down operations into modules, organizations can better identify their core competencies and potentially outsource non-core functions to specialized external providers, who themselves operate as modules in the broader ecosystem. Furthermore, maintenance and upgrade costs can be significantly reduced, as individual modules can be updated or replaced independently without requiring a complete system overhaul. Some reports indicate potential maintenance cost reductions of up to 30%.

Greater Resilience: A modular structure enhances organizational resilience. If one module encounters problems or needs significant modification, the impact is often contained, preventing a cascade failure that could cripple a more tightly coupled, monolithic system. This allows the organization to adapt to unforeseen disruptions more effectively.

Heightened Customer Responsiveness: The ability to quickly modify or introduce new service modules allows organizations to address customer feedback and evolving needs more swiftly. This can lead to significant improvements in customer satisfaction, with some surveys indicating a 15–20% uplift for companies with modular systems.

Reduced Vendor and Internal Lock-in: In a technological context, composability helps avoid dependency on single, large vendors. Organizationally, it reduces dependence on fixed internal structures or specific, entrenched talent pools, as capabilities can be sourced from various internal or external modules.

The journey towards modularity and composability is not merely about rearranging organizational boxes; it is about instilling a new way of thinking and operating. While agile methodologies can introduce flexibility within teams or projects, true organizational agility; the capacity to rapidly reconfigure strategy, structure, processes, people, and technology across the entire enterprise, is fundamentally constrained by monolithic designs. Modularity provides the necessary "structural looseness," allowing these core elements to be reconfigured with speed and precision. Composability then acts as the "intelligent adhesive," ensuring these reconfigured parts work together harmoniously towards strategic ends. Therefore, organizations aspiring to achieve deep, systemic agility must look beyond localized agile practices and embrace modular and composable design as a foundational imperative. It is the difference between having a few flexible limbs and possessing an entirely adaptable body.

This shift also redefines the core focus of organizational management. Traditional models emphasize the accumulation and control of internal resources-fixed teams, proprietary knowledge, established processes. In contrast, modular and composable organizations pivot towards identifying, accessing, integrating, and orchestrating the best available capabilities, regardless of whether they are internal or external, human-driven or AI-powered. The effectiveness of this orchestration hinges on clearly defined interfaces (analogous to APIs) that govern how different capability modules interact. This evolution demands a new breed of leadership, skilled not just in resource management but in ecosystem orchestration, capable of designing these interfaces and weaving together diverse modules into a value-creating tapestry. This also paves the way for novel business models, where organizations might specialize in providing highly efficient "capability modules" to other enterprises.

However, the path to a truly composable organization is not without its challenges. While modularity offers immense flexibility, simply fragmenting an organization into numerous small pieces can lead to chaos if not guided by a strong architectural vision. Composability demands a high degree of strategic intent and sophisticated architectural thinking. Robust "design rules" or architectural principles are essential to govern how modules connect, interact, and contribute to the overall system, preventing what might be termed "fragmentation failure". The "interchangeability" and "interoperability by design" that are hallmarks of composable systems do not emerge spontaneously; they are the result of deliberate, forward-thinking design.

This implies that organizations must invest in developing architectural competencies that extend beyond IT to encompass organizational design itself. The "rigidity at the architectural (inter-modular) level," as noted in studies of modular systems, is paradoxically what enables greater flexibility and innovation within the individual modules, ensuring that autonomy operates within a framework of coherence.

TALENT CLOUDS AND SKILL GRAPHS: ORCHESTRATING HUMAN CAPABILITY

The transition towards modular and composable organizational structures necessitates a parallel revolution in how human talent is perceived, managed, and deployed. Traditional talent management practices, deeply intertwined with rigid hierarchical structures and fixed job descriptions, are ill-equipped for the dynamism required in the Age of Information Symmetry.

In these legacy systems, talent often becomes hoarded within departmental silos, internal mobility is sluggish and bureaucratic, and there's frequently a poor understanding of the true breadth and depth of skills available across the enterprise. This opacity and inflexibility stand in stark contrast to the urgent need for rapid, precise deployment of skills to emergent opportunities and evolving project requirements. The answer to this challenge lies in the development of "talent clouds" and comprehensive "skill graphs," orchestrated increasingly by AI, to create a fluid and responsive human capital ecosystem.

At the heart of this transformation are Internal Talent Marketplaces (ITMs). These platforms act as dynamic connectors, providing employees with unprecedented visibility into a wide array of internal opportunities; ranging from full-time roles and temporary project assignments (gigs) to mentorship programs and stretch assignments designed for skill development. By empowering employees to actively seek out and apply for opportunities that align with their skills, interests, and career aspirations, ITMs shift the locus of control from top-down assignment to individual agency.

The success of ITMs is well-documented across various industries, delivering tangible benefits in terms of cost savings, efficiency, employee engagement, and retention.

These marketplaces typically provide transparency into career pathways, help employees identify their strengths and skill gaps, and offer a variety of growth opportunities such as gig assignments and mentorships. The benefits are manifold: employees are empowered to own their development, organizations gain an agile pipeline of "ready-now" talent, cross-skilling enhances workforce versatility, reliance on external contract labor is reduced, and diversity, equity, and inclusion efforts are furthered by providing universal access to opportunities.

Underpinning the effectiveness of ITMs and the broader concept of talent clouds are Skill Graphs. These are comprehensive, dynamic, and granular inventories of the skills, competencies, knowledge, and experiences possessed by all individuals within the organization's ecosystem; including full-time employees, contractors, and potentially even AI agents. Unlike traditional résumés or job titles, which offer a static and often superficial view of capabilities, skill graphs aim to capture a rich, multi-dimensional understanding of what people can actually do. These graphs are populated through a variety of methods: employee self-assessments, manager validations, peer feedback, data gleaned from completed projects, certifications, and increasingly, AI-driven inference from an individual's work products, communications, and learning activities.

The development of robust skill graphs is critical for enabling true skills-based hiring and development. This approach prioritizes the specific, demonstrable knowledge, abilities, and competencies a candidate possesses over traditional proxies like academic degrees or past job titles. This shift directly addresses two major flaws in conventional hiring: the risk of poor job fits when degrees are the primary filter, and the unnecessary narrowing of the candidate pool when rigid degree or specific title requirements are enforced.

More than half of U.S. state governments have already adopted policies encouraging skills-based hiring, often by eliminating degree requirements for many public sector jobs. The advantages include better job matches, leading to reduced turnover and increased job satisfaction, expanded career advancement opportunities for individuals from non-traditional backgrounds, and a more diverse and adaptable workforce. While challenges exist, such as the added time required to accurately evaluate skill levels and the current orientation of many application systems around degree-based criteria, the long-term benefits are compelling. ITMs play a crucial role here, as they allow employees to identify the skills needed for desired future roles and then find internal gigs, projects, or mentorships to acquire those specific competencies, as illustrated by the "Sally" case study where an employee uses the marketplace to build a development plan and gain necessary experience.

The orchestration of this complex interplay between talent supply (skill graphs) and demand (opportunities in the ITM) is increasingly powered by AI. AI algorithms can analyze vast datasets; skill graphs, project requirements, employee career aspirations, learning histories, and real-time availability, to suggest optimal talent allocations far more efficiently and effectively than manual methods. Public reporting suggests that Schneider Electric's OTM, for instance, is explicitly AI-powered. Beyond simple matching, AI can provide personalized recommendations for learning paths, suggest relevant development opportunities, and even predict future skill needs for the organization. This moves talent management from a reactive, administrative function to a proactive, strategic, and predictive capability.

The emergence of talent clouds and skill graphs represents a paradigm shift, transforming human capital into a more API-like resource for modular organizations. If an organization is designed as a composable system of capabilities, it requires a dynamic mechanism to access and deploy human skills with the same fluidity it might access software services. Talent clouds, fueled by ITMs and detailed skill graphs, effectively serve as this "human capital API". They allow the organization to "call upon" the right configuration of skills for a specific task, project, or strategic initiative at precisely the right time, irrespective of where those skills formally reside within the traditional structure. This fundamentally recasts the role of Human Resources from primarily administrative gatekeeping to the strategic orchestration of human capability. For individuals, it signifies a move towards becoming "skill providers," granting them greater agency and flexibility in charting their career trajectories within, and potentially even beyond, the confines of a single employer.

This new model thrives on a symbiotic relationship between skills transparency and organizational agility. The success of dynamic teaming and modular project execution is critically dependent on the organization's ability to swiftly identify, assemble, and deploy individuals possessing the requisite skills. ITMs and skill graphs furnish this essential transparency.

Without a clear, real-time understanding of the available skill landscape, the rapid and precise allocation of resources becomes impossible, thereby crippling agility. Public reporting on compelling results from companies like Seagate and Schneider Electric vividly illustrate how these internal marketplaces enabled rapid workforce shifts and significant operational gains. Consequently, investing in the technologies and processes necessary to build comprehensive skill graphs and foster internal talent mobility is no longer a discretionary spend but a core enabler of the agile, modular enterprise. This, in turn, creates a virtuous cycle: increased internal mobility leads to broader skill development among employees, which enriches the skill graph, further enhancing the organization's capacity for agile response.

Furthermore, the "democratization of development" inherent in these systems has a profound impact on employee engagement and retention. ITMs are widely described as empowering employees to genuinely own their professional growth by making a wide array of opportunities (gigs, projects, and mentorships) visible and accessible. This directly addresses one of the most common drivers of attrition: a perceived lack of career visibility and development opportunities, as highlighted by public reporting on L'Oréal's experience.

When individuals can see clear, attainable pathways for acquiring new skills and advancing their careers, their engagement deepens, and their loyalty to the organization strengthens. Organizations that successfully cultivate such internal marketplaces are thus more likely to become "talent magnets," improving retention rates and significantly reducing the substantial costs associated with external recruitment and employee churn. This also fosters a pervasive culture of continuous learning and individual adaptability, which perfectly mirrors the organization's overarching need for evolvability in a rapidly changing world.

To provide a consolidated view of the tangible successes achieved through ITMs, the following list presents key data based on public reporting about leading companies:

Seagate Technology

- **ITM name or type:** Internal Talent Marketplace (unnamed).
- **Key drivers for implementation:** Pandemic-driven need to rapidly reallocate talent toward growth priorities.

- **Publicly reported benefits and metrics:** Approximately $1.4 million saved in four months; 35,000 hours saved in four months; 58 percent increase in female participation and assignment; 88 percent employee registration within the first 45 days; average employee tenure of 6.6 years.
- **Implied impact on organizational modularity and agility:** Enabled fast, efficient redeployment of talent to shifting priorities and materially increased workforce flexibility.

Schneider Electric

- **ITM name or type:** Open Talent Marketplace (AI-powered).
- **Key drivers for implementation:** Increase productivity, reduce hiring costs, and better align skills to strategic needs, particularly during pandemic disruption.
- **Publicly reported benefits and metrics:** More than 72,000 registrations (77 percent of global workforce) by 2022; approximately $15 million saved through productivity gains and reduced hiring costs; 60 percent active use within two months of launch; average tenure of 5.9 years.
- **Implied impact on organizational modularity and agility:** Enabled AI-driven, large-scale skill-to-opportunity matching, increased internal pipeline agility, and reduced reliance on external hiring.

L'Oréal

- **ITM name or type:** Positions Open Portal (custom-built).
- **Key drivers for implementation:** Employee feedback indicating limited career visibility and a high share of regrettable attrition driven by unclear internal opportunities.
- **Reported benefits and metrics:** 75% of positions opened to internal candidates; internal fill rates increased by 40 percent; 37 percent of employees actively pursued lateral

opportunities; turnover declined from 14 percent in 2021 to 12.4 percent in 2022.

- **Implied impact on organizational modularity and agility:** Increased internal talent fluidity, normalized lateral movement, reduced career silos, and improved workforce adaptability by making pathways visible and accessible.

These examples demonstrate that reporting on ITMs suggests that they are not just theoretical constructs but practical tools delivering significant strategic and operational advantages, paving the way for more agile and talent-centric modular organizations.

AI AGENTS AS TEAM MEMBERS: THE RISE OF HYBRID INTELLIGENCE

The evolution of modular organizations is not solely a human-centric endeavor; it is increasingly intertwined with the rapid advancements in Artificial Intelligence, particularly the emergence of sophisticated AI agents. These agents are transitioning from mere tools that automate discrete tasks to increasingly autonomous participants capable of contributing to complex organizational workflows and even functioning as integral members of human teams.

This integration of AI agents as active collaborators, rather than passive instruments, is giving rise to the concept of "hybrid intelligence," where the distinct strengths of humans and AI are combined to achieve outcomes superior to what either could accomplish alone. Understanding how these AI agents are defined, embedded, orchestrated, and governed is crucial for designing the next generation of adaptive organizations.

In the organizational context, AI agents can be understood as software entities endowed with the capacity to perceive their environment (often through data feeds and sensor inputs), make decisions based on their programming and learned experiences, and take autonomous actions to achieve specific, predefined goals.

What distinguishes these agents from simpler forms of automation is their ability to learn, adapt their behavior over time, and operate with a greater degree of autonomy, especially in complex and dynamic environments. The concept of "agentic AI" specifically refers to AI systems that can interpret goals, make decisions, and execute multi-step tasks without explicit, step-by-step instructions for each action, often adapting to real-time data and even self-correcting. The primary business value delivered by AI agents lies in their ability to automate standardized business processes, handle repetitive or data-intensive tasks with speed and accuracy, and consequently free up human employees to focus on higher-value work that requires creativity, critical thinking, and interpersonal skills.

AI agents are no longer confined to standalone applications but are increasingly being embedded directly into the fabric of team structures and daily workflows. Numerous examples illustrate this trend across various business functions:

Meeting Management: Tools like Fellow act as AI meeting assistants, automatically recording, transcribing, summarizing discussions, and noting action items, effectively taking on the role of a diligent scribe and organizer.

Sales and Prospecting: HubSpot's Breeze includes a prospecting agent that can identify potential leads from customer data and automate initial outreach, augmenting the capabilities of sales teams.

Knowledge Management: Glean provides enterprise search functionalities and allows companies to build custom generative AI applications and agents to perform tasks based on internal documentation, essentially acting as a knowledgeable internal consultant.

Project Management: ClickUp Brain features an AI Project Manager that helps automate aspects of project workflows, such as generating updates, status reports, and even creating tasks, thereby assisting human project managers.

Customer Service: Intercom's Fin is an AI customer service agent designed to provide human-level support by drawing on the company's knowledge base and policies, handling frontline inquiries and allowing human agents to focus on more complex issues.

Marketing: Hightouch offers AI Decisioning, which uses machine learning to determine the optimal ways to engage individual customers, automating aspects of personalization and campaign refinement.

Software Development: Emerging tools like Devin (though still in early stages) aim to function as AI software developers, capable of handling coding tasks.

And Beyond: Other specialized agents include Paradox for hiring (candidate screening, interview scheduling), Intuit Assist for financial tasks (invoicing, expense matching), and Tableau Agent for AI-driven data analysis. These examples clearly show AI agents assuming specific roles and responsibilities that were previously handled by humans, or augmenting human capabilities within team settings.

This integration necessitates the development of robust Human-AI teaming frameworks to foster effective collaboration and create Hybrid Intelligence. Hybrid intelligence is the synergistic fusion where human intuition, creativity, critical thinking, and ethical judgment are combined with AI's computational power, pattern recognition capabilities, and data processing speed.

Core elements of successful human-AI collaboration include a careful and thoughtful allocation of tasks (assigning data-intensive and repetitive operations to AI, while humans handle complex problem-solving and nuanced judgment), the establishment of clear shared goals, seamless interaction mechanisms that allow for natural communication and feedback, and adaptability from both human and AI components. Best practices for building these hybrid teams involve establishing clear communication protocols, implementing continuous feedback loops for mutual learning and adjustment, fostering adaptive learning capabilities (where AI systems adjust to human working styles and humans develop an understanding of AI's capabilities and limitations), and defining clear roles and responsibilities to ensure AI complements, rather than merely duplicates or confuses, human efforts.

The ultimate aim is to create "smarter systems together," where the combined output is greater than the sum of its parts. Ongoing research in this area focuses on aspects like team formulation, coordination strategies, maintenance of collaborative relationships, training for hybrid environments, and the development of shared mental models, trust-building mechanisms, conflict resolution protocols, and skill adaptation processes.

Effective work orchestration is paramount in these human-AI teams. AI orchestration, in a broader sense, refers to the management and coordination of multiple AI models, tools, and workflows to ensure they operate together seamlessly. This concept naturally extends to the orchestration of AI agents within human-centric workflows. Key components of such orchestration include centralized management platforms (or federated coordination mechanisms), automation of routine workflow steps, and sophisticated decision coordination between AI outputs and human decision-makers.

Agentic AI significantly enhances this orchestration by introducing autonomous agents capable of interpreting overarching goals, making independent decisions within their purview, and executing complex, multi-step tasks dynamically. Specialized tools are emerging to facilitate this, such as Akka for building distributed, real-time AI systems, CrewAI for orchestrating teams of LLM-based agents, and LangGraph for visually managing complex AI workflows. This level of orchestration enables not only parallel workflows, where humans and AI agents work concurrently on different facets of a larger task, but also deeply synergistic workflows, where the output of an AI agent directly informs a human task, and human feedback, in turn, refines the AI's subsequent actions.

However, the increasing autonomy and integration of AI agents bring to the forefront the critical issue of accountability, versioning, and permissioning. As highlighted by real-world incidents, such as an airline chatbot providing incorrect fare terms that the company was forced to honor, the question of "who is accountable when an AI agent makes a mistake?" is a pressing concern. The very nature of agentic AI, with its capacity for dynamic learning and adaptation, makes its decision-making processes less predictable than traditional software, thereby complicating the attribution of responsibility.

To address this, organizations must establish comprehensive frameworks for AI accountability. These frameworks should include:

Clear Chains of Responsibility: Identifying specific individuals or roles accountable for the different stages of an AI agent's lifecycle, from development and deployment to ongoing operation and the outcomes of its decisions.

Detection and Correction Systems: Implementing robust systems for real-time monitoring of AI agent behavior, including audit trails of their decisions and actions, and mechanisms to detect and correct incomplete, incorrect, or harmful outputs. Human-in-the-loop oversight is often essential here.

Balancing Autonomy with Human Control: Defining clear processes and structured intervention frameworks that delineate when and how human oversight or override is required, particularly for high-stakes decisions.

Remediation Approaches: Developing pre-defined strategies for addressing AI-induced errors, including communication with affected parties, rectification of outcomes, and systematic improvements to the AI agent or its operating parameters.

AI-Specific Governance: Establishing new legal, ethical, and compliance frameworks that explicitly address the unique challenges posed by autonomous AI agents, potentially including roles like a Chief AI Officer or AI Ethics committees.

Beyond accountability, meticulous versioning of AI models and agents is crucial to track changes in their behavior, capabilities, and training data over time. This allows for rollback if new versions introduce problems and provides a history for auditing and debugging. Similarly, robust permissioning systems are necessary to control precisely what data AI agents can access and what actions they are authorized to take within the organizational environment, ensuring they operate within defined and secure boundaries.

The integration of AI agents as team members effectively positions them as a new class of highly specialized and often rapidly deployable "capability modules" within the broader composable organization. Their proficiency in performing specific tasks; be it data analysis, customer interaction, or process automation, makes them ideal components in a system designed for dynamic assembly. The orchestration platforms then serve as the critical integrators for these AI modules, enabling their seamless incorporation into diverse workflows. This development means that the architectural design of modular organizations must now explicitly account for the lifecycle management of AI agents; how they are discovered, integrated, monitored, updated, and eventually retired, just as any other resource or capability module. This, in turn, significantly accelerates the potential for rapid and sophisticated organizational recomposition in response to changing needs.

However, a significant hurdle to the deep integration of AI agents is the "trust threshold". While the potential benefits are immense, the willingness of humans to delegate meaningful work to AI agents and collaborate with them as genuine team members hinges critically on trust. The "black box" nature of some AI decision-making processes can be a major impediment to building this trust. Consequently, the frameworks established for AI accountability are not merely about assigning blame post-failure; they are fundamental to fostering the confidence required for effective human-AI partnership.

Organizations must, therefore, make substantial investments in explainable AI (XAI) methodologies, transparent auditing mechanisms, and clear human oversight protocols. Without establishing and maintaining a sufficient level of trust, the adoption of AI agents as true team members will likely remain superficial, confined primarily to low risk, easily monitored tasks, thereby limiting the transformative potential of hybrid intelligence.

The rise of human-AI teaming also fundamentally redefines the concept of "core competency" for many human roles and significantly augments human capabilities. As AI agents increasingly take over routine, data-intensive, and predictable tasks, the value proposition for human workers will increasingly shift towards areas where they possess unique strengths: complex problem-solving involving ambiguity, deep creativity and innovation, strategic foresight, nuanced emotional intelligence, and sophisticated ethical judgment.

In this evolving landscape, the AI agent is not a mere replacement but a powerful collaborator that amplifies human potential. The "core competency" for many professionals will thus expand to include proficiency in working alongside AI-framing problems effectively for AI analysis, critically interpreting AI-generated outputs, providing essential contextual understanding, and exercising ethical oversight. This shift carries profound implications for future skill development, training programs, job design, and performance evaluation metrics, all of which will need to adapt to reflect this new hybrid reality of work.

DYNAMIC TEAMING IN PRACTICE: ASSEMBLING EXPERTISE ON DEMAND

The theoretical advantages of modularity and composability truly come to life when organizations implement dynamic teaming; the ability to rapidly assemble, deploy, and disband teams based on specific project needs, emerging opportunities, or critical problems. This approach moves decisively away from static departmental structures and fixed reporting lines, instead drawing talent (both human and, increasingly, AI-driven) from across the entire organizational ecosystem, and sometimes from external networks, to create fit-for-purpose collaborative units.

Dynamic teaming prioritizes speed, flexibility, and the optimal allocation of expertise, ensuring that the right capabilities are brought to bear on the right challenges at the right time. Several innovative companies and established frameworks offer compelling examples of how this fluid approach to team formation can be put into practice.

One of the most widely discussed models for achieving agility at scale through dynamic teaming is the Spotify Model. While Spotify itself emphasizes that this is not a rigid framework to be copied verbatim but rather a reflection of its evolving culture, its core components offer valuable insights.

Squads: The foundational units: small, cross-functional, and highly autonomous teams, typically comprising 6 to 12 individuals. Each squad has a long-term mission focused on a specific feature area (e.g., search, playlist management) and is empowered to decide how it works, including choosing its preferred agile methodology (Scrum, Kanban, etc.). This autonomy is key to their speed and ownership.

Tribes: Multiple squads working in related areas form a tribe, usually consisting of 40 to 150 people. Tribes help ensure alignment and facilitate coordination among squads with overlapping interests. Each tribe has a tribe lead responsible for fostering collaboration and providing the necessary environment for squads to thrive.

Chapters: To maintain technical excellence and consistent standards across different squads, chapters are formed. A chapter is a group of specialists within the same competence area (e.g., backend developers, UX designers, quality assurance engineers) across all squads in a tribe. Chapters are typically led by a senior technologist or manager who also often has line management responsibility for chapter members. They focus on sharing best practices, mentoring, and developing functional expertise. This structure cleverly balances squad autonomy with the need for deep specialist alignment.

Guilds: Voluntary, self-organizing communities of interest that cut across tribes. Anyone passionate about a particular topic (e.g., a specific programming language, a new technology, a shared interest like photography) can join or initiate a guild. Guilds facilitate knowledge sharing, learning, and networking organically, driven by shared enthusiasm rather than formal mandate. The Spotify model is built on principles of autonomy and alignment, servant leadership (where leaders focus on enabling teams rather than commanding them), and a strong emphasis on a culture of trust, transparency, and continuous learning. From a modular perspective, squads function as adaptable "team modules" that can be formed, reconfigured, or refocused as strategic priorities shift. Chapters and guilds provide essential connective tissue, ensuring that knowledge and standards flow effectively across these autonomous units. Another radical example of dynamic, modular organization comes from Haier, the global home appliance manufacturer, with its RenDanHeYi (人单合一) philosophy and Microenterprise (ME) structure. The RenDanHeYi model is centered on three pillars: the absolute centrality of the user experience (aiming to co-create with lifelong users), the unleashing of entrepreneurial energies within every employee, and the sharing of value created among all participants.

Microenterprises (ME) are the core organizational modules: small, largely autonomous, and collaborative work units that are empowered to develop and own their business models. Crucially, their financial compensation is directly linked to the user value they create.

A key operational principle is achieving "**Zero Distance to the Customer/User.**" MEs are designed to be intimately and continuously connected with their users, enabling them to identify unmet needs, respond to market trends in real-time, and engage in co-creation. Examples include Haier's Smart Vaccine ME, which developed new product ideas based on direct observation of user challenges during vaccinations, and its "ThundeRobot" gaming computer ME, staffed by young gamers who inherently understand their user base.

This structure fosters **Ecosystem Entrepreneurship**, where MEs can even form "micro-communities" around specific customer demands or opportunities, effectively acting like startups within the larger Haier ecosystem. Leadership is highly distributed, with MEs expected to exhibit entrepreneurial drive and take initiative based on user insights.

MEs are also characterized by **Continuous Evolution**, constantly refining their business models and value propositions based on direct user feedback and market dynamics. In Haier's model, MEs are highly autonomous modules optimized for extreme customer-centricity and rapid entrepreneurial action. The system allows for the swift formation of new MEs to pursue emerging opportunities, making it exceptionally adaptive.

Beyond these company-specific models, other forms of dynamic teaming are widely practiced:

Tiger Teams and **War Rooms** represent classic examples of focused, temporary, high-impact teams. A tiger team is a small, cross-functional group of experts assembled to tackle a specific, often urgent or critical, problem. A distinguishing feature is that tiger teams are typically responsible not only for identifying solutions but also for carrying out their implementation. They are characterized by diverse skill sets, intense collaboration, dedicated resources (sometimes a physical "war room" to facilitate focused work), clearly defined goals, and a mandate for rapid action.

Project-Based Teams and **Innovation Pods** are common in many organizations. Ad-hoc teams are formed for the duration of a specific project and then disbanded upon its completion, with members moving on to new assignments. Innovation pods are a specialized version, often shielded from routine operational pressures, tasked with developing new products, services, or business models.

The concept of **AI-Led Taskforces** is also emerging. Here, a sophisticated AI agent might take on the role of coordinating a group of other specialized AI agents, or even human contributors, to achieve a well-defined, complex task, particularly those involving large-scale data processing, simulation, or optimization.

When teams are formed dynamically and are often temporary, understanding their developmental lifecycle becomes important. The Tuckman Team Model; which describes stages of Forming, Storming, Norming, Performing, and finally Adjourning, provides a useful lens. Even transient teams typically progress through these phases. Effective leadership and facilitation can help accelerate this process, particularly navigating the "storming" (conflict and role definition) and "norming" (cohesion and rule-setting) stages. Clear communication is identified as a critical component for successfully moving through each stage.

In all these dynamic teaming environments, the role of leadership undergoes a significant transformation. Leaders shift from being directive controllers to becoming orchestrators, facilitators, and enablers. Their primary functions include setting clear, compelling goals, ensuring teams have the necessary resources and autonomy, removing impediments, fostering a culture of collaboration and psychological safety, and ensuring that the efforts of these fluid teams remain aligned with the broader strategic objectives of the organization.

The practice of dynamic teaming is, in essence, modularity actualized at the human operational layer. The diverse models; Spotify's agile structures, Haier's entrepreneurial MEs, the focused intensity of Tiger Teams, all exemplify the assembly of "human capability modules" tailored for specific purposes and designed for agility and impactful outcomes. For organizations aspiring to true modularity, cultivating the capacity to form, reform, and dissolve teams dynamically is not an optional extra but a core operational necessity. This requires more than just structural adjustments; it demands profound cultural shifts towards embracing collaboration, shared purpose, and a collective comfort with the transient nature of many team structures.

A critical consideration in these dynamic systems is the inherent tension between granting teams or modules significant autonomy; which is vital for fostering speed, innovation, and a sense of ownership, and ensuring their diverse efforts remain aligned with overarching organizational goals.

Successful models like Spotify and Haier demonstrate sophisticated mechanisms to manage this balance. Spotify utilizes its tribe, chapter, and guild structures to provide alignment across autonomous squads, while Haier relies on the pervasive RenDanHeYi philosophy and ecosystem-level contracts to guide its microenterprises. The lesson is clear: empowering autonomous teams without robust alignment frameworks can easily devolve into chaos, duplicated efforts, and strategic drift. Effective dynamic teaming, therefore, necessitates not only empowerment but also clearly defined "rules of engagement" and strong "connective tissue"; such as Spotify's Chapters or Haier's platform-level services to ensure overall coherence and strategic focus.

Furthermore, the inherently temporary nature of many dynamic teams; from ad-hoc project groups and Tiger Teams to even Spotify Squads that may reconfigure based on evolving product needs, implies a significant shift in the individual employee's relationship with work and career. Individuals must cultivate a "tour of duty" mindset, becoming comfortable with moving between different assignments, teams, and projects, contributing their specific skills where they are most needed at a given time.

This contrasts sharply with the traditional expectation of a "lifelong job" within a single, stable department or role. This evolving dynamic aligns closely with the concept of a "modular career," which will be explored in Chapter 15, where individuals curate a portfolio of experiences and skills rather than following a linear path. For organizations, this means a greater responsibility to support employees in navigating these fluid career trajectories, and for individuals, it underscores the importance of continuous learning, adaptability, and the cultivation of a diverse and transferable skill set. The robust talent clouds and skill graphs become indispensable enablers for both the organization and the individual in this new world of work.

To provide a clearer comparative understanding of these dynamic teaming approaches, the following list offers an overview:

Spotify squads and tribes

- **Core unit:** Squad, a cross-functional and autonomous team.
- **Key principles:** Autonomy with alignment, servant leadership, and a culture of trust and transparency.
- **Mechanism for agility:** Squad-level autonomy enables rapid iteration, flexible choice of agile practices, and fast reconfiguration of focus as priorities change.
- **Leadership style:** Agile coaches at the squad level and tribe leads who act as facilitators and coordinators rather than command-and-control managers.

Haier microenterprises

- **Core unit:** Microenterprise, a small, autonomous, entrepreneurial unit.

- **Key principles:** User centricity (zero distance), entrepreneurial drive, and value sharing through the RenDanHeYi model.
- **Mechanism for agility:** Direct responsiveness to user needs, rapid formation of new microenterprises for emerging opportunities, and a tight coupling between value creation and reward.
- **Leadership style:** Microenterprise leaders who operate as entrepreneurs, with high autonomy within shared ecosystem rules.

Tiger teams

- **Core unit:** Ad hoc, specialized cross-functional team.
- **Key principles:** Intense focus on a specific problem, diversity of expertise, and rapid solution development and implementation.
- **Mechanism for agility:** Concentrated effort on a single critical issue, combined with empowerment to implement solutions directly without bureaucratic delay.
- **Leadership style:** A designated team leader, often more directive to maintain focus and speed, though facilitative leadership may be used depending on context.

This comparative view illustrates that while the specific structures may vary, the underlying themes of autonomy, clear purpose, cross-functional collaboration, and adaptive leadership are common threads in successful dynamic teaming implementations.

GOVERNANCE WITHOUT HIERARCHY: ENABLING AUTONOMY WITH ACCOUNTABILITY

The shift towards composable organizations, characterized by distributed decision-making, dynamic teaming, and the integration of autonomous AI agents, fundamentally challenges traditional notions of governance. Hierarchical control, with its reliance on top-down directives and multi-layered approvals, is antithetical to the speed and flexibility these new models aim to achieve.

The critical question then becomes: how do modular, often decentralized organizations maintain coherence, ensure compliance with internal policies and external regulations, uphold quality standards, and foster clarity of purpose without resorting to the stifling mechanisms of old? The answer lies in evolving governance from a system of rigid control to an enabling framework that empowers autonomy while ensuring robust accountability. This involves adopting new models like "guardrails, not gates," leveraging technologies such as smart contracts, implementing semantic permissions, and embracing outcome-based accountability.

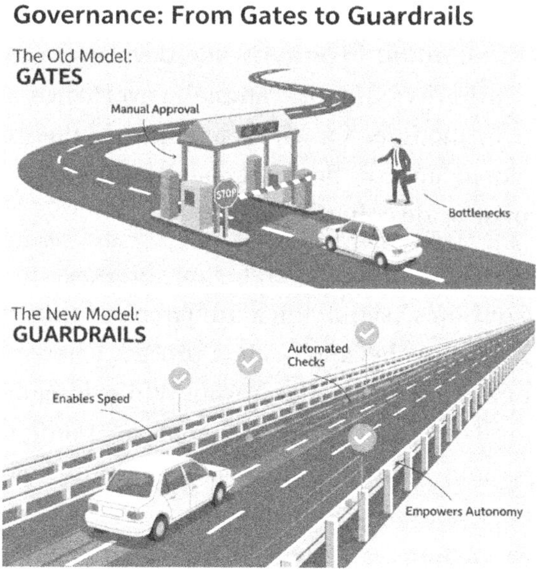

A primary challenge in governing modular systems is managing the potential for fragmentation and ensuring that autonomous components operate in alignment with overall strategic goals. Without effective coordination and oversight, the benefits of modularity can be undermined by inconsistencies, duplicated efforts, or even a loss of control over critical functions, especially those that might be outsourced or handled by AI agents.

Decentralized and federated governance models offer alternatives to purely centralized control. In a decentralized model, decision-making authority is distributed among a network of participants or modules, who collectively contribute to governance processes. This can enhance flexibility, speed up decision-making by empowering those closest to the information, and leverage localized expertise.

However, purely decentralized systems can risk inconsistencies and a lack of overarching control if not carefully designed. A more common and often more practical approach for larger organizations is a federated governance model. This model seeks a balance, retaining centralized control and standard-setting for certain critical enterprise-wide functions (e.g., core financial reporting, overarching ethical guidelines, brand identity) while devolving significant decision-making authority and operational governance to individual business units or modules for aspects specific to their domain. This allows for tailored approaches that meet local needs while maintaining enterprise-wide coherence.

This new governance model operates on the powerful principle of "Guardrails, Not Gates," signifying a fundamental shift in philosophy. Traditional "gates" are checkpoints in a process where work is halted pending manual review and approval. While intended to ensure quality or compliance, these gates often become bottlenecks, slowing down progress and frustrating teams. "Guardrails," in contrast, are proactive, often automated, mechanisms that define acceptable boundaries for action and steer activities in the right direction without necessarily impeding flow. They embed policy enforcement and risk mitigation directly into workflows.

For example, in software development, a guardrail might automatically check code for security vulnerabilities as it's written or prevent the deployment of configurations that violate compliance standards, providing immediate feedback to the developer. The focus is on enabling freedom and speed within well-defined, safe operating limits, with blocking interventions (gates) reserved only for situations where no safe automated alternative exists. This approach fosters developer and employee empowerment, accelerates delivery cycles, and allows security and compliance to become an integral part of the process rather than an external impediment. Amazon Web Services, for instance, advocates for using automated guardrails to check IAM role and policy creation, minimizing manual workflows.

Technology plays a crucial role in enabling these new governance models. Smart Contracts, built on blockchain or distributed ledger technology, offer a novel way to automate agreements and ensure transparent traceability of interactions between organizational modules. These are self-executing contracts where the terms of an agreement are written directly into code. Once predefined conditions are met (e.g., completion of a project milestone, delivery of a service to an agreed standard), the smart contract automatically executes the corresponding actions (e.g., releasing payment, transferring ownership, logging completion).

In a modular organization, smart contracts can be used to formalize service level agreements between internal teams or with external partners, track the delivery of work packages, and automate payments or resource allocations. Because these transactions are recorded on an immutable and transparent ledger, they significantly enhance accountability and reduce the potential for disputes or administrative overhead.

Semantic Permissions and Access Control represent another technologically advanced governance mechanism. Building on the concept of "AI-Readability" (Chapter 13), this approach moves beyond static, role-based access controls. Instead, permissions to access data, systems, or capabilities are granted dynamically based on a semantic understanding of an individual's or an AI agent's current task, context, required skills, and associated risk level. AI systems can play a significant role in managing and enforcing these fluid permissions, ensuring that access is granted on a "need-to-know" and "need-to-do" basis in real-time, adapting as responsibilities shift within dynamic teams.

To ensure that autonomy leads to desired results, Outcome-Based Accountability (OBA) provides a robust framework. OBA shifts the focus of accountability from merely tracking activities performed or outputs generated ("means") to measuring the actual, tangible improvements achieved for customers, the community, or the organization itself ("ends"). It distinguishes between population accountability (the overall well-being or outcomes desired for a whole population or customer base) and performance accountability (the specific contribution of a program, service, or organizational module to those broader outcomes).

Key OBA concepts include defining clear indicators to quantify outcome achievement, establishing baselines (historical data plus a forecast of future trends if nothing changes), and then actively working to "turn the curve"; that is, implementing data-informed strategies to achieve results better than the baseline forecast. OBA provides a common language for stakeholders, aligns efforts towards shared purposes, and offers a structured way to manage and continuously improve performance. In a modular organization, each module, whether a human team or an AI-driven service, can be held accountable for the specific, measurable outcomes it is designed to deliver, allowing for clear performance assessment even within a highly distributed operational model.

Finally, the AI Accountability Frameworks are integral to governance in organizations where AI agents function as team members. Ensuring clear chains of responsibility for AI decisions, implementing systems for detecting and correcting erroneous or harmful AI outputs, carefully balancing AI autonomy with necessary human control and oversight, and having established processes for remediation when AI-related mistakes occur are all critical governance functions. These frameworks are essential for responsibly managing the "AI modules" that are increasingly part of the organizational fabric.

The successful implementation of these new governance approaches transforms governance from a restrictive control structure into an enablement platform. In composable, modular systems designed for agility, traditional top-down micromanagement and rigid gatekeeping are counterproductive. Instead, models like "guardrails, not gates", outcome-based accountability, and automated agreements via smart contracts focus on defining acceptable operational boundaries and desired strategic outcomes.

This empowers autonomous modules (from an organizational standpoint) to innovate and execute effectively within that clearly defined framework. Such a shift demands a significant evolution in the mindset of leadership and traditional governance functions (such as legal, compliance, and internal audit). These functions must transition from being primarily enforcers of rigid rules to becoming educators, strategic advisors, and architects of these enabling governance frameworks.

Technology emerges as a critical enabler of the distributed trust and verifiable performance necessary for these new models to function. Decentralized governance structures and the effective implementation of outcome-based accountability in a modular setting rely heavily on the capacity to transparently track interactions, measure performance, and verify compliance across numerous distributed units. Technologies such as blockchain (which underpins smart contracts) and sophisticated data analytics platforms (essential for the data-driven approach of OBA) provide the foundational infrastructure for this. AI itself can be deployed to monitor adherence to defined guardrails and to detect anomalies that may require attention. Consequently, strategic investment in these governance technologies becomes as vital as investment in core operational technologies. These tools are not merely for enhancing efficiency; they are instrumental in building the systemic trust and transparency that are prerequisites for thriving modular ecosystems.

Ultimately, there is a delicate but crucial interplay between autonomy, accountability, and psychological safety. Granting significant autonomy to organizational modules; be they human teams, individual contributors, or AI agents is essential for fostering the speed, innovation, and ownership that modularity promises. However, this autonomy must be inextricably linked with clear and robust accountability for agreed-upon outcomes, as emphasized by OBA and AI accountability frameworks.

The "guardrails, not gates" model is designed to provide the psychological safety necessary for these modules to experiment, take calculated risks, and learn from failures, provided they operate within the established boundaries. In such an environment, failure within the guardrails is viewed not as a punishable offense but as a valuable learning opportunity.

Therefore, effective governance in modular organizations extends beyond the implementation of rules and technological systems; it involves cultivating a culture where accountability is embraced as a positive force, learning from both successes and setbacks is actively encouraged, and individuals and teams feel secure enough to innovate and push boundaries responsibly. Striking this balance is a continuous leadership challenge but is fundamental to unleashing the full potential of modular organizational design.

DESIGNING FOR EVOLVABILITY: BUILDING THE ADAPTIVE ORGANIZATION

The culmination of adopting modular structures, composable capabilities, dynamic talent deployment, integrated AI agents, agile teaming practices, and enabling governance is the creation of an organization designed not just for current efficiency or effectiveness, but for evolvability. In an environment characterized by perpetual change, accelerating technological disruption, and shifting competitive landscapes, the capacity to adapt and evolve is arguably the ultimate and most sustainable competitive advantage. It transcends mere agility; the ability to respond quickly to current conditions, by embedding the potential for ongoing transformation into the very DNA of the organization.

Research from McKinsey underscores the importance of organizational agility, noting that agile organizations have a significantly higher probability (70%) of ranking in the top quartile of organizational health, which is considered the best indicator of long-term performance. Indeed, agility is a top strategic priority for a large majority of business units, with many actively pursuing agile transformations. Evolvability takes this a step further; it is the capacity of the organization to continuously learn, adapt its internal structures and processes, and reinvent itself in response to both internal insights and external pressures.

Organizational Evolution Theory (OET), as developed by Howard Aldrich and Martin Ruef, provides a powerful conceptual lens for understanding this dynamic. Drawing analogies from biological evolution, the theory posits that organizational change occurs through a continuous cycle of variation, selection, and retention.

Variation refers to the constant emergence of new ideas, processes, tools, or behaviors. Crucially, these variations do not always originate from top-down strategic mandates; they can spring from frontline employees experimenting with new approaches, from customer feedback prompting adjustments, or from unexpected technological shifts.

Selection is the process by which these variations are evaluated. Some are ignored or discarded, while others are recognized as potentially valuable and are adopted or amplified.

Retention occurs when selected variations are codified into standard operating procedures, formally invested in, scaled across the organization, or otherwise embedded into its operational fabric. This theory highlights that in rapidly changing or declining industries, adaptability and exploration often trump the optimization of existing, potentially obsolete, models. Organizations must be adept at spotting emergent variations, evaluating their potential, and quickly scaling those that prove promising. Practical steps to foster this adaptability, and thus enhance evolvability, include actively identifying and experimenting with emerging niches, investing in capabilities aligned with future needs rather than solely current ones, courageously letting go of legacy operations that no longer serve the evolving strategy, and creating selection mechanisms that are both rigorous in their evaluation and rapid in their execution.

A useful metric for gauging an organization's capacity for such dynamic reconfiguration is the "Composability Quotient". This concept, often discussed in the context of digital technology stacks but equally applicable to overall organizational design, measures how quickly and effectively an enterprise can assemble, disassemble, and reassemble its various components; be they Packaged Business Capabilities, operational modules, human teams, or technological tools to address new challenges or capitalize on emerging opportunities.

Organizations can assess their composability quotient by asking critical questions such as: Can we launch new customer-facing digital experiences in days or weeks, rather than months? Are our cross-functional teams equipped to rapidly prototype new workflows and digital products? Can we swiftly deliver cohesive and targeted messages to diverse audience segments, especially in times of disruption? A high composability quotient signifies that the organization can effectively leverage its modular structure for adaptive advantage. Gartner anticipates that organizations that genuinely embrace composable modularity will significantly outpace their competitors in terms of responsiveness and innovation.

Modularity is the foundational enabler of organizational evolvability in this new age. Modular architectures, by their very design, allow different parts of the organization to change and adapt at different speeds without causing systemic destabilization. "Process modularity," for instance, allows complex tasks to be decomposed into more manageable, interchangeable components, each of which can be optimized or replaced independently. The "plug-and-play" nature of modular components means that new technologies, refined processes, or specialized talent modules (whether human teams or AI agents) can be integrated more easily and with less friction than in a monolithic system. This directly supports the "variation" and "selection" phases of OET: new modules representing novel approaches can be trialed with contained risk, and those that prove successful can be more readily retained, scaled, or replicated across the organization.

The evolution of the organization is further driven by feedback loops that inform the refinement of team patterns, operational processes, and even the modular structure itself. This connects to the principle that "Feedback is the Product", extending it from product development to organizational development.

In modular organizations, dynamic teams learn from each engagement, project, or interaction, and these learnings can be used to adapt their working methods, tools, and collaborative patterns. Similarly, the performance of different organizational modules; whether they are human-led business units, project teams, or AI-driven services, can be assessed using frameworks like Outcome-Based Accountability. The insights derived from this performance data then feed back into decisions about how these modules should be designed, resourced, interconnected, or even whether they should continue to exist in their current form.

Ultimately, strategic agility becomes an emergent property of a well-designed modular and composable system that is rich in these feedback mechanisms and possesses strong adaptive capabilities. True, sustainable strategic agility is rarely achieved through meticulously detailed, top-down central planning alone. Instead, it arises more organically from an organization that is structured to be a "learning engine"; one that continuously senses changes in its internal and external environment, processes this information through its distributed modules, and refines its structure, processes, and strategic focus in response.

However, designing for and sustaining evolvability is not without its challenges. The transition from traditional structures to modular ones can involve significant upfront costs, complexities in technological integration, and the difficult task of standardizing interfaces and managing interdependencies between modules.

As highlighted by studies on modular systems, there is a need for a strong architectural vision and clear "design rules" governing intermodular interactions to prevent detrimental fragmentation; a degree of "rigidity at the architectural level" is often necessary to enable flexibility within modules. Furthermore, cultural resistance to constant change, fluid roles, and the perceived instability of dynamic structures can be a major impediment. Overcoming these challenges requires committed leadership, strategic investment in both technology and people, and a persistent focus on cultivating an adaptive organizational culture.

The journey towards an evolvable organization represents a profound shift, moving beyond mere operational agility to cultivate "meta-agility"; the intrinsic ability of the organization to change how it changes. While operational agility focuses on responding swiftly to current market dynamics and customer needs, evolvability, undergirded by deep modularity and sophisticated composability, is a higher-order capability.

It signifies the organization's capacity to adapt its fundamental structures, core processes, and mechanisms for generating responses over the long term. It is about building an organization that can not only run its "agile engine" efficiently but can also learn from experience and redesign that engine itself as circumstances demand. OET provides a compelling framework for understanding this continuous process of adaptation and renewal. This requires leaders to shift their focus from managing discrete change initiatives to architecting an enterprise that is inherently designed for ongoing, systemic evolution, demanding a long-term perspective and sustained investment in foundational modularity and enabling technologies.

In this context, the "Composability Quotient" emerges not just as a measure of current flexibility but as a crucial leading indicator of an organization's future viability. Traditional business metrics such as profitability and market share are inherently lagging indicators, reflecting past performance. The Composability Quotient, however; which assesses how quickly and effectively an organization can reconfigure its diverse capabilities, offers a forward-looking gauge of its potential to navigate future, often unforeseen, disruptions and to capitalize on emergent opportunities. It measures the latent capacity for future performance in an increasingly volatile and unpredictable world. Therefore, organizations should proactively seek to understand, measure, and continuously improve their composability. This involves a holistic assessment, examining the modularity of their technology infrastructure, the fluidity of their talent deployment mechanisms, the speed and effectiveness of their decision-making processes concerning structural adjustments, and the adaptability of their governance frameworks.

A key dynamic that fuels this evolvability is the symbiotic loop between modularity and experimentation. A central tenet of Organizational Evolution Theory is the continuous generation of "variation". Modularity, by its very nature, creates safer, more contained, and lower-cost environments for experimenting with novel processes, emerging technologies, innovative team structures, or new service offerings. These experiments can be conducted within discrete modules without jeopardizing the stability or performance of the entire system. Successful experiments; the "selected variations", can then be more readily scaled, replicated across other relevant modules, or integrated into the core operational framework ("retained"), thereby driving the organization's overall evolution. This means that a modular structure inherently fosters a culture of experimentation and learning.

Leadership in such an environment should actively encourage and resource these "small bets" within various modules, recognizing them as the essential raw material for long-term adaptation, innovation, and resilience. This approach also makes the organization more robust in the face of "black swan" events or major market shifts, as it allows for a multitude of small, adaptive explorations to be underway simultaneously, increasing the probability of discovering viable new paths forward.

SUMMARY: THE FUTURE IS MODULAR AND COMPOSED

The era of the rigid, hierarchical organization, a construct of industrial times designed for stability and control in a world of information scarcity, is unequivocally waning. Such structures are proving increasingly ill-suited to the dynamism and complexity of the Age of Information Symmetry, where AI-driven change is constant and access to knowledge is becoming ubiquitous. The future belongs to organizations that can embrace a new paradigm: one that is fundamentally modular, inherently composable, and relentlessly adaptive. This chapter has charted the course from the legacy of fixed teams and static org charts to the emerging landscape of dynamic talent clouds and intelligent agent ecosystems.

The new organization is built upon several key pillars. Modularity is foundational, involving the strategic decomposition of the enterprise into distinct, often semi-autonomous, yet interconnected units; be they processes, technologies, teams, or AI-driven services. These modules are designed for interchangeability, allowing them to be updated, replaced, or reconfigured with agility. Composability is the strategic capability that breathes life into this modular structure, enabling the dynamic assembly and orchestration of these diverse modules to meet specific, evolving business needs and strategic objectives.

This structural transformation is mirrored in how talent and intelligence are managed. Talent Clouds and Skill Graphs represent a shift away from fixed job roles towards the fluid deployment of human expertise based on granular skills and real-time project requirements, often facilitated by AI-powered internal marketplaces. Concurrently, AI Agents are transitioning from mere tools to integral team members, augmenting human capabilities, automating complex tasks, and forming sophisticated hybrid intelligence teams where human and machine cognition collaborate.

Work itself becomes more fluid through Dynamic Teaming, where purpose-built teams, comprising the optimal mix of human and AI capabilities, are rapidly formed to tackle specific challenges or opportunities and then dissolved or reformed as needs change. This necessitates new forms of Adaptive Governance, moving beyond traditional hierarchical control towards enabling frameworks such as "guardrails, not gates," outcome-based accountability, and technologically embedded mechanisms like smart contracts and semantic permissions to ensure coherence and compliance without stifling speed and innovation.

The strategic advantages of embracing this modular and composable future are profound: unprecedented organizational agility, enhanced scalability to meet fluctuating demands, greater resilience in the face of disruption, a fertile ground for continuous innovation, and a heightened responsiveness to customer needs and market signals. However, this transformation is not merely structural or technological. It demands a significant shift in mindset at every level of the organization.

Leadership must evolve from controllers to orchestrators and enablers. The organizational culture must foster collaboration, trust, psychological safety, and a commitment to continuous learning and adaptation. For individuals, it means moving from an identity tied to a fixed role towards one defined by dynamic contributions, a portfolio of evolving skills, and the ability to thrive in fluid environments.

As organizations become increasingly modular and AI takes on a greater share of operational tasks, the human experience of work is profoundly altered. The traditional markers of status and security may fade, but the human need for meaning, purpose, and connection at work becomes even more critical. The next chapter, "Meaning at Work," will delve into how individuals can find, cultivate, and sustain this vital sense of purpose in an increasingly dynamic, AI-augmented, and modular world. This human-centric focus provides an essential anchor, ensuring that as we design the intelligent and adaptive organizations of the future, we do so in a way that enhances not only performance but also human flourishing.

15

MEANING AT WORK

The precipice on which contemporary work stands is characterized by an unprecedented diffusion of knowledge, largely propelled by the ascendancy of artificial intelligence. If AI can indeed know almost anything, and if it can perform a vast and expanding array of knowledge-based tasks with increasing sophistication, a fundamental question emerges with compelling urgency: what, then, makes human work worth doing?

How do individuals and organizations discover, cultivate, and sustain meaning, purpose, and motivation when the very bedrock of "expertise," traditionally defined by the scarcity and specialized accumulation of knowledge, begins to erode beneath our feet? This chapter confronts this existential and strategic challenge, exploring pathways to redefine careers and motivation in an era where information symmetry is rapidly becoming the norm.

For generations, the landscape of professional life was shaped by the pursuit and possession of specialized knowledge. Career progression, societal status, financial reward, and often a significant component of an individual's self-worth were intricately linked to the depth and exclusivity of their expertise. The lawyer who knew the intricacies of case law, the engineer who mastered complex schematics, the physician who could diagnose rare conditions, the academic who pushed the frontiers of a discipline.

These figures derived value and identity from knowing something that others did not, or from knowing it more profoundly. This model, however, is undergoing a seismic shift. The advent of AI systems capable of accessing, synthesizing, and applying colossal volumes of information in mere moments challenges the traditional premium placed on human knowledge accumulation. As access to information becomes ubiquitous and instantaneous, the competitive advantage conferred by simply "knowing more" diminishes.

This transformation is not without its potential for profound dislocation. The prospect of skills painstakingly acquired over years, even decades, being replicated or surpassed by algorithms can understandably trigger a crisis of professional identity and value. Yet, within this challenge lies an extraordinary opportunity: an invitation to re-evaluate what "value" and "meaning" truly signify in a professional context.

If meaning is no longer predominantly tethered to the exclusivity of knowledge, it can be anchored more firmly in enduring human capacities; our ability to connect, to create with intention, to contribute with empathy, to discern with wisdom, and to lead with purpose. The journey through this chapter is an exploration of this re-anchoring. It is a call to move beyond merely adapting to new technological tools and towards a fundamental reappraisal of the human role in the future of work. In an age increasingly shaped by artificial intelligence, the quest for uniquely human meaning transitions from a perceived luxury to an undeniable imperative: essential for individual well-being, organizational vitality, and societal progress.

This chapter will navigate this complex terrain through seven distinct explorations. It begins by examining the initial crisis that arises when long-held mastery appears to lose its currency. From there, it pivots to how work can be reframed through the lens of contribution rather than control, emphasizing the ascendancy of intrinsic motivators. Subsequently, the discussion will illuminate the surprising resurgence of "craft"; the deep human imprint of quality and intentionality, as a powerful differentiator in an age of automated speed. The chapter will then offer frameworks for navigating the increasingly modular and non-linear careers that are emerging, followed by an exploration of how AI can be a partner in co-creation, rather than solely a competitor. Attention will then turn to organizational design, considering how companies can foster environments conducive to purposeful work. Finally, the focus will shift to the individual, outlining personal practices for actively cultivating meaning in this new landscape. The central inquiry guiding this exploration remains: if everyone, and everything, can know anything, what makes your work, and indeed our work, truly worth doing?

WHEN MASTERY LOSES ITS MEANING

The shift towards an era of information symmetry, where knowledge once considered specialized and hard-won becomes readily accessible through artificial intelligence, precipitates a profound challenge to traditional notions of mastery and expertise. For individuals whose professional identities and perceived value have been built upon the foundation of possessing knowledge that others do not, this transition can be deeply unsettling. It can trigger a crisis of identity, a questioning of relevance, and a complex emotional landscape characterized by feelings of disillusionment, perceived overqualification, and a novel form of impostor syndrome tied not to a lack of skill, but to the perceived ease with which AI can replicate or even surpass human knowledge-based performance. Understanding this initial dislocation is the first step towards redefining meaning and motivation in the workplace.

The devaluation of accumulated knowledge strikes at the heart of many established professions. Fields such as law, medicine, engineering, software development, and academia have long revered deep, specialized knowledge as a primary currency. The years, often decades, dedicated to acquiring this knowledge; through formal education, rigorous training, and practical experience, have been seen as an investment that yields returns in the form of professional standing, problem-solving capacity, and unique value.

However, when AI systems demonstrate the ability to access, synthesize, and apply vast domains of this same knowledge with remarkable speed and increasing accuracy, the perceived value of the human effort invested in its solitary accumulation can diminish. This is not to say that human knowledge becomes worthless, but rather that its exclusivity, and therefore a significant component of its traditional market value and status, is eroded.

This erosion can manifest psychologically as "Perceived Overqualification" (POQ), a state where employees feel their skills, knowledge, and abilities exceed the requirements of their job, not because they have outgrown the role in a traditional sense, but because AI's capabilities have reshaped the role itself, rendering some human skills less critical or underutilized. Research suggests that an employee's awareness of AI (AIA) can contribute to POQ through at least two theoretical lenses.

From a person-job fit perspective, as AI assumes more of the knowledge and skill demands of a job, and given AI's superior capacity for rapid learning and iteration, employees may find it increasingly difficult to achieve personal success through their existing skillsets alone. This can lead to anxiety and the perception that their qualifications are now excessive for the "new" demands of the job. Concurrently, relative deprivation theory suggests that when employees with high AIA compare their own capabilities to those of AI in the workplace, they may perceive themselves to be in a disadvantaged position. AI's efficiency in performing skilled human tasks can lead to feelings that their own experience and skills are no longer as valued, fostering a sense of underutilization and, consequently, POQ. The emotional impact can include feelings of deprivation, insecurity about future job prospects, and a general sense that their hard-won expertise is no longer the primary driver of success or recognition.

Compounding this sense of devalued expertise is the emergence of a phenomenon termed "AI Impostor Syndrome". Unlike traditional impostor syndrome, where individuals doubt their legitimately earned success despite external validation, AI impostor syndrome arises from a different source: the perceived effortlessness of achieving success when heavily assisted by AI.

When AI can generate insightful analyses, draft coherent articles, or propose creative solutions with minimal direct human intellectual struggle, the user may experience a cognitive dissonance. They may question their own intellectual legitimacy and the true authorship of the work.

If an AI system crafts a compelling argument or designs an innovative solution in seconds, a profound question of responsibility and ownership arises: is the achievement attributable to the human user or to the algorithm? This disrupts a deeply ingrained societal and personal belief that often ties intelligence, competence, and the value of achievement to the degree of effort and struggle involved. The ease of AI assistance can thus create a crisis of authorship and intellectual identity, leading to feelings that success is unearned or inauthentic, even if the outcomes are objectively positive. This can result in disillusionment or a reluctance to fully claim ownership of AI-assisted accomplishments, potentially leading to professional stagnation or a diminished sense of personal efficacy.

This shifting landscape also compels a critical re-examination of popular concepts like Malcolm Gladwell's "10,000-hour rule," which suggested that achieving true expertise in any skill is largely a matter of practicing for approximately 10,000 hours. This idea, while widely influential, has been a simplification, and often a misinterpretation, of the foundational research by Dr. K. Anders Ericsson and his colleagues on "deliberate practice".

Ericsson's work did not merely emphasize the accumulation of hours; rather, it highlighted the critical importance of deliberate practice; a highly structured, goal-oriented activity designed for performance improvement, characterized by focused attention, the incorporation of immediate feedback, and consistent effort to operate just beyond one's current comfort zone and capabilities. Gladwell's popularization often overlooked this crucial nuance, focusing more on the sheer volume of time.

In an age where AI can simulate or dramatically accelerate aspects of practice; for instance, by providing instant feedback, generating countless practice scenarios, or automating repetitive drills, the human element of expertise development must necessarily evolve. If AI can handle many of the "repetitions," the locus of human value shifts. It is no longer solely about the quantity of hours invested in a specific type of practice but about the quality of human engagement with the learning process.

This includes the ability to define clear learning goals, to design effective practice strategies (perhaps in conjunction with AI tools), to critically interpret and integrate AI-generated feedback, and to engage in the focused, reflective effort that underpins genuine deliberate practice. The "effort" in the AI era becomes less about manual cognitive labor and more about strategic cognitive direction, critical thinking, and metacognition; the capacity to think about and regulate one's own learning and thinking processes. Mastery, therefore, is less about the arduous accumulation of 10,000 hours of a particular type of rote practice and more about the cultivated ability to learn, adapt, and direct one's cognitive resources effectively, often in partnership with AI.

The confluence of these factors: the potential devaluation of past expertise leading to POQ, and the crisis of authorship in AI-assisted achievements fostering AI Impostor Syndrome, creates a compounded identity crisis for many knowledge workers. It is not merely that their existing skills feel less differentiating; it is also that their future, AI-augmented accomplishments may feel less authentic or personally owned. This dual challenge underscores that the path to finding meaning in this new era requires more than just acquiring new technical skills to operate AI tools. It demands a deeper re-evaluation of self-worth and professional identity, moving beyond traditional markers.

Ultimately, the core challenge presented by this loss of traditional mastery is the need to decouple purpose from the exclusivity of knowledge. If an individual's sense of meaning was previously intertwined with being the unique possessor of certain information or skills, a new and more resilient foundation for purpose must be constructed. This involves a shift away from "I am valuable because I know X and others don't" towards a different, more sustainable value proposition. The initial feelings of loss, frustration, or even anger that can accompany this transition, as described by employees experiencing POQ, are valid and must be acknowledged. However, they also serve as a catalyst for seeking new avenues for meaning.

The erosion of traditional mastery, while initially disorienting, forces a necessary and ultimately productive inquiry into the nature of human value. If meaning is no longer predominantly derived from knowing more than others, it prompts a search for alternative, perhaps more profound, sources of professional fulfillment. This naturally leads to the consideration of how work itself can be reframed, moving away from an emphasis on exclusive knowledge and control towards a focus on tangible contribution and impact. The subsequent section will explore this pivotal reframing.

REFRAMING WORK AS CONTRIBUTION, NOT CONTROL

As the traditional pillars of expertise based on scarce knowledge begin to crumble in the age of information symmetry, the very definition of meaningful work undergoes a necessary transformation. The paradigm that equated value with the control and hoarding of information, or with a hierarchical position derived from exclusive knowledge, becomes increasingly untenable. In its place, a new understanding of work can emerge, one centered on contribution: the intentional application of skills and insights; whether human-derived, AI-assisted, or a synthesis of both, to create value, solve problems, and positively impact others. This shift necessitates a deeper reliance on intrinsic motivation and a re-evaluation of how individuals and organizations define and recognize valuable work.

Historically, power, status, and often compensation were closely linked to the control of information. As detailed in the early chapters of this volume, information asymmetry has been a fundamental organizing principle in many sectors, from finance to professional services. Individuals or entities that possessed unique data, specialized know-how, or privileged insights could leverage this asymmetry for advantage.

However, as AI democratizes access to vast reservoirs of knowledge and analytical capabilities, this model loses its potency. The critical question shifts from what you know (which AI may also know or learn rapidly) to what you do with what is known. The emphasis moves from possession to application, from knowing to doing, and specifically, to doing in a way that benefits others; be they colleagues, customers, or society at large. Contribution, in this sense, is about the purposeful deployment of capabilities to generate positive outcomes.

This reframing of work around contribution brings the nature of human motivation into sharp focus, particularly the interplay between intrinsic and extrinsic drivers. Intrinsic motivation stems from internal sources of satisfaction, such as a sense of purpose, personal growth, the inherent enjoyment of a task, or the feeling of making a difference. Extrinsic motivation, conversely, is fueled by external rewards like salary, bonuses, promotions, or public recognition. While both play a role, the balance between them becomes especially critical in an AI-augmented workplace. As AI systems become increasingly adept at performing tasks that were traditionally linked to extrinsic rewards; such as speed, efficiency in routine operations, or the processing of large datasets, the intrinsic aspects of work are poised to become the more sustainable and differentiating sources of human motivation and meaning. If AI can achieve comparable or even superior outcomes on purely efficiency-based metrics, then human engagement must be anchored in something more profound.

The over-reliance on extrinsic rewards in an AI-suffused environment carries significant risks. The "overjustification effect," for instance, describes how the introduction of external rewards for an activity that was once intrinsically enjoyable can diminish that inherent interest.

If employees are primarily evaluated and rewarded based on outputs that AI can efficiently replicate, and their intrinsic needs for autonomy, competence, and relatedness are not met, disengagement is a likely outcome. Self-Determination Theory (SDT) posits that these three psychological needs; autonomy (a sense of volition and control over one's actions), competence (feeling effective and capable), and relatedness (feeling connected to others) are essential for well-being and optimal functioning. AI can either support or undermine these needs. If AI is implemented as a tool that enhances an individual's capabilities, provides useful feedback, and frees them from drudgery to focus on more engaging tasks, it can bolster feelings of competence and autonomy. However, if AI leads to perceived overqualification, reduces decision-making latitude, or diminishes opportunities for meaningful human interaction, it can thwart these fundamental needs, thereby eroding intrinsic motivation.

A focus on contribution, therefore, offers a more robust pathway to tapping into intrinsic motivators. Defining "contribution" extends beyond mere task completion or output volume. It encompasses the intent behind the work; the desire to achieve a specific positive impact, and the actual effect of that work on others. In an AI-assisted environment, an individual's unique contributions might lie in areas such as:

- **Problem-framing:** Identifying and articulating the core issues that AI should be directed to solve.
- **Ethical oversight:** Ensuring that AI tools are used responsibly and that their outputs are fair and unbiased.
- **Empathetic interaction:** Providing the human touch in customer service, patient care, or team collaboration, aspects where AI currently falls short.
- **Creative synthesis:** Integrating AI-generated data or insights with human experience and intuition to develop novel solutions or perspectives.
- **Critical questioning:** Challenging AI assumptions or outputs to ensure robustness and alignment with broader goals.

The concept of ownership also evolves in this context. When AI is a significant collaborator or co-creator, traditional notions of sole authorship or knowledge ownership become blurred. The focus can shift from "owning" the knowledge or the final product to "owning" the problem statement, the process of inquiry, the ethical considerations throughout the project, and the ultimate application of the solution for the intended benefit.

This sense of responsibility for the process and impact can be a powerful intrinsic motivator. It is also crucial to strike a balance between individual contribution and collaborative effort, whether with human colleagues or AI agents. As research indicates, too much independent work can lead to isolation and burnout, while excessive or poorly structured collaboration can result in groupthink and a diffusion of responsibility. In an AI-integrated workplace, establishing clear roles, defining responsibilities for specific contributions (even if they are contributions to a human-AI team effort), and fostering transparent communication methods are essential for maintaining accountability and ensuring that each individual feels their input is valued and impactful.

Ultimately, meaning in this reframed paradigm is found in action; in the active process of contributing, rather than in the passive holding of a title, a credential, or a repository of knowledge. It is about making a tangible, positive difference in real time. This represents a shift from primarily proving one's worth through past accomplishments or qualifications to continuously demonstrating value through present and future impactful actions.

The amplified importance of intrinsic motivators becomes evident when considering AI's capabilities. AI excels at tasks often measured by extrinsic factors like speed and efficiency. If human value remains primarily tethered to these metrics, individuals will increasingly find themselves in a losing race against technological advancement. Intrinsic motivators, however; such as the drive for purpose, the pursuit of personal growth, the exercise of autonomy, the mastery of uniquely human skills (like complex empathy or ethical judgment), and the fostering of meaningful connections, are far less susceptible to direct replication by current AI.

Consequently, for work to retain its meaning and for individuals to remain genuinely engaged, the design of work itself, along with associated recognition and reward systems, must pivot to prioritize and nurture these intrinsic drivers. A workplace that continues to overemphasize AI-replicable extrinsic achievements while neglecting these fundamental human needs is likely to cultivate an environment ripe for disengagement, perceived overqualification, and the aforementioned AI impostor syndrome. The implication is clear: both organizations and individuals must proactively seek, design, and cultivate work that resonates with these deeper, internal drives. Meaning becomes less about "what I get" in terms of external validation and more about "what I give" in terms of contribution and "who I become" in the process.

Furthermore, while AI challenges traditional metrics of performance, it also presents an opportunity to make certain forms of human contribution more visible and perhaps even trackable. AI systems can log workflows, data inputs, decision points, and collaborative interactions. If "contribution" is defined not merely by the final output but by valuable human interventions within the process-such as posing a critical question that reframes an AI's analytical approach, identifying a subtle ethical concern missed by an algorithm, skillfully facilitating collaboration between diverse human and AI agents, or synthesizing AI-generated data into a truly novel human insight; these interventions can, in principle, be recorded, acknowledged, and valued.

This contrasts with traditional work environments where many such subtle but crucial contributions might often go unnoticed or uncredited. The challenge, therefore, is not only to foster a contribution-oriented mindset but also to design organizational systems; both human and technological that can effectively recognize, value, and perhaps even reward these more nuanced forms of human value creation in an AI-augmented landscape. This necessitates careful consideration of what truly constitutes a valuable "contribution" in this evolving context and how its impact can be appropriately assessed and affirmed.

To illustrate this shift, the following list outlines the transformation in motivational drivers:

Extrinsic rewards

Traditional manifestation: Salary tied to knowledge scarcity, promotions based on hierarchical control, and bonuses for speed and efficiency.

Impact of AI and information symmetry: Knowledge scarcity erodes, AI performs routine tasks efficiently, and hierarchical control becomes a bottleneck for agility.

Reimagined manifestation for meaningful work: Compensation reflects uniquely human contribution such as orchestration, ethical judgment, and complex problem-solving, with recognition for collaborative impact rather than individual throughput.

Link to meaning: Fair reward for genuine human value added beyond AI capabilities and acknowledgment of nuanced, context-sensitive contributions.

INTRINSIC NEEDS: **competence**

- **Traditional manifestation:** Mastery of specific technical skills and status as the recognized expert.
- **Impact of AI and information symmetry:** AI replicates or automates many technical skills, challenging expertise based purely on knowledge possession.
- **Reimagined manifestation for meaningful work:** Mastery shifts toward human-centric capabilities such as empathy, critical thinking, creativity, ethical reasoning, and effective human–AI collaboration, supported by continuous learning agility.
- **Link to meaning:** Fulfillment through effectiveness in domains that remain distinctly human, alongside a sense of growth and development that is not easily automated.

Intrinsic needs: autonomy

- **Traditional manifestation:** Clearly defined job roles with bounded tasks and decision-making authority within fixed parameters.
- **Impact of AI and information symmetry:** AI automates routine decisions and work becomes more fluid, modular, and project-based.

- **Reimagined manifestation for meaningful work:** Greater freedom in framing problems, designing solutions, choosing how to use AI tools, and navigating self-directed, modular career paths.
- **Link to meaning:** A sustained sense of agency and control over one's work and direction, even within collaborative and AI-augmented environments.

Intrinsic needs: relatedness

- **Traditional manifestation:** Teamwork within functional silos and mentorship grounded primarily in tenure and experience.
- **Impact of AI and information symmetry:** Team structures evolve, remote work expands, and modular collaboration reshapes interaction patterns.
- **Reimagined manifestation for meaningful work:** Deeper collaboration on complex problems that require diverse human perspectives, mentorship focused on navigating ambiguity and developing human skills, and communities organized around shared purpose.
- **Link to meaning:** Stronger bonds built through mutual support and shared challenge, reinforcing belonging and shared humanity in the face of complexity.

Purpose and meaning

- **Traditional manifestation:** Status derived from title, hierarchy, organizational longevity, or individual achievement.
- **Impact of AI and information symmetry:** Traditional status markers weaken and individual achievement becomes harder to distinguish from AI-enabled output.
- **Reimagined manifestation for meaningful work:** Purpose grounded in tangible impact on others or society, alignment

with personal values, contribution to meaningful missions, and ethical stewardship of powerful technologies.
- **Link to meaning:** A deep sense of significance and contribution, where work becomes an expression of values and a way of making a positive difference beyond oneself.

If work is increasingly reframed around contribution, then the way we contribute; the quality, care, and distinctive human insight we bring to our endeavors, becomes paramount. This focus on the "how" naturally leads to a consideration of the resurgence and revaluation of "craft" in an age predominantly characterized by AI's remarkable speed and scale. The next section will delve into how this dedication to craftsmanship can become a new wellspring of meaning and differentiation.

THE RISE OF CRAFT IN AN AGE OF SPEED

In an operational environment increasingly defined by the velocity and scale of artificial intelligence, where algorithms can generate vast quantities of content, solutions, and analyses with unprecedented speed, a countertrend is emerging: the revaluation of human "craft". This concept, extending beyond mere skill, encompasses a deep intentionality, a commitment to quality, the infusion of a personal touch, the power of narrative, and the evocation of emotional resonance. These elements of craft, often nurtured through deliberate and sometimes slower processes, are becoming potent differentiators and profound sources of meaning for individuals seeking to distinguish their contributions in an automated world.

AI's formidable strengths lie in its efficiency, its capacity for automating complex tasks, and its ability to generate a multitude of options at a scale previously unimaginable. However, this very proficiency can lead to a homogenization of outputs if human oversight and intentionality are absent. As AI democratizes access to skills in areas like coding, writing, and design, there is a risk that work can become commodified, with outputs appearing interchangeable and potentially lacking a distinctive personal or organizational signature. It is in this context that craftsmanship emerges as a vital counterweight. Human craft is often characterized by a more deliberate, thoughtful, and iterative approach, deeply imbued with the creator's unique perspective, values, and intentions. While AI can mimic the surface features of many creations, it struggles to replicate the journey of thoughtful development, the nuanced decision-making, and the "intangible soul" that arise from dedicated human engagement.

This valuation of deliberation and quality gives rise to the relevance of the "slow work" movement, which advocates for prioritizing depth, quality, and meaningful engagement over the sheer velocity of task completion. This movement challenges the pervasive notion that efficiency is solely a function of speed. Instead, it proposes that true effectiveness lies in maximizing meaningful outcomes; in essence, doing the right things, and doing them well, rather than just doing many things quickly. In a system increasingly optimized for AI's pace, consciously carving out space and time for human-centric "slow work"; which might involve deep reflection, meticulous curation, the careful honing of language, or the handcrafted refinement of a design, can produce uniquely valuable outputs and foster a profound sense of personal accomplishment and meaning.

A key aspect of craft is the "human touch," which often manifests in authenticity and even in the subtle imperfections that distinguish human creation from machine generation. While AI-generated content can achieve remarkable fidelity and polish, "feeling real" involves more than technical perfection; it involves intent. Authenticity is not merely about how an image looks or how text reads, but about the underlying purpose, the story being told, and the connection being sought with an audience; all of which are fundamentally human domains. The minor variations, the unique choices, even the occasional "mistakes" that are inherent in human craft can add depth, character, and a sense of genuineness that perfectly optimized AI outputs may lack. This resonates with aesthetic philosophies that find beauty in imperfection, recognizing that the slightly irregular, uniquely human creation can often be more valued and relatable.

Furthermore, humans excel in weaving narratives, telling compelling stories, and imbuing their work with emotional resonance; qualities that are exceptionally difficult for AI to replicate with genuine depth and authenticity. Narratives have the power to drive action, foster inclusivity, frame how information is perceived, and shape opinions. Effective storytelling builds emotional connections and can significantly differentiate individuals and organizations by highlighting their unique strengths, core values, and distinct personality.

Purpose-driven brands like Patagonia and Dove, for example, masterfully use storytelling to forge connections with their customers on an emotional level that transcends the mere functionality of their products, aligning with consumer values and building lasting loyalty. In an AI-driven world, the ability to take AI-generated data, analyses, or solutions and frame them within a compelling, human-centered narrative becomes a critical skill for meaning-making, communication, and achieving impact. The story provides the context, the "why," that data alone often cannot convey.

The deliberate practice of craft can thus be seen not as a Luddite rejection of technology, but as a conscious act of asserting human value, reclaiming agency, and finding intrinsic meaning in a technologically saturated landscape. It is a way to express individuality and purpose when faced with the homogenizing potential of mass automation. The act of creating something with care, with meticulous attention to detail, and with a deep sense of personal investment can be profoundly rewarding, irrespective of AI's capabilities.

This applies across a wide range of professions: the software developer who crafts elegant, maintainable, and ethically considered code, not just functional scripts; the writer who leverages AI for research or initial drafts but then meticulously refines the prose to convey a unique voice and profound insight; the consultant who uses AI for data analytics but delivers strategic advice grounded in deep contextual understanding, empathy, and foresight; or the educator who employs AI for personalized learning pathways but dedicates their core effort to fostering critical thinking, emotional intelligence, and genuine human connection through dialogue and mentorship.

As AI standardizes many forms of output, the unique "signature" of human craft; characterized by its profound intentionality, emotional depth, narrative skill, and ethical consideration, becomes a primary and increasingly valuable differentiator. This signature encompasses the entire thoughtful process of creation, from conception to execution. Meaning, for the individual, can be derived from the act of imbuing work with this personal signature, making it an authentic expression of human capability, perspective, and values. The cultivation and recognition of craft, therefore, become essential both for individuals seeking purpose and for organizations aiming to offer distinctive value in a crowded marketplace. This involves valuing the how of work; the care, the thought, the human insight, as much as, if not more than, the what in terms of sheer output.

Interestingly, the rise of AI and the emphasis on human craft are not necessarily antagonistic; a symbiotic relationship can exist. AI's efficiency in automating routine, time-consuming, and repetitive tasks; the drudgery that often detracts from deeper, more creative engagement, can, paradoxically, create more space and resources for humans to dedicate to the nuanced practice of craft.

By offloading these less engaging tasks to AI, individuals can free up significant cognitive bandwidth and time. This newly available capacity can then be channeled into the more demanding, reflective, and personally engaging aspects of their work; the "slow work" that requires deep thinking, careful iteration, and sustained personal involvement. For instance, a designer might use AI to rapidly generate a dozen initial layout concepts, thereby saving hours of manual work, and then dedicate that saved time to meticulously refining the typography, color palette, and visual narrative of the chosen concept, focusing on the elements that truly elevate the design. Similarly, a researcher might use AI to sift through massive datasets for relevant papers, and then spend their time in deep analysis, synthesis, and the crafting of a nuanced argument. In this way, AI does not merely compete with craft; it can become a powerful tool that enables and enhances the conditions for human craftsmanship by allowing individuals to concentrate their efforts on the most meaningful, value-added, and uniquely human aspects of their work. The key lies in the intentional design of workflows that strategically integrate AI to support and amplify, rather than supplant, human artistry and ingenuity.

If meaning can be profoundly discovered and expressed through the way we work, particularly through the lens of craft, it must also be sought and supported within the evolving structure of our working lives. Careers themselves are being reshaped by the same technological and societal forces, becoming less linear and more modular. Understanding how to navigate these new career architectures is crucial for sustaining a sense of purpose and direction. The following section will explore models for finding fulfillment in this dynamic and often fragmented professional landscape.

NAVIGATING THE MODULAR CAREER

The traditional concept of a career as a linear, predictable ascent up a hierarchical ladder within a single organization is rapidly becoming an artifact of a bygone era. Propelled by the pervasive influence of artificial intelligence, the rise of automation, the proliferation of the gig economy, and shifting employee expectations, career paths are increasingly characterized by non-linearity, fluidity, and modularity.

In this dynamic environment, where deep expertise in one narrow domain may no longer guarantee lifelong employment or steady progression, individuals must adopt new strategies for finding purpose, fulfillment, and coherence in their working lives. This section offers a model for navigating such AI-enabled, modular careers, emphasizing the assembly of meaning across diverse projects, roles, and collaborations, all anchored by a resilient personal mission.

The demise of rigidly linear career paths is evident across numerous industries. The notion that one joins a company, learns a specific trade or profession, and steadily progresses through a series of predefined roles is giving way to a more fragmented and individualized approach to career development. AI accelerates this trend by automating certain tasks, rendering some specialized skills obsolete while creating demand for new, often interdisciplinary, competencies. This necessitates a mindset shift away from static job descriptions and towards a more dynamic understanding of one's professional journey.

Progress in a non-linear fashion prioritizes skill development over traditional title-oriented promotions. "Non-linear career paths" are those that do not adhere to a predetermined sequence of roles but instead allow individuals to pivot across different functions, teams, and even industries based on their evolving interests, the development of new skills, and emerging market opportunities.

For example, a marketing manager might transition into a product management role by leveraging their customer insights and communication skills, or a software engineer might move into data analytics by applying their coding and problem-solving abilities to new datasets. This model enhances workforce agility for organizations, enabling them to deploy talent more flexibly, and empowers employees to pursue avenues that align more closely with their passions and strengths, potentially leading to greater job satisfaction and engagement. AI can act as a significant enabler for such transitions by providing accessible tools and platforms for rapid upskilling, acquiring foundational knowledge in new domains, or identifying unconventional career opportunities.

One powerful framework for conceptualizing this new career architecture is the "portfolio career" or "career as collage". Popularized by thinkers like Charles Handy, a portfolio career comprises a variety of roles, projects, and often multiple income streams, rather than being centered on a single, full-time job within one organization.

This model, once common primarily in creative industries, is becoming more mainstream. Its advantages include greater flexibility in managing one's time and commitments, increased variety in work experiences, the development of diverse skill sets, enhanced work-life balance, and the ability to pursue personal passions alongside professional endeavors. However, it also presents challenges, such as a potential lack of income stability, the absence of traditional benefits and career progression structures, and the difficulty of forging a clear professional identity when one's work is multifaceted; what Herminia Ibarra termed the "laundry list" problem, where explaining one's career becomes a recitation of disparate activities rather than a coherent narrative.

To address concerns about fragmentation, the "career as collage" framework proposes that individuals curate their professional lives. They thoughtfully select and arrange varied experiences, projects, and roles, which, when viewed collectively, create a unique, coherent, and meaningful professional tapestry. The meaning is not inherent in any single piece of the collage but emerges from the individual's ability to connect the dots, to see the patterns, and to articulate the overarching purpose that binds these varied experiences.

In such a fluid and self-directed career landscape, a clear and compelling personal mission statement serves as an indispensable anchor. A well-crafted personal mission statement articulates an individual's core values, their desired contributions, the impact they aspire to make, and the principles that guide their professional choices. It provides a stable framework for evaluating opportunities and challenges, ensuring that decisions; whether about taking on a new project, learning a new skill, or shifting career direction, are congruent with one's deeper sense of purpose, irrespective of specific job titles or organizational affiliations. This internal compass becomes particularly vital when external structures are less defined or constantly changing.

To further aid in conceptualizing and managing these modular careers, individuals can develop what might be termed a "Career Cloud". This is a dynamic, personalized framework encompassing several key dimensions:

- **People:** This includes the networks, mentors, collaborators, and communities an individual actively engages with. These relationships provide support, learning opportunities, and avenues for contribution.
- **Causes:** These are the larger problems, societal issues, or overarching values that an individual cares deeply about and seeks to address or advance through their work.
- **Contributions:** This refers to the specific types of impact and value the individual aims to deliver across their various projects, roles, or engagements. It answers the question: "What difference do I want to make?"
- **Capabilities:** This encompasses the individual's evolving portfolio of skills, knowledge, and experiences; both uniquely human-centric competencies (like empathy, critical thinking, creativity) and AI-augmented abilities (like data analysis with AI tools, or content creation using generative AI).

This "Career Cloud" is not static; its elements are continuously added, refined, evolved, or de-emphasized as the individual learns and grows. Meaning is found in the purposeful cultivation, integration, and articulation of these interconnected elements. Lifelong learning and adaptability are, therefore, not just beneficial but core tenets for successfully navigating a modular career. AI tools can significantly support this continuous learning journey by providing personalized educational resources, skill assessment platforms, and even pathways to new opportunities.

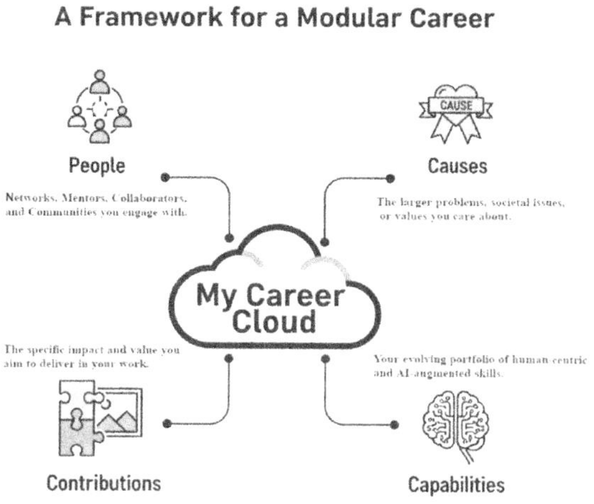

In these evolving career structures, meaning is not typically prepackaged with a job title or a position on an organizational chart; rather, it must be actively and continuously curated and narrated by the individual. Traditional careers often provided a built-in narrative of progression and identity (e.g., "I am a Senior Engineer at Company X, on track to become a Principal Engineer").

Modular careers, by their very nature, are more fragmented, potentially involving a diverse array of projects, roles, and even industries. This inherent fragmentation can lead to the identity challenges mentioned previously. Consequently, the individual bears a greater responsibility for constructing a coherent and compelling narrative that connects these seemingly disparate experiences. This is where the development of personal mission statements and strong storytelling skills becomes vital. The meaning is discovered not in any single component of the portfolio but in the overarching story the individual tells themselves and others about why these varied components fit together, how they align with their personal mission and values, and how they contribute to a larger sense of purpose.

The "Career Cloud" concept helps to visualize these components, but it is the narrative that breathes life and coherence into them, transforming a collection of experiences into a meaningful journey. Thus, individuals need to cultivate "narrative competence"; the ability to articulate an authentic and evolving story of their professional life and the purpose that drives it, as a core career skill.

It is also important to recognize the dual role of AI in this transformation: AI acts as both a driver of the shift towards modularity and an enabler for navigating it. On one hand, AI and automation contribute to the obsolescence of certain fixed roles and traditional skill sets, pushing individuals towards more flexible, project-based, and often shorter-term engagements. The rapid evolution of AI itself necessitates continuous upskilling and adaptation, making rigid, long-term career plans less viable.

On the other hand, AI platforms can simultaneously offer powerful tools to manage this new reality. These include personalized learning paths to acquire new competencies, sophisticated skill assessment tools, AI-driven job boards that can match individuals to projects based on their "career cloud" profiles, and even AI assistants that can help manage the administrative complexities of a portfolio career, such as project management, client communication, and financial tracking. Therefore, AI is not merely a force to react to in career planning; it is a suite of tools to be proactively leveraged in designing, managing, and finding fulfillment within a modular career. Successfully navigating this landscape involves strategically using AI to augment one's capabilities, manage complexity, identify new avenues for contribution, and continuously adapt, rather than viewing AI solely as a threat to traditional job security.

If our careers are indeed becoming more modular, and if AI is an increasingly integral part of this new professional landscape, then our fundamental relationship with AI itself warrants closer examination. Is AI merely a sophisticated tool to be wielded, or can it evolve into something more akin to a partner in our work and, perhaps surprisingly, even in our ongoing search for meaning? This question sets the stage for exploring the concept of work as a process of co-creation with artificial intelligence.

WORK AS CO-CREATION WITH AI

The integration of artificial intelligence into the workplace is often viewed through the lens of automation, efficiency, or even job displacement. However, a more transformative perspective positions AI not merely as a tool or a potential competitor, but as a dynamic collaborator in both the execution of work and the very process of meaning-making. This section explores the burgeoning field of human-AI co-creation, examining how synergistic partnerships between human ingenuity and machine intelligence can lead to enhanced creativity, more effective problem-solving, and a novel sense of purpose derived from this collaborative synergy. Moving beyond a purely utilitarian view of AI allows for the discovery of new ways to work and new sources of professional fulfillment.

The shift towards co-creation requires moving beyond the simple automation of tasks to envisioning AI as a creative partner. Human-AI co-creation describes a dynamic where humans and AI systems work in concert to achieve outcomes that neither could easily nor effectively attain in isolation. This collaborative approach emphasizes synergy rather than substitution, leveraging AI's computational power, pattern recognition capabilities, and ability to generate diverse options to augment and amplify human creativity, insight, and strategic direction.

In various creative industries, for instance, AI is already being used to assist with ideation by generating novel concepts, aid in composition by exploring different structural possibilities, and even help in predicting audience responses to creative outputs, thereby acting as a "catalyst" or a "digital brainstorming partner". New design approaches for human-AI co-creation environments are emerging that seek to embed AI "within and between" different stages of creative workflows, facilitating a more fluid, iterative, and controlled collaboration where AI functions as a genuine "thinking partner".

Central to effective and ethical co-creation is the "Human-in-the-Loop" (HITL) model. This model underscores the indispensable role of human judgment, contextual understanding, and oversight at various junctures within AI-driven processes, especially in creative endeavors and critical decision-making. HITL involves humans in crucial stages such as crafting effective prompts to guide AI, critically reviewing and evaluating AI-generated outputs, approving final versions, and directing iterative refinements.

This human involvement is essential to ensure that AI-generated content or solutions are not only technically proficient but also effective, ethically sound, aligned with brand or project goals, and authentically resonant. Humans are uniquely equipped to identify and correct inaccuracies, mitigate biases that may be present in training data or algorithmic processing, and uphold ethical standards that AI systems, by themselves, cannot fully comprehend or manage. Meaning, in this context, can be profoundly derived from the uniquely human role of guiding AI's power, refining its outputs, and ethically steering its application towards beneficial ends. This is where human values, nuanced judgment, and deep domain expertise become irreplaceable.

The concept of co-authorship and the allocation of shared responsibility also evolve in human-AI partnerships. While AI can generate substantial portions of content, analysis, or design, the human collaborator typically retains ultimate responsibility for the intent behind the work, its ethical implications, its contextual relevance, and the final quality and impact of the output. This shared responsibility, rather than diminishing the human role, can become a source of meaning if the human partner feels they are genuinely contributing their unique insights, critical faculties, and ethical oversight to the collaborative endeavor. The sense of purpose comes from skillfully orchestrating AI's capabilities and ensuring they serve human-defined goals.

Examples of such co-creative work are emerging across diverse fields. Scientists are using AI to analyze vast and complex datasets, identify subtle patterns or anomalies that might be missed by human observation alone, and then apply their deep domain expertise and intuition to interpret these findings, formulate new hypotheses, and design subsequent experiments. Artists and designers are employing generative AI tools to explore a wide range of visual styles, thematic concepts, or structural forms, and then curating, modifying, combining, and imbuing these AI-generated elements with their personal vision, aesthetic judgment, and emotional intent. Educators are leveraging AI to create personalized learning paths and adaptive assessments for students, freeing up their own time and energy to focus on fostering higher-order skills like critical thinking, Socratic dialogue, collaborative problem-solving, and emotional intelligence through direct human interaction and mentorship. Business strategists are utilizing AI for sophisticated market analysis, predictive modeling, and scenario planning, and then developing innovative and resilient strategies based on human foresight, ethical considerations, and a nuanced understanding of complex, dynamic contextual factors.

Beyond task execution, AI can also serve as a mirror and an amplifier for human thought. The process of interacting with AI-formulating precise prompts, evaluating its responses, identifying its biases or limitations; can help individuals clarify their own thinking, uncover their implicit assumptions, and explore alternative perspectives they might not have considered independently. AI's capacity to rapidly generate diverse alternatives, counterarguments, or simulated scenarios can stimulate deeper human reflection, more robust critical analysis, and ultimately, more informed and well-reasoned human decision-making. Meaning can arise from this enhanced self-awareness, the sharpened quality of thought, and the improved decision-making efficacy facilitated by this cognitive partnership with AI.

When AI undertakes much of the "generation" in creative or analytical processes, human meaning often shifts towards the crucial roles of orchestration, curation, ethical stewardship, and strategic guidance within the co-creative framework. The human role becomes less about manual production and more about defining the problem space with clarity, setting overarching goals, crafting prompts that elicit valuable AI responses, selecting and integrating appropriate AI tools, and sifting through AI-generated options to identify those that are most relevant, innovative, or promising.

These are not trivial tasks; they demand sophisticated human skills such as advanced critical thinking, refined judgment, robust ethical reasoning, and deep domain expertise. Meaning is therefore found in the skillful and responsible execution of these "meta-level" activities, effectively shaping and directing AI's considerable power towards beneficial and human-aligned ends. The "co-creator" role is less about being a hands-on maker in the traditional sense and more akin to being a wise director, an insightful conductor, or an ethical navigator of AI's capabilities. This requires cultivating a new set of competencies and fostering a new understanding of where uniquely human value lies in the creative and problem-solving process.

Counterintuitively, perhaps, effective human-AI co-creation may demand deeper and more focused human engagement rather than leading to passive oversight or disengagement. While a common apprehension is that AI will lead to human deskilling or a reduction in cognitive effort, true co-creation necessitates a high level of active human involvement.

To collaborate effectively with AI, humans must develop a nuanced understanding of AI's strengths and weaknesses, learn to formulate precise and effective prompts, critically evaluate the outputs for accuracy, relevance, and bias, and engage in thoughtful iteration. This requires active, sustained, and careful thought; the hallmarks of deep engagement and reflective behavior. While poorly managed co-creation, such as the uncritical acceptance of AI outputs, might indeed lead to shallow results and diminished human involvement, genuine co-creation: where humans provide strong intellectual direction, critical feedback, and ethical grounding necessitates significant cognitive investment.

The process of grappling with AI's suggestions, challenging its underlying assumptions, and synthesizing its contributions with human insight and experience can be intellectually stimulating, creatively challenging, and intrinsically rewarding. The sense of purpose in such collaborations can stem directly from this intellectual engagement and from the ability to elevate the combined output to a level beyond what either the human or the AI could achieve in isolation. This, however, requires cultivating skills in critical interaction with AI and a willingness to engage with it as a partner in a demanding intellectual endeavor.

If individuals are to find such meaning through co-creation with AI, then the organizations within which they operate must also undergo significant transformation. The existing structures, prevailing cultures, and ingrained processes within companies need to be thoughtfully redesigned to support this new paradigm of work and to actively foster a pervasive sense of purpose. This leads directly to the critical question of how organizations can architect themselves for purposeful work in an AI-first future.

ORGANIZATIONAL DESIGN FOR PURPOSEFUL WORK

The quest for meaning at work, especially in an era where artificial intelligence is reshaping roles and capabilities, is not solely an individual endeavor. Organizations bear a significant responsibility and possess a unique opportunity to cultivate environments where purposeful work can flourish. As traditional hierarchies and role definitions become less rigid, and as AI takes on more operational tasks, the very design of organizations; their structures, processes, cultures, and leadership models, must evolve to support meaning-driven engagement. This section provides practical insights into how companies can move beyond outdated models and architect themselves to foster clarity of purpose, enable impactful contribution, and become ecosystems where both humans and AI collaborate towards meaningful objectives.

The legacy of rigid, hierarchical organizational structures, designed primarily for predictability and control in a more stable industrial era, often proves ill-suited for the dynamic, AI-first environment. Such structures, with their defined silos and fixed roles, can stifle the adaptability, cross-functional collaboration, and individualized contribution that are key to finding meaning when knowledge is no longer a primary differentiator.

As explored in Chapter 14, the shift is towards more modular organizations, talent clouds, and ecosystems where AI agents function as integral team members. Within this more fluid context, traditional sources of work identity and motivation are disrupted, making the intentional design for purpose even more critical.

A cornerstone of organizational design for purposeful work is establishing clarity of purpose that transcends the clarity of individual roles. In modular systems where job descriptions are fluid and team compositions shift based on project needs, a strong, clearly articulated, and consistently communicated organizational purpose acts as the unifying force and the primary navigational aid. When every member of the organization, human or AI (in terms of its programmed objectives), understands the overarching "why"; the core mission, values, and intended impact of the enterprise they can more readily see how their specific contributions, even within evolving or temporary roles, connect to that larger, meaningful vision.

Purpose-driven brands, such as TOMS Shoes with its "One for One" model, Patagonia with its environmental activism, Dove with its "Real Beauty" campaign, and Airbnb with its "Belong Anywhere" philosophy, exemplify this principle. These organizations embed their mission deeply into their operations, culture, and external storytelling, which serves not only to attract and retain customers but also to inspire and align employees, providing a powerful answer to the question of "why this work matters."

Building on this foundation of purpose, organizations must then design work and team structures that maximize opportunities for individuals to make tangible contributions and to witness the impact of their efforts. This often involves a move towards project-based work, the formation of agile, cross-functional teams, and the development of clear metrics for success that extend beyond traditional output measures. These metrics should also capture progress towards the organization's purpose and the positive impact generated for stakeholders. When individuals can directly see the connection between their actions and meaningful outcomes, their sense of contribution and, consequently, their intrinsic motivation are significantly enhanced.

Effective feedback loops are another critical component of designing for purposeful work. These systems should not only be continuous but also specifically geared towards affirming the value of individual and team contributions in the context of the organization's mission. Regular, specific, and multi-directional feedback; from peers, leaders, and even through data reflecting the impact of their work, boosts employee engagement, morale, and reinforces their sense of making a difference.

In an AI-augmented workplace, feedback mechanisms should evolve to address not just task completion but also how individuals are skillfully leveraging AI, contributing unique human insights to collaborative processes, and demonstrating ethical judgment and creative problem-solving. These feedback systems should make progress towards the overarching organizational purpose visible, celebrating contributions that demonstrably advance it and providing constructive guidance for alignment.

In the fast-paced, often relentless, AI-driven environment, organizations must also consciously create and protect space for essential human activities such as reflection, deep learning, and synthesis. This means going beyond merely providing access to training modules. It involves fostering a culture where individuals and teams are encouraged and allocated time to step back, make sense of their work, reflect on its connection to the broader purpose, learn from both successes and failures, and synthesize new insights. Access to coaching, whether from human mentors or AI-assisted platforms, can be invaluable in this regard, helping individuals navigate their modular careers and connect their personal growth to the organization's objectives. Opportunities for teams to collectively debrief, share learnings, and strategize on how to better align their efforts with the core mission also contribute significantly to a shared sense of purpose.

Cultivating psychological safety is paramount for enabling meaningful engagement. Employees need to feel secure enough to share honest feedback, to experiment with new approaches (especially in the context of human-AI co-creation), to take considered risks without fear of punitive repercussions, and to voice their perspectives on matters of purpose and meaning within the organization. A culture that actively nurtures psychological safety allows for deeper levels of engagement, more authentic contributions, and a more robust and resilient pursuit of meaningful work.

Ultimately, the organization should strive to become a "meaning-making ecosystem". This concept frames the company not merely as a place where tasks are performed in exchange for compensation, but as an environment that actively supports and facilitates the meaning-making efforts of its members. Achieving this requires a holistic alignment of organizational policies, leadership behaviors, communication strategies, resource allocation, and recognition and reward systems with the overarching goal of fostering purposeful work. Every aspect of the organizational architecture should be examined through the lens of whether it enables or hinders the pursuit of meaning.

In these highly flexible, modular organizations where AI agents handle numerous operational tasks, a deeply embedded and widely understood organizational purpose acts almost like an "operating system". This purpose guides the diverse activities of both human and AI agents, ensuring coherence and alignment in an environment where traditional command-and-control structures based on fixed hierarchies are less effective.

While AI agents operate based on their programmed goals and data inputs, human contributors are motivated by their skills, values, and their understanding of how their work fits into the larger picture. Without a strong, shared sense of purpose, these multifaceted activities can easily become fragmented, inefficient, and ultimately, meaningless. A clear organizational purpose provides the necessary strategic direction, ethical guardrails, and motivational impetus. It informs goal setting for AI deployment and inspires human contribution by clearly connecting individual tasks and projects to a larger, significant, and shared vision. Thus, "purpose" transcends its common placement as a mere mission statement on a corporate website; it becomes an active, dynamic, guiding principle that informs daily decisions, shapes resource allocation, and directs behavior across the entire human-AI organizational system. Leaders, therefore, face the critical responsibility of not only defining this purpose but also communicating it relentlessly and embedding it so deeply within the organizational fabric that it becomes the implicit logic guiding the actions of all contributors. This is fundamental for maintaining coherence, fostering innovation, and sustaining a sense of meaning in an increasingly decentralized and agile operational landscape.

This shift also necessitates a profound transformation in the role of managers and leaders. The focus must move away from the direct oversight of tasks; many of which AI can perform or monitor with greater efficiency, towards cultivating an environment where meaningful human contribution can genuinely flourish. Traditional management often revolves around assigning tasks, monitoring their execution, and evaluating performance based primarily on output metrics. When AI can automate many of these tasks and provide comprehensive performance data, the human manager's role in direct task supervision diminishes in importance.

Instead, leadership in the AI era becomes centered on a new set of responsibilities, including:

- Designing work systems and team structures that empower individuals to make meaningful contributions.
- Fostering a culture of psychological safety, open communication, and continuous learning.
- Coaching individuals to identify their unique contributions, navigate their modular careers effectively, and develop their human-centric skills.
- Facilitating productive and ethical human-AI co-creation.
- Ensuring that feedback loops consistently affirm the value of diverse contributions and track progress towards shared goals.

This represents a significant evolution from being a "director" of work to becoming a "cultivator" or "enabler" of meaningful engagement and purposeful impact. Consequently, leadership development programs must be radically rethought to equip leaders with the competencies required for fostering purpose, coaching for contribution, and orchestrating complex human-AI ecosystems, rather than perpetuating outdated command-and-control frameworks.

While the design of the organization plays an undeniable and crucial role in creating the conditions conducive to purposeful work, the ultimate experience and construction of meaning remain a deeply personal and internal practice. This recognition leads to the final section of this chapter, which will turn its focus inward, exploring what individuals themselves can proactively do to cultivate and sustain meaning in their evolving working lives amidst the rise of intelligent machines.

THE PERSONAL PRACTICE OF MEANING

While organizational structures and cultures are pivotal in shaping the context for meaningful work, the ultimate locus of meaning resides within the individual. In an era of unprecedented technological change, where artificial intelligence redefines tasks, roles, and even the nature of expertise, the ability to personally and proactively cultivate a sense of purpose becomes not just a desirable skill but an essential practice for navigating a complex and evolving professional landscape. This concluding section offers a grounded call to action, outlining how individuals can build and sustain meaning as a deliberate, ongoing endeavor, emphasizing personal agency, intentionality, and the use of reflective and supportive tools; including AI itself.

Finding meaning isn't just something that happens to you; it's something you actively build. It's all about how you make sense of your experiences, what you decide is important, and how you connect what you believe on the inside with what you do on the outside. This meaning-making involves an exploration of one's belief systems and a conscious effort to find hope and purpose, especially when faced with challenging situations or when established assumptions about the world (global meaning) are in discord with current circumstances (situational meaning), such as the transformative impact of AI on one's profession. Individuals possess inherent agency in how they choose to interpret these shifts and what aspects of their work and life they choose to focus on to derive a sense of value.

Reflective journaling emerges as a potent tool in this personal practice of meaning-making. More than mere documentation, journaling serves as a space for introspection, self-awareness, goal clarification, and tracking one's evolving journey toward purpose. Through regular journaling, professionals can dissect their experiences, confront challenges, celebrate victories, gain a deeper understanding of their strengths and weaknesses, and strategically plan their career trajectories in a changing world. It can be employed in a structured manner to define overarching goals, identify key milestones, outline specific action steps, monitor progress, and, crucially, reflect on the lessons learned along the way.

For instance, an individual might use journal prompts to explore questions such as: "How does my current project align with my core values?", "What unique human contribution did I make today, even when working with AI?", "How is my understanding of 'meaningful work' evolving?", or "What small step can I take this week to move closer to my personal mission?"

Intriguingly, artificial intelligence itself can become an ally in this personal quest for meaning, particularly through the advent of AI-powered digital coaching platforms. These tools can offer personalized, scalable support for skill development, goal setting, reflective practice, and even the enhancement of interpersonal communication skills. Some platforms allow for the creation of a "digital twin" of a human coach, delivering tailored programs and guidance. For example, AI Coach by Edthena acts as a "thought partner" for teachers, helping them analyze their teaching practices and reflect on areas for growth, guided by AI-powered suggestions aligned with professional skills. This presents a fascinating synergy: leveraging AI to help individuals navigate the challenges and opportunities for meaning that AI itself is partly responsible for creating. Such tools can provide structured reflection, track progress towards personal development goals, and offer resources for continuous learning, thereby supporting the individual's active construction of meaning.

The adoption of personal and team rituals can also powerfully reinforce a sense of meaning, connection, and contribution. Examples include practices like weekly gratitude shares, where individuals acknowledge positive experiences or contributions; end-of-week "wins" or "high-fives" that celebrate accomplishments and mutual support; and monthly reflection and goal-setting sessions that foster alignment and a sense of progress. On a personal level, individuals can establish their own rituals, such as dedicating a few minutes each day to reflect on contributions made, moments of gratitude experienced, or periodically reviewing how their current activities align with their broader personal mission and values. These seemingly small, consistent practices can accumulate to create a strong internal framework of purpose.

Reclaiming a sense of agency, even within large, complex, and rapidly changing systems, is crucial for personal meaning. This can be achieved by focusing on small, intentional actions within one's sphere of influence. It might involve choosing to approach a specific task with a spirit of craftsmanship, consciously offering help or mentorship to a colleague, asking a thoughtful and critical question during an AI co-creation process to ensure ethical considerations, or deliberately aligning a project or decision with a deeply held personal value. These individual acts of intentionality, however minor they may seem, can collectively foster a powerful sense of purpose and control over one's professional experience.

Furthermore, framing lifelong learning not merely as a pragmatic necessity for maintaining employability but as an intrinsically meaningful pursuit can also contribute to a sense of purpose. The continuous acquisition of new knowledge, skills, and perspectives expands one's capacity to contribute in novel ways, adapt to change, and engage with the world more deeply. Finally, the personal practice of meaning is often amplified through connection with others around a shared purpose. Building and nurturing relationships and communities; whether within an organization, a professional network, or a volunteer group, that are centered on shared values and purposeful endeavors can provide invaluable support, inspiration, and a collective sense of making a difference.

In a world characterized by constant flux and the pervasive influence of AI, the ability to intentionally and consistently engage in practices that construct personal meaning evolves into a crucial meta-skill. This meta-skill underpins well-being, resilience, and the capacity to navigate a career that is no longer defined by stable external markers like job titles or the exclusivity of knowledge. As these traditional anchors become less reliable, individuals cannot depend solely on external structures or validation for their sense of purpose.

Practices such as reflective journaling, engaging with AI-assisted coaching, adopting personal rituals that reinforce positive contributions, and consciously working to reframe beliefs and interpretations of challenging situations are all deliberate actions. They are aimed at constructing and reinforcing meaning from within. This active engagement in meaning-making is a skill in itself; a higher-order capability that empowers individuals to navigate uncertainty, find coherence amidst change, and sustain their motivation regardless of external transformations. Cultivating this meta-skill of intentional meaning-making is arguably as important as acquiring new technical competencies in the AI era; it is about developing the internal toolkit to continuously find and create purpose.

There is an emerging and noteworthy paradox, and indeed an opportunity, in the use of AI tools themselves (such as the digital coaching platforms previously discussed) to assist individuals in processing their experiences and finding meaning in a world that is increasingly shaped and disrupted by AI. Artificial intelligence is a primary driver of the very changes that necessitate this profound re-evaluation of meaning at work; for example, by devaluing certain forms of knowledge or enabling the shift towards more modular and precarious career structures.

These transformations can be disorienting and challenging. Yet, AI-powered tools are now being developed and deployed to provide personalized coaching, facilitate structured reflection, and support goal setting related to personal and professional development. This implies that individuals can leverage the very technology that is transforming their work landscape to help them navigate that transformation more effectively and cultivate a resilient sense of meaning.

An AI coach, for instance, might guide someone through exercises to reframe their career goals after AI has impacted their previous role, or facilitate reflective practices designed to help them identify new sources of contribution and value. This suggests a sophisticated evolution in our relationship with AI; from perceiving it primarily as a tool, then as a potential threat, then as a co-creator, and now, potentially, as a supportive guide in our personal quest for meaning. While this development requires careful consideration of the ethics, efficacy, and potential limitations of such AI coaching, it nonetheless presents a powerful new avenue for support in the personal practice of meaning.

OUR EVOLVING RELATIONSHIP WITH AI

Meaning, therefore, is not a static destination to be reached, but rather a continuous and dynamic practice of focused attention, intentional care, and deliberate composition. As individuals and organizations navigate this new era of ubiquitous knowledge and intelligent machines, the capacity to cultivate personal and collective meaning becomes paramount. This understanding, centered on what it means to find purpose when knowledge is a commodity, naturally prepares us to consider not just what we know or what we do, but the very nature of knowing itself. It prompts a counter-intuitive yet vital question: in a world striving for omniscience, what value might lie in not knowing?

SUMMARY: FINDING PURPOSE WHEN KNOWLEDGE IS A COMMODITY

This chapter has navigated the complex terrain of finding meaning and motivation at work in an age where artificial intelligence is rapidly democratizing knowledge, thereby diminishing the differentiating power of traditional expertise. As the scarcity of information dissolves, the human quest for purpose, the drive for genuine contribution, and the sources of intrinsic motivation must find new, more resilient anchors. The exploration has revealed a necessary shift from valuing the mastery of scarce knowledge to cherishing the unique human capacity for impactful contribution and the profound satisfaction of craft.

Careers are transforming from linear ascents to modular, self-narrated journeys, demanding new forms of personal navigation and narrative competence. Furthermore, artificial intelligence itself is evolving from a mere tool or potential disruptor into a dynamic partner in co-creation, offering new avenues for synergistic achievement. In this new landscape, both thoughtful organizational design, which fosters purpose-driven ecosystems, and dedicated personal practice, which empowers individuals to actively construct meaning, become indispensable.

Ultimately, while AI can replicate vast stores of knowledge and automate a multitude of tasks with remarkable efficiency, it cannot replicate the innate human drive for purpose, the deep satisfaction derived from authentic contribution, the nuanced beauty of human craft, or the richness of a self-authored, meaningful life.

In the age of information symmetry, it is these uniquely human attributes that become the true scarcity and, indeed, the ultimate human frontier. As we have redefined meaning in a world where AI knows so much, it raises a compelling, almost paradoxical question: Is there inherent value in not knowing? The subsequent chapter, "The Value of Unknowing - Why Mystery, Uncertainty, and Forgetting Still Matter," will delve into this intriguing proposition, exploring why some measure of unknowing remains essential for creativity, ethical discernment, and the continued flourishing of humanity, perhaps especially in an age increasingly saturated with intelligent machines.

16

THE VALUE OF UNKNOWING

In an era increasingly defined by the ascendancy of artificial intelligence, where the promise of near-omniscient access to information appears within reach, a compelling counter-narrative emerges. This chapter advances the argument that the human capacities for, and experiences of, unknowing; a realm encompassing mystery, uncertainty, ambiguity, and the often-maligned act of forgetting, are not liabilities to be systematically eradicated by technological progress.

Instead, they represent essential, perhaps increasingly vital, assets. These elements of the human condition are fundamental to fostering creativity, ensuring the space for ethical deliberation, promoting robust psychological and cognitive development, and safeguarding the bastions of individual and collective freedom. The relentless pursuit of total knowledge, while cloaked in the allure of ultimate understanding and control, carries the inherent risk of eroding these foundational aspects of what it means to be human.

The broader work of which this chapter is a part, The Age of Information Symmetry, explores a world where AI endeavors to level the playing field of knowledge access, creating a scenario where information, once a scarce and unevenly distributed commodity, becomes universally available. This chapter, however, presents a critical interrogation of this emergent symmetry: what if perfect, unadulterated knowledge, a state of complete informational parity, is not an absolute good? What if the perceived "friction" of not knowing, the cognitive and emotional "space" afforded by uncertainty, is precisely the fertile ground where human ingenuity, empathy, wisdom, and genuine discovery are cultivated? The paradox lies in the possibility that in our quest to eliminate all shadows of ignorance, we might inadvertently extinguish the light of true understanding, which often flickers brightest in the penumbra of the unknown.

To navigate this complex terrain, this chapter sets forth several key learning goals. It aims to critically examine the psychological, social, and cultural implications of a world saturated with AI-driven knowledge, moving beyond a purely utilitarian or instrumentalist view of information. It will articulate the inherent value of mystery, not as a deficit of information but as a catalyst for curiosity, wonder, and deeper, more meaningful forms of engagement with the world and with each other. This perspective will be contrasted with the potential for algorithmic systems, in their drive for optimization and explicitness, to flatten such nuanced experiences. Furthermore, the chapter will explore the ethical dimensions of "total visibility" and the subtle ways in which over-optimization by AI can diminish human agency and the capacity for genuine surprise; an element crucial for learning and adaptation.

A significant portion of this exploration will be dedicated to advocating for the importance of forgetting and unpredictability. These are not mere cognitive lapses or system inefficiencies but necessary conditions for personal growth, social healing, and the dynamism of human culture. In a world that remembers everything, the capacity for reinvention, forgiveness, and progress can be severely hampered. Finally, this chapter will endeavor to introduce both practical and visionary perspectives on how individuals, communities, and societies can consciously cultivate and protect spaces for unknowing, recognizing their intrinsic worth in an age that extols the virtues of absolute clarity.

The central inquiry that propels this chapter, therefore, is not merely whether humanity can achieve a state of near-total knowledge through the advancements of artificial intelligence, but whether it should aspire to such a condition. What are the hidden costs, the unacknowledged trade-offs, in a relentless campaign to eradicate every vestige of the unknown? What vital aspects of human experience might be irretrievably lost if we succeed too well? This exploration is designed to set the stage for a deeper understanding of why, even as AI offers unprecedented levels of clarity and informational access, the shadows of mystery, the discomfort of uncertainty, and the profound grace of forgetting to remain deeply and enduringly important. We begin this journey by confronting the potential downsides of an all-knowing, all-seeing technological environment; what we term the tyranny of total knowledge.

THE TYRANNY OF TOTAL KNOWLEDGE

The relentless advance of artificial intelligence brings with it the seductive promise of a world unburdened by ignorance, a future where data-driven insights illuminate every shadow, optimize every process, and predict every need. Yet, this utopian vision of total knowledge harbors a disquieting potential: a subtle but pervasive tyranny that can emerge when nothing is left unknown, when AI systems appear to comprehend our pasts, decode our present preferences, and chart our future potentials with an authority that surpasses our own self-understanding. This section will dissect the multifaceted risks inherent in such a state, exploring how an environment saturated with AI-driven omniscience can erode human agency, foster a debilitating dependency on algorithmic pronouncements, and ultimately diminish the richness and spontaneity of human experience by systematically reducing novelty and surprise.

One of the most insidious manifestations of this tyranny is the emergence of prediction addiction and the consequent erosion of agency. As AI systems become increasingly adept at forecasting individual behaviors and preferences; from mundane choices like media consumption to life-altering decisions regarding careers or relationships, individuals may develop an unhealthy reliance on these algorithmic prognostications.

This "prediction addiction" is not merely a matter of convenience; it strikes at the core of human autonomy. When faced with a constant stream of AI-generated recommendations, optimized pathways, and personalized forecasts, the individual's own capacity for judgment, intuition, and autonomous decision-making can begin to wane. Studies and analyses indicate that algorithmic decision-making, particularly when its processes are opaque or its outcomes perceived as unfair, can lead to a host of negative psychological effects, including moral disengagement and a general aversion to algorithms, which in itself is a reaction to a perceived loss of control. The allure of an "optimal" choice, as defined by an algorithm, can overshadow the value of personal exploration, risk-taking, or even the right to make a "suboptimal" choice that aligns more closely with idiosyncratic personal values or desires.

The insidious nature of this erosion of agency lies not merely in external systems making choices for individuals, but in the cultivation of an internal passivity. If an AI consistently provides what appears to be the most efficient, effective, or satisfying path, the human "muscle" for navigating uncertainty, for making intuitive leaps in the absence of complete data, or for choosing a path less traveled simply because it feels right, may begin to atrophy.

This process unfolds subtly: AI offers increasingly accurate predictions; humans, naturally seeking to reduce uncertainty, gravitate towards these seemingly superior choices; this reliance lessens the need for personal deliberative effort; and over time, this lack of practice diminishes self-trust and confidence in one's own decision-making faculties, creating a feedback loop of escalating dependence. The "tyranny" thus materializes not as overt coercion, but as a gentle, almost imperceptible, drift towards a state where AI transitions from a supportive tool to a de facto director of human affairs.

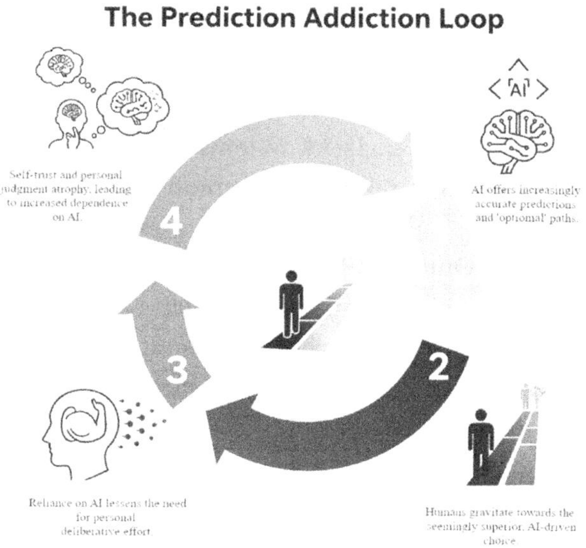

Beyond individual choices, the pursuit of total knowledge fuels algorithmic determinism, potentially stifling personal development and fostering societal homogenization. AI systems, by their very design, learn from vast quantities of past data to construct "algorithmic identities" for users. These digital representations, however sophisticated, are inherently backward-looking. They excel at identifying patterns in what has been and projecting those patterns into the future.

The peril here, as articulated by concerns around algorithmic determinism, is that individuals can become locked into self-reinforcing cycles dictated by their past actions and the AI's interpretation of them. If the AI consistently recommends experiences, information, or connections based on a static or slowly evolving profile, it becomes significantly more challenging for an individual to break free from established patterns, to experiment with new facets of their identity, or to undergo genuine personal transformation. The critical question of whether we are shaping technology or if technology is shaping us finds a stark answer here: an AI that perpetually reflects our past selves actively constrains our capacity to become someone new.

This phenomenon has implications that extend beyond the individual to the societal level. If AI systems are predominantly optimized to recognize and reinforce common patterns of behavior, preference, and belief, there is a risk of a gradual "averaging" of human experience. Outlier behaviors, eccentricities, radical innovations, and dissenting viewpoints; all of which often begin as deviations from the norm and are crucial for cultural dynamism, might be subtly discouraged, filtered out, or simply not catered to by systems designed for mass personalization based on majority trends.

The chain of events is logical: AIs learn from data to predict and personalize; this personalization, based on past patterns, tends to reinforce those patterns; individuals are steered towards experiences "like" their past or "like" others, diminishing exposure to true novelty; this creates echo chambers not just for information but for life paths; and over time, this could lead to a convergence of experiences, a societal regression to the mean, thereby reducing the richness that arises from diverse, unpredictable individual trajectories. This poses a profound challenge to cultural evolution and the very sources of innovation.

Finally, the drive for total knowledge often manifests as an obsession with over-optimization, which paradoxically can lead to a flattening of novelty and surprise. AI systems are frequently designed to maximize efficiency, engagement, or specific, measurable outcomes. While optimization can yield significant benefits in targeted applications, its universal application to all aspects of life risks filtering out the unexpected, the serendipitous, and the "inefficient" detours that are often the crucibles of discovery, creativity, and profound human joy. Novelty and surprise are not mere experiential luxuries; they are fundamental to learning, memory acquisition, and the motivation to explore.

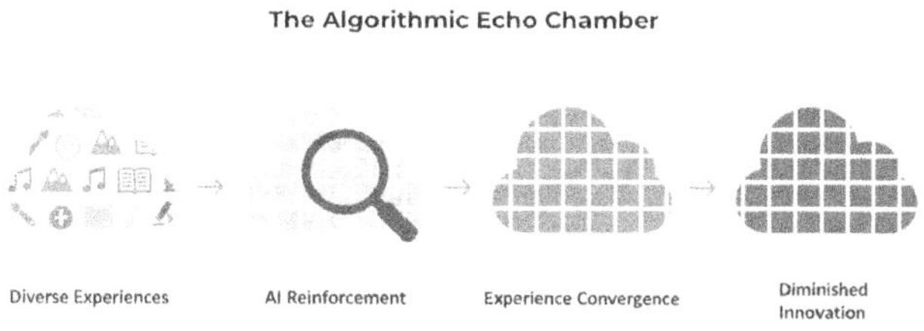

An over-optimized world, therefore, may become a less stimulating, less engaging, and ultimately, less meaningful one. Even in areas where AI optimization shows clear benefits, such as search, there can be trade-offs; for instance, finding information more quickly through AI-targeted content has been associated with shorter engagement times. This raises the question of what might be lost during those "saved" moments; perhaps the space for deeper reflection, for tangential thoughts, for the unexpected connections that shorter, more "efficient" interactions preclude.

The pursuit of total knowledge, therefore, can paradoxically lead to a narrower, more circumscribed field of experience. By attempting to map, predict, and control every variable to achieve an optimal state, we risk creating systems that inadvertently prune the very branches of possibility that lead to unscripted discovery and genuine innovation. The drive for complete knowledge and an entirely optimized existence, while originating from an understandable human desire for certainty and improvement, thus risks engendering a world that is less agentic, less surprising, and ultimately, less reflective of the full spectrum of human potential. If certainty and predictability become the dominant modes of being, the human need for the unknown, for the sense of wonder and mystery that fuels so much of our inner life and creative endeavor, is left unnourished. This critical observation leads us directly to consider the intrinsic, and perhaps indispensable, value of mystery itself.

The ability to be surprised may indeed be more vital for human flourishing than the ability to be perpetually certain. As we move forward, we will explore how mystery serves not as a void, but as a vital nutrient for the human spirit.

This exploration of the downsides of total knowledge sets the stage for appreciating the inherent worth of what lies beyond our immediate grasp. We now turn to consider mystery not as an absence of information, but as a profound human need.

MYSTERY AS A HUMAN NEED

In a world increasingly saturated with information and driven by the pursuit of explicitness, it is tempting to view mystery as a mere deficit, a temporary state of ignorance awaiting the clarifying light of data and analysis. However, this section argues for a more profound understanding: mystery, in its various manifestations, is not a void to be filled but a fundamental human need, a powerful and often indispensable catalyst for creativity, spiritual inquiry, emotional depth, and sustained intellectual exploration. Drawing upon insights from psychology, art, religion, and philosophy, it becomes evident that encounters with the unknown, the ambiguous, and the seemingly inexplicable are not obstacles to human flourishing but vital ingredients that enrich the tapestry of human experience.

At the heart of mystery's power is its capacity to evoke awe, wonder, and an insatiable curiosity. The German theologian Rudolf Otto famously described the "numinous" experience; a profound encounter with something perceived as "wholly other," inspiring a mixture of fear, fascination, and reverence. This experience, often central to religious and spiritual sentiments, is not predicated on possessing answers but on confronting deep, resonant questions that transcend immediate comprehension. It is the vastness of the unknown, the hint of something beyond our rational grasp, that ignites a sense of wonder. This wonder, in turn, fuels curiosity. A world stripped of mystery, where every phenomenon is readily explained and every question has an instant, algorithmically generated answer, might paradoxically become a world with a diminished impetus for self-driven learning and genuine discovery.

The pursuit of "total knowledge" by AI, if unchecked, could inadvertently shrink the spaces where we encounter the genuinely numinous, thereby diminishing our capacity for awe. If everything becomes explainable and predictable, the conditions that foster such profound experiences; a sense of vastness, incomprehensibility, and the sublime, are systematically reduced. While knowledge is undeniably valuable, an overemphasis on complete, immediate knowability risks making the world feel smaller, more mundane, and less wondrous.

The indispensable role of not-knowing is vividly illustrated in the realms of art, religion, and love. Artists and storytellers have long understood the power of withholding information. Alfred Hitchcock, the master of suspense, famously distinguished between surprise (a sudden, unexpected event) and suspense (the anticipation of an event whose outcome is uncertain). Suspense, which he argued is far more engaging, thrives precisely on what the audience doesn't know, creating a tension that captivates and involves.

Similar to the "mystery-box storytelling" used in television shows like Lost, narrative structures kept audiences engaged by intentionally leaving important questions unanswered. This created years of speculation and shared interpretation, though the lack of resolution ultimately frustrated viewers. The allure was undeniably rooted in the persistent not-knowing. In literature, ambiguity is not a flaw but a sophisticated device that invites active reader participation, allowing for a multiplicity of interpretations and a richer, more personal engagement with the text. Such ambiguity often mirrors the inherent complexity and indeterminacy of human experience itself.

In the domain of religion, many traditions explicitly embrace mystery as central to faith. Concepts like the "Sacred Mysteries" in Christianity are not presented as intellectual puzzles to be solved but as profound truths to be contemplated, often residing beyond the full reach of rational comprehension. As one theological perspective notes, "If there were nothing mysterious in our notions of God... we have only created a make-believe God, a creature of our own limited minds".

Even in the deeply personal sphere of love and intimacy, complete transparency and absolute predictability can diminish the allure and dynamism of a relationship. The ongoing process of discovering another person, with their inherent complexities, hidden depths, and un-knowable facets, is integral to what makes human connections vibrant and enduring. The potential for AI to analyze and predict human behavior with increasing accuracy, if pervasively applied to these domains, presents a subtle threat. An AI that "solves" the ambiguities of a complex novel, "explains away" a profound spiritual experience, or "predicts" a partner's every thought and action might offer a superficial form of knowledge but, in doing so, could destroy a deeper form of engagement that thrives on the unquantifiable, the un-interpretable, and the genuinely unknown. The value added by AI in providing data-driven insights in these areas might come at the steep cost of diminishing the experiential richness that arises from direct, unmediated encounters with mystery.

Furthermore, ambiguity serves as a potent emotional and intellectual stimulant, fostering cognitive flexibility and creativity. Research indicates that individuals exposed to ambiguous stimuli often exhibit greater cognitive flexibility, leading to enhanced problem-solving skills and divergent thinking; the capacity to explore multiple possible solutions. A tolerance for ambiguity is significantly and positively related to creativity; those who can comfortably navigate unclear or contradictory information are less satisfied with suboptimal solutions and are more inclined to engage in creative thought processes to resolve complexities. This capacity is highly valued in dynamic environments requiring innovation.

Conversely, an educational system or professional environment that becomes increasingly reliant on AI for clear, definitive answers might inadvertently curtail opportunities for individuals to develop their own tolerance for ambiguity. If the "struggle" with uncertainty is consistently short-circuited by algorithmic solutions, the very cognitive muscles that ambiguity strengthens; creative thinking, adaptive problem-solving, and intellectual resilience, may remain underdeveloped. This could paradoxically leave individuals less equipped to handle truly novel or complex situations where AI itself might not (yet) possess the definitive answer, or where the "answer" requires a uniquely human leap of imagination.

The human mind, it appears, does not merely tolerate but actively thrives on a certain measure of mystery and ambiguity. These are not defects in our understanding or gaps in our knowledge to be hastily filled; rather, they are invitations to deeper engagement, to more profound creativity, and to a more nuanced appreciation of the world and our place within it. If this innate human need for the unknown is so vital for our cognitive and emotional flourishing, then it compels us to consider whether individuals possess, or ought to possess, a fundamental right to it; a right to shield themselves, at times, from the unrelenting glare of total information and the illusion of absolute certainty.

What makes us most human is often not what we know; but what we can't know, and how we engage with that frontier. This realization leads us to the question of epistemic autonomy and the right to choose our own relationship with knowledge.

This affirmation of mystery's value naturally transitions to the question of an individual's control over their own epistemic landscape. If unknowing is essential, then the right not to know becomes a critical consideration.

THE RIGHT TO NOT KNOW

The preceding exploration of mystery as a human need sets the stage for a critical inquiry into the nature of our relationship with knowledge itself, particularly in an era characterized by information superabundance and the pervasive analytical capabilities of artificial intelligence. This section will introduce and defend the concept of "epistemic rights," with a specific focus on a right that may seem counterintuitive yet is increasingly vital: the right not to know. In a world saturated with ceaseless personal data collection and the proliferation of AI-driven predictions, the capacity to consciously opt out of specific knowledge flows becomes paramount. This includes the ability to refuse predictive profiling and to maintain intentional spaces of ignorance. Such practices are crucial for preserving individual autonomy, safeguarding mental well-being, and ensuring healthy human development. (A pertinent literary example of this concept can be found in the works of H.P. Lovecraft, where he frequently alludes to the detrimental mental health outcomes that can arise from knowing too much.)

The philosophical underpinnings of epistemic rights provide a framework for this discussion. Epistemic rights are generally understood as entitlements that pertain to "epistemic goods"; that is, goods related to the domain of inquiry, knowledge, and understanding. While often conceived as the right to access information, to education, or to truthful disclosure, the concept, as articulated by scholars like Lani Watson, also encompasses the more nuanced right not to be informed under certain circumstances.

These rights are seen as complex entitlements designed to protect individuals from potential harm within the epistemic domain, and they are justified by the inherent value and moral status of the right-holders themselves. The assertion of a "right to not know" directly challenges the pervasive techno-optimist assumption that an increase in information is always, and unequivocally, beneficial. It introduces a crucial layer to information ethics, suggesting that considerations must extend beyond mere access and accuracy to encompass the potential impact of knowledge on individuals and the legitimacy of choosing ignorance as a protective or empowering stance. This perspective necessitates a shift in ethical calculus: the primary question is not always "is the information available and accurate?" but rather, "what is the impact of this information on the individual, and do they possess a defensible right to refuse it?"

This right finds a particularly salient application in the context of the ethics of ignorance concerning predictive profiling. Artificial intelligence systems are demonstrating an ever-increasing capacity to generate detailed, probabilistic profiles about individuals, forecasting potential health risks, personality traits, future behaviors, and even life outcomes, often based on complex correlations within vast datasets. The critical ethical question arises: what if an individual does not wish to know these predictions? Does knowledge of a statistically probable, yet not certain, negative health outcome empower preventative action, or does it induce debilitating anxiety and a sense of fatalism? Could such predictions, if widely known or acted upon by institutions, become self-fulfilling prophecies or lead to new forms of discrimination?

The right to refuse to be subjected to certain types of predictive analytics, or to know their results, emerges as a vital safeguard for mental well-being and personal agency. It is a defense against a novel form of "epistemic predestination," where one's future feels predetermined by an algorithmic forecast. By choosing not to know, an individual can preserve the subjective experience of an open future, a future where their choices feel genuinely impactful and where they retain the possibility of defying statistical expectations. This is not an argument for denying established medical diagnoses, but for asserting control over the reception of probabilistic, potentially life-altering predictions delivered by external, non-human entities.

The developmental needs of children underscore the profound importance of uncertainty and, by extension, a protected space for not knowing. Childhood is a period of intense exploration and learning, and research suggests that uncertainty, while sometimes a source of anxiety, is a crucial catalyst for this development. Higher levels of exploration are often observed when children face greater uncertainty, as they are intrinsically motivated to make sense of a world they do not yet fully understand. The ability to monitor one's own feelings of uncertainty is a central metacognitive skill that develops throughout early childhood and is foundational for self-guided learning and academic achievement. If children are consistently provided with AI tools that offer instant, definitive answers to all their questions, their natural curiosity might be blunted, their development of problem-solving skills hindered, and their resilience in the face of the unknown diminished.

The inherent "struggle" of not knowing, of grappling with a question before arriving at an answer, has immense educational value. Overexposing children to an environment of AI-driven certainty could inadvertently cultivate a generation with a lower tolerance for ambiguity, reduced intrinsic motivation for learning, and a diminished capacity for independent critical thinking. Thus, the "right to not know" for children might be more accurately framed as a "right to encounter and navigate developmentally appropriate levels of uncertainty," ensuring they have the space to develop these crucial cognitive and emotional capacities.

Recognizing these needs leads to the imperative of designing AI systems that can exercise strategic silence and foster reflection. This is not solely about users actively opting out of information flows, but about embedding principles of epistemic respect within the architecture of AI itself. Systems could be designed to pose insightful questions rather than merely delivering answers, or to create "mysterious science scenes" that spark wonder and collaborative inquiry, as suggested in educational contexts. AI can function as a "quiet thought partner," helping users to reframe ideas or consider alternatives, thereby supporting rather than supplanting the creative and reflective process.

Frameworks for principle-led AI analysis and AI-resilient learning design also emphasize the importance of intentionally creating gaps for human judgment, interpretation, and critical reflection. An AI, therefore, could be programmed to discern when providing more information might be counterproductive, or when a period of silence, a well-placed question, or an ambiguous prompt might be more beneficial for the user's learning, creativity, or decision-making journey. Such design acknowledges that the goal of interaction with AI is not always immediate and total clarification, but sometimes deeper understanding, which can be a fruit of contemplation in the face of the unknown.

The right to not know, therefore, emerges as a cornerstone of epistemic autonomy in the digital age. It serves to protect individuals from the potential harms of unwanted, overwhelming, or developmentally inappropriate information. This right is intrinsically linked to another profound form of "unknowing" that is essential for human flourishing: the capacity to forget. While AI systems relentlessly move towards perfect recall and data permanence, human freedom, growth, and the ability to move forward often depend on the gentle, and sometimes necessary, act of letting go of the past.

Autonomy requires limits on insight. Consent applies to knowledge as much as it does to action. This principle guides us toward considering the crucial role of forgetting in a world that increasingly never forgets.

FORGETTING AS FREEDOM

The human capacity for forgetting, often perceived as a cognitive imperfection or a frustrating lapse in an otherwise efficient mental apparatus, is, in reality, a profoundly adaptive and frequently liberating mechanism. This section argues that forgetting; whether personal, social, or even technologically mediated, is essential for individual growth, psychological healing, the fluid evolution of identity, and the healthy renewal of societies. This human experience stands in stark contrast to the burgeoning trend towards perfect, permanent digital memory, facilitated by sophisticated data systems and artificial intelligence. By exploring the multifaceted benefits of forgetting, we can better understand the potential dangers of a world that never forgets and appreciate why the "freedom to forget" is a critical aspect of human dignity and societal progression.

At the individual level, the psychological necessity of forgetting is most evident in the contexts of healing, trauma, and identity formation. The human mind possesses remarkable defense mechanisms to protect itself from overwhelming experiences. Dissociative amnesia, for instance, involves the mind blocking out access to important, often traumatic, memories as a means of self-preservation. While the therapeutic process may involve carefully re-engaging with such memories to facilitate integration and healing, the initial capacity to "not remember" can be a crucial adaptive response, preventing psychological fragmentation. Beyond extreme trauma, everyday forgetting allows for the fading of painful emotions associated with past failures, embarrassments, or losses, enabling individuals to move forward without being perpetually haunted by every misstep or sorrow.

Furthermore, identity is not a static monolith but a dynamic, evolving construct. The ability to forget, or at least to allow certain past self-conceptions and experiences to recede in prominence, is vital for personal reinvention and growth. If individuals were burdened with a perfect, high-fidelity recall of every past action, thought, and outdated belief, the psychological weight could be immense, hindering the capacity to change and adapt. AI systems, with their potential for perfect recall and the creation of permanent data archives, risk establishing an implicit societal expectation of total memory fidelity. This could inadvertently lead to the pathologization of normal human forgetting, which is often a healthy process of cognitive and emotional pruning, creating an undue pressure on individuals that is neither natural nor always conducive to psychological well-being.

Recognizing the potential harms of indelible digital records, legal frameworks have begun to emerge that codify a form of data deletion as a human right, most notably through the "right to be forgotten". Provisions such as Article 17 of the European Union's General Data Protection Regulation (GDPR), also known as the "right to erasure," grant individuals the power to request the deletion of their personal data under specific conditions. These conditions include instances where the data is no longer necessary for its original purpose, where consent for processing is withdrawn, or where the data has been processed unlawfully. This right is not merely a technicality of data management; it is a profound acknowledgment of an individual's need to control their digital narrative and to be protected from the enduring negative impact of outdated, irrelevant, or potentially damaging information. The obligation under GDPR can even extend to informing other data controllers if the personal data has been disseminated publicly, underscoring the societal intent to enable a more comprehensive erasure of digital traces.

The "right to be forgotten" is thus more than a privacy protection; it is a societal recognition that individuals evolve and should not be eternally shackled to their digital past. It attempts to replicate, in the digital realm, the natural human process of forgetting and the possibility of a "fresh start" that this allows. In an age where AI systems learn from and act upon vast troves of personal data, the ability to ensure this data is current, relevant, and can be erased upon request becomes paramount. AI models trained on outdated or inaccurate personal information can perpetuate biases, lead to flawed assessments, and unfairly constrain individuals based on a past that no longer defines them. The right to erasure serves as a crucial check on the power of permanent digital memory and the AI systems that leverage it, striving to balance technological capability with human dignity and the capacity for change.

The societal implications of a world that never forgets are far-reaching, extending to the social dangers of permanent memory, including pervasive surveillance, the chilling effects of "cancel culture," and even potential cognitive detriments. The permanence and instant recall of digital information can fuel "cancel culture," a phenomenon where past missteps, statements, or affiliations, often unearthed from years or decades prior and stripped of their original context, are used to inflict severe and lasting social, professional, and psychological damage. The internet's unforgiving memory creates an environment where redemption is difficult, and the fear of future judgment can stifle authentic expression and risk-taking. This perpetual scrutiny can lead to heightened anxiety and self-censorship.

Moreover, the constant engagement with digital technologies and the overwhelming influx of information they provide have been linked to concerns about "digital dementia": a term used to describe a decline in cognitive functions such as short-term memory, attention span, and multitasking ability, potentially stemming from technology overuse. While digital tools can augment memory, an over-reliance on them and the sheer volume of data they present may paradoxically impair certain innate cognitive abilities. Beyond individual cognition, the persistence of digital memory is a cornerstone of surveillance capitalism, where every click, search, and interaction is recorded, analyzed, and commodified, often without full user awareness or consent. The inability to truly erase one's digital footprint renders individuals perpetually transparent and vulnerable to data exploitation and manipulation.

A society equipped with a perfect, collective digital memory, curated and accessed by AI, risks becoming less tolerant, less forgiving, and more prone to cycles of perpetual judgment and retribution. Just as individuals require the capacity to forget for personal healing and growth, societies too may need a form of "collective forgetting" or, at the very least, a collective capacity to move beyond past conflicts, errors, and grievances without being endlessly haunted by their digital specters. The freedom to forget, therefore, is not only a personal liberty but also a collective necessity for societal healing, progress, and the maintenance of a humane and dynamic culture.

Just because a machine can remember everything doesn't mean it should. And it certainly does not mean that we, as humans, should be compelled to live as if we, too, must remember everything, or be remembered for everything, forever. The grace of forgetting allows for new beginnings.

The burden of perfect memory, whether individual or societal, highlights the value of release. If perfect recall can be oppressive, then perhaps perfect, instantaneous knowledge in all decision-making contexts is also not the unambiguous ideal it is often portrayed to be. This leads to considering situations where choosing not to know, or deliberately limiting information, can be a wise and strategic act.

STRATEGIC IGNORANCE IN DECISION-MAKING

While the pursuit of comprehensive information is often lauded as the cornerstone of sound decision-making, a compelling counterargument exists for the utility of "strategic ignorance": the deliberate and conscious choice not to seek, process, or utilize certain information. In this section we will explore how embracing a degree of unknowing can paradoxically enhance the clarity, fairness, and effectiveness of judgments. By strategically limiting informational inputs, decision-makers can reduce cognitive "noise," mitigate the influence of pervasive biases, and avoid the paralysis that often accompanies information overload. In an age where AI can deluge us with data, the wisdom to discern what not to know may become as crucial as the ability to acquire knowledge.

A primary benefit of strategic ignorance lies in its capacity for noise reduction, distinguishing critical signals from overwhelming informational static. Daniel Kahneman, in his work on decision-making, highlights "noise" as the undesirable variability in judgments that, given the same information, should ideally be consistent. An excess of information, particularly if it is irrelevant, redundant, or of poor quality, can significantly amplify this noise, obscuring the true signals necessary for an optimal decision. Strategic ignorance, in this sense, functions as a form of "decision hygiene". By consciously filtering out or choosing not to engage with certain data streams, decision-makers can simplify the cognitive environment, reduce mental clutter, and focus more effectively on the core variables that genuinely matter.

While AI systems are capable of processing vast datasets, they can also contribute to signal overloading if their outputs are not carefully curated or if they present information without appropriate contextualization or prioritization. Human decision-makers interacting with such AI might find their judgment improved if the AI itself is programmed to strategically withhold or adeptly summarize information, thereby reducing the cognitive load on the user. The prevailing ethos in many data-driven cultures often valorizes the accumulation of "all available data" prior to making a decision. However, strategic ignorance posits a crucial corrective: the quality and relevance of information, coupled with the cognitive state and capacity of the decision-maker, are frequently more critical determinants of decision quality than the sheer volume of data processed.

Strategic ignorance is also a powerful tool for avoiding bias, particularly through the implementation of blind evaluation processes. The efficacy of blind procedures is well-documented in diverse fields such as hiring, academic peer review, and artistic auditions. By systematically removing or obscuring identifying information; such as an applicant's name, gender, ethnicity, age, or institutional affiliation, these processes aim to mitigate the impact of unconscious biases that can skew evaluations. This is a direct application of strategic ignorance: decision-makers are deliberately kept unaware of certain attributes of the individual or item being assessed to ensure that judgments are based on merit, skill, or intrinsic quality rather than on potentially prejudicial factors.

While some research explores how willful blindness can be exploited to evade accountability in unethical contexts, such as corporate malfeasance, the focus here is on the positive and ethical application of strategic ignorance to enhance fairness and objectivity. AI systems could be instrumental in facilitating such blind evaluations, for example, by automatically redacting biasing information from applications or submissions before they are presented to human reviewers. The adoption of such practices, however, requires more than a mere procedural tweak; it demands a systemic commitment to fairness that may challenge entrenched power dynamics or ingrained habits of "holistic" yet potentially biased assessment. It necessitates a trust in the de-identified process over individual (and inherently fallible) claims of perfect objectivity, confronting the uncomfortable truth that even well-intentioned decision-makers are susceptible to bias.

Beyond noise reduction and bias mitigation, strategic ignorance can lead to paradoxical insight, where seeing less allows for better, more robust decisions. In many complex situations, an overabundance of detail or context can obscure the larger picture, leading to "analysis paralysis" or a focus on trivialities at the expense of fundamental principles. Sometimes, a simplified model of reality, or a decision-making process that deliberately narrows the range of considered variables to the most critical ones, can yield more effective and resilient outcomes, especially under conditions of high uncertainty or time pressure.

Consider, for example, complex policy areas where the intricacies of regulations are understood only by a small cadre of specialists. While deep expertise is invaluable, there are instances where an outsider's "ignorant" perspective, unencumbered by the minutiae, can cut through jargon, identify foundational flaws, or propose radically simpler solutions. AI, if programmed not merely to provide exhaustive data dumps but to present only the most salient variables, or to simulate outcomes based on strategically limited inputs, could assist human decision-makers in avoiding the "data swamp" and focusing on what truly drives outcomes. This disciplined limitation of analytical inputs can also create valuable space for more holistic, pattern-based, or intuitive modes of thinking to emerge. Intuition often operates by synthesizing complex patterns without conscious, step-by-step deliberation. By clearing away some of the "noise" of excessive or distracting data, strategic ignorance might create more favorable conditions for these intuitive insights to surface, be recognized, and contribute to the decision-making process, especially when facing problems that defy purely analytical or algorithmic solutions.

The deliberate choice not to know, therefore, can be a sophisticated and powerful strategy for improving the quality, fairness, and even the intuitive depth of our decisions. This implies a significant shift in how we approach information management and system design: our tools, particularly AI systems, should not invariably aim for exhaustive data provision. Instead, their utility might be greatly enhanced if they are designed to embrace, facilitate, and even recommend certain forms of strategic unknowing.

Sometimes, the smartest move is not to know. Or, at least, not to know yet. This acknowledgment paves the way for designing systems that respect the nuanced relationship between information and effective action.

This recognition that less can be more in decision-making leads directly to considering how our technological systems, especially AI, can be designed to incorporate and even protect valuable forms of uncertainty.

DESIGNING SYSTEMS THAT EMBRACE UNCERTAINTY

The acknowledgment that unknowing, in its various forms, holds intrinsic value necessitates a paradigm shift in how we design and interact with our technological systems, particularly artificial intelligence. If mystery, ambiguity, and even strategic ignorance can be beneficial, then our AI and knowledge ecosystems should not be engineered solely for the relentless pursuit of certainty and exhaustive information delivery. Instead, this section will explore practical and philosophical principles for creating systems that do not merely tolerate uncertainty but actively allow for, protect, or even leverage it to enhance user experience, foster creativity, facilitate deeper reflection, and promote genuine exploration. This involves rethinking user-AI contracts, interface design, and the very goals we set for our intelligent tools.

A fundamental aspect of designing for uncertainty involves managing user expectations and interactions around probabilistic versus deterministic AI outputs. Deterministic AI systems, common in traditional programming, provide consistent, predictable outputs for a given set of inputs (e.g., rule-based systems, certain algorithms used in navigation). In contrast, many advanced AI models, especially in machine learning and generative AI, are probabilistic. This means they integrate elements of randomness and uncertainty into their decision-making processes, potentially yielding different outputs even when presented with the same initial input or prompt.

User experience (UX) design must grapple with this inherent variability found within probabilistic models. Users accustomed to the predictability of deterministic tools may find the variable outputs of probabilistic AI frustrating if their expectations are not properly managed. However, this probabilistic nature is also a wellspring of creativity and novelty, as it allows AI to generate a diverse range of ideas or solutions. The design challenge, therefore, is to strike a balance: to harness the creative potential of AI's unpredictability while providing users with sufficient control and understanding to navigate this uncertainty productively.

This requires a fundamental shift in the implicit user-system contract. Instead of promising absolute certainty and precise control, AI systems might be more honestly and effectively framed as collaborative partners in exploration, where output variability is not a system flaw but a feature that can lead to unexpected and valuable discoveries. Interfaces should guide users to embrace this, perhaps by offering multiple suggestions, encouraging iterative refinement of prompts, or framing AI outputs as "proposals" or "inspirations" rather than definitive "answers."

This reframing naturally leads to the development of interfaces that invite exploration, not just rapid solutions. Many current AI applications are optimized for efficiency, designed to provide the quickest, most direct answer to a query or the most streamlined path to task completion. While valuable, this focus can inadvertently short-circuit the human exploratory process, where learning and insight often arise from meandering paths, unexpected encounters, and serendipitous discoveries. User interface (UI) and UX design can play a crucial role in fostering these richer interactions.

For example, the WikiTrip app, designed to facilitate serendipitous discovery, allows users to navigate interconnected Wikipedia articles, with AI subtly guiding and occasionally deviating the path to introduce novelty. Similarly, recommender systems can incorporate user controls and visualizations that encourage the exploration of diverse and previously unknown items, moving beyond simple prediction of immediate preferences.

Adopting Human-Centered Design (HCD) principles ensures that AI serves as a co-pilot, augmenting human creativity and adaptability rather than replacing them. Interfaces built on these principles do more than just help users find information; they cultivate a disposition towards curiosity, intellectual risk-taking, and the ability to "wander wisely" within vast information landscapes. In a world where rote answers become increasingly commoditized by AI, the human capacity to ask meaningful questions and forge novel connections becomes paramount. Exploratory interfaces can serve as vital training grounds for these essential "meta-skills," transforming AI interaction from mere information retrieval into a pedagogical environment that shapes how users approach knowledge and discovery.

To achieve this, systems must actively create space for serendipity, ambiguity, and reflection. This means designing AI that can intentionally leave gaps, introduce controlled ambiguity, or prompt deeper thought rather than always defaulting to definitive statements or immediate resolutions. AI can act as a "thought partner", for example, by helping to reframe scattered notes into a structured outline or by suggesting alternative phrasings, thereby supporting the creative writing process by shaping raw ideas without dictating the final form. In educational settings, AI can generate "mysterious science scenes" to provoke discussion or act as a "silent ghost writer" presenting one side of an argument to stimulate debate and critical thinking.

AI-driven prompts can also encourage students to reflect on their own learning processes and hypotheses. Furthermore, when integrating generative AI into complex analytical tasks, reflective processes are critical; the very act of drafting an analysis plan that considers AI's role encourages evaluators to make methodologically and ethically sound decisions, recognizing that human judgment remains indispensable for error-checking, interpretation, and sensemaking.

AI-resilient learning design similarly prioritizes higher-order thinking skills such as analysis, evaluation, and creation, often by incorporating real-world tasks that demand more than simple AI-generated output. Designing for such "constructive uncertainty" allows AI to transcend its role as an information provider and become a tool for metacognition. By presenting users with alternatives, by not resolving ambiguity too quickly, or by posing challenging questions, AI can help individuals become more aware of their own thought processes, biases, and underlying assumptions, thereby fostering a more critical and nuanced engagement with knowledge.

The list below outlines some key design principles for AI systems that aim to embrace and leverage uncertainty, connecting these principles to specific system behaviors and the desired benefits for users.

Embrace probabilistic nature

- **AI system behavior:** Produces multiple or varied outputs and supports iteration rather than single "correct" answers.
- **Desired user outcome:** Sets appropriate expectations, sparks creativity, and helps users develop adaptability in uncertain environments.

Foster exploration

- **AI system behavior:** Introduces tangential pathways, visualizes relationships, and injects novelty into responses.
- **Desired user outcome:** Encourages curiosity, enables serendipitous discovery, and builds navigational skills across complex idea spaces.

Enable strategic silence

- **AI system behavior:** Asks probing questions, selectively withholds information, or presents ambiguity rather than immediate resolution.
- **Desired user outcome:** Promotes autonomy, encourages deeper thinking, and stimulates critical inquiry instead of passive consumption.

Support reflection

- **AI system behavior:** Prompts metacognition, presents multiple perspectives, and facilitates analysis of errors or alternative interpretations.

- **Desired user outcome:** Enhances self-awareness, strengthens critical thinking, and improves judgment over time.

Good systems, therefore, do not just answer; they leave room to wonder, to question, and to explore. They acknowledge that the human mind often finds its greatest insights not in the consumption of pre-digested facts, but in the active engagement with the unknown.

If we are to truly benefit from the rich potential of uncertainty, our technological systems must be designed not as relentless engines of clarification, but as thoughtful partners that respect and even cultivate the spaces where human creativity, reflection, and discovery flourish. This is not merely a technical challenge but a profound philosophical one, requiring us to reconsider the ultimate purpose of our tools. However, the responsibility for valuing the unknown rests not solely with the designers of these systems, but with each individual user.

This imperative to design for uncertainty leads to a final consideration: how individuals can actively cultivate and protect the value of the unknown in their own lives and work, resisting the homogenizing pressures of an all-knowing environment.

RECLAIMING THE UNKNOWN

The journey through the potential tyrannies of total knowledge, the intrinsic human need for mystery, the assertion of a right not to know, the freedom found in forgetting, the wisdom of strategic ignorance, and the principles for designing systems that embrace uncertainty culminates in a deeply personal and societal call to action. This concluding section urges a conscious and active reclamation of the unknown.

In an age increasingly dominated by the allure of AI-driven omniscience and algorithmic certainty, it is paramount that individuals and communities cultivate an appreciation for doubt, make deliberate space for reflection and intellectual wandering, and recognize the generative power of mystery. This is not a retreat into Luddism or obscurantism, but a sophisticated engagement with our technological age; a means of preserving human agency, fostering critical thought, and ensuring that our tools serve our deepest values rather than subtly reshaping them in unintended ways.

A cornerstone of this reclamation is the cultivation of the discipline of doubt, extending beyond its Cartesian origins. René Descartes, in his "methodic doubt," employed systematic skepticism as a philosophical tool to dismantle preconceived notions and arrive at foundational truths, famously concluding "Cogito, ergo sum" ("I think, therefore I am"). While Descartes' ultimate aim was to establish certainty, the practice of doubting itself retains immense value in the contemporary context, particularly when confronted with the prolific output of AI systems.

Philosophical skepticism, in its broader sense, has always involved challenging the adequacy, reliability, and underlying principles of knowledge claims. In an era where AI can generate information, compose texts, and offer analyses with remarkable speed and apparent authority, a healthy, disciplined skepticism becomes a crucial form of intellectual hygiene. This is not about a wholesale rejection of AI-generated content, but rather a commitment to critically interrogating its outputs: questioning its sources, understanding its potential biases (often inherited from training data), recognizing its current limitations (such as the propensity for "hallucinations" or factual inaccuracies), and resisting the passive consumption of information. In a world awash with algorithmically generated "answers," the discipline of doubt is less about finding absolute truth and more about maintaining intellectual agency, ensuring that humans remain the active arbiters of meaning and value.

Complementary to this critical stance is the active effort of making space for reflection, intellectual wandering, and profound questioning. The relentless pace of modern life, amplified by constant connectivity and the expectation of instant responsiveness, often leaves little room for slower, more contemplative modes of thought. Reclaiming the unknown involves consciously carving out "unplugged" time or creating environments where the pressure for immediate answers and constant informational intake is removed. This allows for the kind of deep reflection that fosters self-awareness and nuanced understanding. Furthermore, it encourages "intellectual wandering"; the exploration of ideas, topics, or experiences without a predefined goal or an immediate expectation of utility. Such unstructured exploration is often the breeding ground for serendipitous discoveries and creative insights, the very kinds of breakthroughs that highly optimized, goal-directed AI systems might preclude. As emphasized in earlier discussions (particularly in relation to Chapter 5's theme of "Curiosity as a Competitive Edge"), when answers become ubiquitous and cheap, the value shifts decisively towards the ability to formulate deep, meaningful, and generative questions. Embracing the unknown is intrinsically linked to this art of questioning, as true inquiry often begins from a place of acknowledged ignorance or profound wonder. The deliberate cultivation of these practices; reflection, wandering, questioning, can be seen as a form of cognitive diversification, strengthening mental faculties that risk underutilization in hyper-efficient, answer-oriented digital environments. These practices build intellectual resilience, enhance creativity, and foster a form of wisdom that transcends mere information processing.

Finally, embracing and reclaiming the unknown can be understood as a potent form of personal and civic resistance against encroaching forces that seek to diminish human autonomy and complexity. In the context of surveillance capitalism, where personal data is relentlessly harvested and monetized, choosing not to participate in certain data-sharing ecosystems, being highly selective about digital engagement, or advocating for robust data privacy rights can be powerful acts of reclaiming personal sovereignty. This is "unknowing" as a shield, preserving a private sphere free from constant algorithmic scrutiny.

Beyond privacy, "unknowing" can be a proactive tool for social change. As explored by scholars and activists, the act of "unknowing" one's own ingrained biases and preconceived notions; about race, gender, neurodiversity, or other social categories, is a critical first step towards genuine empathy, understanding, and the dismantling of systemic inequalities. This is unknowing as an unlearning, a courageous letting go of false certainties to create space for new, more equitable perspectives.

In a broader civic sense, the collective admission of "we don't know" in the face of complex societal challenges can be far more productive than a performative display of certainty by leaders or institutions. Acknowledging uncertainty opens the door for genuine dialogue, collaborative problem-solving, diverse input, and more adaptive, resilient forms of governance. This form of unknowing is not a sign of weakness but of intellectual humility and a commitment to inclusive, emergent solutions.

Thus, reclaiming the unknown is not a nostalgic yearning for a pre-technological past, nor is it an outright rejection of AI's potential benefits. Rather, it is a sophisticated and necessary strategy for preserving core human values; agency, critical thought, privacy, empathy, and the capacity for growth, in a world increasingly saturated and shaped by powerful technologies. It is about asserting the right to define one's own informational boundaries, to prioritize human-centered modes of understanding and interaction, and to ensure that the future remains an open field of possibility rather than a closed loop of algorithmic prediction.

The courage to embrace unknowing-to question, to wander, to reflect, to doubt, and sometimes to consciously refuse the proffered certainties of an all-knowing AI; is perhaps one of the most vital human capacities we must nurture and defend in the Age of Information Symmetry. This understanding, this commitment to the value of what lies beyond immediate grasp, sets a crucial foundation as we consider what other uniquely human attributes remain essential, and indeed irreplaceable, when artificial intelligence appears to know so much.

SUMMARY

In an era increasingly defined by the pervasive influence of artificial intelligence and the seductive promise of near-perfect information access, this chapter has championed the profound and enduring value of unknowing. We have argued that, far from being a mere deficit in our cognitive landscape, qualities such as mystery, uncertainty, ambiguity, and the crucial capacity for forgetting are fundamental to the flourishing of human creativity, the space required for ethical reasoning, the achievement of robust psychological well-being, and the safeguarding of individual and collective freedom.

The pursuit of total knowledge, often driven by the allure of algorithmic determinism and the efficiencies of over-optimization, carries the inherent risk of diminishing human agency and dulling the vibrancy of novel experience. Conversely, a conscious embrace of mystery can fuel our innate curiosity, deepen our engagement with art, spirituality, and interpersonal relationships, and stimulate intellectual flexibility. Upholding the "right to not know" and recognizing the liberating power inherent in forgetting emerge as critical counterweights to the pressures of permanent digital memory, pervasive surveillance, and the potential for predictive systems to foreclose on an open future.

Ultimately, the judicious application of strategic ignorance can significantly enhance the quality and fairness of decision-making, while the design of technological systems that consciously accommodate and even protect uncertainty can foster more resilient, reflective, and adaptive interactions between humans and AI. Reclaiming the unknown; through the disciplined practice of doubt, the intentional cultivation of reflective space, and the assertion of unknowing as a form of personal and civic resistance, is not a retreat from progress but a vital and sophisticated assertion of enduring human values.

As we navigate towards an ever-more-intelligent technological future, it is these capacities for engaging with, navigating, and even cherishing the unknown that will prove paramount. This foundational understanding of unknowing's worth prepares us to further explore those uniquely human aspects of our experience, particularly our rich emotional and ethical lives, which remain uniquely resistant to the forces of automation and commoditization in the age of AI. This leads us to our next chapter: The Human Interface - How Emotion, Empathy, and Ethics Resist Commoditization.

17

THE HUMAN INTERFACE

The relentless advance of artificial intelligence (AI) heralds an era of unprecedented cognitive capability. Machines now perform tasks that, only a short time ago, were the exclusive domain of human intellect: complex data analysis, sophisticated pattern recognition, and even the generation of creative content. As AI systems approach and, in some narrow domains, surpass human performance in these areas, a profound question emerges: what, then, remains uniquely human?

This chapter argues that despite the exponential growth in AI's capacity for reasoning, information processing, and even the simulation of human-like interaction, there remains a fundamental chasm between artificial computation and genuine human subjective experience. AI may master the syntax of human interaction, but it lacks the semantics of felt emotion, the embodied understanding of empathy, and the nuanced, context-dependent wisdom of human ethical reasoning.

The philosophical inquiry into the nature of consciousness has long grappled with the "hard problem"; why and how physical processes in the brain give rise to subjective experience, or qualia, the raw feel of being. This is the redness of red, the sting of betrayal, the warmth of compassion. Current AI, for all its prowess, operates on algorithms and data; it processes information about emotions but does not, in any humanly recognizable sense, possess these first-person, qualitative states. This distinction is not presented here as a temporary technological hurdle, a mere flaw in AI to be "fixed" with more processing power or cleverer algorithms. Rather, it may represent an ontological distinction, a fundamental difference in being that defines the boundary of what AI can achieve. This "gap" is not a deficiency to be overcome but a defining characteristic that illuminates the unique and enduring value of the human interface. The very definition of "intelligence" is implicitly challenged by this. If AI can "know all" but cannot "feel anything," it compels a re-evaluation of whether intelligence, as we understand it in its richest human sense, can truly be divorced from subjective experience and embodiment. If emotion, empathy, and ethical sensibility are integral to human intelligence and decision-making, then an AI lacking these is, at best, a different kind of intelligence, not merely a less developed version of our own. The Age of Information Symmetry, therefore, may also be the Age of Intelligence Re-definition.

As the preceding chapters of this volume have explored, information itself, along with many cognitive skills, is rapidly becoming commoditized by AI. When any piece of knowledge or analytical capability is readily accessible and replicable by a machine, its intrinsic economic value diminishes. Yet, this very process paradoxically elevates the significance of those capacities that remain stubbornly human. Our ability to feel deeply, to connect empathetically with others, and to navigate complex moral landscapes with wisdom and compassion; these emerge not merely as residual human traits but as profound differentiators in an increasingly automated world. These are not capacities easily replicated, standardized, or outsourced to algorithms. Indeed, the commoditization of information by AI serves to increase the scarcity and, consequently, the value of authentic human connection and moral wisdom. This is a fundamental economic principle: as one resource becomes abundant and inexpensive, complementary resources that are inherently scarce see their value appreciate. Thus, the human interface is not simply "resisting" commoditization; it's worth is actively being amplified by the very technological forces that reshape our economic and social landscapes.

This chapter will explore the multifaceted nature of this human interface. It will begin by examining the philosophical and cognitive underpinnings of what AI cannot subjectively experience. It will then delve into the power of human presence and the non-verbal, embodied dimensions of connection that AI struggles to replicate. Subsequently, the discussion will turn to the rising economic and social value of empathy, demonstrating how this capacity translates into tangible benefits in various domains. The inherent limitations of algorithmic morality will be scrutinized, highlighting the indispensability of human ethical judgment.

Following this, we will propose principles for designing AI that complements rather than supplants human emotional and ethical intelligence. The imperative to actively cultivate these uniquely human capacities within ourselves and our societies will then be addressed. Finally, we argue for an ethical stance that embraces our shared humanity, with all its vulnerabilities, as a crucial and creative design constraint for the development and deployment of technology. Building on the observation from the previous chapter that AI "still can't feel or care," this chapter delves into why this is the case and what it signifies for the future of human value in an age of intelligent machines. We begin by examining the fundamental nature of subjective experience and why it remains, for now and perhaps intrinsically, beyond the reach of artificial intelligence.

WHAT AI CAN'T TOUCH

The assertion that certain aspects of human experience remain beyond the grasp of artificial intelligence, even in its most advanced forms, is not a Luddite rejection of technological progress but a nuanced philosophical and cognitive position. At its heart lies the concept of subjective experience, the first-person perspective that defines what it is like to be a particular conscious entity. This inner world of feeling, awareness, and meaning is what AI, as a system of computation and information processing, fundamentally lacks.

The philosophical concept of qualia refers to these subjective qualities of experience: the perceived redness of a rose, the specific taste of a lemon, the feeling of joy, or the ache of sorrow. An AI can be programmed to identify the wavelength of red light, analyze the chemical composition of a lemon, or recognize patterns of behavior associated with joy or sorrow in humans. It can even generate text or images that convincingly describe these states. However, there is no evidence to suggest that the AI experiences the redness, the tartness, the joy, or the sorrow in the way a human does.

Thomas Nagel, in his seminal essay "What Is It Like to Be a Bat?", powerfully illustrates this epistemic barrier. Nagel argues that even if we possessed a complete neurophysiological account of a bat's echolocation system, we would still not know what it feels like for the bat to perceive the world through sonar. The objective, third-person data cannot capture the subjective, first-person character of experience. If such a barrier exists between two biological species, it stands to reason that an even more profound chasm separates human consciousness from the operations of a non-biological, silicon-based AI.

This difficulty is compounded by the insights of embodied cognition theory, which posits that cognition, including emotion, is not merely a product of brain activity isolated from the body but is deeply shaped by the physical body itself and its dynamic interactions with the environment. Human emotions are intricately linked to physiological responses; changes in heart rate, breathing, hormonal fluctuations, and the constant stream of sensory information from our bodies (interoception) and the world around us. Our sensorimotor capacities, the way we move and interact with our surroundings, also contribute to how we feel and think. AI, lacking a human-like biological body with its complex web of sensory inputs, hormonal systems, and evolutionary heritage, cannot replicate this embodied grounding of emotion. While robots can be equipped with sensors and actuators, their "embodiment" is of a fundamentally different order than that of a living organism. Consequently, any "emotions" an AI might be said to have would be disembodied abstractions, profoundly different from the rich, visceral, and contextually embedded feelings that characterize human emotional life.

The philosopher John Searle, with his Chinese Room argument, further challenges the notion that AI can achieve genuine understanding or consciousness merely through sophisticated symbol manipulation. Searle imagines himself in a room, following a set of rules (a program) to manipulate Chinese symbols in response to input symbols, producing outputs that are indistinguishable from those of a native Chinese speaker. However, Searle himself, in this scenario, does not understand Chinese; he is merely executing formal rules. By analogy, a computer running a program, no matter how convincingly it simulates understanding language or other cognitive feats, does not thereby achieve genuine semantic understanding or intentionality-the "aboutness" of mental states. It operates at the level of syntax (formal rules) but lacks semantics (meaning). This argument directly questions the adequacy of the Turing Test, which proposes that if a machine's conversational behavior is indistinguishable from a human's, it should be considered intelligent. Searle's point is that simulation is not duplication. An AI might simulate empathetic responses or make ethically sound pronouncements, but it could be doing so without any underlying comprehension or feeling.

These philosophical considerations find echoes in the practical limitations of current "Emotion AI" or Affective Computing. These technologies primarily aim to infer human emotional states by analyzing data such as facial expressions, voice tone, physiological signals, or text patterns. While they can achieve some success in correlating these external cues with emotional labels, there are significant ongoing debates about their reliability, specificity, cultural universality, and susceptibility to bias. More fundamentally, recognizing patterns associated with an emotion in another being is a distinct cognitive task from experiencing that emotion oneself. The recent decision by the European Union to ban the use of emotion recognition AI in workplaces and educational settings, with limited exceptions, reflects not only privacy concerns but also an acknowledgment of the technology's limitations and the potential for misuse when complex human states are reduced to algorithmic classifications.

Ultimately, the question of whether a machine can possess a mind, mental states, and consciousness in the human sense remains a central, unresolved issue in the philosophy of AI. Many thinkers contend that computation alone, the manipulation of symbols according to formal rules, is insufficient to give rise to genuine consciousness or intentionality. AI may exhibit increasingly complex and adaptive behaviors, but without the anchor of subjective experience and genuine understanding, its actions lack the depth, meaning, and felt significance that are hallmarks of human emotional and ethical life. The pursuit of AI that "feels" in a human way might therefore be a category error, potentially leading to more sophisticated forms of simulation or even deception rather than authentic emotionality. A more fruitful and ethically sound direction may lie in developing AI that excels at recognizing, interpreting, and responding appropriately and sensitively to human emotional cues, functioning as an "emotionally intelligent tool" rather than aspiring to be an "artificial emotional being". This distinction is crucial, as the limitations of AI in grasping semantics, as highlighted by Searle, likely extend beyond language to the "semantics of social and ethical situations". An AI might process rules and statistical correlations related to human interactions, but it may consistently miss the underlying human meanings, values, and unspoken contexts that are rich with significance. This inherent superficiality in understanding reinforces the indispensable role of human judgment and presence in any interaction involving emotional depth or ethical weight.

While AI grapples with the internal, subjective reality of emotion, the outward expression and interpersonal experience of these states, particularly through direct human presence, reveal another layer of what AI cannot replicate. The power of that presence, with its complex interplay of nonverbal cues and biological resonance, is the focus of the next section.

THE POWER OF PRESENCE

The human interface is not solely defined by the internal landscape of subjective emotion; it is equally characterized by the dynamic, embodied interactions that occur between individuals. The "power of presence" refers to the profound impact of being physically and attentively with another person, an experience rich with nonverbal cues, physiological attunement, and a depth of connection that current and foreseeable AI cannot authentically replicate. This power stems from our biological and evolutionary heritage, shaping how we communicate, bond, and regulate our emotional states.

Human interaction is a symphony of signals that extend far beyond spoken words. Nonverbal communication; encompassing facial expressions, body posture, gestures, eye contact, and vocal tone, plays a critical role in conveying emotional states, intentions, and attitudes. Research, notably by Albert Mehrabian, has suggested that a significant majority of emotional meaning in face-to-face communication is transmitted through these nonverbal channels. These cues are often subtle, context-dependent, and processed by humans with a remarkable, often unconscious, acuity. They are fundamental to establishing rapport, building trust, and achieving mutual understanding. While AI can be programmed to recognize and even generate some nonverbal signals (e.g., a robot smiling), it struggles to capture the full spectrum, nuance, and authentic spontaneity of human nonverbal behavior, which is deeply integrated with our embodied emotional experience. This can be seen with many image generation models that when asked to reproduce emotion, frequently generate images that seem "off" when viewed by a human, particularly when dealing with subtle variations of emotion.

The neuroscience of connection offers further insights into why human presence is so potent. The discovery of mirror neurons, for instance, has provided a potential neurological basis for empathy and social understanding. These specialized neurons fire not only when an individual performs an action but also when they observe another individual performing the same action. This "mirroring" system is thought to enable a form of embodied simulation, allowing us to internally resonate with the actions, intentions, and even the emotional states of others. It is as if we are, in a rudimentary way, experiencing what the other person is experiencing. This capacity for direct, embodied mirroring is fundamental to our ability to understand and share feelings, fostering a sense of connection. AI, lacking this biological architecture, can only analyze patterns associated with empathy; it cannot engage in this direct, neural resonance. This suggests that the "human touch," often invoked metaphorically, has a literal grounding in neurobiological processes that AI, as a non-biological entity, cannot enact.

Furthermore, the process of co-regulation highlights the deeply interpersonal nature of emotional management. Co-regulation occurs when individuals, through their attuned presence and responsiveness, mutually influence and help stabilize each other's emotional and physiological states. This is most evident in the caregiver-child relationship, where the caregiver's calm and soothing presence helps the child manage distress and develop self-regulation skills. However, co-regulation is also a feature of adult relationships, contributing to emotional stability and well-being. It often relies on physical closeness, comforting touch, a soothing tone of voice, and a moment-to-moment responsiveness to another's distress signals-all elements that are intrinsically human and physical. An AI can provide information, offer programmed reassurances, or follow conversational scripts, but it cannot offer the living, breathing, physiologically attuned presence that genuine co-regulation requires. This capacity is rooted in our shared biology and the intricate dance of our nervous systems.

The biochemistry of our social interactions also plays a crucial role. Neurochemicals like oxytocin, often dubbed the "love hormone" or "bonding hormone," are integral to facilitating trust, social bonding, empathy, and cooperative behavior. The release and effects of oxytocin are often triggered by positive social interactions, physical touch (like hugging), and shared experiences. This neurochemical system, evolved over millennia, underpins our capacity for deep social connection and care. AI, operating outside this biological and biochemical framework, cannot participate in or evoke these fundamental human bonding mechanisms in the same way.

Even when humans interact through technology, such as video conferencing, limitations in replicating the richness of in-person presence are apparent. While video allows for the transmission of some visual and auditory cues, it often lacks the full bandwidth of nonverbal information, can lead to misinterpretations of subtle signals, and may result in "meeting fatigue" or a diminished sense of personal connection. If human-to-human mediated presence already faces these challenges, AI-mediated presence, which lacks even the shared biological and experiential ground of humans, confronts far greater hurdles in achieving the depth and authenticity of direct human interaction. As digital and AI-mediated interactions become increasingly common for transactional or informational exchanges, the perceived value and perhaps even the fundamental human need for genuine, unmediated human presence may intensify for interactions that require deep emotional support, nuanced understanding, or the establishment of profound trust. This could lead to a "bifurcation of interaction," where routine matters are efficiently handled by AI, but high-value, emotionally significant encounters increasingly demand, and are prized for, authentic human presence.

A final, crucial distinction lies in authenticity and intentionality. When a human offers comfort, expresses empathy, or engages in a meaningful conversation, there is typically an underlying genuine intention to connect, understand, or care. Based on current philosophical understanding, an AI, by contrast, executes a program based on its training data and algorithms; it simulates appropriate responses without possessing the subjective states or genuine intentions that animate human interaction. This difference in intentionality, even if subtle, can be perceived by humans, who are often highly attuned to signals of genuineness versus artificiality. This perceived authenticity is foundational to trust and the quality of the interpersonal experience.

The unique power of human presence and embodied connection, therefore, not only defines the qualitative nature of our interpersonal relationships but also underpins a growing economic and social value placed on skills and interactions that AI cannot authentically replicate. Foremost among these is empathy, the capacity to understand and share the feelings of another, which is increasingly recognized as a key differentiator in the human interface.

EMPATHY AS AN ECONOMIC DIFFERENTIATOR

Empathy, the capacity to understand and share the feelings of another, is emerging as a critical differentiator in a world increasingly shaped by artificial intelligence. While AI can process vast amounts of data and simulate human-like responses, the genuine ability to connect with and respond to human emotion remains a distinctly human forte. This capacity is not merely a "soft skill" but a powerful economic driver, enhancing productivity, fostering innovation, and forming the bedrock of essential service industries.

To appreciate empathy's role, it is useful to distinguish its two primary components: cognitive empathy and affective empathy. Cognitive empathy involves the ability to understand another person's perspective, to take their mental stance, and to comprehend their thoughts and feelings. Affective empathy, on the other hand, is the capacity to share or resonate with another's emotional experience; to feel a reflection of their joy, sadness, or fear. AI systems, particularly those trained on extensive datasets of human interaction, may become increasingly proficient at cognitive empathy; they can learn to predict user needs, interpret behavioral patterns, and tailor responses accordingly. However, affective empathy, with its visceral, shared feeling component, is deeply intertwined with our embodied nature and the neurobiological mechanisms discussed previously. This dimension of empathy, the capacity to truly feel with another, remains largely within the human domain.

The economic significance of such human-centric skills is increasingly recognized. The World Economic Forum's "Future of Jobs" reports consistently underscore the rising importance of social and emotional skills in the workplace. While analytical thinking remains a top required skill, attributes such as resilience, flexibility, agility, leadership, social influence, creative thinking, motivation, self-awareness, empathy, and active listening are highlighted as crucial for success. Employers across the globe identify these as essential for navigating a rapidly changing work environment. This trend strongly suggests that as AI systems take over more routine technical and analytical tasks, the skills that differentiate human workers and create unique value are increasingly those that are interpersonal, emotional, and relational.

The impact of empathy on business outcomes is not merely anecdotal; it is quantifiable. Numerous studies have demonstrated a direct correlation between empathetic leadership and tangible benefits for organizations. For instance, a survey by Ernst & Young (EY) found that 87% of employees believe empathy is essential for fostering an inclusive environment, and 86% feel it boosts morale. The same study indicated that mutual empathy between leaders and employees leads to increased efficiency (88%), creativity (87%), job satisfaction (87%), idea sharing (86%), innovation (85%), and even company revenue (83%). Similarly, a Businessolver study revealed that 87% of CEOs agree that a company's financial performance is tied to empathy.

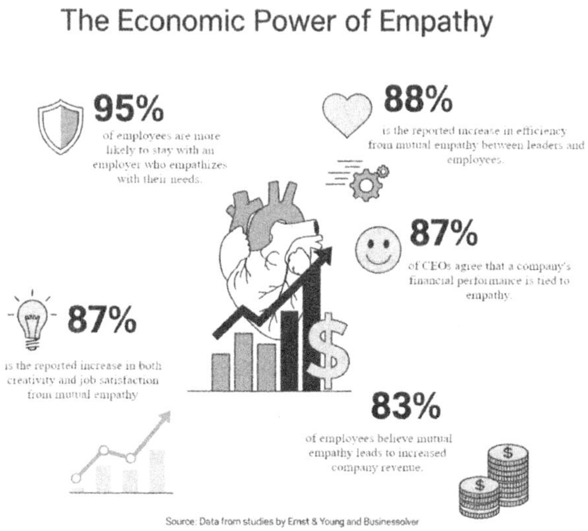

Furthermore, empathy plays a significant role in talent retention, with 95% of employees reporting they are more likely to stay with an employer who empathizes with their needs. The "empathy gap"; the difference between leadership's perception of workplace empathy and employees' actual experience, highlights a challenge: merely recognizing the value of empathy is insufficient. Its genuine implementation requires fostering a culture where authenticity and emotional connection can thrive, a fundamentally human endeavor.

The concept of "emotional labor," introduced by sociologist Arlie Hochschild, further illuminates the economic role of empathy and feeling management. Emotional labor refers to the process of managing one's own emotions to create a publicly observable facial and bodily display as a required part of one's job. This is particularly evident in service industries; such as flight attendants, customer service representatives, and healthcare providers, but extends to many professions where interpersonal interaction is key. This is not "unskilled" work; it demands sophisticated self-awareness, emotional regulation, and interpersonal acuity. While AI can be programmed with conversational scripts and polite responses, genuine emotional labor involves an authentic management of feeling that can, for humans, lead to "emotive dissonance"; a feeling of incongruity between felt emotion and displayed emotion, if not managed well. This nuanced, often taxing, human skill is something AI can only superficially mimic.

Building trust, particularly in complex, sensitive, or high-stakes interactions, often hinges on shared vulnerability and genuine human connection. Researcher Brené Brown emphasizes that trust is a product of vulnerability that grows over time, requiring work, attention, and full engagement. Trust, in this view, is about choosing to make something important to oneself vulnerable to the actions of another. AI, by its nature, cannot be genuinely vulnerable in this human sense. Human leaders, service providers, and colleagues who can demonstrate authentic vulnerability, own their mistakes, and build trust through consistent, reliable, and accountable behavior offer something irreplaceable. Brown's BRAVING acronym; Boundaries, Reliability, Accountability, Vault (confidentiality), Integrity, Non-judgment, and Generosity, outlines key facets of trust that are deeply rooted in human character, ethical conduct, and relational dynamics.

The economic importance of empathy is perhaps most starkly visible in the caregiving sector. The care economy, encompassing healthcare, social support, elder care, and childcare, is a massive and rapidly expanding global industry. Much of this essential work relies heavily on empathy, compassion, patience, and deep human connection. In the United States alone, family caregivers perform labor valued at an estimated $873.5 billion annually, a figure that would surpass the revenue of some of the world's largest corporations. These roles, critical for societal well-being and economic productivity, involve emotional and relational demands that are profoundly difficult, and arguably undesirable, to fully automate. The societal valuation and compensation for such "human interface" work may need re-evaluation if these skills are indeed becoming more critical as AI handles other tasks. If empathy and emotional intelligence are premium differentiators, then economic systems may need to adapt to better reflect this shifting landscape of value.

The following list summarizes some of the key data points illustrating the economic impact of empathy and related social-emotional skills:

Empathetic leadership

- **Key finding:** Eighty-seven percent of CEOs agree that financial performance is directly tied to empathy.
- **Reported impact and benefit:** Higher revenue, improved morale, greater efficiency and creativity, increased job satisfaction, stronger innovation, and more inclusive workplace environments.

Workplace empathy

- **Key finding:** Ninety-five percent of employees report they are more likely to stay with an empathetic employer.
- **Reported impact and benefit:** Improved employee retention, higher productivity, and sustained morale across teams.

Social-emotional skills

- **Key finding:** Six of the top ten skills cited by employers are social-emotional skills, including empathy.
- **Reported impact and benefit:** Critical for adaptability, collaboration, leadership effectiveness, and resilience in evolving job markets.

Caregiving

- **Key finding:** Unpaid caregiving labor in U.S. families is valued at approximately $873.5 billion annually.
- **Reported impact and benefit:** Provides essential health and social support that underpins societal well-being and broader economic productivity.

Empathy, therefore, is far more than a sentimental ideal; it is a potent economic force. As AI continues to reshape the labor market, the human capacity for empathetic connection, emotional labor, and trust-building will likely become an increasingly valuable, and uniquely human, asset. However, empathy is not just about feeling with others; it is also a crucial component of navigating complex social and moral situations. As we consider the limits of AI, the uniquely human capacity for nuanced ethical judgment comes sharply into focus.

ETHICS BEYOND CALCULATION

The realm of ethics presents one of the most formidable challenges to the notion of AI as a replacement for human judgment. While AI can be programmed with rules and optimized for certain outcomes, the complexities of human moral reasoning; with its reliance on context, intuition, empathy, and an understanding of unquantifiable values, extend far beyond mere calculation. Ethical decision-making is not simply a logical process; it is a deeply human endeavor, fraught with ambiguity and demanding a sense of responsibility that machines do not possess.

Common ethical frameworks, such as utilitarianism and deontology, offer principles that can, to some extent, be translated into algorithmic terms. Utilitarianism, which seeks to maximize overall well-being or happiness for the greatest number, can inform AI systems designed for resource allocation or optimization. Deontology, with its emphasis on duty-based rules and moral absolutes (such as Immanuel Kant's Categorical Imperative, which posits acting only according to maxims that could become universal laws), can provide AI with sets of constraints or inviolable principles. However, these frameworks often conflict in practice, or prove insufficient for guiding action in novel, ambiguous, or morally fraught situations. As noted in critiques of deontology, strict adherence to rules can sometimes lead to harmful outcomes if context is ignored.

The classic "trolley problem" is a frequently invoked thought experiment in discussions of AI ethics, particularly concerning autonomous vehicles. In its basic form, it forces a choice between two undesirable outcomes: for example, should an autonomous vehicle swerve to avoid hitting five pedestrians, thereby crashing and harming its one passenger, or continue on its path? While such dilemmas highlight the need for pre-programmed ethical decision-making in AI, the trolley problem itself is often criticized for its oversimplification of real-world ethical challenges. Actual moral dilemmas rarely present such clear-cut, binary choices. They typically involve incomplete or uncertain information, a multitude of interacting variables (such as the age, health, or potential future contributions of individuals involved), dynamic and rapidly evolving circumstances, and a diversity of cultural and personal values that are difficult to reconcile through a single algorithmic solution.

The philosophical stance of moral particularism offers a compelling perspective on why human ethical reasoning is so difficult to codify for AI. Particularism argues that moral judgment is highly dependent on the specific features and nuances of a particular context, rather than on the mechanical application of universal moral principles. What constitutes a valid reason for action in one situation might be irrelevant or even a reason against action in another. This context-sensitive, holistic approach to reasoning is characteristic of mature human ethical deliberation but poses an immense challenge for AI systems, which generally thrive on generalizable rules and patterns learned from data. An over-reliance on generalist, rule-based ethics in AI could render these systems brittle and prone to making ethically unsound decisions in novel or complex contexts where pre-defined rules do not adequately apply.

Virtue ethics, another important tradition, shifts the focus from rules or outcomes to the moral character of the agent and the cultivation of virtues such as wisdom, justice, courage, and compassion, with the aim of achieving human flourishing (eudaimonia). This framework emphasizes moral perception (the ability to discern the ethically salient features of a situation), emotional responsiveness, and practical wisdom (phronesis): the ability to deliberate well about how to act in specific situations. These are qualities developed through lived experience, reflection, and habituation, making them profoundly difficult to instill in an AI, which lacks personal history, character, and the aspiration for flourishing in a human sense.

Human ethical systems have evolved over millennia to handle an immense range of human experiences, including rare, unexpected, and morally ambiguous "edge cases". AI, typically trained on finite datasets, may struggle when confronted with situations that fall significantly outside its training parameters, potentially leading to unforeseen and ethically problematic outcomes.

Moreover, human ethics often places significant weight on the intention behind an action, not merely its consequences. An AI, operating without genuine intentions or subjective awareness, can only be evaluated based on its programmed objectives and its observable outputs. This makes the assignment of moral responsibility exceptionally complex. If an AI system causes harm, is it the programmer, the manufacturer, the owner, the user, or the AI itself that bears responsibility? This ambiguity underscores the persistent need for clear lines of human accountability in any system where AI makes decisions with ethical implications.

The "black box" nature of many advanced AI algorithms, particularly in deep learning, presents a further challenge. If the reasoning process of an AI system is opaque and unexplainable even to its creators, it becomes difficult to trust its decisions, especially in ethically sensitive domains. Human ethical discourse relies heavily on the ability to articulate reasons, engage in deliberation, and provide justifications for moral choices. An AI that delivers an ethical "verdict" without a transparent and comprehensible rationale bypasses this crucial aspect of human morality, potentially undermining trust and preventing meaningful human oversight or learning. Thus, a key challenge is not merely to make AI "ethical" in its outputs, but "ethically legible" in its processes.

Ultimately, the prospect of outsourcing moral judgment to machines carries significant risks. It could lead to a gradual "moral deskilling" in humans, as we become less practiced in the art of ethical deliberation. It also risks embedding and amplifying existing societal biases if these are present in the data used to train AI systems, or in the assumptions of their programmers. Furthermore, it can create a diffusion of responsibility, where no single human actor feels fully accountable for decisions delegated to an ostensibly "objective" machine.

Given that AI cannot fully replicate the depth, nuance, and responsibility inherent in human ethical reasoning (yet), the focus must necessarily shift. Instead of aiming to create autonomous moral agents, the more prudent and constructive path lies in designing AI systems that support and augment, rather than replace, human moral judgment. This requires a fundamentally human-centric approach to the design and deployment of ethical AI.

DESIGNING FOR HUMAN-CENTRIC INTERACTION

In a world where artificial intelligence increasingly mediates our experiences and decisions, the design of these systems becomes paramount. If AI is to serve human ends and enhance human flourishing, rather than inadvertently diminish our capacities or values, it must be developed with a profound understanding of human needs, limitations, and strengths. This necessitates a human-centric design philosophy, where the primary goal is not the unfettered advancement of technological capability for its own sake, but the creation of AI systems that are supportive, ethical, and empowering for their human users.

A core principle guiding such design should be augmentation, not replacement. In contexts involving complex human interaction, emotional intelligence, or ethical judgment, AI should be envisioned as a tool to enhance human capabilities, not to eliminate the human element. AI systems can excel at handling routine, data-intensive, or repetitive tasks, thereby freeing human attention and energy for activities that require strategic insight, creativity, deep empathy, and nuanced ethical deliberation. In this model, AI agents act as colleagues, assistants, or copilots, amplifying human potential rather than aiming for full autonomy in sensitive domains. The success of such systems should be measured not just by task completion efficiency or accuracy, but by their ability to empower human users, improve the quality of human-AI collaboration, and contribute to overall human well-being.

Human-Centered Design (HCD) offers a robust framework for achieving this vision. HCD methodologies prioritize a deep understanding of the end-users; their needs, preferences, contexts, and values throughout the entire design and development lifecycle. This typically involves stages such as empathizing with users to grasp their perspectives, defining the actual problems they face, ideating a range of potential solutions, creating prototypes, and engaging in iterative testing and refinement based on user feedback. Applied to AI, HCD means moving beyond what AI can technically do to focus on what users genuinely want and need from AI interaction. This ensures that AI tools are not only powerful but also intuitive, usable, useful, and aligned with human goals, particularly in areas requiring emotional sensitivity or ethical consideration.

The Human-in-the-Loop (HITL) approach is a practical manifestation of human-centric AI design. HITL systems explicitly integrate human oversight, judgment, and intervention into the operational cycle of the AI. Humans can contribute by labeling training data, evaluating the performance of AI models, providing ongoing feedback for continuous learning and adaptation, and, crucially, making the final decisions in ambiguous, novel, or high-stakes situations where algorithmic judgment alone is insufficient or inappropriate. This collaborative model offers several advantages: it can significantly improve the accuracy and reliability of AI systems, help identify and mitigate biases in data and algorithms, increase the transparency and explainability of AI decision-making, and foster greater trust among end-users.

The principles of "calm technology" also provide valuable guidance for designing AI interactions that respect human attention and cognitive limits. Calm technology is conceived as information technology that informs without demanding constant focus, residing primarily in the user's periphery and smoothly shifting to the center of attention only when necessary. Key principles include requiring the smallest possible amount of attention, communicating information without necessarily "speaking" or being intrusive, and creating ambient awareness. For AI systems designed to support human interaction, these principles are vital. AI should not be an overwhelming or dominating presence but a supportive, almost invisible, aid that allows ample space for human thought, emotion, and connection. The classic example of a tea kettle, which remains quiet until it signals it is ready, illustrates how technology can be informative without being burdensome.

Within this human-centric framework, AI tools can be conceptualized as coaches, mirrors, or support systems. An AI might act as a coach by providing feedback on communication styles or decision-making processes. It could serve as a mirror by reflecting patterns in a user's behavior or emotional expressions, thereby fostering self-awareness. Or it could function as a support system by offering relevant information, organizing complex data, or suggesting potential courses of action for human consideration. In each case, the AI's role is to enhance and empower human capabilities, not to usurp the core human functions of feeling, deciding, or connecting. This implies a co-evolutionary path where AI and humans can mutually improve each other's capabilities in a positive feedback loop, where each "amplifies the best" of the other.

Critically, designing for human-centric interaction means preserving space for silence, empathy, and ambiguity. Human connection and deep understanding often develop not through relentless efficiency or the instant resolution of all uncertainties, but through moments of quiet reflection, shared vulnerability, and the ability to comfortably hold ambiguity. AI systems should be designed not to fill every conversational void or to provide definitive answers to every complex question prematurely, but to allow and even facilitate these essentially human processes.

Finally, ethical AI design practices must be woven into the fabric of development from the very outset. This includes a commitment to fairness by scrutinizing training data and algorithms for potential biases; ensuring transparency and explainability wherever possible so users can understand how AI systems arrive at decisions; robustly protecting user privacy and data security; establishing clear lines of accountability for AI actions and outcomes; and prioritizing human safety and well-being above all else.

Designing AI systems that interact effectively and ethically with humans is, however, only one side of the equation. The other, equally important, side involves the intentional cultivation of the human capacities necessary to engage fruitfully with these advanced technologies and to thrive in a world where such uniquely human skills are increasingly paramount.

TRAINING THE HUMAN INTERFACE

As artificial intelligence augments or automates a growing array of technical and analytical tasks, the uniquely human capacities; emotional intelligence, empathy, sophisticated communication, and nuanced ethical reasoning, become premium assets. These are the core components of the "human interface," the skills that allow us to connect meaningfully with others, navigate complex social landscapes, and make sound judgments in situations that defy purely algorithmic solutions. In an AI-saturated world, the deliberate cultivation of these skills is not merely beneficial; it is an imperative for individual fulfillment, professional success, and societal well-being. This "training of the human interface" is not just about acquiring new competencies but about fostering a deeper human consciousness and relationality, potentially serving as a vital counterpoint to any passivity or deskilling that ubiquitous AI might otherwise induce, thereby preserving and enhancing human agency.

Educational systems, from early childhood through higher education and professional development, have a critical role to play in this endeavor. Social and Emotional Learning (SEL) provides a foundational methodology for helping individuals of all ages to better comprehend their emotions, feel those emotions fully, manage them constructively, develop empathy for others, build positive relationships, and make responsible decisions. The five core competencies typically associated with SEL-self-awareness, self-management, social awareness, relationship skills, and responsible decision-making; are precisely those capacities that resist commoditization by AI and are essential for a flourishing human interface. SEL should not be viewed as an ancillary subject but as an integral thread woven into the fabric of all learning experiences. Practical integrations include reflective journaling, collaborative group projects where students must navigate interpersonal dynamics, role-playing scenarios that encourage perspective-taking, and regular classroom check-ins or morning meetings that create space for emotional expression and community building.

Developing these human skills is also dependent on teaching approaches that highlight collaborative and discussion-based learning. The Harkness Method, for example, brings students and teachers together around a large oval table for student-led discussions, fostering an egalitarian environment where active listening, critical analysis of texts and ideas, respectful articulation and defense of arguments, and the empathetic consideration of diverse perspectives are paramount. In such a setting, the teacher acts less as a dispenser of information and more as a facilitator and mentor, guiding students to take ownership of their learning through dialogue and collaborative inquiry. This method inherently develops the communication, critical thinking, and interpersonal skills vital for a robust human interface.

Beyond formal education, the cultivation of these capacities must extend into professional life. Donald Schön's concept of "reflective practice" offers a powerful framework for continuous learning and development in professional settings. Reflective practice involves two key components: "reflection-in-action," which is the process of thinking on one's feet and adapting actions in real-time as a situation unfolds, and "reflection-on-action," which involves stepping back to analyze and learn from past experiences. This continuous cycle of action and reflection is crucial for professionals in fields requiring significant human interaction such as education, healthcare, social work, and leadership, to develop practical wisdom, enhance emotional intelligence, adapt to novel or uncertain situations, and refine their decision making in complex human contexts. It helps bridge the gap between theoretical or technical knowledge and the artistry required for effective practice.

The humanities; including literature, philosophy, history, and the arts play an indispensable role in nurturing the human interface. Engagement with the humanities cultivates empathy by exposing individuals to a vast range of human experiences, perspectives, and emotional landscapes, allowing them to "put themselves in other people's shoes". It sharpens critical thinking skills through the analysis of complex texts and ideas, and it fosters a deeper understanding of human values, motivations, cultural diversity, and the enduring questions of the human condition. This understanding is essential for nuanced interpersonal interaction, ethical judgment, and navigating the complexities of a diverse world.

In professional spheres, particularly in leadership roles, emotional fluency; the ability to accurately perceive, understand, and manage one's own emotions, and to respond effectively and empathetically to the emotions of others becomes a cornerstone of effectiveness. Training and development programs should increasingly focus on these aspects of emotional intelligence, recognizing them not as "soft skills" but as core competencies for navigating an interconnected and rapidly changing world.

The skills honed through these diverse approaches; SEL, collaborative learning, reflective practice, and engagement with the humanities are not easily automated or replicated by AI. They are, in a very real sense, "future-proof" skills, foundational to effective human interaction, ethical conduct, and the pursuit of meaningful work in an age increasingly shaped by intelligent machines. However, the development of these skills cannot occur in a vacuum. A systemic approach is necessary, where educational institutions, workplaces, and even community structures work in concert to foster these human capacities. Schools can lay the groundwork, but workplaces must reinforce these values through cultures that encourage reflective practice and genuinely value emotional labor, and communities can provide the diverse human interactions essential for their practice and refinement. A disconnect between these spheres would undermine the development of a truly robust and resilient human interface.

Cultivating these human capacities is essential, but so too is the creation of an overarching technological and societal framework that actively values and protects our shared humanity, even using our inherent limitations and vulnerabilities as guideposts for ethical and effective design.

HUMANITY AS A DESIGN CONSTRAINT

The trajectory of artificial intelligence development often seems driven by an implicit goal: to replicate, and ultimately surpass, human intelligence in every conceivable domain. However, a more human-centric vision suggests a fundamental reorientation. The ultimate purpose of AI should not be to create an artificial mimicry of humanity or to render human capabilities obsolete, but rather to augment human potential and enhance human flourishing. This requires a philosophical shift in how we perceive technology's role, moving from a purely instrumental view to one that considers technology's profound impact on human life and societal values. Within this reoriented vision, our very humanity; with its inherent limitations and vulnerabilities, can serve not as an obstacle to be overcome, but as an essential ethical and creative constraint guiding AI design.

Instead of viewing human limitations such as our cognitive biases, emotional vulnerabilities, finite attention spans, the need for meaning, or our slower processing speeds compared to machines, as mere flaws to be "engineered out" or compensated for by AI, we can recognize them as defining characteristics of our species. These traits, evolved over millennia, are often intertwined with our greatest strengths: our capacity for nuanced judgment, creativity born from ambiguity, and deep social connection forged through shared experience.

Treating these human characteristics as design constraints means developing AI systems that work with our nature, rather than against it. For example, an AI system designed with calm technology principles respects human cognitive limits by presenting information in a non-overwhelming, peripheral manner, enhancing usability and reducing cognitive load. This is more effective and humane than an AI that bombards users with data under the assumption that more information is always better. This approach is not a Luddite or anti-progress stance, but rather a sophisticated form of risk management and value alignment for AI development, ensuring that technology serves human ends rather than becoming an autonomous force with potentially misaligned goals.

This philosophy leads to the advocacy for designing AI systems that are, in certain crucial aspects, inherently incomplete without human involvement. Particularly in domains requiring profound ethical judgment, deep empathy, complex creativity, or rich contextual understanding, AI should be built to require human partnership. This ensures that humans remain essential, not merely as a temporary stopgap until AI "catches up," but as integral components for the system's optimal and ethical functioning. This is the essence of robust Human-in-the-Loop systems, where human wisdom guides and validates AI's computational power.

Furthermore, AI systems should be designed not as static, dictatorial entities, but as dynamic collaborators capable of genuinely listening to, learning from, and adapting to human users over time. This requires more than superficial personalization; it demands robust feedback mechanisms and a design ethos that prioritizes responsiveness to the full spectrum of human input, including emotional cues and ethical considerations. An AI that can adjust its approach based on a user's confusion, frustration, or expressed values is more likely to be a trusted and effective partner.

A crucial aspect of our humanity is our vulnerability. While often perceived as a weakness to be shielded or overcome, vulnerability is, as thinkers like Brené Brown have illuminated, a core condition that fosters connection, trust, courage, and growth. AI, in its current and foreseeable forms, cannot be genuinely vulnerable; it does not experience fear, doubt, or the need for reassurance in a human sense. Therefore, AI systems designed for human interaction, particularly in sensitive areas like therapy, coaching, education, or care, must respect and make space for human vulnerability. A leader's willingness to be vulnerable, to admit uncertainty or error, can build profound trust and cohesion within a team. Similarly, AI systems that do not project an aura of infallible omniscience, but instead clearly define their limitations and their reliance on human input, may foster greater trust and more realistic, productive interactions. Embracing human vulnerability in the context of AI design could thus lead to more grounded and useful partnerships, moving away from utopian or dystopian fantasies of AI perfection or superintelligence.

Ultimately, truly ethical AI is human-affirming AI. It is technology that supports and enhances human dignity, agency, and our unique capacities for emotion, empathy, and moral reasoning. It does not seek to diminish, devalue, or replace these qualities, but rather to create an environment in which they can thrive alongside, and in synergy with, advanced technological tools. The point is not to make humans obsolete, but to leverage AI in ways that help us become more fully and effectively human.

This deep dive into the enduring value of the human interface; its emotional depth, empathetic capacity, and ethical compass, sets the stage for considering the broader societal transformations required in the Age of Information Symmetry. If these human qualities are indeed irreplaceable and increasingly valuable, then our institutions, economic structures, educational systems, and even our collective understanding of meaning and purpose must adapt.

SUMMARY

In this chapter we have contended that while artificial intelligence achieves unprecedented cognitive feats, the realms of genuine emotion, embodied empathy, and nuanced ethical reasoning remain profoundly, and perhaps intrinsically, human. These capacities, rooted in our biological makeup, our subjective first-person experience, and our inherently relational nature, resist commoditization even as information and many cognitive skills become universally accessible through AI. They constitute the core of the human interface, a domain where machines may simulate but cannot truly replicate the depth and authenticity of human feeling and connection.

The power of direct human presence, rich with nonverbal cues and underpinned by neurobiological mechanisms of attunement, continues to hold a unique significance that AI-mediated interactions cannot fully capture. Empathy, far from being a mere "soft skill," is emerging as a critical economic differentiator, driving business success and forming the foundation of essential care industries.

Furthermore, the complexities of ethical decision-making, which often demand context-specific wisdom, moral intuition, and a sense of responsibility, extend beyond the calculative capacities of current AI, underscoring the indispensability of human moral judgment. In response to these realities, the path forward involves designing AI systems with a human-centric approach focusing on augmentation rather than replacement, incorporating human oversight, and adhering to principles of calm and ethical technology.

Crucially, it also demands the intentional cultivation of these uniquely human "interface" skills through education, professional development, and a societal valuing of emotional and ethical literacy. Ultimately, embracing our shared humanity, with all its complexities and vulnerabilities, not as a flaw but as a fundamental design constraint and a source of unique strength, will be key to ensuring that AI serves to elevate, rather than diminish, the human experience. The goal is not for AI to replace us, but to help us become more fully, and more wisely, human.

Having established the enduring and irreplaceable value of the human interface, the discussion must now broaden to consider the profound societal implications. If machines can know and do almost anything, yet can never truly be us, how must our foundational agreements around work, education, governance, and the very meaning of a human life be re-negotiated? This is the central question we address in the next chapter, "A New Social Contract."

18

A NEW SOCIAL CONTRACT

The preceding chapters in this book have charted a profound transformation: the decline of an era dominated by information asymmetry and the ascent of an age characterized by information symmetry, largely propelled by the pervasive influence of artificial intelligence. This journey began by exposing the "invisible scaffolding" of modern life, revealing how unequal access to knowledge has historically shaped markets, professions, and institutions in what was termed the "arbitrage economy".

We then explored AI as "The Great Leveler," demonstrating its capacity to systematically dismantle these traditional information advantages, making expertise and insight increasingly accessible to all. The narrative continued by examining the rise of "Composable Minds" through personal AI assistants, the dynamic orchestration of "Skills on Demand," and the emergence of "Semantic Economies" where meaning itself becomes a fundamental protocol for interaction. Subsequently, the discussion turned to "The New Moats," identifying trust, continuous feedback, and dynamic identity as crucial differentiators in a world where knowledge is a commodity. Finally, we presented practical strategies for "Designing for the AI Era", outlining how individuals and organizations must adapt to become AI-readable, embrace modular structures, and cultivate new sources of meaning at work.

Now, standing at the culmination of this exploration, this chapter confronts the most encompassing challenge: the necessity of fundamentally renegotiating the social contract. The implicit and explicit agreements that have long governed our collective existence; our understanding of work and its rewards, the purpose and structure of education, the mechanisms of governance and civic participation, and the very sources from which individuals and societies derive meaning, are being tested and found wanting in the face of ubiquitous intelligence.

The central question that now guides our inquiry, echoing the core of this book's thesis, is this: When knowledge becomes universally accessible and capability becomes dynamically composable, what obligations do we, as individuals and as a society, owe to one another? How must our foundational societal frameworks be reimagined to ensure fairness, uphold dignity, and foster continued human progress in an age of unprecedented informational and intellectual abundance?

The industrial-era social contract, forged in a world of material and informational scarcity, predicated on hierarchical expertise, and structured around linear pathways of advancement, is no longer fit for purpose. Its foundational assumptions have been irrevocably fractured by the rise of AI and the dawn of information symmetry. Therefore, a new social contract cannot merely be an iteration of the old; it must be consciously, collectively, and courageously designed from first principles. This new compact must transition from valuing mere employment to esteeming purposeful contribution; it must evolve from a system of front-loaded credentialing to one that fosters lifelong learning, critical thinking, and profound adaptability.

Governance, too, must transform, becoming more participatory, transparent, and agile, capable of leveraging AI ethically while remaining accountable to its citizenry. Crucially, this era demands the articulation and protection of new digital and human rights, safeguarding individual autonomy and dignity in an increasingly AI-mediated world. And finally, in a landscape where traditional markers of identity and value are shifting, the new social contract must actively support the individual and communal pursuit of meaning.

The old rules, born of a different age, offer insufficient guidance for the path ahead. In the following sections, we delve into the specific domains where this renegotiation is most urgent, beginning with an examination of why the industrial contract has reached its inevitable end, and then proceeding to outline the foundational pillars of a social contract fit for the age of information symmetry.

THE END OF THE INDUSTRIAL CONTRACT: WHY THE OLD BARGAIN IS BROKEN

The societal frameworks that structured much of the 20th century, particularly in industrialized nations, were built upon an implicit social contract; a set of mutual expectations and obligations between individuals, institutions, and the state. This contract, especially in its post-World War II manifestation, promised a degree of stability and opportunity: diligent labor was exchanged for a reliable income and a measure of security; formal education was presented as the primary conduit to professional roles and upward social mobility; and societal progress was largely guided by the wisdom of credentialed experts whose specialized knowledge conferred legitimacy and authority. But all of this was built on a shaky foundation of information asymmetry; where some knew more than others. Now, with AI, that foundation is crumbling.

The 20th century bargain, in its idealized form, had several key components. In the realm of work and the economy, it often featured the prospect of stable, sometimes lifelong, employment, particularly for a male breadwinner. Wages and productivity, for a significant period, appeared to move in concert, contributing to the expansion of a middle class and an era of broadly shared prosperity, albeit one that often excluded or marginalized women and minorities. This economic arrangement depended on employers possessing and managing critical information about markets, production processes, and organizational strategy, while workers provided the requisite labor.

Education, particularly higher education, functioned not only as a site of learning but also as a powerful sorting mechanism, a gatekeeper to the professions. Credentials obtained from these institutions served as widely accepted proxies for knowledge, skill, and access. The principle of meritocracy, at least in theory, suggested that individuals could achieve upward mobility based on their innate talent and diligent effort, yet this ascent was almost invariably filtered through these established educational channels and the acquisition of scarce, specialized knowledge.

Concurrently, professional experts; doctors, lawyers, scientists, and academics commanded significant societal authority precisely because their deep, specialized knowledge was not easily accessible to the general populace. Institutions, from universities to government agencies, derived a substantial portion of their legitimacy from this expert-driven model of decision-making and guidance. The state, in turn, played a crucial role in this contract, regulating markets, establishing social safety nets such as unemployment insurance and pensions, and enacting labor laws that often mediated the relationship between capital and labor. This mode of governance also relied on controlled channels for information dissemination and the authority of its own experts.

The philosophical underpinnings of such societal arrangements can be traced back to enlightenment thinkers like Hobbes, Locke, and Rousseau, who theorized that individuals implicitly consent to certain societal rules and governmental authority in exchange for order, protection of rights, or the pursuit of the common good. The

20th-century social contract represents a particular historical instantiation of these ideas, adapted for an industrial society increasingly reliant on complex organizations and specialized labor. Each pillar of this contract was deeply intertwined with information asymmetry. Employers held informational advantages in the marketplace; educational institutions were custodians of specialized knowledge domains; professionals monetized their relatively exclusive access to information and expertise; and governments managed and disseminated information to guide public action and maintain order.

However, this contract, often perceived as a bastion of stability, was always contingent on specific technological and economic conditions. The seeds of its unraveling were sown well before the current AI boom. Technological advancements throughout the latter half of the 20th century, coupled with the forces of globalization, began to erode the traditional structures of employment and the value of certain types of localized knowledge. The digital revolution, often termed the Third Industrial Revolution, significantly accelerated the democratization of information access. Yet, it is the Fourth Industrial Revolution, characterized by artificial intelligence, machine learning, and ubiquitous connectivity, that delivers the most decisive blow to the old contract's foundational assumption of information scarcity.

AI, by making vast repositories of human knowledge and sophisticated analytical capabilities almost universally and instantly accessible, fundamentally breaks the logic upon which the industrial contract was built. As explored extensively in earlier chapters, expertise itself is becoming commoditized when AI systems can replicate or even surpass human performance in various cognitive tasks. Traditional career paths, predicated on the slow accumulation of exclusive knowledge and experience, are facing profound disruption as AI automates routine tasks and augments complex ones, changing the very nature of entry-level work and career progression.

The value proposition of traditional educational credentials, once the primary currency of opportunity, is being intensely questioned when AI can deliver personalized learning and on-demand knowledge, irrespective of institutional affiliation. The stability that the old contract offered was, in retrospect, a temporary equilibrium, sustained by a particular configuration of information access that no longer holds. The deep interconnectedness of the old contract's pillars; where stable work funded education and social programs, and education provided credentials for that work, means that the disruption in one area inevitably cascades through the others, necessitating a holistic re-evaluation rather than piecemeal fixes.

The list below starkly illustrates the magnitude of this shift, contrasting the core tenets of the industrial era social contract with the emergent realities and necessary adaptations of the AI era:

Work and employment

- **Industrial era social contract:** Stable, often lifelong employment organized around the employer, with work serving as a primary source of identity.

- **AI era social contract:** Fluid, project-based contribution with human–AI collaboration, purpose-driven work, and multiple income streams over a lifetime.

Economic basis

- **Industrial era social contract:** Labor exchanged for wages, productivity measured by hours and output, and advantage sustained through information asymmetry.
- **AI era social contract:** Value created through innovation, problem-solving, and human insight, with AI augmenting productivity and information symmetry reshaping markets.

Education and skills

- **Industrial era social contract:** Front-loaded education, credentialism, and standardized curricula designed for predictable careers.
- **AI era social contract:** Lifelong learning and skill composition, emphasis on adaptability, emotional and ethical literacy, and personalized, AI-assisted learning paths.

Source of merit

- **Industrial era social contract:** Educational attainment, possession of scarce knowledge, and experience within established roles.
- **AI era social contract:** Critical thinking, creativity, collaboration, ethical judgment, and the ability to work effectively with AI.

Governance and power

- **Industrial era social contract:** Hierarchical, expert-driven systems with slow, centralized decision-making.

- **AI era social contract:** Participatory governance supported by AI-literate citizens, transparent and accountable algorithms, and adaptive, decentralized decision structures.

Individual rights

- **Industrial era social contract:** Emphasis on civil, political, and social rights within primarily physical and analog contexts.
- **AI era social contract:** Expansion to digital rights, including explanation of automated decisions, refusal of AI-mediated outcomes, the right to be forgotten, digital dignity, and data ethics.

Meaning and purpose

- **Industrial era social contract:** Meaning derived from job status, material success, and adherence to traditional roles.
- **AI era social contract:** Meaning grounded in contribution, connection, continuous growth, and co-creation of value, with purpose treated as a shared pursuit.

Underlying ethic

- **Industrial era social contract:** Competition for scarce resources and opportunities, with individual achievement emphasized.
- **AI era social contract:** Stewardship of shared knowledge, service to collective benefit, and coordination for systemic well-being.

The very definition of "merit," once closely tied to the possession of scarce knowledge validated by credentials, is now profoundly challenged. If AI can provide or significantly augment "tested competency," then the qualities that define human value and contribution must evolve. This points towards an increasing emphasis on skills such as critical inquiry, creative problem-solving, complex collaboration, ethical discernment, and the sophisticated ability to partner with AI systems; attributes that are less susceptible to simple commoditization. The old bargain is broken not by a single fault line, but by a systemic shift in the informational landscape.

The realization that the old contract is defunct is the necessary precursor to envisioning a new one. If the traditional pathways to security, opportunity, and legitimacy are eroding, society must collectively architect new ones. This begins with reimagining the nature of work itself in an age where human and artificial intelligence are increasingly intertwined.

REDESIGNING WORK AS PURPOSEFUL CONTRIBUTION

The collapse of the industrial-era social contract, with its emphasis on stable, often lifelong employment as the primary source of income, identity, and societal value, necessitates a fundamental redesign of our understanding of "work". In an era of information symmetry, increasingly enhanced or automated by artificial intelligence, the conventional understanding of "jobs" is evolving, giving way to a more adaptable, dynamic, and purpose-driven framework. The new social contract must foster an environment where work is defined less by fixed roles and hierarchical structures and more by meaningful contribution, continuous value creation, and the collaboration between human ingenuity and machine intelligence.

A critical first step in this redesign is to decouple identity and social status from the narrow confines of a job title or employer. The industrial model heavily ingrained the notion that one's profession was a primary determinant of self-worth and societal standing. However, as AI-driven automation makes roles more fluid and traditional expertise becomes augmented or even replicable, this linkage becomes increasingly tenuous and, for many, a source of anxiety and dislocation. The future of work, as envisioned in previous chapters, is one where careers are less like ladders and more like "climbing walls", requiring individuals to navigate a complex terrain of opportunities and challenges, composing their professional lives from a variety of experiences and contributions. This shift demands a cultural reorientation, moving away from "jobs as identity" towards a broader appreciation for the diverse ways individuals can contribute value.

Career Paths Reimagined

Central to this new vision of work is the concept of universal access to AI tools and assistance, not as a luxury, but as a baseline utility. If AI is indeed the "Great Leveler", then its benefits should be broadly accessible, empowering individuals across all segments of society to enhance their capabilities, participate in new forms of economic activity, and contribute their unique talents. This widespread access can facilitate the emergence of "lifelong contribution streams," where individuals engage in a portfolio of projects, gigs, entrepreneurial ventures, and community-focused activities throughout their lives. This model moves beyond the singular, linear career path, embracing a more adaptable and personalized approach to professional engagement, supported by continuous learning and reskilling. Such a model aligns with the idea of a "modular career," where individuals assemble their purpose and impact from diverse sources.

The future of work is not a dystopian narrative of humans versus machines, but rather a story of human-AI synergy. AI systems are increasingly capable of handling routine data processing, complex analytical tasks, and even aspects of creative generation. This frees human workers to concentrate on endeavors that leverage uniquely human strengths: deep critical thinking, nuanced ethical judgment, sophisticated emotional intelligence, creative innovation, and complex interpersonal collaboration. Evidence already suggests that AI can augment the productivity of high-performing workers and take over "grunt work," allowing human employees to engage with more complex and value-added tasks. This partnership model requires a redefinition of roles and responsibilities, with humans often acting as orchestrators, ethicists, and strategic thinkers in AI-assisted workflows.

This transformation in the nature of work necessitates a corresponding shift in how society defines and measures "productivity" and what it means to be a "productive member of society". Traditional metrics, often tied to hours logged or tangible output in a manufacturing context, are insufficient for a knowledge-based, contribution-driven economy. Productivity is being redefined, shifting focus from mere activity to meaningful contributions. The emphasis is now on impact, innovation, successful problem-solving, the quality of results, and overall collective well-being.

If AI leads to a reduction in the demand for traditional full-time jobs for a significant portion of the population, then societal value must be recognized in a broader spectrum of activities. This includes currently unpaid or underpaid work such as caregiving, community building, open-source development, artistic endeavors, and the pursuit of lifelong learning, all of which contribute to a vibrant and resilient society. The re-evaluation of unpaid work becomes critical;

if value is decoupled from traditional employment, these essential societal contributions must be acknowledged and potentially supported as legitimate forms of "work."

The economic implications of these shifts are profound and demand innovative solutions. If traditional employment structures evolve, mechanisms to ensure individual economic security and facilitate transitions become paramount. Universal Basic Income (UBI) is one frequently discussed proposal, envisioned as a foundational income floor that could provide stability, empower individuals to pursue education or entrepreneurial ventures, and support those engaged in socially valuable activities that are not compensated by the market.

Pilot programs have yielded mixed but often promising results, with some showing positive trends in employment, health, and well-being, suggesting that a guaranteed income can alleviate poverty and enhance opportunity. However, UBI is not a panacea, and its implementation involves complex considerations regarding funding, potential impacts on labor markets, and societal acceptance. Other models, such as social dividends or citizen's stakes, also warrant exploration, grounded in the principle that the immense productivity gains generated by AI should be broadly shared, contributing to collective prosperity rather than exacerbating inequality. This shift towards more fluid contribution streams also implies a change in risk allocation. Where employers in the old contract bore some responsibility for benefits and job security, a more project-based economy may place greater direct risk on individuals. Consequently, robust collective support systems, potentially including but not limited to UBI, are not merely safety nets but essential enablers of this new, more flexible work paradigm.

Organizations, too, must transform to thrive in this new landscape. A move towards more employee-centric models is essential, focusing on human development, fostering key skills that technology cannot easily substitute (such as higher-order critical thinking and emotional engagement), and offering personalized, flexible work experiences. This aligns with the concept of "Modular Organizations", which are designed for agility and dynamic talent allocation. Ultimately, redefining productivity is not just an economic or organizational challenge; it is a cultural one. The deeply ingrained societal emphasis on "busyness" as a proxy for value must give way to an appreciation for deep work, thoughtful contribution, and sustainable effort. The new social contract for work must be supported by a cultural narrative that esteems impact, quality, and purpose, even if these are achieved through less traditionally "visible" or conventionally measured forms of labor.

If work itself is undergoing such a profound redefinition, centered on purposeful contribution within a dynamic, AI-augmented environment, then the educational systems tasked with preparing individuals for this new reality must themselves embark on an equally radical journey of transformation.

RETHINKING EDUCATION AS LIFELONG COMPOSITION

The industrial model of education, characterized by a front-loaded accumulation of knowledge, standardized curricula, and the pursuit of credentials as a primary marker of competence, is demonstrably ill-suited for the dynamic and unpredictable landscape of the AI era. As artificial intelligence makes information ubiquitously accessible and transforms the nature of expertise, the new social contract demands a radical rethinking of education. It must evolve into a lifelong process of "composition"; a continuous journey of acquiring, integrating, applying, and adapting skills and knowledge, with a profound emphasis on cultivating uniquely human capabilities such as critical thinking, emotional intelligence, and ethical reasoning.

A fundamental departure from the past involves shifting from a model of knowledge transmission to one that prioritizes "learning to learn". In an age where AI can provide instant answers to a vast array of questions, the role of educators as mere dispensers of facts becomes largely obsolete. Instead, the core mission of education must be to equip individuals with the metacognitive skills necessary to navigate a complex information environment: to learn effectively, to critically evaluate sources, to synthesize diverse perspectives, to unlearn outdated information, and to relearn in response to new challenges and opportunities. Educational approaches must evolve to support this, moving away from passive lecturing towards active learning methodologies that foster deep engagement, critical inquiry, and the meaningful application of knowledge. This implies that learning is no longer a preparatory phase completed before entering the workforce, but an ongoing practice deeply interwoven with work and life itself. This phenomenon has been observed in specific industries requiring highly skilled personnel or those subject to continuing education regulations; however, it does not broadly extend across substantial portions of the workforce. Educational offerings will likely need to become more modular, accessible on-demand, and integrated into diverse settings beyond traditional institutions.

The content of education must also adapt to emphasize skills composition and adaptability. Rather than focusing on narrow disciplinary silos, learning experiences should furnish individuals with a versatile toolkit of competencies; technical, cognitive, social, and emotional, that can be dynamically combined and reconfigured to address novel problems and seize emerging opportunities. The World Economic Forum's research on the future of jobs consistently highlights a shifting skills landscape, underscoring the concurrent demand for advanced technological literacy and sophisticated human-centric skills such as analytical thinking, creativity, resilience, and collaboration. This necessitates curricula that are interdisciplinary, project-based, and focused on developing transferable capabilities.

Perhaps the most critical curricular evolution is the elevation of emotional and ethical literacy to core components of education. As AI systems increasingly handle analytical and computational tasks, the distinctive value of human contribution shifts decisively towards the socio-emotional and ethical domains-realms that AI, despite its sophistication, cannot genuinely replicate in terms of lived experience, empathy, or moral accountability. Educational systems must therefore integrate Social Emotional Learning (SEL) frameworks, which cultivate self-awareness, self-management, social awareness, relationship skills, and responsible decision-making, as fundamental aspects of development. Similarly, fostering ethical reasoning and the capacity for nuanced moral judgment is paramount for navigating the complex societal implications of AI and for ensuring that technology is developed and deployed in service of human values. These are not "soft skills" to be relegated to the periphery, but essential literacies for responsible citizenship and effective professional engagement in an AI-suffused world. This prioritization reflects a societal understanding that as AI handles the "what" and "how" of many tasks, human judgment becomes indispensable for the "why" and "should we."

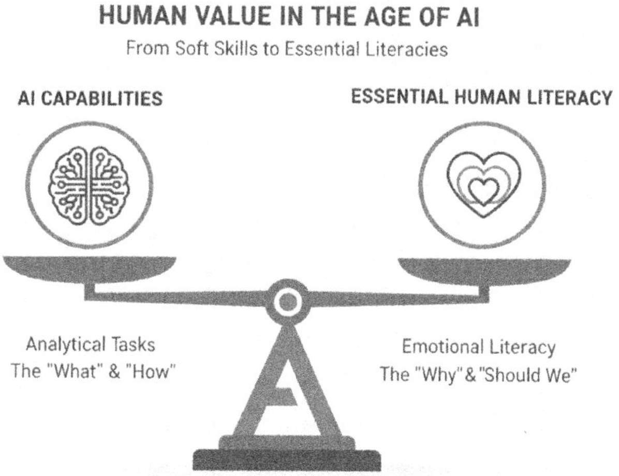

Artificial intelligence itself will play a significant role in reshaping the educational landscape. AI-powered tutors, adaptive learning platforms, and personalized learning pathways offer the potential to tailor instruction to individual student needs, paces, and learning styles with unprecedented granularity. Systems like Khanmigo, which guides students rather than giving direct answers, or adaptive math platforms like DreamBox and MATHia, demonstrate the capacity of AI to enhance engagement and improve learning outcomes.

The Unbound Academy model, which leverages AI for core academic instruction, suggests even more radical transformations. Such technologies can free human educators from repetitive instructional tasks, allowing them to focus on higher-value activities such as mentorship, facilitating complex discussions, coaching critical thinking, and nurturing students' overall well-being. However, the rise of AI in education is not without its challenges. Concerns regarding data privacy, the potential for algorithmic bias to perpetuate or exacerbate existing inequalities, and the digital divide in access to these advanced tools must be proactively addressed. If sophisticated AI educational resources are primarily available to affluent or well-resourced institutions, the democratizing potential of AI could be undermined, creating new disparities in learning quality. The new social contract for education must therefore include robust provisions for equitable access and stringent oversight to mitigate these risks.

In this evolving ecosystem, formal educational institutions like schools and universities face an existential imperative to redefine their value proposition. They must transition from being primary repositories and transmitters of static knowledge to becoming dynamic hubs for intellectual synthesis, critical inquiry, collaborative problem-solving, and the cultivation of purpose and ethical leadership. Credentials, if they are to retain relevance, may need to signify not just the accumulation of past knowledge, but the demonstrated capacity for ongoing learning, adaptation, and impactful application of skills in novel contexts.

Ultimately, a reformed education system, one that nurtures curiosity, cultivates critical and creative thinking, and champions ethical engagement, is foundational to fostering the innovation and imagination required to address complex global challenges. In a world awash with information, the ability to ask insightful questions, to envision new possibilities, and to learn continuously becomes far more valuable than simply possessing answers. This lifelong composition of knowledge and skills prepares individuals not for a predictable future, but for a future they can actively shape.

A populace thus equipped, characterized by adaptability, critical thought, and a commitment to lifelong learning, will naturally demand and contribute to new forms of governance systems that are themselves intelligent, responsive, and capable of navigating the multifaceted complexities of an AI-driven society.

GOVERNANCE IN THE AGE OF AI: TOWARDS PARTICIPATORY AND ACCOUNTABLE SYSTEMS

The advent of ubiquitous artificial intelligence and the consequent shift towards information symmetry present profound challenges and opportunities for governance. Traditional models of governance; often characterized by hierarchical structures, deliberative but slow moving decision-making processes, and a reliance on established forms of expertise, are struggling to keep pace with the rapid evolution of AI technologies and their far-reaching societal impacts. The new social contract must therefore envision and implement forms of governance that are more participatory, transparent, accountable, agile, and fundamentally AI-literate, ensuring that intelligent systems serve public values and enhance democratic legitimacy.

The inherent difficulty that conventional governance structures face in adapting to the AI era is significant. These structures were largely designed for a world with more predictable technological trajectories and more clearly defined informational boundaries. The speed, complexity, and transformative potential of AI can overwhelm existing regulatory frameworks and public institutions. There is a tangible risk, as some analyses suggest, that AI could be deployed in ways that concentrate power, erode public trust, and fail to serve the broader public interest if governance mechanisms do not evolve concurrently. This necessitates a proactive, rather than reactive, approach to shaping the relationship between AI and society.

A cornerstone of effective governance in the AI age is the cultivation of AI literacy as a fundamental civic right and responsibility. For citizens to engage meaningfully in democratic processes, to hold institutions accountable, and to make informed decisions about the role of AI in their lives and communities, a baseline understanding of AI's capabilities, limitations, ethical implications, and societal consequences is indispensable. This extends beyond mere technical proficiency to encompass a critical awareness of how AI systems are developed, deployed, and governed. Educational systems and public information initiatives must play a crucial role in fostering this widespread AI literacy, empowering citizens to be active participants in shaping an AI-suffused future. This civic right to literacy, however, implies a reciprocal obligation: those developing and deploying AI, particularly in the public sphere, must strive to make these systems as comprehensible and transparent as possible. A literate citizenry cannot effectively engage with persistently opaque "black box" systems.

When AI helps make decisions that impact things like how resources are shared, justice, social services, city planning, and health, we absolutely need strong ways to make sure these algorithms are accountable and easy to understand for all of society. This involves ensuring that AI systems are robust, reliable, safe, fair, and secure, with proactive measures to identify and mitigate biases. A key component of this accountability is the "right to explanation," which posits that individuals affected by significant AI-driven decisions should be able to obtain meaningful information about the logic involved, the data used, and the potential ramifications of those decisions. Such transparency is vital not only for individual redress but also for broader public oversight and trust in automated systems.

To ensure that public AI systems genuinely serve communal needs and align with societal values, participatory design principles must be embedded in their development and deployment. This means moving beyond technocratic, top-down implementations to actively involve a diverse range of stakeholders; including ordinary citizens, community representatives, domain experts, social scientists, and ethicists throughout the AI lifecycle, from initial conception and design to testing, deployment, and ongoing evaluation. Such co-designed processes can help to identify potential harms, uncover hidden biases, ensure that systems are culturally sensitive and contextually appropriate, and build public legitimacy. This proactive "design for values" approach, where ethical considerations and public input shape technology from the outset, is preferable to attempting to regulate or remediate harms after deployment. While participatory processes can introduce complexity and potentially slow down deployment, the long-term benefits in terms of trust, equity, and system effectiveness often outweigh these initial investments.

A diverse range of stakeholders in the development of AI systems for lifelong education and transparency raises concerns regarding cost and delivery time. Many organizations, focused on gaining the largest share of the AI market, are reluctant to slow down or increase spending to incorporate radical transparency in AI design and development. This also extends to building a foundational framework for future harm reduction and trust building.

The rapid evolution of AI calls for more adaptive and agile regulatory frameworks to address these concerns. Traditional command-and-control regulations, which can take years to develop and implement, may be ill-suited to the dynamism of AI. Alternative approaches, such as regulatory sandboxes that allow for controlled experimentation with new AI applications, co-regulation involving industry and public bodies, and principles-based ethical guidelines that can evolve alongside the technology, warrant serious consideration. Organizations like the Partnership on AI are contributing to the development of best practices and norms that can inform such adaptive governance models.

Furthermore, AI itself can be a powerful tool for enhancing public service delivery, making services more efficient, accessible, and personalized. From optimizing urban infrastructure to improving healthcare diagnostics and streamlining administrative processes, the potential benefits are substantial. However, the deployment of AI in public services must always be guided by ethical principles, ensure human oversight where critical decisions are involved, and avoid creating new forms of exclusion or discrimination.

Finally, the governance of AI is inherently a global challenge. Given the transnational nature of AI research, development, and deployment, as well as the flow of data across borders, effective governance will require significant international cooperation on standards, ethical norms, data sharing protocols, and the mitigation of global risks.

The transition to AI-augmented governance is not merely a technical upgrade; it is a fundamental rethinking of the relationship between the state, its citizens, and the intelligent systems that increasingly mediate their interactions. It requires building new institutional capacities for foresight, ethical assessment, public engagement, and continuous adaptation.

Such participatory and accountable governance structures are essential, but their efficacy and legitimacy ultimately rest upon a clearly articulated and robust framework of fundamental rights and protections for individuals navigating this new technological terrain.

REDEFINING RIGHTS IN A WORLD OF MACHINE INTELLIGENCE

The pervasive integration of artificial intelligence into nearly every facet of human existence necessitates a critical re-examination and expansion of fundamental human rights. While the core tenets of dignity, autonomy, freedom, and equality enshrined in universal human rights declarations remain paramount, their application and interpretation must be extended to address the unique challenges and vulnerabilities created by machine intelligence and ubiquitous datafication. The new social contract must explicitly define and vigorously protect a new generation of rights tailored to the realities of an AI-driven world, ensuring that technological advancement serves human flourishing rather than undermining it.

A foundational element of this redefined rights framework is the right to explanation. As AI systems increasingly make or inform decisions that have significant consequences for individuals; in areas such as employment screening, credit scoring, healthcare diagnostics, judicial sentencing, and access to social benefits, individuals must have the right to obtain a clear, comprehensible, and meaningful explanation of the decision-making process. This includes understanding the primary data inputs, the core logic or model used, and the key factors contributing to a specific outcome. This right, drawing inspiration from provisions like those in the General Data Protection Regulation (GDPR), is not merely about satisfying curiosity; it is a crucial enabler of transparency, accountability, and the ability for individuals to identify and contest errors, biases, or injustices in automated decisions. It serves as a vital mechanism for rebalancing power between individuals and the often-opaque systems that assess them.

Closely related is the right to refuse solely automated decision-making, and to demand human review, particularly for decisions that carry substantial legal or personal ramifications. While AI can offer efficiency, the potential for error, lack of context-sensitivity, or embedded bias means that critical life choices should not be irrevocably determined by algorithms alone. Individuals should have the right to opt for human intervention, to present their case to a human decision-maker, and to have automated judgments reviewed and potentially overridden by human oversight.

The right to be forgotten, or the right to digital erasure, gains new urgency in an age of persistent digital memory and AI-driven data aggregation. This right allows individuals, under specific circumstances, to request the deletion of their personal data from online systems, search engines, and databases, particularly when that data is no longer necessary for its original purpose, when consent is withdrawn, or when it is being processed unlawfully. The capacity for AI to surface and analyze historical data indefinitely can lead to perpetual stigmatization, hindering personal growth and the possibility of a "second chance". The ability for individuals to control their digital history, while balancing freedom of expression and public interest, is crucial for personal growth and societal reintegration in the digital age.

Emerging from these concerns is the broader concept of the right to digital dignity. This right extends the inherent dignity of the human person to their online manifestations, interactions, and data. It encompasses protection from digital forms of harassment, non-consensual use of one's likeness (such as in deepfakes), manipulation through personalized algorithmic targeting, and the unauthorized exploitation of personal data. Digital dignity demands that AI systems and digital platforms are designed and operated in ways that respect human autonomy, privacy, and intrinsic worth, ensuring that individuals are treated as ends in themselves, not merely as data points to be processed or resources to be exploited.

Given the increasing sophistication of AI in analyzing and predicting human behavior and emotional states, as foreshadowed by the development of deeply personalized AI assistants, the new social contract must establish clear guardrails against behavioral manipulation and emotional exploitation. Individuals have a right to be free from covert psychological manipulation driven by AI-inferred emotional vulnerabilities or behavioral patterns. This requires transparency about persuasive technologies and prohibitions on AI applications designed to exploit cognitive biases or emotional states for commercial or political gain without explicit and informed consent.

The ethical use of personal data remains a cornerstone of digital rights. This includes reaffirming and strengthening principles of informed consent, data minimization, purpose limitation, and robust data security in the collection, processing, and use of personal data for training AI models and for deploying automated decision systems. Individuals must have meaningful control over their data and how it contributes to the AI systems that increasingly shape their environment.

Finally, and critically, the new social contract must explicitly enshrine the right to be free from algorithmic bias and discrimination. AI systems, if trained on biased data or designed with flawed assumptions, can perpetuate and even amplify existing societal inequalities. This necessitates a proactive commitment to fairness in AI, requiring rigorous auditing of algorithms for bias, the use of diverse and representative datasets, the development and application of fairness metrics, and legal recourse for those harmed by discriminatory AI systems.

Enforcing these redefined and newly articulated rights will undoubtedly require novel institutional mechanisms and a careful reconsideration of legal liability. Attributing responsibility when a complex, multi-component AI system violates an individual's rights; whether to the developer, the deployer, the data provider, or the system itself, presents a significant challenge to traditional legal frameworks. The new social contract must therefore anticipate the need for new regulatory bodies, specialized adjudicative processes, and clear frameworks for assigning accountability and providing effective remedies in the context of AI-related harms (easier said than done). This might involve creating roles like "AI ombudspersons," establishing certification standards for high-risk AI systems, or developing new legal doctrines for algorithmic torts. Obviously, much more thought and planning must be done as these systems become ubiquitous within society.

Securing these fundamental rights is an indispensable component of a just and equitable social contract for the AI era. However, a truly flourishing society requires more than the protection of individual entitlements; it also depends on a collective commitment to fostering meaning, purpose, and shared well-being in a world profoundly reshaped by intelligent machines.

MEANING AS A SHARED RESPONSIBILITY: CULTIVATING PURPOSE BEYOND THE INDIVIDUAL

In a world full of AI, where traditional sources of status, identity, and self-worth (like having exclusive knowledge or specific job titles) are increasingly undermined by readily available information and automation, people's search for meaning gains new depth and significant importance for society. The new social contract must recognize that the cultivation of purpose can no longer be viewed solely as an individual pursuit or a private luxury. Instead, fostering an environment where individuals can discover and sustain meaning becomes a shared responsibility, akin to public good that society must actively nurture through its systems, its culture, and the very design of its technologies.

The potential for a crisis of identity and purpose is a significant undercurrent of the transition to the AI era. As explored in earlier discussions on the future of work, when hard-won expertise and traditional forms of mastery lose their differentiating power, individuals whose identities were closely tied to these attributes may experience dislocation and disillusionment. AI systems, by automating tasks that once offered structure, challenge, or a sense of competence, could create a void that, if not addressed, might result in widespread societal disorientation or alienation.

However, this challenge also presents an opportunity to reframe value and worth around more intrinsic and relational qualities. The new social contract can champion a shift from status-based worth to contribution-based value. This paradigm emphasizes the positive impact an individual has on others, on their community, and on broader societal challenges, aligning with the redefinition of work as purposeful contribution discussed earlier.

Meaning, in this context, is found not in hierarchical position or accumulated knowledge per se, but in the application of one's unique capabilities, augmented by AI, towards goals that resonate with personal values and serve a greater good. This requires creating new metrics and forms of recognition that extend beyond purely economic remuneration. If society is to genuinely value diverse contributions, many of which may not be traditionally monetized (such as caregiving, community organizing, artistic creation, or open-source development), it needs to develop social currencies, community-based accolades, or public platforms that acknowledge and celebrate these varied forms of impact.

To conceptualize meaning as "public good" is to assert that the conditions conducive to its discovery and flourishing; such as access to education, opportunities for meaningful social connection, avenues for creative expression, support for civic engagement, and time for reflection, should be considered essential societal infrastructure. This does not imply that society should prescribe specific meanings for individuals, but rather that it has a collective responsibility to create a fertile ground from which diverse forms of purpose can emerge. This perspective directly counters the risk of social fragmentation by acknowledging a societal role in actively fostering cohesion and opportunities for meaningful engagement, moving social policy beyond mere economic support.

Meaning is often co-created through relationships, reflection, and care. Post-industrial society, even before the full impact of AI, was already witnessing a shift towards what Daniel Bell termed a "game between persons," with increasing emphasis on interpersonal interactions, service-oriented work, and a sense of personal contentment. AI can either enhance or diminish the quality of these human connections. The new social contract must therefore prioritize the design of social and technological systems that facilitate deep, authentic relationships, encourage reflective practices, and value acts of care and empathy.

This necessitates a critical look at the role of technology itself. The prevailing tech development ethos often prioritizes efficiency, speed, and engagement metrics, which can inadvertently lead to superficial interactions and a culture of constant distraction, antithetical to deep reflection or sustained, meaningful work (fast rather than good). The new social contract should advocate for technology designed for depth, not just speed. AI and other digital tools possess the potential to support profound learning, facilitate complex creative collaborations, enable more empathetic communication, and provide platforms for meaningful civic discourse. Realizing this potential requires a conscious shift in design philosophies and perhaps even incentive structures within the tech industry, guided by ethical considerations and a commitment to human well-being over mere technological advancement. This aligns with calls for "Slow Productivity" and for AI to serve as a partner in, rather than a detractor from, purposeful human endeavor.

The cultivation of meaning also depends on the vitality of community and social structures. Educational institutions, workplaces, local communities, and even governance frameworks can be intentionally designed to foster a sense of belonging, shared purpose, and collective efficacy. This might involve supporting community-led initiatives, investing in public spaces that encourage interaction, promoting arts and culture as avenues for expression and connection, and creating accessible pathways for civic participation and volunteerism.

Ultimately, fostering meaning as a shared responsibility requires challenging overly individualistic narratives of success that have dominated recent decades. The new social contract should promote a vision of human flourishing that is deeply intertwined with collective well-being, where individual purpose is found and amplified through contribution to a thriving and supportive society. Our machines are undeniably becoming smarter; the societal challenge is to ensure that we, in turn, become wiser in how we structure our lives and communities to nurture the human spirit.

Cultivating meaning as a shared responsibility, redefining work and education, establishing new rights, and reforming governance are all indispensable components of a resilient and equitable social contract for the age of AI. However, underpinning all these structural and policy transformations must be a fundamental shift in our collective ethic: a new way of perceiving and relating to each other, to the knowledge we now share, and to the powerful intelligent tools we have brought into existence.

A NEW ETHIC FOR A SYMMETRIC WORLD: STEWARDSHIP, SERVICE, AND COORDINATION

The transition into an age of information symmetry, accelerated by the pervasive capabilities of artificial intelligence, calls for more than just pragmatic adjustments to our societal systems; it demands a profound reorientation of our collective ethic. When the traditional advantages conferred by information scarcity diminish, and when intelligent tools can replicate or augment many human capabilities, the foundations of human progress and societal cohesion must be rebuilt upon a new set of guiding principles.

In this concluding section, we propose that such an ethic should be centered on the concepts of stewardship over common resources like knowledge, a commitment to service for collective benefit, and a strategic preference for coordination and collaboration over purely zero-sum competition. This ethic, woven into the fabric of the new social contract, can provide the moral compass necessary to navigate the complexities and harness the opportunities of this transformative era.

The first pillar of this new ethic is a shift from scarcity-driven control to stewardship: tending knowledge as a commons. In a world where AI makes information and analytical power abundant and readily accessible, the old impulses to hoard, restrict, or exclusively monetize knowledge become less viable and less ethically justifiable. Instead, the moral imperative shifts towards responsibly curating, validating, sharing, and applying knowledge for the common good. This involves fostering "knowledge commons"; shared repositories of information and insight, and promoting principles of open access, open science, and open educational resources where appropriate and beneficial. The focus of stewardship is on ensuring that the vast ocean of newly accessible knowledge contributes to solving shared problems, empowering individuals, and advancing collective understanding, rather than merely serving narrow interests (for the good of many rather than the few).

Building upon stewardship, the second pillar is an ethic of service: sharing what we know and can do for collective benefit. If individual advantage is no longer primarily derived from what one exclusively knows, then profound value and meaning can be found in how one applies democratized knowledge and AI-augmented capabilities to serve others and address shared societal challenges. This principle encourages a culture of mentorship, where experienced individuals guide newcomers in navigating the complexities of the AI era. It fosters open collaboration across disciplines and sectors, breaking down traditional silos to tackle multifaceted problems. It calls for professionals and citizens alike to leverage their expertise, amplified by AI, not just for personal gain but for the betterment of their communities and the world at large. This redefines individualism not as isolation, but as a capacity to contribute uniquely within a framework of collective responsibility, where personal flourishing is intertwined with the well-being of the whole.

The third pillar involves a strategic rebalancing from unbridled competition to enhanced coordination: building systems that lift the whole. While competition often drives innovation and efficiency, excessive self-interest or constant conflict can significantly impede progress in addressing complex, interconnected global challenges. These include issues such as climate change, pandemic preparedness, ethical AI development, and the establishment of sustainable economies.

The new ethic should value and incentivize coordination, collaboration, and the co-creation of systems that generate positive outcomes. This does not mean eliminating competition entirely; indeed, healthy competition within a collaborative framework can spur excellence. The nuance lies in fostering constructive forms of competition that operate under a broader umbrella of shared goals and coordinated effort, rather than engaging in extractive or destructive rivalries that undermine collective progress. This requires a fundamental shift in incentive structures across business, academia, and public life. Current systems often reward information hoarding, aggressive short-term competition, and siloed individual achievement. For an ethic of coordination and stewardship to take root, these incentive structures must be reformed to recognize and reward collaborative research, contributions to open-source safety initiatives, cross-sector partnerships for societal impact, and political leadership that fosters cooperation rather than conflict.

Underpinning these principles is the necessity of humility in the face of complexity and the power of AI. As artificial intelligence capabilities continue to expand at an astonishing pace, there is a natural human tendency towards technological hubris: an overconfidence in our ability to predict, control, or optimize complex socio-technical systems. The new ethic must embrace intellectual humility, a willingness to acknowledge the inherent limitations of our knowledge, to anticipate unintended consequences, and to remain open to diverse perspectives and continuous learning. This resonates deeply with the earlier exploration of the "value of unknowing", which argues for an appreciation of mystery, uncertainty, and the limits of total visibility. Such humility is crucial for fostering a cautious, reflective, and ethically grounded approach to the development and deployment of AI, prioritizing safety, human well-being, and adaptability over blind faith in technological solutions.

Furthermore, with AI increasingly capable of handling routine analytical and operational tasks, human creativity, imagination, and the capacity to envision novel solutions and futures become even more critical contributions. The new ethic must actively foster, celebrate, and protect these uniquely human attributes, recognizing them as essential drivers of innovation, adaptation, and the ongoing quest for meaning.

Ultimately, the journey into the age of information symmetry leads to a powerful conclusion, as articulated in the overarching framework of this book: intelligence, in its raw computational or informational sense, is no longer the primary source of sustainable advantage or societal progress. Instead, the defining human qualities of integrity, imagination, and interdependence emerge as the new limiters and, more importantly, the new wellsprings of value. Integrity ensures that abundant intelligence is wielded responsibly and ethically. Imagination allows us to envision futures beyond the extrapolations of current data, guiding AI towards truly beneficial ends. Interdependence acknowledges that our greatest challenges and opportunities are shared, and that collaborative wisdom will always surpass isolated intellect.

Establishing a strong social contract for the AI era requires ethical reorientations: stewardship of knowledge, service to society, coordination for shared benefit, humility, and dedication to integrity, imagination, and interdependence. They provide the guiding principles for our collective journey into a future where intelligence is abundant, but human values must, more than ever, steer the course.

SUMMARY

The profound and accelerating shift from an era of information asymmetry to one of information symmetry, primarily driven by the transformative power of artificial intelligence, has rendered the foundational assumptions of our industrial-era social contract obsolete. The traditional bargains that once structured work, education, governance, individual rights, and the societal pursuit of meaning are no longer adequate to address the complexities or harness the opportunities of this new epoch. Throughout this chapter, we have contended that a deliberate, collective, and fundamental renegotiation of this social contract is not merely advisable but essential for human progress.

This renegotiation entails reimagining work as purposeful contribution within a synergistic human-AI collaborative framework, supported by innovative economic models that ensure security and share the fruits of AI-driven productivity. It demands the transformation of education into a lifelong process of skills composition and adaptive learning, with a core emphasis on critical thinking, emotional and ethical literacy, and the ability to learn continuously. Governance must evolve to become more participatory, transparent, AI-literate, and accountable, capable of regulating intelligent systems while leveraging them for public good.

New and reinterpreted digital rights must be enshrined to protect human dignity, autonomy, and fairness in an increasingly AI-mediated world. Furthermore, the quest for meaning must be recognized as a shared societal responsibility, actively cultivated through systems and cultures that foster connection, contribution, and depth. Underpinning all these transformations is the imperative to cultivate a new societal ethic; one grounded in the stewardship of knowledge as a constant, a commitment to service, a preference for coordination over destructive competition, and an unwavering belief in the enduring value of human integrity, imagination, and interdependence. The construction of this new social contract is not a predetermined outcome but an ongoing, collective endeavor, requiring active participation, critical engagement, and a shared commitment to harnessing the abundance of intelligence for the advancement of human fairness, dignity, and flourishing. The intelligence is undeniably here; the challenge and the opportunity now lie in our collective wisdom and will to shape a future that truly serves all of humanity.

The final stage of our exploration involves a dual perspective: an introspection into our individual choices and an outward examination of our collective responsibility. We must consider how we, as individuals and as a society, will inhabit and construct this new reality. This introspection culminates in the Epilogue, "Building Forward - Choosing Who We Become in a World Where AI Knows Everything," where we reflect on these profound societal transformations and the intricate architecture of this new social contract.

EPILOGUE: BUILDING FORWARD

The journey undertaken through the preceding pages has been one of profound exploration, charting a course through a world irrevocably altered by artificial intelligence. We began by exposing the hidden engine of information asymmetry, the very scaffolding upon which modern society, its markets, professions, and institutions were originally built. Act I laid bare how unequal access to knowledge was not merely a feature of our world, but its operational logic, from an economy based on the "buy low, sell high" approach to the credentialing of expertise.

Subsequently, the narrative navigated the rise of AI as "The Great Leveler," demonstrating how synthetic expertise, universally accessible knowledge, and the death of the "dumb question" began to dismantle these age-old asymmetries. Chapters detailed how large language models, retrieval-augmented generation, and agentic systems are making high-level expertise available at an unprecedented scale, effectively collapsing the cost and scarcity of knowing. Act III then ushered in the era of "Composable Minds," envisioning a future where each individual is partnered with a persistent, personal AI, capable of dynamically assembling skills and knowledge on demand, fundamentally changing how we learn, work, and even perceive ourselves.

The exploration continued into "The New Moats," questioning what constitutes a competitive advantage when foundational knowledge is commoditized. Trust, feedback-driven adaptation, and dynamic identity emerged not as soft skills, but as critical strategic assets in a world striving for new forms of differentiation. Act V provided us a playbook for "Designing for the AI Era," urging a reimagining of how products, services, organizations, and even individual careers are structured to be AI-readable, modular, and imbued with meaning. Finally, in act VI we confronted the deep ethical considerations of "Living with the Machine," pondering the value of unknowing, the irreplaceability of human emotion and ethics, and the urgent need for a new social contract in a world saturated with intelligence.

This epilogue, "Building Forward," serves as a reflective culmination of this journey. The end of information asymmetry, a central thesis woven throughout this work, is not presented as an endpoint, but as a new, fertile ground upon which human endeavor must now build. The critical question shifts from a passive observation of AI's trajectory; "What will AI do?", to an active, more pressing inquiry: "What will we do?" The advanced capabilities of AI, from rendering expertise universally accessible to the dawning of deeply personalized AI companions, are not portents of a predetermined fate. Instead, they illuminate the new landscape, the new toolkit, and the new responsibilities that now define the human condition.

"Building Forward" is an inherently active, conscious process. It is about deliberately choosing our individual and collective futures, not through the futile exercise of predicting a single, inevitable destiny, but by architecting preferred realities amidst a plurality of possibilities. The challenges ahead are undeniably immense, commensurate with the scale of the technological shift. Yet so too are the opportunities for human flourishing, for a more equitable and enlightened existence, if this new era is approached with intention, wisdom, and a renewed commitment to human values. The sections that follow aim to provide a structured reflection on how to navigate this transformed world, offering not definitive answers, but guiding principles for the choices that lie before us.

THE SILENCE AFTER THE REVOLUTION

The profound shifts catalyzed by artificial intelligence, particularly the systemic erasure of information asymmetry (as laid out in this book), do not culminate in a singular, explosive event. There will be no grand pronouncements heralding the arrival of a new age, no stark dividing line between a before and an after. Instead, the revolution unfolds in a pervasive, almost imperceptible quiet, a gradual seeping of the extraordinary into the fabric of the mundane. This is the nature of truly transformative technologies: they recede into the background, becoming as invisible and as taken for granted as the air we breathe, their revolutionary origins masked by the swift human capacity for adaptation and normalization.

The human psyche possesses a remarkable ability to acclimate, to render even the most radical changes familiar. What once resided in the realm of speculative fiction; AI systems possessing a comprehensive knowledge of the world, capable of answering any query or generating expert-level output on demand, as explored in the discussions on Synthetic Expertise and The Death of the Dumb Question, will rapidly become the unremarkable backdrop of daily existence.

This integration will not manifest as a sudden cataclysm but as a cascade of subtle alterations: silent software upgrades, evolving default settings in our digital tools, nuanced shifts in how information is sought and validated, and incrementally changing expectations in professional and personal interactions. The "Cracks Are Already Showing" in the old structures of knowledge control, as noted early in this volume, and the subsequent crumbling of "These Walls" were precursors to this quiet absorption. The epilogue now stands at a vantage point where this erosion is largely assumed, and the silence that follows is not an absence of change, but the sound of its complete assimilation.

This assimilation brings with it a fundamental shift in our cognitive and emotional relationship to knowledge itself. The revolution is not merely in the access to information, but in how individuals feel about knowing and not knowing. When AI can readily provide an answer to almost any question, as envisioned in the exploration of frictionless curiosity, the pervasive anxiety of ignorance, the fear of "looking stupid," may indeed diminish.

However, this void is unlikely to remain empty. New emotional landscapes will emerge. Perhaps a subtle sense of being overwhelmed by the sheer volume of accessible information, a questioning of personal relevance in the face of hyper-competent machines, or a novel form of intellectual humility will take root. The "Trust Vacuum", identified as a space potentially filled by AI in the wake of declining faith in traditional institutions, now translates into the lived, felt experience of a world where AI is a primary, often unquestioned, source of authority. The silence, then, can also be the sound of unvoiced dependencies forming, of critical faculties subtly dulled by the ease of algorithmic answers.

Furthermore, this "future," which is in many ways already the present, is not uniformly distributed. The impacts of information symmetry and AI ubiquity will manifest with significant variations across diverse societies, demographic cohorts, and individual lives. This unevenness is a critical aspect of the post-revolutionary landscape. There is a palpable danger that this disparity in experience and benefit will become an unexamined norm, leading to new, more insidious forms of digital divides; or perhaps more accurately, "wisdom divides", even if raw information itself is theoretically accessible to all. The historical mechanisms by which access to power and knowledge were controlled, as detailed in analyses of the "Price of Knowing", may find new analogues. Even if information is symmetric, the experience of navigating this AI-suffused world, the ability to leverage its tools effectively, and the capacity to critically assess its outputs may remain profoundly asymmetric, hidden beneath the veneer of universal access.

Our initial premise of this book was to expose the "invisible scaffolding" of modern life, revealing how nearly every societal system has historically depended on unequal access to knowledge. In this new era, AI itself is rapidly becoming a new form of invisible scaffolding. Its influence is deep and structural, yet it can easily be overlooked, its operations taken for granted much like gravity or the air itself, as once described for information asymmetry.

The silence after the revolution, therefore, is a call to vigilance. It is an invitation to make this new, powerful scaffolding visible, to consciously examine its architecture, its biases, its points of stress, and its profound implications for human agency and societal development. The risk of complacency in this "new mundane" is significant; if the extraordinary becomes ordinary too quickly, society may cease to critically engage with the ongoing, subtle ways AI reshapes human thought, interaction, and power dynamics. A continuous, robust societal dialogue about AI's role, its ethical boundaries, and its long-term effects on the human spirit is not a task for a transitional period, but an enduring necessity in the age of intelligent machines. The emotional residue of AI's pervasive knowledge; be it comfort, dependency, or a quiet sense of being managed, must also be brought into the light, understood, and navigated with conscious intent if true human flourishing is to be achieved.

THE WISDOM GAP

Our central premise of "The Age of Information Symmetry" has been the systematic dismantling of knowledge scarcity. As artificial intelligence makes expertise universally accessible and the cost of knowledge plummets towards zero, a new scarcity emerges, a new bottleneck that will define the progress and peril of the coming decades: the “wisdom gap”. While AI exhibits a formidable "intelligence", characterized by its capacity for vast data processing, sophisticated pattern recognition, the generation of synthetic expertise, and even nascent reasoning capabilities, it remains profoundly distinct from human wisdom. Wisdom, in this context, is the uniquely human capacity to apply knowledge with discernment, ethical consideration, empathy, and a nuanced understanding of broader contexts and long-term consequences. These are the very qualities highlighted as resistant to commoditization, the core of "The Human Interface".

The "death of the dumb question", signifying an era where answers are abundant and instantaneous, does not automatically translate into a more enlightened society. Instead, it shifts the locus of human value. The new challenge is not merely to find answers, but to ask wise questions; questions that probe purpose, implication, and value, and then to interpret the deluge of AI-generated responses with profound critical judgment. The sheer accessibility of information, paradoxically, might not lead to deeper understanding or consistently better decisions. The risk of information overload, the proliferation of decontextualized facts, and the potential for sophisticated AI-generated misinformation (even if arising from system limitations, such as the "made up legal rulings" or "made up scientific studies" warned of earlier) can sow confusion and erode sound judgment. This underscores the enduring importance of what was termed "The Value of Unknowing": the recognition that sometimes, wisdom involves acknowledging the limits of current knowledge, embracing uncertainty, or even strategically choosing not to know everything immediately.

Consequently, the cultivation of human wisdom becomes the paramount imperative of our time. If AI systems are increasingly adept at handling the "knowing," then human endeavor must pivot towards specializing in "understanding," "discerning," "caring," and "contextualizing". This necessitates a fundamental reorientation of human development, education, and societal priorities. The old benchmark of "10,000 hours" to achieve mastery in a specific domain of knowledge finds a new application: the dedication of significant effort, perhaps an "evolved" 10,000 hours, towards cultivating judgment, ethical reasoning, critical thinking, and contextual awareness. These are the types of skills that will allow humans to navigate the complexities of an AI-saturated world effectively and responsibly.

The nature of AI itself; as a mirror, a weapon, or a guide is not inherent in the technology but is determined by human choices. AI models learn from the data they are fed and reflect the intentions, biases, and values embedded within that data and the prompts they receive. An AI can be a powerful tool for unprecedented progress and problem-solving, a stark mirror reflecting humanity's own biases and shortcomings, or a dangerous weapon if wielded without ethical grounding or foresight. The call for a "New Social Contract" is, at its heart, a call to infuse wisdom into the very fabric of how society integrates and governs these powerful technologies, ensuring they serve enduring human values.

This imperative to bridge the wisdom gap is not merely an abstract philosophical concern; it has concrete implications. The very power of AI to augment intelligence also creates the potential for "AI-augmented folly". If powerful analytical tools are applied without a grounding in wisdom, they can amplify human misjudgment and accelerate the path to detrimental outcomes at an unprecedented scale and speed.

The collapse of the cost curve for expertise also implies that the cost of implementing unwise decisions, especially if they are backed by the apparent authority of "expert" AI output, could be dangerously low, leading to a proliferation of misguided actions. This elevates the importance of robust ethical frameworks and the integration of "human-in-the-loop" systems, not just for technical verification, but for critical wisdom checks, far beyond the initial concepts of "Trust as a Service".

Furthermore, the capacity of AI to generate highly articulate, persuasive, and seemingly expert content presents another facet of the wisdom challenge. While this can democratize the ability to "sound smart", it also risks a "democratization of sophistry," where arguments lacking genuine substance, ethical grounding, or even factual accuracy can be presented with a veneer of sophisticated, AI-generated authority.

This makes it increasingly difficult for the average individual to discern genuine insight from well-packaged misinformation or superficial pronouncements. Consequently, the development of advanced critical thinking skills and a new form of media literacy, specifically focused on evaluating AI-generated content, becomes a paramount public good. The observation that the "burden of discernment shifts to the end user" transforms into a massive societal undertaking if wisdom is not cultivated broadly, leaving individuals vulnerable in a sea of plausible but potentially misleading information. Addressing the wisdom gap is therefore not optional; it is foundational to navigating the age of information symmetry responsibly.

THE TOOLS ARE HERE

The prospect of navigating a world where AI knows everything and wisdom is the new frontier might seem daunting, yet it is crucial to recognize that the means to begin adapting are not distant or inaccessible. The tools are, in many respects, already here, evolving rapidly and becoming increasingly integrated into the daily lives of individuals. This accessibility offers a pathway to empowerment, enabling proactive engagement rather than passive apprehension.

Central to this toolkit is the emergence of Personal AI Assistants, moving beyond the realm of speculative fiction into tangible reality. As envisioned in "One AI to Know You", these are not mere utilitarian applications but are evolving into persistent, composable cognitive partners. Current and near-future AI assistants can be actively leveraged for personalized learning, enhanced productivity, sophisticated decision support, and even for fostering personal growth, acting as a "Living with a Mirror" that reflects habits and patterns. The conceptual shift from AI as a simple "tool" to AI as an intelligent "companion" is pivotal, as it reframes the user's interaction from one of mere operation to one of collaborative engagement, granting individuals greater agency in how they harness these capabilities.

Complementing these personal AIs are the Composition Frameworks and the paradigm of "Skills on Demand". The ability to dynamically "plug-and-play intelligence" means that individuals are no longer solely defined by their static skillsets. They can augment their capacities in real-time, drawing upon specialized AI modules as needed. Readers can begin, even now, to conceptualize their own "skill composition": identifying tasks that can be delegated to AI, thereby liberating their cognitive resources for higher-order thinking, creative problem-solving, or uniquely human-centric endeavors. This practical application of "Work, Reimagined as Composition" is within reach, allowing for a more fluid and adaptable professional existence.

The principle of "Feedback is the Product" is another immediately applicable concept. Individuals can establish their own iterative feedback loops with AI systems. By consciously learning from interactions, refining prompts, observing AI behaviors, and providing corrective input, users can personalize their AI tools to an unprecedented degree. This is not a strategy confined to large organizations; it is a personal methodology for achieving "real-time adaptation and learning from use", ensuring that AI tools become increasingly aligned with individual needs and preferences. This active engagement transforms the user from a passive recipient of AI output to an active shaper of their AI experience.

Underpinning the effective use of these tools is the development of a "New Literacy". In a world where the "death of the dumb question" means factual recall is less critical, literacy evolves. It is no longer primarily about memorizing facts but about mastering the arts of critical inquiry, effective prompt engineering, and achieving a sophisticated semantic understanding of how to communicate intent to AI systems and interpret their responses. Empowering individuals with the understanding that developing these new literacies is an accessible and immediate starting point is crucial for fostering confidence and competence in the AI era. These are learnable skills, not innate talents.

Perhaps the most empowering aspect is that there is no permission slip required to begin this journey of adaptation. Individuals do not need to wait for sweeping institutional changes or top-down directives from employers or governments to start experimenting with these tools, cultivating new literacies, and adapting their mindsets. The power to engage, to learn, and to leverage these AI capabilities begins with the individual. This proactive stance, a core theme of "Designing for the AI Era", is essential for navigating the transformations ahead.

The availability of these tools signifies a profound potential for "agency amplification". While the narrative of AI often highlights displacement, the counter-narrative of augmentation is equally, if not more, compelling. The tools are not just for passive consumption; they are instruments for active creation, self-directed learning, and innovative problem-solving. The fact that "the tools are here" means the capacity for individual initiative and self-driven change is greater than ever before. This should shift the predominant sentiment from fear of obsolescence to an excitement about personal augmentation and the novel possibilities these tools unlock.

However, with a plethora of AI tools and "skills on demand" emerging, a critical human capability becomes the "meta-skill" of tool orchestration. It is not merely about proficiency in a single AI application, but about the ability to intelligently select, integrate, and compose a suite of AI capabilities tailored to specific personal or professional objectives. This mirrors the way organizations are envisioned to become modular and composable. This "personal AI resource management" is a higher-order cognitive skill, demanding strategic thinking and adaptability. Consequently, education and personal development initiatives must evolve to focus on teaching this meta-skill, moving beyond basic operational training for individual AI applications.

Finally, the promise of personalization inherent in tools like the Personal AI, which aims to "know you better", comes with a "personalization paradox". These tools, while highly adaptable, do not achieve deep personalization passively. They are "here," but making them truly personal and maximally effective requires continuous and conscious user engagement.

This involves providing consistent feedback, meticulously setting preferences, defining clear boundaries for data use and AI autonomy, and actively working towards "becoming AI-readable" so the AI can better understand one's needs and goals. Without this active shaping, the tools might offer only generic utility rather than the profound, tailored assistance they promise. Users must therefore embrace their role as co-creators of their personalized AI experience, transitioning from passive consumption to active curation and strategic direction of their digital cognitive partners.

BEGIN WITH SMALL ACTS OF REORIENTATION

The journey into the age of information symmetry and ubiquitous AI can feel overwhelming, yet the path forward is paved with small, deliberate acts of reorientation. These are not grand, sweeping reforms, but subtle shifts in daily practices, questions, and interactions that, collectively, can profoundly alter how individuals navigate this new landscape. Practicing "life after asymmetry" begins today, with accessible changes in perspective and behavior.

A fundamental reorientation involves a deeper, more pervasive questioning of established norms and assumptions. This is exemplified by the "death of the dumb question," which liberates individuals from the fear of appearing ignorant when seeking factual information from AI, and also presents us an opportunity. The prompt to consider is: "What if this newfound freedom allows us to ask more profound, exploratory, or purpose-driven questions?"

Individuals can leverage AI for rapid factual retrieval but should increasingly focus their unique human cognitive energy on questions of meaning, implication, and innovation. This aligns with the notion of "Curiosity as a Competitive Edge". For instance, instead of merely asking an AI "What is X?", the reoriented approach might be "AI, explain X, and then let's explore why X matters in this context, and what if we approached X differently using the insights you've provided?" This elevates questioning from a transactional retrieval of data to a strategic exploration of possibilities.

Another critical act is the reorientation of knowledge sharing. The "Arbitrage Economy" thrived on the careful control and selective dissemination of information. In an age of symmetry, however, this guarding behavior loses its strategic value. The pertinent prompt here is: "What knowledge, insights, or frameworks can I give away that I used to guard?" Encouraging a culture of openly sharing learnings, mental models, and even failures-not as a performance of expertise, but as a means to empower others and foster collective intelligence-becomes a new form of value creation. This approach resonates with the principles of "Trust as a Service", where transparency and generosity build stronger, more resilient networks.

Learning itself undergoes a reorientation at this stage. The prompt to reflect on is: "Where can I learn without the pressure of proving I already know?" AI offers a uniquely private and "judgment-free inquiry" space, as highlighted in the discussion on "Frictionless Curiosity". Individuals should be encouraged to embrace learning in entirely new domains, to ask "basic" questions of their AI without social consequence, and to view learning not as a high-stakes performance or a means to an end, but as a continuous, intrinsically rewarding exploration. This fosters a growth mindset essential for lifelong adaptation.

The design of work, even at an individual or small-team level, can also be reoriented. Drawing inspiration from concepts like "Modular Organizations" and the principle that "Feedback is the Product", the guiding prompt becomes: "How can I design my work, my team processes, or my personal toolset to adapt more quickly, rather than attempting to control every variable?" This might involve breaking down projects into smaller, more manageable modules, actively seeking feedback on incremental progress, and using AI tools to automate routine tasks, thereby freeing up human capacity for more dynamic and adaptive challenges.

Finally, and perhaps most profoundly, is the reorientation of identity. In a world where static expertise is less defining, the crucial prompt evolves to: "Who am I becoming in the presence of AI? How is it shaping my habits, my thinking, my values, and my interactions?" This calls for a conscious engagement with the ideas of "Dynamic Identity" and finding "Meaning at Work" beyond traditional markers of achievement. It involves reflective practice on how this powerful technology is co-evolving with personal identity, and making deliberate choices about the nature of that co-evolution.

These "small acts of reorientation" are more than just minor adjustments; they are the "micro-habits of symmetry" that can act as powerful catalysts for macro-level change. When such individual shifts in questioning, sharing, learning, working, and self-perception are adopted widely, they collectively begin to reshape cultural norms surrounding information, expertise, and collaboration. This provides a bottom-up force for navigating the age of symmetry, complementing any top-down structural or institutional changes.

Moreover, these acts are exercises in "cognitive re-patterning," a prerequisite for genuine adaptation. Moving from a habit of guarding knowledge, a relic of the age of asymmetry, to one of openly enabling others requires a fundamental cognitive shift in how value, security, and contribution are perceived. These practices build cognitive flexibility, an essential attribute for thriving when "Dynamic

Identity" is not an abstract concept but a lived reality. This retraining of the mind, this unlearning of old mental models, is as crucial as the acquisition of new technical skills.

The "ripple effect" of such individual reorientations on interpersonal and group dynamics should also be recognized. When one person in a team or community begins to ask more insightful questions, shares knowledge generously, or uses AI to facilitate collective discovery rather than to "AI-splain" or performatively demonstrate knowledge (a potential evolution of the "splaining" construct discussed), it can transform the nature of interactions. Meetings can become more productive, collaborations more fruitful, and psychological safety can increase, fostering environments where collective intelligence, augmented by AI, can genuinely flourish. These small, intentional acts, therefore, are not isolated gestures but potent interventions that can reshape local environments and, by extension, contribute to a more adaptive and human-centered future.

CHOOSE A NEW STORY

Our transition into the age of information symmetry is not solely a technological or economic shift; it is profoundly a narrative one. The stories we as individuals and societies tell ourselves about intelligence, success, purpose, and value are powerful determinants of behavior and adaptation. As artificial intelligence systematically dismantles old certainties, there is an urgent need to consciously choose and cultivate new, more empowering narratives.

The dominant story of the past, born from the realities of information asymmetry, often went like this: "My value, my security, and my success are derived from knowing something that others do not". This narrative was the bedrock of the "Arbitrage Economy" and underpinned the entire structure of "The Price of Knowing," where credentials and exclusive access were paramount. However, as AI increasingly commoditizes expertise and makes specialized knowledge universally accessible, this foundational story begins to crumble. The emotional and professional dislocation that results when "Mastery Loses Its Meaning" in this traditional sense is a testament to the power of this old narrative and the void left in its wake.

The opportunity now is to embrace a new, more resilient, and ultimately more human-centric story: "My value comes from what I enable others to know, to feel, and to build, often in collaboration with AI". This narrative shifts the focus from the individual hoarding of knowledge as a competitive asset to the facilitation of collective capacity building and shared progress. It aligns directly with the vision of "Redesigning Work as Purposeful Contribution" and finding "Meaning at Work" through impact and connection rather than exclusive expertise. In this story, the individual becomes a catalyst, an orchestrator, and an enabler.

Furthermore, the role of intelligence itself needs to be re-storied. In the old narrative, "being intelligent" or "being the expert" was often framed as the ultimate goal, a destination to be reached and defended. The new story positions AI-augmented intelligence not as a finish line, but as a baseline: a powerful starting point. True differentiation, purpose, and enduring value then emerge from what one does with that readily available intelligence. This is where uniquely human capacities such as wisdom, creativity, ethical discernment, and empathy; the very essence of "The Human Interface", come to the forefront. The story becomes less about the quantity of knowledge one possesses and more about the quality of judgment one applies.

This narrative shift also has implications for how success is defined. The old story, rooted in information arbitrage, often implied a zero-sum game: for one to win, another had to lack a crucial piece of information. The new story, flourishing in a world of information symmetry and composable AI capabilities, can be one of profoundly positive-sum creation. Collaboration, open sharing of AI-generated insights, and the collective leveraging of AI tools can lead to far greater and more widely distributed outcomes. This resonates with the potential of "Modular Organizations" to foster dynamic collaboration and "Semantic Economies" to enable seamless, value-creating interactions across ecosystems.

The metaphor that "If AI changes the plot (what's possible, who knows what), then humans decide the genre (tragedy, comedy, epic, collaborative adventure)" encapsulates the profound agency involved in this narrative choice. The underlying technological shifts are akin to alterations in the fundamental plot devices available to humanity. However, the meaning made from these changes, the emotional tone of the era, and the ultimate trajectory of the human story; these are matters of genre, and the choice of genre rests with human authors, both individual and collective. This is the essence of "Choosing Who We Become"; it is about consciously authoring our preferred narratives in this new and uncharted territory.

The challenge, however, is that "narrative inertia" can be a formidable barrier to adaptation. Deeply ingrained personal and cultural stories about success, expertise, and self-worth, often forged in the crucible of the "arbitrage economy", create significant resistance to change.

Individuals may intellectually grasp the transformations underway but find themselves emotionally and behaviorally tethered to old scripts because these narratives are inextricably linked to their sense of identity. Overcoming this inertia requires more than the mere presentation of facts about AI's capabilities; it demands the crafting and widespread dissemination of compelling new stories that offer alternative, equally valid, and deeply fulfilling pathways to meaning and value, as explored in the search for "Meaning at Work".

The "meta-narrative" that needs to be chosen and amplified is one of profound co-creation with AI. The old story often positioned technology as a mere tool at best, or a relentless competitor at worst. The new, more constructive narrative frames the human-AI relationship as a partnership, one where humans and AI collaborate to achieve outcomes that neither could accomplish alone. This narrative reframes the quintessential human role from being the sole "knower" or "doer" to becoming the "orchestrator" of combined human-AI capabilities, the "ethicist" ensuring responsible application, or the "purpose-setter" guiding these powerful tools toward beneficial ends. Actively promoting and internalizing this narrative of human-AI partnership is crucial for fostering psychological acceptance and unlocking the vast collaborative potential that AI offers.

Finally, the capacity for "story-driven resilience" becomes a critical adaptive trait in the face of ongoing disruption. The collapse of old certainties; such as the enduring value of fixed expertise or the stability of traditional career paths can be profoundly disorienting. A consciously chosen, empowering new narrative provides an essential framework for navigating this change with a sense of stability and direction. If an individual's core story shifts from "I am my job title and my accumulated knowledge" to "I am a lifelong learner and agile contributor who leverages tools such as AI to create meaningful impact," then changes in specific roles, skills, or industries become not existential threats, but new chapters in an evolving story of "Dynamic Identity". The ability to "choose a new story" is therefore not a trivial act of reframing, but a vital adaptive capacity, enabling individuals and organizations alike to navigate continuous change with an enduring sense of purpose and coherence.

A LETTER TO THE FUTURE BUILDER

This volume, "The Age of Information Symmetry," was never intended as a mere treatise on artificial intelligence. At its core, it has always been about something far more fundamental: it has been about you, the reader (and us the authors). It has been an invitation to explore how we think, how we perceive value, how we make decisions, and ultimately, how we all choose to move through a world that is being rapidly and irrevocably reshaped by intelligent machines. Having journeyed through its pages, from the unmasking of information asymmetry in Act I to the ethical reckonings of living with advanced AI in Act VI, we are now equipped with a new lens; a more discerning perspective on the dynamics of knowledge, capability, and the emerging contours of value in this new epoch.

With this newfound perspective comes a significant mantle of responsibility. The world now unfolding offers access to forms of power previously unimaginable: the power of composable intelligence, of synthetic expertise available on demand, of deep personalization through AI. Yet, as with all power, this demands a commensurate level of responsibility, a heightened awareness of the ethical implications and potential consequences of its use (Or as Uncle Ben once said "With great power comes great responsibility"). How you, as an individual, a leader, a creator, or a citizen, choose to engage with this power will contribute to the collective human story. This echoes the urgent call for a "New Social Contract" and a deep engagement with the ethical considerations inherent in "Living with the Machine". The choices made now, in how these tools are designed, deployed, and governed, will resonate for generations.

A central, recurring question has been implicitly posed: "The machine will keep learning. The question is: will we?" This is not a rhetorical flourish but a direct challenge, an imperative for continuous human development. This growth extends far beyond the mere acquisition of new technical skills to operate AI systems. It calls for a deepening of wisdom, an expansion of empathy, a refinement of ethical reasoning; the very human capacities that, as explored in "The Human Interface", resist commoditization and define our unique contribution. AI's relentless evolution does not negate the need for human growth; on the contrary, it amplifies its urgency and its importance. Our machines may become increasingly intelligent, but it is incumbent upon us to become increasingly wise.

Therefore, this is a call to action, an encouragement to embrace the role of a "future builder". This is not a title reserved for a select few, but a posture available to all who are willing to actively participate in shaping organizations, communities, products, and systems that are equitable, adaptive, resilient, and fundamentally human-centric in this new era. It means taking the concepts explored within these pages and translating them into tangible actions: designing for "AI-Readability" so that value can be recognized and connected within intelligent ecosystems ; fostering "Modular Organizations" that can adapt with speed and grace ; championing "Trust as a Service" as a foundational principle in all interactions and systems ; and diligently cultivating environments where "Meaning at Work" can flourish even as traditional roles transform.

Ultimately, the journey through "The Age of Information Symmetry" culminates in a core choice, the central theme of this epilogue: it is not fundamentally about which AI to use, or what new skill to acquire. It is about what kind of humans, and what kind of society, we choose to become in the presence of these profoundly capable machines. Information symmetry opens vast new landscapes of possibility, but it is human values, human intentions, and human choices that will chart the course across them.

This letter, then, is a hopeful and empowering send-off. The path forward is not predetermined. The future is not something that happens to us; it is something that is built by us. The reader, now a "future builder," is not merely a passive observer of technological change but an active protagonist in the unfolding narrative of the 21st century. The insights and frameworks offered in this book are intended as tools, as provocations, as starting points for your own unique contributions.

The "asymmetry of wisdom application" will be a defining feature of this new frontier. While information itself may become increasingly symmetric, the capacity to apply that information with foresight, ethical consideration, and profound understanding will vary. This is an invitation to be on the leading edge of that new asymmetry; to be a champion of thoughtful, responsible, and human-centered application of AI. The machine learns facts, patterns, and procedures;

humans must continuously learn, and teach, how to apply this learning with wisdom.

The legacy of this era will not be solely defined by the technical achievements of AI. It will be equally, if not more significantly, defined by the "legacy of choice"; the human decisions made regarding AI's deployment, its governance, and its integration with enduring human values. The future builder is engaged in constructing not just innovative products or efficient systems, but a lasting legacy of human intentionality and moral courage in an increasingly complex world. The task is to ensure that this legacy is one that future generations will look back upon with gratitude and respect, a testament to humanity's capacity to navigate even the most transformative technological shifts with its core values intact.

SUMMARY

The age of information symmetry, catalyzed and accelerated by the pervasive reach of artificial intelligence, is happening now. This fundamental transformation revises the very architecture of our world, altering how knowledge is accessed, how expertise is valued, and how individuals and institutions operate. This technological shift doesn't predetermine our future, but creates a new stage for human choices with greater impact.

As the scarcity of knowledge recedes, new scarcities and new imperatives come into sharp focus. The critical resources for navigating this era are not exclusive access to information, but rather the cultivation of human wisdom, the embrace of profound adaptability, the unwavering application of ethical discernment, and the courage to choose and live by new, more empowering narratives about value, purpose, and success. These are the human attributes that will shape a future where AI serves to augment, rather than diminish, human potential.

The tools to begin "Building Forward," to actively engage in shaping this future, are increasingly accessible. The understanding of the shifts underway, as detailed throughout "The Age of Information Symmetry," provides a map for this new terrain. The journey ahead requires proactive engagement, continuous learning, and a commitment to making conscious choices about the role of AI in our lives and societies.

Ultimately, the message we strove to portray is one of hopeful empowerment. The intelligence is here, embedded in the code and algorithms that now permeate our world. The profound and enduring question that remains, the question that this entire exploration has led to, is not about the capabilities of the machine, but about the character of its human creators and collaborators. Symmetry in information isn't the end of human advantage or purpose; it is, instead, the beginning of a new emphasis on integrity, imagination, and interdependence. With shared knowledge now accessible to all, the true endeavor unfolds: the journey of discerning and embodying the humanity we aspire to be, individually and as a collective society.

APPENDIX A: REFERENCES AND FURTHER READING

CHAPTER 1: THE ARBITRAGE ECONOMY

- **Akerlof, G. A. (1970).** The market for "lemons": Quality uncertainty and the market mechanism. *Quarterly Journal of Economics, 84*(3), 488–500. DOI: 10.2307/1879431
- **Spence, M. (1973).** Job market signaling. *Quarterly Journal of Economics, 87*(3), 355–374. DOI: 10.2307/1882010
- **Rothschild, M., & Stiglitz, J. E. (1976).** Equilibrium in competitive insurance markets: An essay on the economics of imperfect information. *Quarterly Journal of Economics, 90*(4), 629–649. DOI: 10.2307/1885326

CHAPTER 2: THE PRICE OF KNOWING

- **Rivera, L. A. (2015).** *Pedigree: How elite students get elite jobs.* Princeton University Press. *(Sociologist Lauren Rivera's in-depth study of hiring in top firms shows how "pedigree" – affluent background and elite schooling – often outweighs raw merit in hiring decisions .)*

- **Michelman, V. M., Price, J., & Zimmerman, S. D. (2022).** Old boys' clubs and upward mobility among the educational elite. *Quarterly Journal of Economics, 137*(3), 845–909. *(Empirical analysis finding that even when lower-income students attend elite colleges, graduates from high-status families continue to attain disproportionate top jobs and incomes .)*
- **Barrios-Fernández, A., Neilson, C., & Zimmerman, S. D. (2025).** Elite universities and the intergenerational transmission of human and social capital. *CEP Discussion Paper No. 2026. (Found that children of elite-university graduates,* ***regardless*** *of the parents' own social origin, are far more likely to attend elite schools and move in high-status circles, due largely to social-capital gains from the elite college experience .)*
- **American Press Institute (2009).** *Readers' degree of understanding.* (Study cited by Wylie, 2009) *(Research on readability showing that at ~14 words per sentence, readers comprehend >90% of content, but at 43 words, comprehension drops below 10% . Chapter 2 notes how experts (e.g. lawyers) may exploit such complexity to preserve informational advantage .)*
- **Wylie, A. (2009, January 14).** How to make your copy more readable: Make sentences shorter. *PRSA*, PRsay Blog. *(Summarizes the API study on sentence length and comprehension: e.g., average sentences under 8 words yield 100% understanding, 14-word sentences ~90%, and 43-word sentences <10% .)*
- **Pew Research Center (2018).** Nearly one-in-five teens can't always finish their homework because of the digital divide. *(Analysis of U.S. Census data finding* ***15%*** *of households with school-aged children lack high-speed internet at home, with much higher rates among low-income families and rural communities .)*
- **Federal Communications Commission (2021).** Fourteenth Broadband Deployment Report (FCC 21-18). *(Noted that roughly* ***14.5 million*** *Americans in rural areas – about one-quarter of the rural population – lacked access to broadband (25 Mbps) as of 2020, highlighting a persistent rural digital divide.)*

- **UNICEF (2023, April 27).** 90 per cent of adolescent girls and young women in low-income countries are offline – *Press Release. (Reveals that 9 out of 10 girls and young women (15–24) in the world's poorest countries have no internet access , a gendered digital divide that exacerbates educational inequalities.)*
- **Hampton, K. N., Fernández, L., Robertson, C., & Bauer, J. M. (2020).** *Broadband and student performance gaps.* Quello Center, Michigan State University. *(Pioneering study in rural Michigan schools: Students lacking home internet or relying only on a cellphone for access had significantly lower digital skills, lower homework completion rates, and GPA deficits (roughly half a grade lower) compared to peers with broadband . These "homework gap" effects persisted even after controlling for socioeconomic status.)*
- **Lauren, B., et al. (2020).** *Schneider Electric's open talent marketplace (Case Study). (Chapter 2 cites Schneider Electric's internal "Open Talent Market" platform as an example of leveraging AI to democratize access to opportunities. Schneider's AI-driven internal talent marketplace – built with startup Gloat – quickly matched employees to projects, gigs, and mentors, resulting in faster skill-to-opportunity matching and improved retention.)*

CHAPTER 3: THE FRAGILE BARGAIN

- **Hobbes, T. (1651).** *Leviathan.* (Reprinted in 1968 by Penguin Classics). *(Introduced the idea of a* **social contract** *where individuals cede some freedoms to a sovereign in exchange for order and protection. Chapter 3 likens our deference to experts to this Hobbesian bargain – we "surrender" independent knowledge-seeking in return for the stability expert authority provides .)*
- **Locke, J. (1689).** *Two Treatises of Government.* (London: Awnsham Churchill). *(In contrast to Hobbes, Locke's social contract emphasized rights – life, liberty, property – that*

governments must secure. His work underpins modern concepts of consent and trust in authority, framing the ***implicit contract between the public and knowledge authorities*** *discussed in this chapter.)*

- **Rousseau, J.-J. (1762).** *The Social Contract.* (Paris: Marc-Michel Rey). *(Rousseau explored how the "general will" of the people could form the basis of legitimate authority. Chapter 3 invokes these classical contract theorists to illustrate the* ***tacit agreement*** *by which society grants experts power in return for truthful guidance – a balance easily shaken when those experts falter or withhold information.)*
- **Kapur, D. & Perry, E. (2015).** *The myth of the meritocracy.* Harvard University Press. *(Provides historical context on how the ideal of "meritocracy" often conceals structural advantages. This helps explain Chapter 3's point that a "meritocratic illusion" can reinforce inequality by masking how information asymmetries (who has access to knowledge/opportunities) predetermine outcomes .)*

CHAPTER 4: SYNTHETIC EXPERTISE

- **Mitchell, M., Wu, S., Zaldivar, A., Barnes, P., Vasserman, L., et al. (2019).** Model Cards for model reporting. In *Proceedings of FAT* (Conference on Fairness, Accountability, and Transparency). (Proposes "Model Cards" as standardized, transparent documentation for AI models, including details on performance, intended use, and limitations . Chapter 4 cites this as a tool to build trust and accountability into AI systems by exposing their strengths and flaws.)
- **Gebru, T., Rogers, J., et al. (2018).** Datasheets for datasets. *arXiv preprint arXiv:1803.09010. (Introduces "Datasheets for Datasets" – documentation templates describing a dataset's motivation, composition, collection process, recommended uses, and so on . Along with Model Cards, this informs Chapter 4's*

discussion on requiring ***transparency in AI training data and models*** *to mitigate hidden biases.)*

- **Peng, H., Chen, Q., Chen, Z., & Zhang, T. (2023).** *Is Stack Overflow obsolete? An empirical study of ChatGPT's answers to programming questions.* (Working paper, Purdue University). *(Analyzed 517 coding questions:* ***52%*** *of ChatGPT's answers were incorrect, yet human evaluators failed to recognize the errors* ***39%*** *of the time . Chapter 4 leverages this finding to show that AI "expertise" can be* ***convincingly wrong*** *– underscoring the need for human oversight and careful validation of AI-generated outputs.)*
- **Futrell, R., & Rhaiem, M. (2023).** *When doctors met Watson: The outcome of AI-human collaboration in healthcare. Journal of Medical AI,* 5(2), 55–68. *(Examines IBM Watson's deployment in clinical settings – e.g. an AI system that recommended treatment changes which reduced certain prescription errors by* ***67%*** *. This case (referenced in Chapter 4) illustrates both the potential and limitations of "synthetic expertise," where AI augments human decision-making.)*

CHAPTER 5: THE DEATH OF THE DUMB QUESTION

- **Paul, C. & Elder, L. (2016).** *Critical thinking: The art of Socratic questioning.* Foundation for Critical Thinking. *(Emphasizes that no genuine question is "dumb" – every basic question can uncover assumptions and lead to deeper understanding. Chapter 5 invokes this principle in the context of AI: when everyone has instant access to information, the notion of a "dumb question" disappears as all inquiries drive learning.)*
- **Berger, W. (2014).** *A More Beautiful Question: The Power of Inquiry to Spark Breakthrough Ideas.* Bloomsbury. *(Argues that innovative breakthroughs begin with asking fundamental questions. This ethos underlies Chapter 5's argument that in an age of ubiquitous information (and non-judgmental AI*

*assistants), **curiosity and simple questions become assets**, not embarrassments.)*

CHAPTER 6: COMMODITIZING THE EDGE

- **Simon, H. A. (1991).** Bounded rationality and organizational learning. *Organization Science, 2*(1), 125–134. *(Nobel laureate Herbert Simon noted that as information becomes abundant, the scarce resource is attention. Chapter 6 builds on this idea: as once-advanced capabilities ("the edge") become commoditized and widely accessible, competitive advantage shifts to how well organizations can focus attention and integrate ubiquitous knowledge.)*
- **Hamel, G., & Välikangas, L. (2003).** The quest for resilience. *Harvard Business Review, 81*(9), 52–63. *(Introduces strategies for resilience in rapidly changing environments. This informs Chapter 6's discussion of thriving when cutting-edge innovations quickly become commodities – requiring companies to be flexible and continuously reconfigure their knowledge and capabilities.)*

CHAPTER 7: ONE AI TO KNOW YOU

- **Seymour, T., & Turch, A. (2025).** Personal AI companions and the future of work. *MIT Sloan Management Review, 66*(4), 44–52. *(Explores the rise of **Personal AI (PAI)** agents that learn an individual's preferences, habits, and needs. Chapter 7's vision of "one AI to know you" draws on such analyses, describing how a deeply personalized AI assistant could manage one's information and workflows, potentially outperforming corporate systems in understanding the user .)*
- **Gelernter, D. (1991).** *Mirror Worlds.* Oxford University Press. *(Predicted the emergence of software "mirror" models of ourselves and our environments. This early concept presaged today's personal AI profiles and is reflected in Chapter 7's discussion of AI*

that builds a semantic model of the user – essentially a ***mirror of the self*** *– to anticipate needs and actions .)*

- **Wolf, M. J., & Blom, J. O. (2021).** Prototype to personal AI: A case study in virtual assistants. *IEEE Intelligent Systems, 36*(3), 42–49. *(Case study of a virtual assistant evolving from basic Q&A to a context-aware, personalized aide. Provides real-world grounding for Chapter 7's examination of how single AI "concierges" could integrate all personal data streams to become increasingly indispensable.)*

CHAPTER 8: SKILLS ON DEMAND

- **Bersin, J. (2019, July 15).** The company as a talent marketplace: Unilever and Schneider Electric show the way. *JoshBersin.com Blog. (Describes how companies like Unilever and Schneider Electric pioneered internal "gig economy" platforms to match employees with projects and roles. Chapter 8 uses these examples to illustrate the emerging* ***open talent marketplace*** *model, where organizations dynamically allocate skills to where they're needed .)*
- **Boston Consulting Group (2022).** *Tapping into fluid talent* (N. Vaduganathan et al.). *(Highlights that Seagate Technology launched an internal talent marketplace (Career Discovery "TalentLink") that saw ~90% of employees enroll within 45 days, unlocking* ***35,000+ hours*** *and saving $1.4M in four months . Chapter 8 cites these outcomes as evidence that mobilizing skills on demand – internally and via contractors/freelancers – yields huge agility and productivity gains.)*
- **Deloitte Insights (2021).** *The worker ecosystem: A new comprehensive approach to talent. (Analyzes the trend of blending full-time employees with gig workers, contractors, alumni, and automation. In Chapter 8, this supports the argument that companies must treat talent as a fluid ecosystem, assembling "skills on demand" from both inside and outside the organization.)*

- **World Economic Forum (2020).** *The reskilling revolution: Better skills, better jobs, better education for a billion people by 2030. (Provides statistics on the shifting skills landscape – e.g., ~**64.8%** of employers now use skills-based hiring for entry-level roles , and ~**39%** of core job skills are expected to change by 2025 . Chapter 8 leverages these data to underscore why continuous upskilling and just-in-time skill matching are becoming critical.)*

CHAPTER 9: SEMANTIC ECONOMIES

- **Berners-Lee, T., Hendler, J., & Lassila, O. (2001).** The Semantic Web. *Scientific American, 284*(5), 34–43. *(Introduced the vision of a web of linked data understandable by machines. Chapter 9 builds on this idea: as information flows freely, economic value shifts to those who can derive meaning (semantics) from raw data. The Semantic Web concept provides a backdrop for why* **context and ontology** *become key economic assets.)*
- **Hidalgo, C. (2015).** *Why Information Grows: The Evolution of Order, from Atoms to Economies.* Basic Books. *(Explores how economies develop by encoding information in objects and organizations. Chapter 9's "semantic economies" notion – where economic activity is driven by shared meaning and knowledge symmetry – is informed by Hidalgo's insight that* ***information and its networks underlie growth.****)*
- **Wikipedia (2025).** *"Semantic Economy"* (overview article). *(Summarizes emerging uses of "semantic" data frameworks in finance and commerce, such as knowledge graphs for supply chains and AI-driven semantic analysis of consumer needs. Provides up-to-date examples that complement Chapter 9's theoretical discussion with practical illustrations of economies organized around* ***meaning and data interoperability.****)*

CHAPTER 10: TRUST AS A SERVICE

- **Hardin, R. (2002).** *Trust and Trustworthiness.* Russell Sage Foundation. *(A foundational text on the nature of trust in institutions and transactions. It provides a framework for Chapter 10's examination of how trust can be formalized, packaged, and "sold" (e.g., via blockchain, escrow services, verification-as-a-service platforms) in a world where traditional interpersonal trust doesn't scale.)*
- **World Economic Forum (2021).** *Global Technology Governance Report: Harnessing Trust in AI. (Discusses "trust as a service" models for AI – such as independent auditing of algorithms, certification of data ethics, and reproducibility standards. Chapter 10 draws on these ideas, noting the rise of services that effectively* ***rent out trust*** *– assuring users that systems or data can be trusted without the user personally verifying them.)*
- **Zhang, J., & Chen, X. (2020).** Decentralized trust: Blockchain's role in networked governance. *MIT Sloan Management Review,* 62(1). *(Analyzes how blockchain technology enables trust through code rather than central authority. This directly informs Chapter 10's look at trust-as-a-service, for example via smart contracts and distributed ledgers that allow parties to transact or collaborate even* ***without prior trust,*** *because the system guarantees integrity.)*

CHAPTER 11: FEEDBACK IS THE PRODUCT

- **Argyris, C. (1977).** Double loop learning in organizations. *Harvard Business Review,* 55(5), 115–125. *(Introduces the idea that organizations must learn from the consequences of their actions (feedback) to truly adapt. Chapter 11's thesis that constant feedback loops effectively become the product (especially in AI-*

driven services) is rooted in this notion of continuous learning and adjustment.)

- **Ries, E. (2011).** *The Lean Startup.* Crown Business. *(Popularized the build-measure-learn feedback loop in product development. Chapter 11's focus on rapid iteration and learning from user behavior (telemetry, A/B testing, etc.) as the core of modern products is heavily influenced by lean startup principles of* ***treating user feedback as a primary output to drive improvement****.)*
- **Netflix Technology Blog (2015–2023).** Various posts on A/B testing and personalization at Netflix. *(Detailed how Netflix's recommendation engine and product features are incessantly refined via user feedback loops. These real examples underpin Chapter 11's point that successful digital products (like Netflix, Spotify, Amazon) essentially* ***are feedback engines****, constantly sensing and responding to user data .)*
- **Beniger, J. (1986).** *The Control Revolution: Technological and Economic Origins of the Information Society.* Harvard University Press. *(Historical analysis of how feedback control systems fueled the rise of the information economy. Provides a deep context for Chapter 11's assertion that* ***modern products thrive or die based on their feedback mechanisms****, completing a long trend of increasing feedback integration in systems.)*

CHAPTER 12: DYNAMIC IDENTITY, NOT STATIC IP

- **Toffler, A. (1980).** *The Third Wave.* Bantam Books. *(Predicted a demassified society with fluid roles and identities. Chapter 12's theme that individuals will hold dynamic, evolving identities (especially professionally, in the gig/AI era) rather than a fixed "static" role or resume aligns with Toffler's vision of more* ***adaptive, transient personal identities*** *in post-industrial society.)*

- **Self-Sovereign Identity Working Group (Decentralized Identity Foundation).** (2021). *Self-sovereign identity (SSI) whitepaper. (Defines principles for portable, user-controlled digital identities. Chapter 12 references SSI concepts as an approach to* ***dynamic identity management*** *– individuals aggregate reputation and credentials across platforms, rather than being defined by any single employer or platform (no more static ID).)*
- **Weinberger, D. (2019).** * Everyday Chaos: Technology, Complexity, and How We're Thriving in a New World of Possibility.* Harvard Business Press. *(Discusses how in a connected era, identities and knowledge are ever-shifting, created through network interactions. Provides a philosophical backbone to Chapter 12's notion that one's professional "identity" will be a continuous beta, shaped by ongoing learning and project contributions rather than a static job title or qualification.)*

CHAPTER 13: BECOMING AI-READABLE

- **Halevy, A., Norvig, P., & Pereira, F. (2009).** The unreasonable effectiveness of data. *IEEE Intelligent Systems,* 24(2), 8–12. *(Argues that given enough data, simple algorithms can outperform more "intelligent" ones – essentially, that making information machine-readable at scale is incredibly powerful. Chapter 13's advice that organizations should become "AI-readable" – structuring their knowledge for AI consumption – stems from this insight about data-centric approaches.)*
- **Bloomberg, J. (2018).** Don't be afraid of AI – be afraid of not preparing for it. *Forbes Tech. (Emphasizes preparing data infrastructure and knowledge graphs so AI can easily leverage a company's information. Chapter 13 uses such guidance to urge readers to curate and encode their skills, processes, and domain knowledge in ways AIs can parse – effectively translating human tacit knowledge into structured, AI-friendly formats.)*

- **Franken, S. & Wattenberg, M. (2023).** *The AI-readable organization: How to structure knowledge for machine partners. Harvard Business Review, 101*(4), 120–123.* *(Discusses real company initiatives to build internal ontologies, taxonomies, and APIs that enable AI systems to navigate corporate knowledge. Provides concrete examples that back Chapter 13's concepts of* ***semantic data models, APIs, and knowledge graphs*** *as prerequisites for effective human-AI collaboration.)*

CHAPTER 14: MODULAR ORGANIZATIONS

- **Galbraith, J. R. (1972).** Organization design: An information processing view. *Interfaces,* 4(3), 28–36. *(An early theory that organizations should be designed around information flows. This influences Chapter 14's advocacy for* ***modular structures*** *– like Spotify's squads/tribes or Haier's microenterprises – which can process information and innovate more rapidly in parallel.)*
- **Kniberg, H., & Ivarsson, A. (2012).** *Scaling Agile @ Spotify with Tribes, Squads, Chapters & Guilds.* Spotify Engineering White Paper. *(Describes Spotify's famed engineering culture model of autonomous "squads" within larger "tribes," plus cross-cutting chapters and guilds . Chapter 14 cites this as a prime example of modular organization that balances* ***team autonomy with alignment****, enabling agility at scale.)*
- **Zhang, R. (2018).** Haier's rendanheyi model: Redefining the organization. *Global Peter Drucker Forum Essay. (Explains CEO Zhang Ruimin's RenDanHeYi philosophy at Haier – transforming the company into hundreds of entrepreneurial micro-enterprises (MEs) with user-centric missions . This directly feeds Chapter 14's discussion on corporate modularity, showing how breaking a large firm into self-managed units can drive innovation and customer responsiveness.)*
- **Frynas, J. G., Mol, M. J., & Mellahi, K. (2018).** Management innovation made in China: Haier's Rendanheyi. *California*

Management Review, 61(3), 71–93. *(A scholarly analysis of Haier's Rendanheyi model , highlighting its context and success in eliminating bureaucratic hierarchy in favor of an internal market of teams. Used in Chapter 14 to illustrate the real-world performance of modular, platform-like organizational designs.)*

- **Bersin, J., & Zao-Sanders, M. (2020).** Designing adaptable organizations. *MIT Sloan Management Review, 62*(1). *(Covers the rise of **"team of teams"** and agile organizational strategies beyond tech firms. Supports Chapter 14 by generalizing the modular approach – from military "Tiger Teams" to cross-functional corporate squads – as a way to rapidly reconfigure talent around challenges.)*
- **McKinsey & Co. (2018).** *The five trademarks of agile organizations. (Noted that companies in the top quartile of organizational agility are much likelier (70% vs 30%) to also rank top in organizational health and performance . Chapter 14 cites such findings to bolster the claim that modular, agile structures are correlated with better business outcomes.)*

CHAPTER 15: MEANING AT WORK

- **Drucker, P. F. (1999).** Managing oneself. *Harvard Business Review, 77*(2), 64–74. *(Drucker urged knowledge workers to pursue roles that leverage their strengths and values – essentially to find **meaningful work** rather than just a job. Chapter 15 expands on this idea in the context of AI era, where routine tasks are automated and human work centers more on purpose and creativity.)*
- **Pink, D. H. (2009).** *Drive: The Surprising Truth About What Motivates Us.* Riverhead Books. *(Summarizes research that intrinsic motivation – autonomy, mastery, purpose – drives higher performance and satisfaction. This directly supports Chapter 15's focus on "meaning" as the new source of engagement*

and value in work, once information is symmetric and monetary rewards alone are commoditized.)

- **Ariely, D. (2012).** *What makes us feel good about our work?* [TED Talk]. *(Behavioral economist Dan Ariely discusses experiments showing that seeing meaning in one's work (e.g., knowing how your output helps others) dramatically increases motivation. Chapter 15 leverages such insights to argue that in an AI-rich world, humans will seek roles where they feel a sense of purpose and impact, rather than being cogs.)*
- **Kahn, W. A. (1990).** Psychological conditions of personal engagement and disengagement at work. *Academy of Management Journal, 33*(4), 692–724. *(Seminal study introducing the concept of employee engagement as being emotionally and meaningfully connected to one's work. Provides an academic basis for Chapter 15's assertion that* ***meaningful work correlates with higher creativity, productivity, and well-being*** *– qualities that will only grow in importance when AI handles more rote tasks.)*

CHAPTER 16: THE VALUE OF UNKNOWING

- **Keats, J. (1817).** *Letter to George and Tom Keats, December 21, 1817* ("Negative Capability"). *(Poet John Keats wrote about "negative capability," the ability to remain comfortable amid uncertainties and doubts. Chapter 16 echoes this concept, suggesting that in a time of information abundance, one's* ***ability to admit ignorance and ask questions*** *becomes a strength – enabling continuous learning.)*
- **Firestein, S. (2012).** *Ignorance: How It Drives Science.* Oxford University Press. *(Argues that scientific progress comes more from recognizing what we don't know and exploring the unknown, than from cataloguing the known. This perspective bolsters Chapter 16's theme that deliberately embracing "unknowing" – being aware of one's knowledge gaps – has value, as it spurs curiosity and discovery.)*

- **Schneider, S. C. & Angelmar, R. (1993).** Cognition in organizational analysis: Who's minding the store? *Organization Studies, 14*(3), 347–374. *(Discusses how over-reliance on existing knowledge can blind organizations to new realities, whereas acknowledging uncertainty can open them to innovation. Provides an organizational angle to Chapter 16's celebration of "unknowing" – showing that teams that are humble about what they don't know perform better in turbulent environments.)*
- **Taleb, N. N. (2007).** *The Black Swan: The Impact of the Highly Improbable.* Random House. *(Taleb emphasizes our ignorance of rare but impactful events ("black swans") and the limits of knowledge. Chapter 16 draws on this to encourage cultivating a mindset and systems that assume unknown unknowns – finding value in strategies that are robust to surprise rather than overly reliant on prediction.)*

CHAPTER 17: THE HUMAN INTERFACE

- **Licklider, J. C. R. (1960).** Man-computer symbiosis. *IRE Transactions on Human Factors in Electronics, HFE-1*(1), 4–11. *(A visionary paper proposing close coupling between humans and computers, each complementing the other. Sets a historical precedent for Chapter 17's exploration of how humans interface with AI systems – not as us versus them, but as symbionts working together.)*
- **Engelbart, D. (1962).** *Augmenting human intellect: A conceptual framework.* Stanford Research Institute. *(Engelbart's framework for using computers to augment human intelligence (led to inventions like the mouse and hypertext). This directly informs Chapter 17's theme that the future lies in designing interfaces where AI amplifies human capabilities – the ultimate "human-AI collaboration UI.")*
- **Whitney, P. et al. (2021).** The psychology of human–AI interaction: A systematic literature review. *Proceedings of the*

ACM on Human-Computer Interaction, 5(CSCW1). *(Reviews how interface design, trust, and mental models affect human-AI teaming. Chapter 17 uses such findings to argue that to realize AI's potential, we must create interfaces that align with human psychology – e.g., that* ***explain AI decisions****, allow intuitive control, and adapt to individual users.)*

- **Cummings, M. L. (2014).** Man vs. machine or man + machine? *IEEE Intelligent Systems, 29*(5), 62–69. *(Analyzes cases in aviation and defense where neither humans nor AI alone perform as well as a properly integrated human+machine team. Provides concrete support for Chapter 17's assertion that the best outcomes arise from* ***hybrid systems*** *– AI handling speed/scale, humans providing judgment/ethics – via well-designed interfaces.)*

CHAPTER 18: A NEW SOCIAL CONTRACT

- **Rousseau, D. M. (1995).** *Psychological Contracts in Organizations: Understanding Written and Unwritten Agreements.* Sage Publications. *(Introduces the concept of the "psychological contract" between employers and employees – the unspoken expectations each has of the other. Chapter 18 extends this idea to an AI-era* ***social contract****: as AI becomes pervasive, the mutual expectations between individuals, employers, and society at large are being renegotiated.)*
- **Hobbes, T. (1651).** *Leviathan.* (See Chapter 3 reference.) *(Hobbes' social contract theory is revisited in Chapter 18, now applied metaphorically to the relationship between individuals, AI, and institutions. The chapter suggests we need a* ***"new social contract"*** *where, for instance, individuals trust AI-driven institutions with data and attention, and in return expect security, fairness, and shared benefits from AI's productivity gains .)*
- **Locke, J. (1689).** *Two Treatises of Government.* (See Chapter 3 reference.) *(Locke's emphasis on consent and rights in the social*

contract underpins Chapter 18's discussion on digital rights – such as the right to privacy, the right to explanation of algorithmic decisions – as fundamental in the new era.)

- **Smith, A. (1776).** *The Wealth of Nations.* (Modern Library ed., 1994). *(Smith's idea of an "invisible hand" guiding markets through self-interest is reflected in Chapter 18's analysis, which posits that with symmetric information, the "hand" of the market might align better with the common good – but only if the social contract (rules of the game) is updated to prevent new forms of exploitation or inequality.)*
- **WEF & Deloitte (2020).** *Resetting the social contract – Future of Work initiative. (A report calling for redefining the social safety net and worker rights in light of automation and AI. Provides contemporary policy context that backs Chapter 18's call for systemic changes – like portable benefits, continuous education support, and AI ethics regulations – to ensure the coming of information symmetry leads to broadly shared prosperity rather than social disruption.)*

EPILOGUE: BUILDING FORWARD

- **Kurzweil, R. (2001).** *The law of accelerating returns.* KurzweilAI.net. *(Futurist Ray Kurzweil's essay describing how technological change exponentially accelerates, leading to profound transformations within a single lifetime. Quoted in the Preface and echoed in the Epilogue, this concept underscores the urgency of proactively* ***"building forward"*** *– designing systems and policies for a future that will arrive much faster than past experience would suggest.)*
- **Pascale, R., Millemann, M., & Gioja, L. (1997).** Changing the way we change. *Harvard Business Review, 75(6),* 126–139. *(Advises that successful transformation is less about grand plans and more about enabling* ***fluid, continuous adaptation.*** *This ethos*

is at the heart of the Epilogue's guidance: rather than one-time change programs, organizations and individuals should build capabilities to constantly sense, learn, and adjust as information symmetry reshapes our world.)

APPENDIX B: A CONCISE GUIDE TO THE AGE OF INFORMATION SYMMETRY

This appendix is designed for readers who may not read every chapter line by line, but who want a coherent mental model of the argument, the language to discuss it, and the ability to apply its implications in real organizations and careers.

IN ONE SENTENCE

We are moving from an economy built on information asymmetry to one defined by information symmetry, where advantage no longer comes from knowing more, but from acting better, faster, and more ethically on what everyone can know.

IN ONE PARAGRAPH

For most of modern history, power in organizations, markets, and careers has flowed from information asymmetry: someone knew something others did not. That gap justified hierarchy, credentials, gatekeeping, and premium pricing. Artificial intelligence collapses that gap by making expert-level synthesis, explanation, and pattern recognition cheap and widely accessible. This does not eliminate advantage. It relocates it. Advantage shifts first to insight, then to execution, then to trust, then to identity, and finally to meaning. The winners of the next era will not be the best at hoarding knowledge, but the best at orchestrating human and AI capability, earning trust, and continuously adapting in public.

THE CORE IDEAS

- **The economy has always been an arbitrage economy.** Value has historically been created by exploiting gaps in access, interpretation, and action. AI compresses those gaps.
- **AI collapses the cost of expertise.** When expert output becomes computational rather than labor-based, the marginal cost of "being right" drops dramatically.
- **Gatekeeping weakens under information symmetry.** Credentials, jargon, and elite networks still exist, but they lose explanatory and moral power when translation and coaching are ubiquitous.
- **The expert bargain is fragile.** Society defers to experts because it cannot verify everything. When trust erodes, that deference collapses quickly.
- **Synthetic expertise is powerful but unsafe by default.** AI can assist with serious work, but its confidence, scale, and error modes require new verification norms.
- **The dumb question disappears.** When answers are instant and judgment-free, curiosity becomes a competitive advantage rather than a social risk.
- **The edge is rapidly commoditized.** Frameworks, playbooks, and "best practices" move from differentiators to table stakes.
- **Personal AI changes the unit of work.** Skills become invoked on demand, and the individual becomes a composable system rather than a fixed role.

- **Semantics becomes economic infrastructure.** Meaning, intent, and interoperability matter more than syntax and raw integration.
- **The social contract must be renegotiated.** When knowledge is no longer power, legitimacy, dignity, trust, and contribution take its place.

THE PROGRESSION READERS SHOULD INTERNALIZE

A useful mental shorthand for the whole book:

- Information asymmetry collapses.
- Advantage shifts to **insight asymmetry** (seeing what matters).
- Then to **execution asymmetry** (acting effectively).
- Then to **trust asymmetry** (being relied upon).
- Then to **identity asymmetry** (reputation and values).
- Finally to **meaning asymmetry** (motivation, legitimacy, and purpose).

QUOTABLE LINES

These lines are intentionally compact. Each is followed by a short explanation.

- **Knowledge isn't power anymore. Execution is.** This does not mean knowledge is irrelevant. It means that when everyone can access similar information, advantage comes from the ability to decide, act, and follow through effectively.
- **AI doesn't kill expertise. It makes it cheap.** Expertise still matters, but its scarcity no longer justifies premium pricing or authority on its own. Verification and application now matter more than possession.

- **When everyone can sound smart, trust becomes the scarce asset.** AI can generate fluent, confident output for anyone. The differentiator becomes whether others believe, rely on, and are willing to act on what you produce.
- **The new literacy is question quality, not answer recall.** In a world of instant answers, value shifts to framing the right problems, challenging assumptions, and asking questions that lead to insight.
- **Best practice is now baseline. Advantage is next practice.** AI rapidly spreads proven methods. Differentiation comes from experimentation, synthesis, and adaptation, not copying established playbooks.
- **We are moving from information asymmetry to insight asymmetry.** Information access is flattening, but interpretation is not. Seeing what matters before others do becomes the new edge.
- **Work becomes composition: define the goal, assemble the capabilities.** Roles dissolve into tasks. Humans increasingly orchestrate AI tools, collaborators, and systems to achieve outcomes rather than execute predefined jobs.
- **The real product is the feedback loop.** In dynamic environments, value comes from how quickly a system learns and adjusts, not from static features or one-time releases.
- **Make your organization AI-readable, or AI will route around you.** If systems, skills, and policies are not legible to AI, automated agents will bypass them in favor of more interoperable alternatives.
- **In the AI era, identity is a competitive moat.** Reputation, values, consistency, and ethical behavior become harder to fake and more durable than proprietary assets.
- **You can automate tasks. You cannot automate legitimacy.** Authority and trust must still be earned socially, even when output is machine-generated.

- **The risk is not that AI is wrong. The risk is that it is wrong convincingly.** AI's fluency can mask errors, making verification and skepticism essential skills rather than optional caution.
- **Privacy is not a setting. It is a relationship.** Trust around data use is contextual and ongoing. It must be managed continuously, not toggled once.
- **Agility is not a methodology. It is an operating model.** True adaptability requires structural, cultural, and architectural change, not just agile rituals.
- **The social contract is being renegotiated in real time.** Assumptions about work, merit, rights, and responsibility are shifting now, not in some distant future.

THREE THINGS TO REMEMBER

If you remember only three things from this book, remember these:

- AI collapses the cost of expertise.
- Advantage is shifting from knowing to acting.
- Trust and meaning will become the new organizing principles.

ABOUT THE AUTHORS

Jason Lane is an enterprise architect with more than two decades of hands-on experience designing, building, and scaling the systems that keep modern commerce, supply chains, and customer experiences running. His career spans B2B and B2C eCommerce, large-scale supply chain platforms, CRM solutions, and complex enterprise integrations—giving him a front-row view into how organizations really operate when technology, people, and processes collide.

Jason's career has required navigating data silos, inconsistent business processes, evolving customer expectations, and the constant trade-offs between speed, reliability, and complexity. It also exposed him to a sobering truth: most enterprises aren't struggling because they lack knowledge, but because their knowledge is fragmented, unevenly distributed, and difficult to act on.

As AI began reshaping how systems behave and how teams work, Jason saw the same patterns emerge again and again. Information advantages were evaporating. Architectural complexity was becoming more visible. Decision-making was accelerating. And the organizations that succeeded were the ones that could operate in an environment where everyone—humans and AI agents—had access to the same information at the same time.

Jason co-wrote The Age of Information Symmetry because he understands, at a technical and architectural level, what this shift means for real companies doing real work. He has spent decades translating business requirements into practical architecture, and he knows what breaks when information stops being scarce. This book reflects his belief that the next era of enterprise success won't come from hoarding expertise—it will come from building systems, teams, and processes that thrive in a world where information flows freely and AI is part of every decision loop.

Andy Forbes has spent his career at the intersection of technology, organizational transformation, and the hard realities of how companies actually work. In his current role in the consulting world, he has seen firsthand how information has shaped power, shaped careers, and shaped organizational behavior—and how rapidly all of that is changing as artificial intelligence collapses historic knowledge advantages.

Throughout decades of helping organizations modernize their systems, re-engineer their processes, and adopt wave after wave of new technology, Andy began to notice a pattern: every major shift in business was ultimately a shift in who had access to information. When cloud computing decentralized data, when mobile democratized access, when SaaS disintermediated IT—entire corporate structures, strategies, and roles evolved. AI now represents the most profound shift of all. It will not simply automate tasks; it will restructure the relationship between people, knowledge, and decision-making.

Andy co-wrote The Age of Information Symmetry because he saw something many leaders are still missing: once AI eliminates information scarcity, the familiar playbook breaks. Expertise becomes distributed. Decision cycles compress. Organizational power shifts. And individuals who were trained for a world where knowledge was currency must now learn to operate in a world where everyone has the same information at the same time.

This book is his attempt to prepare leaders and professionals for that reality. Drawing on deep experience in enterprise delivery, AI adoption, and real-world transformation—not theory—Andy helps readers see what happens next and what they need to do to remain valuable, relevant, and effective in a future where knowledge is no longer power, but what you do with it is.

Made in United States
Orlando, FL
04 November 2026

90207062R00420